D0564807

IB STUDY GUIDES

Physics

FOR THE IB DIPLOMA

Standard and Higher Level

Tim Kirk

OXFORD
UNIVERSITY PRESS

Great Clarendon Street, Oxford OX2 6DP

Oxford University Press is a department of the University of Oxford.
It furthers the University's objective of excellence in research, scholarship, and education by publishing worldwide in

Oxford New York

Auckland Bangkok Buenos Aires Cape Town Chennai
Dar es Salaam Delhi Hong Kong Istanbul Karachi Kolkata
Kuala Lumpur Madrid Melbourne Mexico City Mumbai Nairobi
São Paulo Shanghai Taipei Tokyo Toronto

Oxford is a registered trade mark of Oxford University Press
in the UK and in certain other countries

© Tim Kirk 2003, 2007

The moral rights of the author have been asserted

Database right Oxford University Press (maker)

First published 2003

Second edition 2007

All rights reserved. No part of this publication may be reproduced, stored in a retrieval system, or transmitted, in any form or by any means, without the prior permission in writing of Oxford University Press, or as expressly permitted by law, or under terms agreed with the appropriate reprographics rights organization. Enquiries concerning reproduction outside the scope of the above should be sent to the Rights Department, Oxford University Press, at the address above

You must not circulate this book in any other binding or cover and you must impose this same condition on any acquirer

British Library Cataloguing in Publication Data

Data available

ISBN 978-0-19-915141-7

10 9 8 7 6

Acknowledgements

We are grateful to the following to reproduce the following copyright material.

p78 (bottom) After J.N. Howard 1959; Proc. R.R.E 47, 1459: and R.M. Goody and G.D. Robinson, 1951; Quart. J. Roy Meteorol. Soc. 77, 153;
p167 (bottom right) Cold Spring Harbor Laboratory Archives;
pp153 (left) ECHOS/Nonstock Inc./Photolibrary;
pp153 (right) Jeff Corwin/Getty Images.

We have tried to trace and contact all copyright holders before publication. If notified the publishers will be pleased to rectify any errors or omissions at the earliest opportunity.

Printed in Great Britain by Bell and Bain Ltd., Glasgow

Paper used in the production of this book is a natural, recyclable product made from wood grown in sustainable forests. The manufacturing process conforms to the environmental regulations of the country of origin.

Mixed Sources
Product group from well-managed forests and other controlled sources
www.fsc.org Cert no. TT-COC-002769
© 1996 Forest Stewardship Council

Introduction and acknowledgements

Many people seem to think that you have to be really clever to understand Physics and this puts some people off studying it in the first place. So do you really need a brain the size of a planet in order to cope with IB Higher Level Physics? The answer, you will be pleased to hear, is 'No'. In fact, it is one of the world's best kept secrets that Physics is easy! There is very little to learn by heart and even ideas that seem really difficult when you first meet them can end up being obvious by the end of a course of study. But if this is the case why do so many people seem to think that Physics is really hard?

I think the main reason is that there are no 'safety nets' or 'short cuts' to understanding Physics principles. You won't get far if you just learn laws by memorising them and try to plug numbers into equations in the hope of getting the right answer. To really make progress you need to be familiar with a concept and be completely happy that you understand it. This will mean that you are able to apply your understanding in unfamiliar situations. The hardest thing, however, is often not the learning or the understanding of new ideas but the getting rid of wrong and confused 'every day explanations'.

This book should prove useful to anyone following a pre-university Physics course but its structure sticks very closely to the recently revised International Baccalaureate syllabus. It aims to provide an explanation (albeit very brief) of all of the core ideas that are needed throughout the whole IB Physics course. To this end each of the sections is clearly marked as either being appropriate for everybody or only being needed by those studying at Higher level. The same is true of the questions that can be found at the end of the chapters.

I would like to take the opportunity to thank the many people that have helped and encouraged me during the writing of this book. In particular I need to mention David Jones from the IB curriculum and assessment offices in Cardiff and Paul Ruth who provided many useful and detailed suggestions for improvement – unfortunately there was not enough space to include everything. The biggest thanks, however, need to go to Betsan for her support, patience and encouragement throughout the whole project.

Tim Kirk
October 2002

Second edition

In addition to thanking, once again, Betsan, my family, friends and colleagues, for all their help and support, I would also like to put on record my sincere thanks to the many students and teachers who commented so favourably on the first edition of this book. I am particularly grateful to all of those who took time to pass on their suggestions for improvements in this second edition.

This book is dedicated to the memory of my father, Francis Kirk.

Tim Kirk
November 2007

CONTENTS

(*Italics denote topics which are exclusively Higher Level.*)

The realm of physics – range of magnitudes of quantities in our universe

ORDERS OF MAGNITUDE – INCLUDING THEIR RATIOS

As stated in the introduction to the IB Physics diploma programme, Physics seeks to explain nothing less than the Universe itself. In attempting to do this, the range of the magnitudes of various quantities will be huge.

If the numbers involved are going to mean anything, it is important to get some feel for their relative sizes. To avoid 'getting lost' among the numbers it is helpful to state them to the nearest **order of magnitude** or power of ten. The numbers are just rounded up or down as appropriate.

Comparisons can then be easily made because working out the ratio between two powers of ten is just a matter of adding or subtracting whole numbers. The diameter of an atom, 10^{-10} m, does not sound that much larger than the diameter of a proton in its nucleus, 10^{-15} m, but the ratio between them is 10^5 or 100000 times bigger. This is the same ratio as between the size of a railway station (order of magnitude 10^2 m) and the diameter of the Earth (order of magnitude 10^7 m).

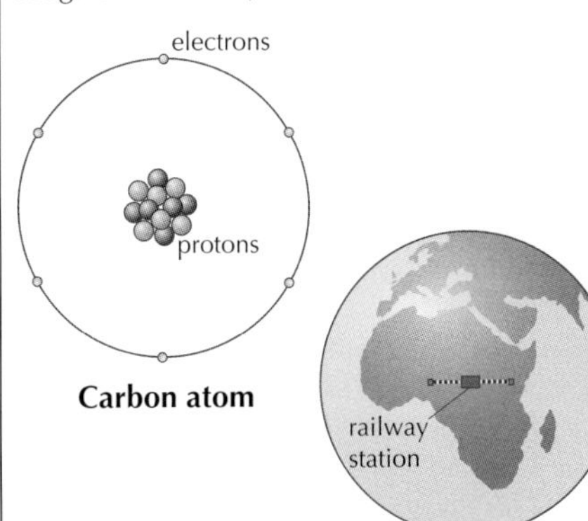

For example, you would probably feel very pleased with yourself if you designed a new, environmentally friendly source of energy that could produce 2.03×10^3 J from 0.72 kg of natural produce. But the meaning of these numbers is not clear – is this a lot or is it a little? In terms of orders of magnitudes, this new source produces 10^3 joules per kilogram of produce. This does not compare terribly well with the 10^5 joules provided by a slice of bread or the 10^8 joules released per kilogram of petrol.

You do NOT need to memorise all of the values shown in the tables, but you should try and develop a familiarity with them. The ranges of magnitudes (first and last value) for the fundamental measurements of mass, length and time need to be known.

RANGE OF MASSES

Mass / kg	
10^{52}	total mass of observable Universe
10^{48}	
10^{44}	
	mass of local galaxy (Milky Way)
10^{40}	
10^{36}	
10^{32}	
	mass of Sun
10^{28}	
10^{24}	mass of Earth
10^{20}	total mass of oceans
	total mass of atmosphere
10^{16}	
10^{12}	
10^{8}	laden oil supertanker
10^{4}	elephant
	human
10^{0}	
	mouse
10^{-4}	
10^{-8}	grain of sand
	blood corpuscle
10^{-12}	
	bacterium
10^{-16}	
10^{-20}	
	haemoglobin molecule
10^{-24}	
10^{-28}	proton
	electron
10^{-32}	

RANGE OF LENGTHS

Size / m	
10^{26}	radius of observable Universe
10^{24}	
10^{22}	
	radius of local galaxy (Milky Way)
10^{20}	
10^{18}	
10^{16}	distance to nearest star
10^{14}	
10^{12}	
	distance from Earth to Sun
10^{10}	
	distance from Earth to moon
10^{8}	
	radius of the Earth
10^{6}	
10^{4}	deepest part of the ocean / highest mountain
10^{2}	tallest building
10^{0}	
10^{-2}	length of fingernail
10^{-4}	thickness of piece of paper
	human blood corpuscle
10^{-6}	wavelength of light
10^{-8}	
10^{-10}	diameter of hydrogen atom
10^{-12}	wavelength of gamma ray
10^{-14}	
	diameter of proton
10^{-16}	

RANGE OF TIMES

Time / s	
10^{20}	
	age of the Universe
10^{18}	
	age of the Earth
10^{16}	
10^{14}	age of species-Homo sapiens
10^{12}	
10^{10}	
	typical human lifespan
10^{8}	
	1 year
10^{6}	
	1 day
10^{4}	
10^{2}	
10^{0}	heartbeat
10^{-2}	
	period of high-frequency sound
10^{-4}	
10^{-6}	
10^{-8}	passage of light across a room
10^{-10}	
10^{-12}	
	vibration of an ion in a solid
10^{-14}	
	period of visible light
10^{-16}	
10^{-18}	
	passage of light across an atom
10^{-20}	
10^{-22}	
	passage of light across a nucleus
10^{-24}	

RANGE OF ENERGIES

Energy / J	
10^{44}	energy released in a supernova
10^{34}	
10^{30}	
	energy radiated by Sun in 1 second
10^{26}	
10^{22}	
	energy released in an earthquake
10^{18}	
10^{14}	energy released by annihilation of 1 kg of matter
10^{10}	energy in a lightning discharge
10^{6}	energy needed to charge a car battery
10^{2}	
	kinetic energy of a tennis ball during game
10^{-2}	
	energy in the beat of a fly's wing
10^{-6}	
10^{-10}	
10^{-14}	
10^{-18}	energy needed to remove electron from the surface of a metal
10^{-22}	
10^{-26}	

The SI system of fundamental and derived units

FUNDAMENTAL UNITS

Any measurement and every quantity can be thought of as being made up of two important parts:

1. the number and
2. the units.

Without **both** parts, the measurement does not make sense. For example a person's age might be quoted as 'seventeen' but without the 'years' the situation is not clear. Are they 17 minutes, 17 months or 17 years old? In this case you would know if you saw them, but a statement like

length = 4.2

actually says nothing. Having said this, it is really surprising to see the number of candidates who forget to include the units in their answers to examination questions.

In order for the units to be understood, they need to be defined. There are many possible systems of measurement that have been developed. In science we use the International System of units (SI). In SI, the **fundamental** or **base** units are as follows

Quantity	SI unit	SI symbol
Mass	kilogram	kg
Length	metre	m
Time	second	s
Electric current	ampere	A
Amount of substance	mole	mol
Temperature	kelvin	K
(Luminous intensity	candela	cd)

You do not need to know the precise definitions of any of these units in order to use them properly.

DERIVED UNITS

Having fixed the fundamental units, all other measurements can be expressed as different combinations of the fundamental units. In other words, all the other units are **derived units**. For example, the fundamental list of units does not contain a unit for the measurement of speed. The definition of speed can be used to work out the derived unit.

$$\text{Since speed} = \frac{\text{distance}}{\text{time}}$$

$$\text{Units of speed} = \frac{\text{units of distance}}{\text{units of time}}$$

$$= \frac{\text{metres}}{\text{seconds}} \text{ (pronounced 'metres per second')}$$

$$= \frac{\text{m}}{\text{s}}$$

$$= \text{m s}^{-1}$$

Of the many ways of writing this unit, the last way (m s^{-1}) is the best.

Sometimes particular combinations of fundamental units are so common that they are given a new derived name. For example, the unit of force is a derived unit – it turns out to be kg m s^{-2}. This unit is given a new name the newton (N) so that 1N = 1 kg m s^{-2}.

The great thing about SI is that, so long as the numbers that are substituted into an equation are in SI units, then the answer will also come out in SI units. You can always 'play safe' by converting all the numbers into proper SI units. Sometimes, however, this would be a waste of time.

There are some situations where the use of SI becomes awkward. In astronomy, for example, the distances involved are so large that the SI unit (the metre) always involves large orders of magnitudes. In these cases, the use of a different (but non SI) unit is very common. Astronomers can use the astronomical unit (AU), the light-year (ly) or the parsec (pc) as appropriate. Whatever the unit, the conversion to SI units is simple arithmetic.

$1 \text{ AU} = 1.5 \times 10^{11} \text{ m}$

$1 \text{ ly} = 9.5 \times 10^{15} \text{ m}$

$1 \text{ pc} = 3.1 \times 10^{16} \text{ m}$

There are also some units (for example the hour) which are so common that they are often used even though they do not form part of SI. Once again, before these numbers are substituted into equations they need to be converted.

The table below lists the SI derived units that you will meet.

SI derived unit	SI base unit	Alternative SI unit
newton (N)	kg m s^{-2}	–
pascal (Pa)	kg m^{-1} s^{-2}	N m^{-2}
hertz (Hz)	s^{-1}	–
joule (J)	kg m^2 s^{-2}	N m
watt (W)	kg m^2 s^{-3}	J s^{-1}
coulomb (C)	A s	–
volt (V)	kg m^2 s^{-3} A^{-1}	WA^{-1}
ohm (Ω)	kg m^2 s^{-3} A^{-2}	VA^{-1}
weber (Wb)	kg m^2 s^{-2} A^{-1}	V s
tesla (T)	kg s^{-2} A^{-1}	Wb m^{-2}
becquerel (Bq)	s^{-1}	–
gray (Gy)	m^2 s^{-2}	J kg^{-1}
sievert (Sv)	m^2 s^{-2}	J kg^{-1}

PREFIXES

To avoid the repeated use of scientific notation, an alternative is to use one of the list of agreed prefixes given in the IB data booklet. These can be very useful but they can also lead to errors in calculations. It is very easy to forget to include the conversion factor.

For example, 1 kW = 1000 W. 1 mW = 10^{-3} W (in other words, $\frac{1\text{W}}{1000}$)

Uncertainties and error in experimental measurement

ERRORS – RANDOM AND SYSTEMATIC (PRECISION AND ACCURACY)

An experimental error just means that there is a difference between the recorded value and the 'perfect' or 'correct' value. Errors can be categorised as **random** or **systematic**.

Repeating readings does not reduce systematic errors.

Sources of random errors include

- The readability of the instrument
- The observer being less than perfect
- The effects of a change in the surroundings.

Sources of systematic errors include

- An instrument with **zero error**. To correct for zero error the value should be subtracted from every reading.
- An instrument being wrongly **calibrated**
- The observer being less than perfect in the same way every measurement

An **accurate** experiment is one that has a small systematic error, whereas a **precise** experiment is one that has a small random error.

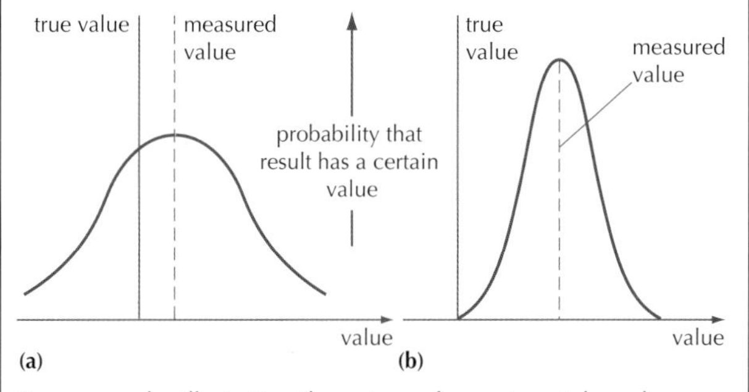

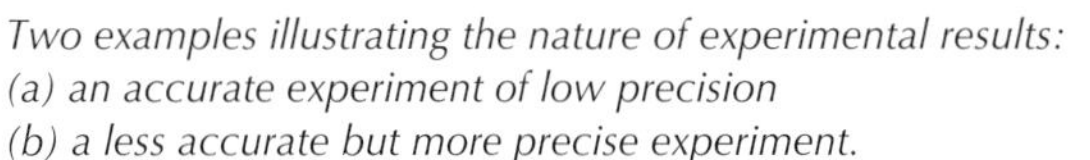

Two examples illustrating the nature of experimental results:
(a) an accurate experiment of low precision
(b) a less accurate but more precise experiment.

GRAPHICAL REPRESENTATION OF UNCERTAINTY

In many situations the best method of presenting and analysing data is to use a graph. If this is the case, a neat way of representing the uncertainties is to use **error bars**. The graphs below explains their use.

Since the error bar represents the uncertainty range, the 'best-fit' line of the graph should pass through ALL of the rectangles created by the error bars.

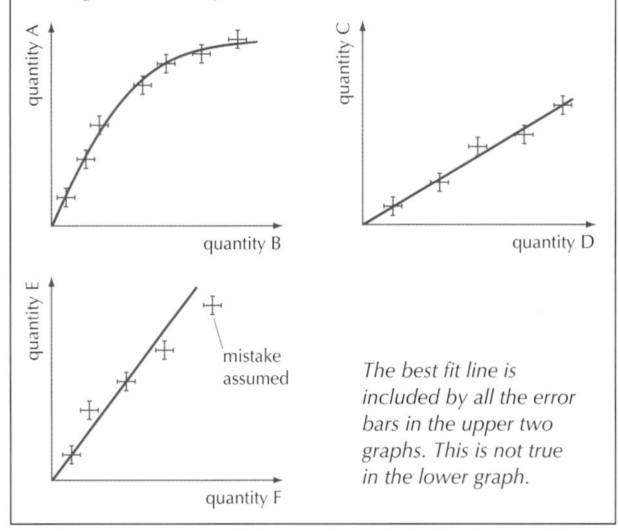

The best fit line is included by all the error bars in the upper two graphs. This is not true in the lower graph.

Systematic and random errors can often be recognised from a graph of the results.

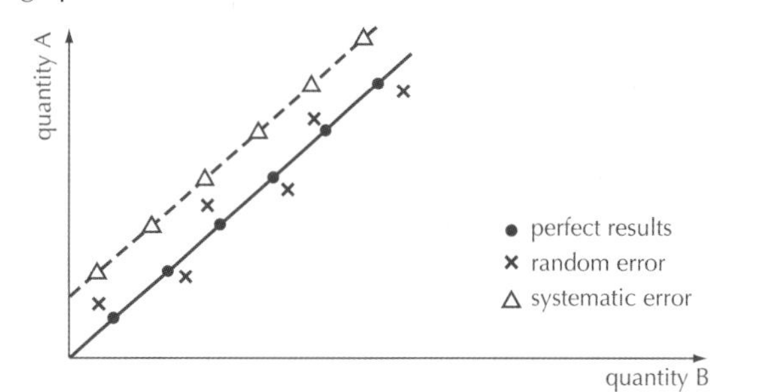

Perfect results, random and systematic errors of two proportional quantities.

ESTIMATING THE UNCERTAINTY RANGE

An **uncertainty range** applies to any experimental value. The idea is that, instead of just giving one value that implies perfection, we give the likely range for the measurement.

1. Estimating from first principles

All measurement involves a readability error. If we use a measuring cylinder to find the volume of a liquid, we might think that the best estimate is 73 cm^3, but we know that it is not exactly this value (73.000 000 000 00 cm^3).

Uncertainty range is ± 5 cm^3.
We say volume = 73 ± 5 cm^3.

Normally the uncertainty range due to readability is estimated as below.

Device	Example	Uncertainty
Analogue scale	Rulers, meters with moving pointers	± (half the smallest scale division)
Digital scale	Top-pan balances, digital meters	± (the smallest scale division)

2. Estimating uncertainty range from several repeated measurements

If the time taken for a trolley to go down a slope is measured five times, the readings in seconds might be 2.01, 1.82, 1.97, 2.16 and 1.94. The average of these five readings is 1.98 s. The deviation of the largest and smallest readings can be calculated (2.16 – 1.98 = 0.18; 1.98 – 1.82 = 0.16). The largest value is taken as the uncertainty range. In this example the time is 1.98 s ± 0.18 s

SIGNIFICANT DIGITS

Any experimental measurement should be quoted with its uncertainty. This indicates the possible range of values for the quantity being measured. At the same time, the number of **significant digits** used will act as a guide to the amount of uncertainty. For example, a measurement of mass which is quoted as 23.456 g implies an uncertainty of ± 0.001 g (it has five significant digits), whereas one of 23.5 g implies an uncertainty of ± 0.1 g (it has three significant digits).

A simple rule for calculations (multiplication or division) is to quote the answer to the same number of significant digits as the LEAST precise value that is used.

For a more complete analysis of how to deal with uncertainties in calculated results, see page 8.

Estimation

ORDERS OF MAGNITUDE

It is important to develop a 'feeling' for some of the numbers that you use. When using a calculator, it is very easy to make a simple mistake (e.g. by entering the data incorrectly). A good way of checking the answer is to first make an estimate before resorting to the calculator. The multiple-choice paper (paper 1) does not allow the use of calculators.

Approximate values for each of the fundamental SI units are given below.

1 kg	A packet of sugar, 1 litre of water. A person would be about 50 kg or more
1 m	Distance between one's hands with arms outstretched
1 s	Duration of a heart beat (when resting – it can easily double with exercise)
1 amp	Current flowing from the mains electricity when a computer is connected. The maximum current to a domestic device would be about 10 A or so
1 kelvin	1K is a very low temperature. Water freezes at 273 K and boils at 373 K. Room temperature is about 300 K
1 mol	12 g of carbon–12. About the number of atoms of carbon in the 'lead' of a pencil

The same process can happen with some of the derived units.

1 m s^{-1}	Walking speed. A car moving at 30 m s^{-1} would be fast
1 m s^{-2}	Quite a slow acceleration. The acceleration of gravity is 10 m s^{-2}
1 N	A small force – about the weight of an apple
1 V	Batteries generally range from a few volts up to 20 or so, the mains is several hundred volts
1 Pa	A very small pressure. Atmospheric pressure is about 10^5 Pa
1 J	A very small amount of energy – the work done lifting a apple off the ground

POSSIBLE REASONABLE ASSUMPTIONS

Everyday situations are very complex. In physics we often simplify a problem by making simple assumptions. Even if we know these assumptions are not absolutely true they allow us to gain an understanding of what is going on. At the end of the calculation it is often possible to go back and work out what would happen if our assumption turned out not to be true.

The table below lists some common assumptions. Be careful not to assume too much! Additionally we often have to assume that some quantity is constant even if we know that in reality it is varying slightly all the time.

Assumption	Example
Friction is negligible	Many mechanics situations – but you need to be very careful
No heat lost	Almost all thermal situations
Mass of connecting string etc is negligible	Many mechanics situations
Resistance of ammeter is zero	Circuits
Resistance of voltmeter is infinite	Circuits
Internal resistance of battery is zero	Circuits
Material obeys Ohm's law	Circuits
Machine 100% efficient	Many situations
Gas is ideal	Thermodynamics
Collision is elastic	Only gas molecules have perfectly elastic collisions
Object radiates as a perfect black body	Thermal equilibrium e.g. planets

CALCULUS NOTATION

If one wants to mathematically analyse motion in detail, the correct way to do this would be to use a branch of mathematics called calculus. A knowledge of the details of calculus is **not** required for the IB physics course. However sometimes it does help to be able to use calculus shorthand.

Symbol	Pronounced	Meaning	Example
Δx	'Delta x'	= 'the change in x'	Δt means 'the change in time'
δx	'Delta x'	= 'the small change in x'	δt means 'the small change in time'
$\frac{\Delta x}{\Delta t}$	'Delta x divided by delta t' or 'Delta x over delta t'	= 'the AVERAGE rate of change of x' This average is calculated over a relatively large period of time	$\frac{\Delta s}{\Delta t}$ often means the average speed
$\frac{\delta x}{\delta t}$	'Delta x divided by delta t' or 'Delta x over delta t	= 'the AVERAGE rate of change of x' This average is calculated over a small period of time	$\frac{\delta s}{\delta t}$ means the average speed
$\frac{dx}{dt}$	'Dee x by dee t' or Dee by dee t of x	= 'the INSTANTANEOUS rate of change of x' This value is calculated at one instant of time	$\frac{ds}{dt}$ means the instantaneous speed

Graphs

PLOTTING GRAPHS – AXES AND BEST FIT

The reason for plotting a graph in the first place is that it allows us to identify trends. To be precise, it allows us a visual way of representing the variation of one quantity with respect to another. When plotting graphs, you need to make sure that all of the following points have been remembered

- The graph should have a title. Sometimes they also need a key.
- The scales of the axes should be suitable – there should not, of course, be any sudden or uneven 'jumps' in the numbers.
- The inclusion of the origin has been thought about. Most graphs should have the origin included – it is rare for a graph to be improved by this being missed out. If in doubt include it. You can always draw a second graph without it if necessary.
- The final graph should, if possible, cover more than half the paper in either direction.
- The axes are labelled with both the quantity (e.g. current) AND the units (e.g. amps).
- The points are clear. Vertical and horizontal lines to make crosses are better than 45 degree crosses or dots.
- All the points have been plotted correctly.
- Error bars are included if appropriate.
- A best-fit trend line is added. This line NEVER just 'joins the dots' – it is there to show the overall trend.
- If the best-fit line is a curve, this has been drawn as a single smooth line.
- If the best-fit line is a straight line, this has been added WITH A RULER
- As a general rule, there should be roughly the same number of points above the line as below the line.
- Check that the points are randomly above and below the line. Sometimes people try to fit a best-fit straight line to points that should be represented by a gentle curve. If this was done then points below the line would be at the beginning of the curve and all the points above the line would be at the end, or vice versa.
- Any points that do not agree with the best-fit line have been identified.

MEASURING INTERCEPT, GRADIENT AND AREA UNDER THE GRAPH

Graphs can be used to analyse the data. This is particularly easy for straight-line graphs, though many of the same principles can be used for curves as well. Three things are particularly useful: the **intercept**, the **gradient** and the **area under the graph**.

1. Intercept

In general, a graph can intercept (cut) either axis any number of times. A straight-line graph can only cut each axis once and often it is the ***y*-intercept** that has particular importance. (Sometimes the *y*-intercept is referred to as simply 'the intercept'.) If a graph has an intercept of zero it goes through the origin. **Proportional** – note that two quantities are proportional if the graph is a straight line THAT PASSES THROUGH THE ORIGIN.

Sometimes a graph has to be 'continued on' (outside the range of the readings) in order for the intercept to be found. This process is known as **extrapolation**. The process of assuming that the trend line applies between two points is known as **interpolation**.

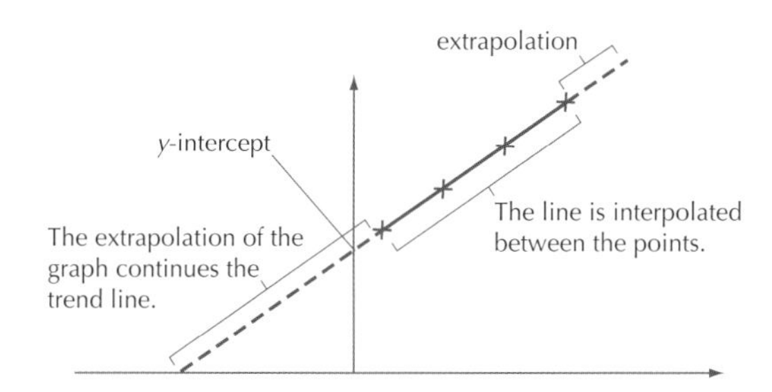

2. Gradient

The gradient of a straight-line graph is the increase in the *y*-axis value divided by the increase in the *x*-axis value.

The following points should be remembered

- A straight-line graph has a constant gradient.
- The triangle used to calculate the gradient should be as large as possible.
- The gradient has units. They are the units on the *y*-axis divided by the units on the *x*-axis.
- Only if the *x*-axis is a measurement of time does the gradient represent the RATE at which the quantity on the *y*-axis increases.

The gradient of a curve at any particular point is the gradient of the tangent to the curve at that point.

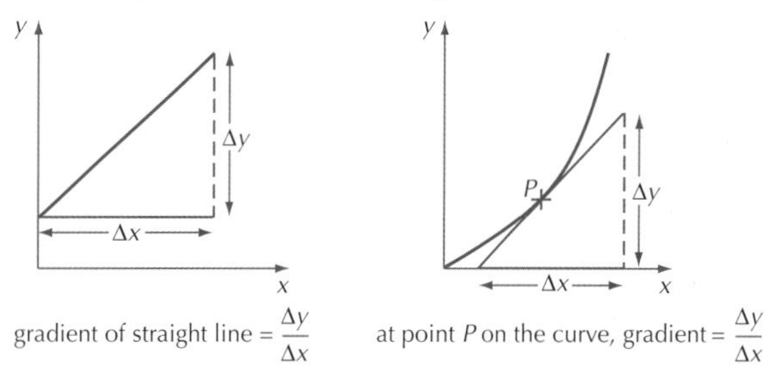

gradient of straight line = $\frac{\Delta y}{\Delta x}$ at point P on the curve, gradient = $\frac{\Delta y}{\Delta x}$

3. Area under a graph

The area under a straight-line graph is the product of multiplying the average quantity on the *y*-axis by the quantity on the *x*-axis. This does not always represent a useful physical quantity. When working out the area under the graph

- If the graph consists of straight-line sections, the area can be worked out by dividing the shape up into simple shapes.
- If the graph is a curve, the area can be calculated by 'counting the squares' and working out what one square represents.
- The units for the area under the graph are the units on the *y*-axis multiplied by the units on the *x*-axis.
- If the mathematical equation of the line is known, the area of the graph can be calculated using a process called **integration**.

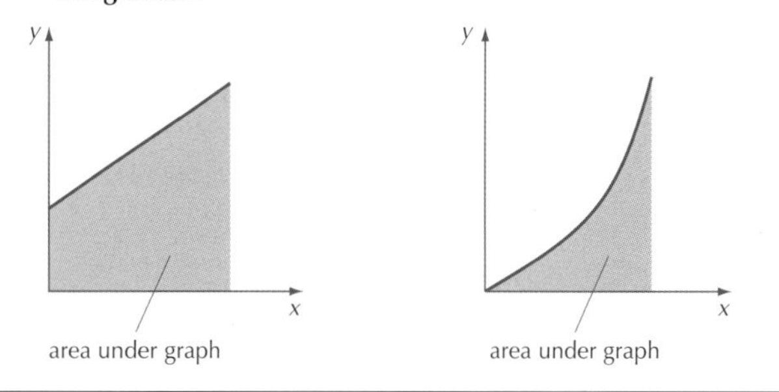

Graphical analysis and determination of relationships

EQUATION OF A STRAIGHT-LINE GRAPH

All straight-line graphs can be described using one general equation

$$y = mx + c$$

y and x are the two variables (to match with the y-axis and the x-axis).
m and c are both constants – they have one fixed value.

- c represents the intercept on the y-axis (the value y takes when $x = 0$)
- m is the gradient of the graph.

In some situations, a direct plot of the measured variable will give a straight line. In some other situations we have to choose carefully what to plot in order to get a straight line. In either case, once we have a straight line, we then use the gradient and the intercept to calculate other values.

For example, a simple experiment might measure the velocity of a trolley as it rolls down a slope. The equation that describes the motion is $v = u + at$ where u is the initial velocity of the object. In this situation v and t are our variables, a and u are the constants.

You should be able to see that the physics equation has exactly the same form as the mathematical equation. The order has been changed below so as to emphasis the link.

$$v = u + at$$
$$y = c + mx$$

By comparing these two equations, you should be able to see that if we plot the velocity on the y-axis and the time on the x-axis we are going to get a straight-line graph.

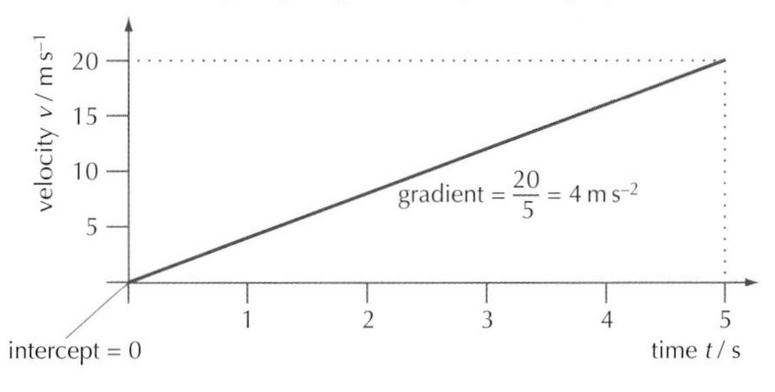

The comparison also works for the constants.

- c (the y-intercept) must be equal to the initial velocity u
- m (the gradient) must be equal to the acceleration a

In this example the graph tells us that the trolley must have started from rest (intercept zero) and it had a constant acceleration of 4.0 m s^{-2}.

CHOOSING WHAT TO PLOT TO GET A STRAIGHT LINE

With a little rearrangement we can often end up with the physics equation in the same form as the mathematical equation of a straight line. Important points include

- Identify which symbols represent variables and which symbols represent constants.
- The symbols that correspond to x and y must be variables and the symbols that correspond to m and c must be constants.
- If you take a variable reading and square it (or cube, square root, reciprocal etc) – the result is still a variable and you could choose to plot this on one of the axes.
- You can plot any mathematical combination of your original readings on one axis – this is still a variable.
- Sometimes the physical quantities involved use the symbols m (e.g. mass) or c (e.g. speed of light). Be careful not to confuse these with the symbols for gradient or intercept.

Example 1
The gravitational force F that acts on an object at a distance r away from the centre of a planet is given by the equation

$F = \frac{GMm}{r^2}$ where M is the mass of the planet and the m is mass of the object.

If we plot force against distance we get a curve (graph 1).

We can restate the equation as $F = \frac{GMm}{r^2} + 0$

and if we plot F on the y-axis and $\frac{1}{r^2}$ on the x-axis

we will get a straight-line (graph 2).

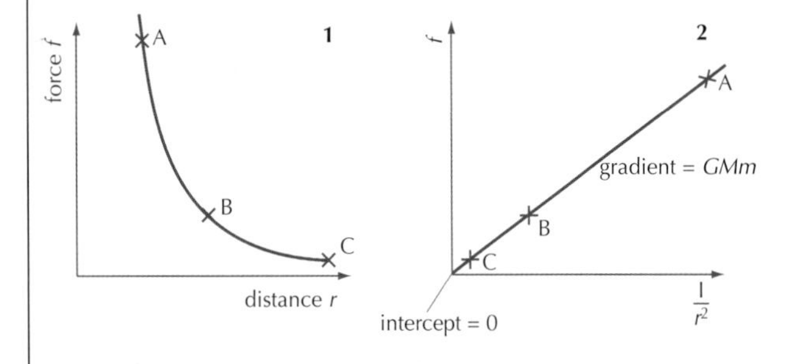

Example 2
If an object is placed in front of a lens we get an image. The image distance v is related to the object distance u and the focal length of the lens f by the following equation.

$$\frac{1}{u} + \frac{1}{v} = \frac{1}{f}$$

There are many possible ways to rearrange this in order to get it into straight–line form. You should check that all these are algebraically the same.

$$v + u = \frac{uv}{f} \quad \text{or} \quad \frac{v}{u} = \frac{v}{f} - 1 \quad \text{or} \quad \frac{1}{u} = \frac{1}{f} - \frac{1}{v}$$

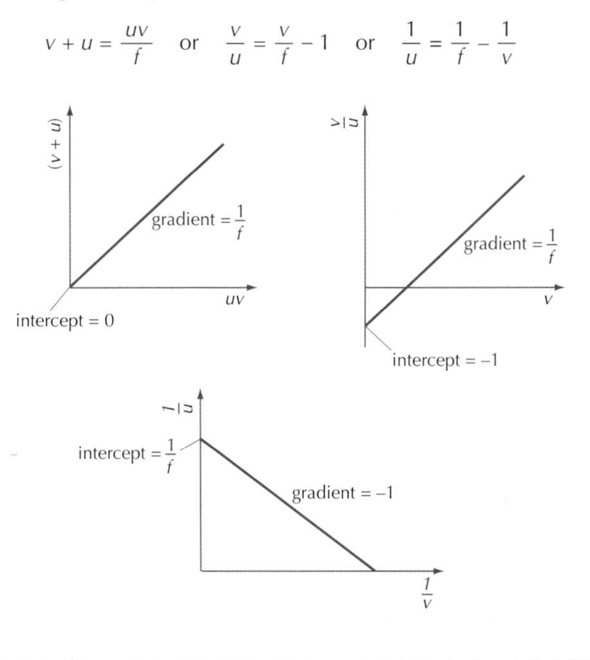

Graphical analysis – logarithmic functions

LOGS – BASE TEN AND BASE e

Mathematically,

If $a = 10^{b}$

Then $\log(a) = b$

[to be absolutely precise $\log_{10}(a) = b$]

Most calculators have a 'log' button on them. But we don't **have** to use 10 as the base. We can use any number that we like. For example we could use 2.0, 563.2, 17.5, 42 or even 2.7182818284590452353602874714. For complex reasons this last number IS the most useful number to use! It is given the symbol e and logarithms to this base are called **natural logarithms**. The symbol for natural logarithms is ln (x). This is also on most calculators.

If $p = e^{q}$

Then $\ln(p) = q$

The powerful nature of logarithms means that we have the following rules

$\ln(c \times d) = \ln(c) + \ln(d)$

$\ln(c \div d) = \ln(c) - \ln(d)$

$\ln(c^{n}) = n \ln(c)$

$\ln\left(\frac{1}{c}\right) = -\ln(c)$

These rules have been expressed for natural logarithms, but they work for all logarithms whatever the base.

The point of logarithms is that they can be used to express some relationships (particularly power laws and exponentials) in straight line form. This means that we will be plotting graphs with logarithmic scales.

A normal scale increases by the same amount each time.

1 2 3 4 5 6 7 8 9 10 11

A logarithmic scale increases by the same ratio all the time.

10^0	10^1	10^2	10^3
1	10	100	1000

POWER LAWS AND LOGS (LOG-LOG)

When an experimental situation involves a power law it is often only possible to transform it into straight-line form by taking logs. For example, the time period of a simple pendulum, T, is related to its length, l, by the following equation.

$T = k\, l^{p}$

k and p are constants.

A plot of the variables will give a curve, but it is not clear from this curve what the values of k and p work out to be. On top of this, if we do not know what the value of p is, we can not calculate the values to plot a straight-line graph.

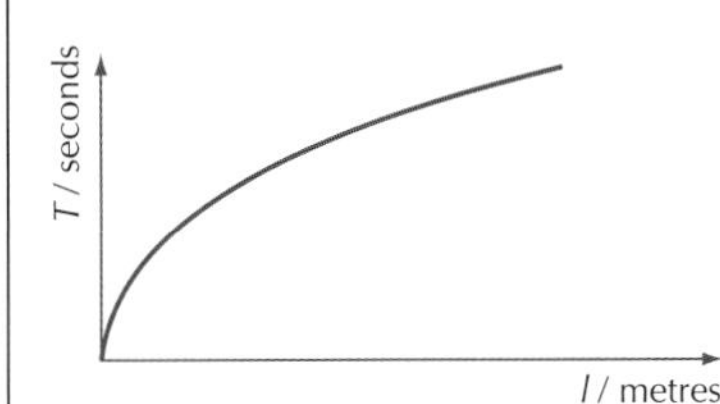

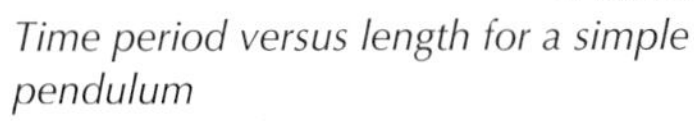
Time period versus length for a simple pendulum

The 'trick' is to take logs of both sides of the equation. The equations below have used natural logarithms, but would work for all logarithms whatever the base.

$\ln(T) = \ln(k\, l^{p})$

$\ln(T) = \ln(k) + \ln(l^{p})$

$\ln(T) = \ln(k) + p \ln(l)$

This is now in the same form as the equation for a straight-line

$y = c + mx$

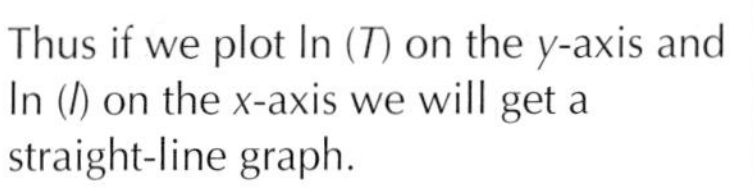
Thus if we plot ln (T) on the y-axis and ln (l) on the x-axis we will get a straight-line graph.

The gradient will be equal to p

The intercept will be equal to $\ln(k)$ [so $k = e^{(\text{intercept})}$]

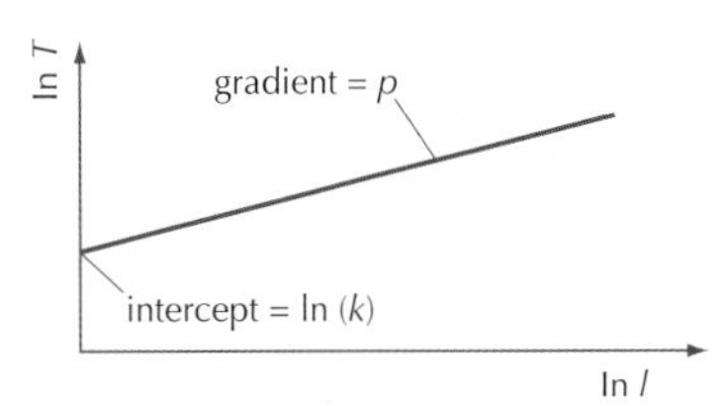

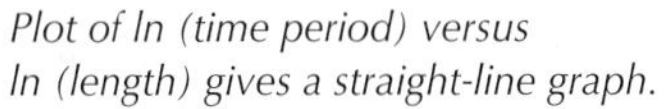
Plot of ln (time period) versus ln (length) gives a straight-line graph.

Both the gravity force and the electrostatic force are inverse-square relationships. This means that the force ∝ (distance apart)$^{-2}$. The same technique can be used to generate a straight-line graph.

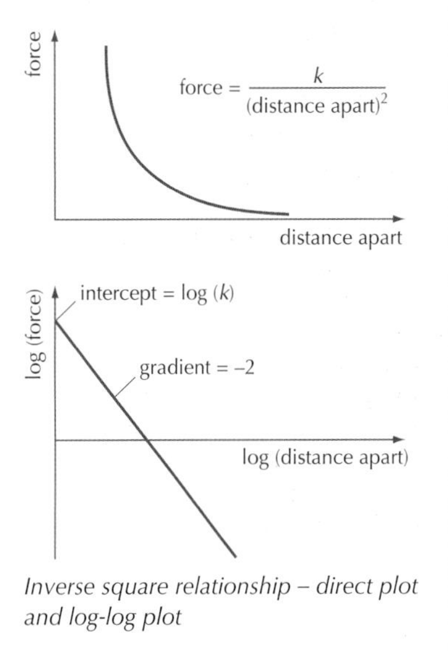

Inverse square relationship – direct plot and log-log plot

EXPONENTIALS AND LOGS (LOG-LINEAR)

Natural logarithms are very important because many natural processes are exponential. Radioactive decay is an important example. In this case, once again the taking of logarithms allows the equation to be compared with the equation for a straight line.

For example, the count rate R at any given time t is given by the equation

$R = R_0\, e^{-\lambda t}$

R_0 and λ are constants.

If we take logs, we get

$\ln(R) = \ln(R_0\, e^{-\lambda t})$

$\ln(R) = \ln(R_0) + \ln(e^{-\lambda t})$

$\ln(R) = \ln(R_0) - \lambda t \ln(e)$

$\ln(R) = \ln(R_0) - \lambda t$ [$\ln(e) = 1$]

This can be compared with the equation for a straight-line graph

$y = c + mx$

Thus if we plot ln (R) on the y-axis and t on the x-axis, we will get a straight line.

Gradient = $-\lambda$

Intercept = $\ln(R_0)$

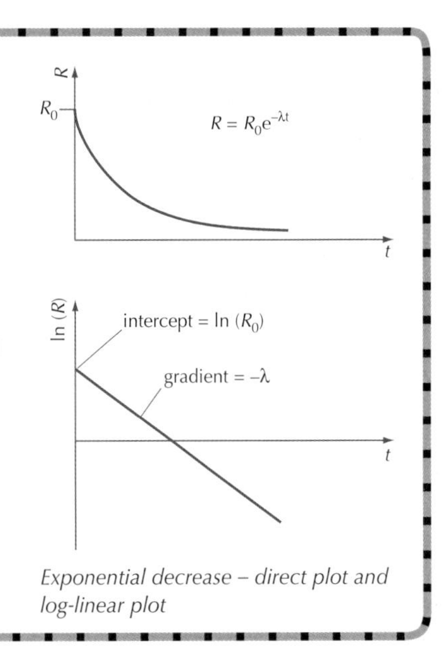

Exponential decrease – direct plot and log-linear plot

Uncertainties in calculated results

MATHEMATICAL REPRESENTATION OF UNCERTAINTIES

For example if the mass of a block was measured as 10 ± 1 g and the volume was measured as 5.0 ±0.2 cm^3, then the full calculations for the density would be as follows.

Best value for density = $\frac{\text{mass}}{\text{volume}} = \frac{10}{5} = 2.0 \text{ g cm}^{-3}$

The largest possible value of density = $\frac{11}{4.8} = 2.292 \text{ g cm}^{-3}$

The smallest possible value of density = $\frac{9}{5.2} = 1.731 \text{ g cm}^{-3}$

Rounding these values gives density = $2.0 \pm 0.3 \text{ g cm}^{-3}$

We can express this uncertainty in one of three ways – using **absolute**, **fractional** or **percentage uncertainties**.

If a quantity p is measured then the absolute uncertainty would be expressed as $\pm\Delta p$.

Then the fractional uncertainty is

$$\frac{\pm\Delta p}{p},$$

which makes the percentage uncertainty

$$\frac{\pm\Delta p}{p} \times 100\%.$$

In the example above, the fractional uncertainty is ±0.15 or ±15%.

Thus equivalent ways of expressing this error are

density = $2.0 \pm 0.3 \text{ g cm}^{-3}$

OR density = $2.0 \text{ g cm}^{-3} \pm 15\%$

Working out the uncertainty range is very time consuming. There are some mathematical 'short-cuts' that can be used. These are introduced in the boxes below.

MULTIPLICATION, DIVISION OR POWERS

Whenever two or more quantities are multiplied or divided and they each have uncertainties, the overall uncertainty is approximately equal to the **addition** of the **percentage** (fractional) uncertainties.

Using the same numbers from above,

$\Delta m = \pm 1$ g

$$\frac{\Delta m}{m} = \pm\left(\frac{1 \text{ g}}{10 \text{ g}}\right) = \pm 0.1 = \pm 10\%$$

$\Delta v = \pm 0.2 \text{ cm}^3$

$$\frac{\Delta V}{V} = \pm\left(\frac{0.2 \text{ cm}^3}{5 \text{ cm}^3}\right) = \pm 0.04 = \pm 4\%$$

The total % uncertainty in the result = ± (10 +4)%
= ± 14 %

14% of $2.0 \text{ g cm}^{-3} = 0.28 \text{ g cm}^{-3} \approx 0.3 \text{ g cm}^{-3}$

So density = $2.0 \pm 0.3 \text{ g cm}^{-3}$ as before.

In symbols, if $P = Q \times R$ or if $P = \frac{Q}{R}$

Then $\frac{\Delta P}{P} = \frac{\Delta Q}{Q} + \frac{\Delta R}{R}$ [note this is ALWAYS added]

Power relationships are just a special case of this law.

If $P = R^n$

Then $\frac{\Delta P}{P} = n\left(\frac{\Delta R}{R}\right)$

For example if a cube is measured to be 4.0 ± 0.1 cm in length along each side, then

% Uncertainty in length = $\pm\left(\frac{0.1}{4.0}\right) = \pm 2.5\ \%$

Volume = $(\text{length})^3 = (4.0)^3 = 64 \text{ cm}^3$

% Uncertainty in [volume] = % uncertainty in [$(\text{length})^3$]
= 3×(% uncertainty in [length])
= 3×(± 2.5 %)
= ± 7.5 %

Absolute uncertainty = 7.5% of 64 cm^3
= 4.8 cm^3 ≈ 5 cm^3

Thus volume of cube = 64 ± 5 cm^3

OTHER MATHEMATICAL OPERATIONS

If the calculation involves mathematical operations other than multiplication, division or raising to a power, then one has to find the highest and lowest possible values.

Addition or subtraction

Whenever two or more quantities are added or subtracted and they each have uncertainties, the overall uncertainty is equal to the **addition** of the **absolute** uncertainties.

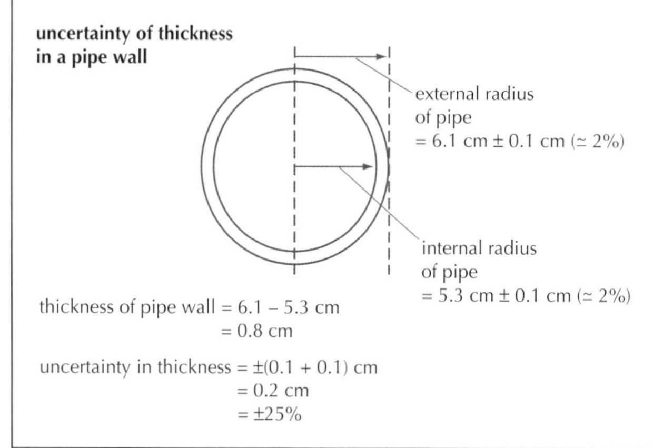

Other functions

There are no 'short-cuts' possible.

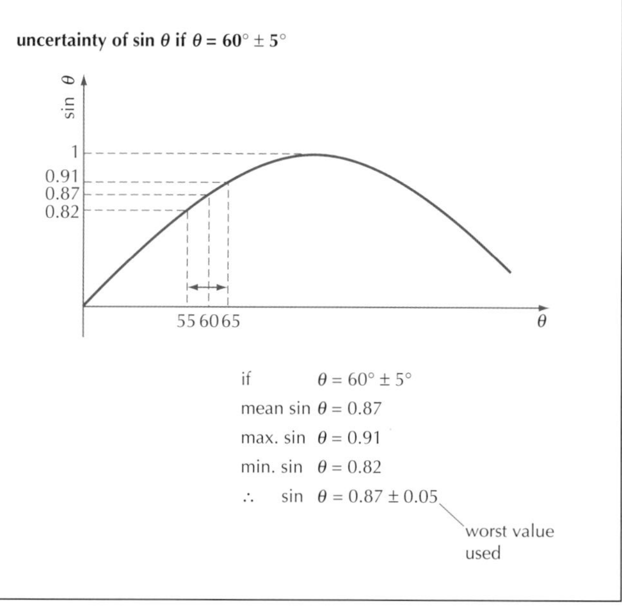

Uncertainties in graphs

ERROR BARS

Plotting a graph allows one to visualise all the readings at one time. Ideally all of the points should be plotted with their error bars. In principle, the size of the error bar could well be different for every single point and so they should be individually worked out.

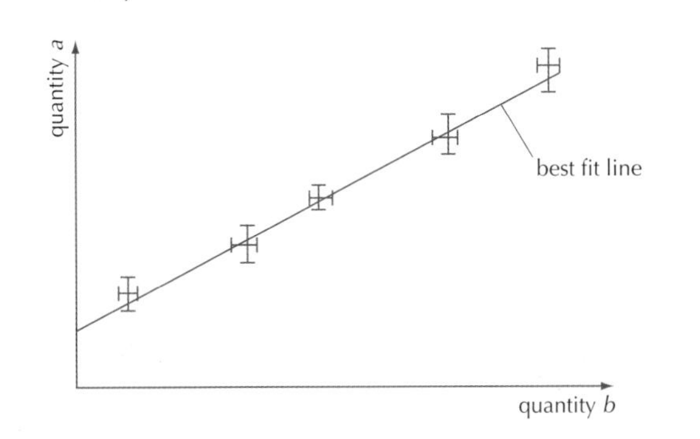

In practice, it would often take too much time to add all the correct error bars, so some (or all) of the following short cuts could be considered.

- Rather than working out error bars for each point – use the worst value and assume that all of the other error bars are the same.
- Only plot the error bar for the 'worst' point i.e. the point that is furthest from the line of best fit. If the line of best fit is within the limits of this error bar, then it will probably be within the limits of all the error bars.
- Only plot the error bars for the first and the last points. These are often the most important points when considering the uncertainty ranges calculated for the gradient or the intercept (see below).
- Only include the error bars for the axis that has the worst uncertainty.

UNCERTAINTY IN SLOPES

If the gradient of the graph has been used to calculate a quantity, then the uncertainties of the points will give rise to an uncertainty in the gradient. Using the steepest and the shallowest lines possible (i.e. the lines that are still consistent with the error bars) the uncertainty range for the gradient is obtained. This process is represented below.

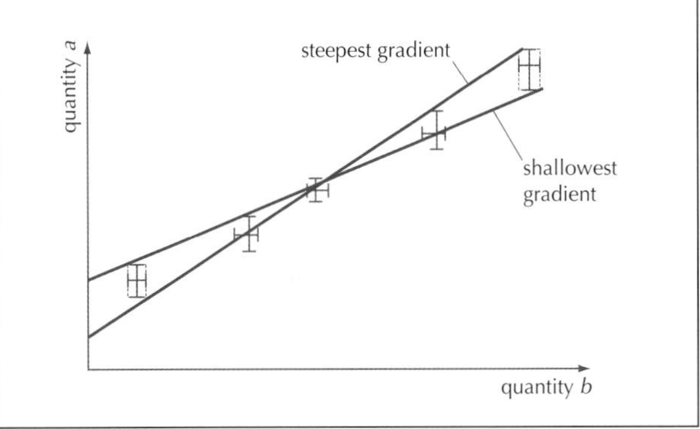

UNCERTAINTY IN INTERCEPTS

If the intercept of the graph has been used to calculate a quantity, then the uncertainties of the points will give rise to an uncertainty in the intercept. Using the steepest and the shallowest lines possible (i.e. the lines that are still consistent with the error bars) we can obtain the uncertainty in the result. This process is represented below.

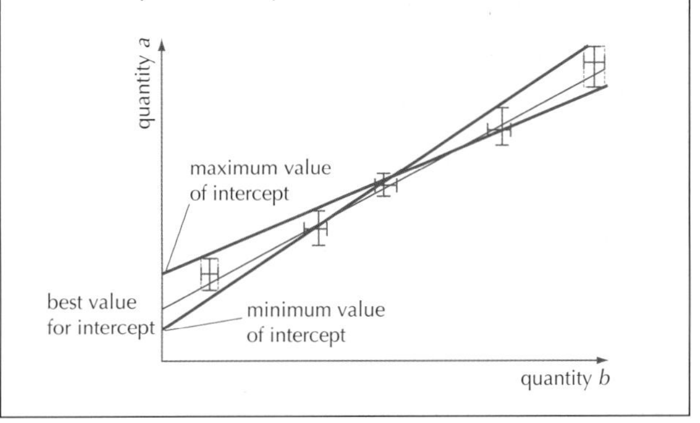

Vectors and scalars

DIFFERENCE BETWEEN VECTORS AND SCALARS

If you measure any quantity, it must have a number AND a unit. Together they express the **magnitude** of the quantity. Some quantities also have a direction associated with them. A quantity that has magnitude and direction is called a **vector** quantity whereas one that has only magnitude is called a **scalar** quantity. For example, all forces are vectors.

The table lists some common quantities. The first two quantities in the table are linked to one another by their definitions (see page 13). All the others are in no particular order.

Vectors	**Scalars**
Displacement ⟷	Distance
Velocity ⟷	Speed
Acceleration	Mass
Force	Energy (all forms)
Momentum	Temperature
Electric field strength	Potential or potential difference
Magnetic field strength	Density
Gravitational field strength	Area

Although the vectors used in many of the given examples are forces, the techniques can be applied to all vectors.

COMPONENTS OF VECTORS

It is also possible to 'split' one vector into two (or more) vectors. This process is called **resolving** and the vectors that we get are called the **components** of the original vector. This can be a very useful way of analysing a situation if we choose to resolve all the vectors into two directions that are at right angles to one another.

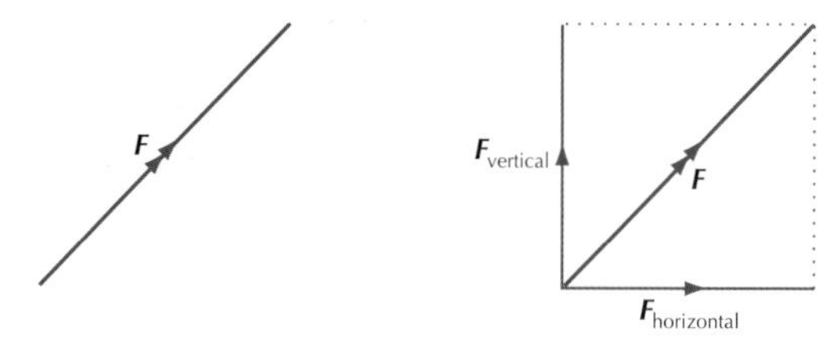

Splitting a vector into components

These 'mutually perpendicular' directions are totally independent of each other and can be analysed separately. If appropriate, both directions can then be combined at the end to work out the final resultant vector.

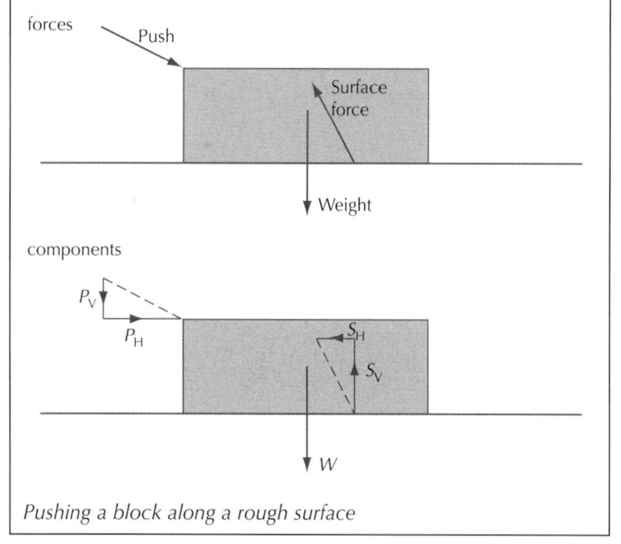

Pushing a block along a rough surface

REPRESENTING VECTORS

In most books a bold letter is used to represent a vector whereas a normal letter represents a scalar. For example ***F*** would be used to represent a force in magnitude AND direction.The list below shows some other recognised methods.

$\vec{F}$, $\bar{F}$ or $\underset{\sim}{F}$

Vectors are best shown in diagrams using arrows:

- the relative magnitudes of the vectors involved are shown by the relative length of the arrows
- the direction of the vectors is shown by the direction of the arrows.

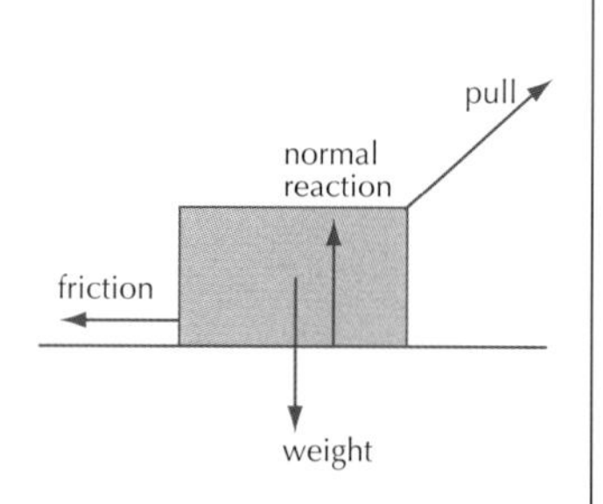

ADDITION / SUBTRACTION OF VECTORS

If we have a 3N and a 4N force, the overall force (resultant force) can be anything between 1N and 7N depending on the directions involved.

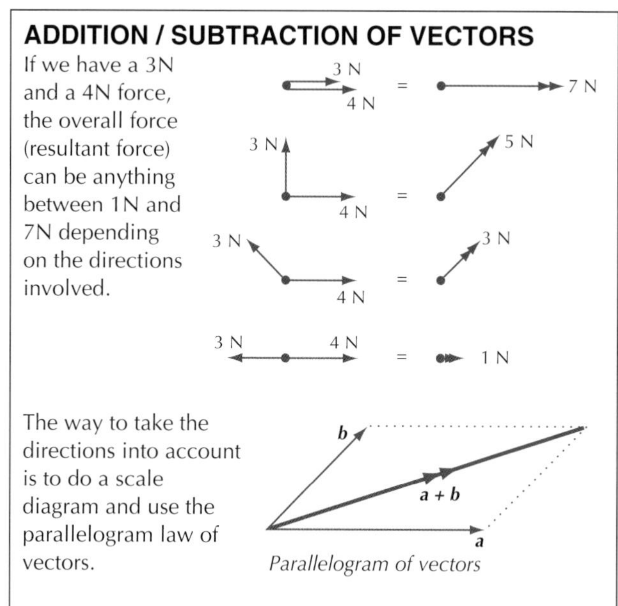

The way to take the directions into account is to do a scale diagram and use the parallelogram law of vectors.

Parallelogram of vectors

This process is the same as adding vectors in turn – the 'tail' of one vector is drawn starting from the head of the previous vector.

TRIGONOMETRY

Vector problems can always be solved using scale diagrams, but this can be very time consuming. The mathematics of trigonometry often makes it much easier to use the mathematical functions of sine or cosine. This is particularly appropriate when resolving. The diagram below shows how to calculate the values of either of these components.

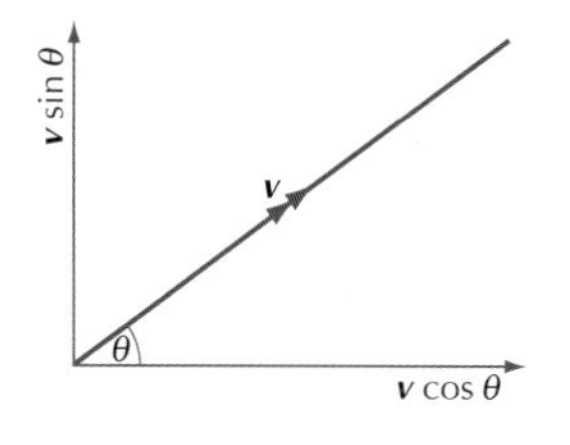

See page 16 for an example.

IB QUESTIONS – PHYSICS AND PHYSICAL MEASUREMENT

1 Which one of the following quantities is a scalar?

A Weight

B Distance

C Velocity

D Momentum

2 Which one of the following is a fundamental unit in the International System of units (S.I.)?

A newton

B ampere

C joule

D pascal

3 Gravitational field strength may be specified in N kg^{-1}. Units of N kg^{-1} are equivalent to

A m s^{-1}

B m s^{-2}

C kg m s^{-1}

D kg m s^{-2}

4 Which one of the following is a scalar quantity?

A Electric field

B Acceleration

C Power

D Momentum

5 A motor car travels on a circular track of radius, a, as shown in the figure. When the car has travelled from P to Q its displacement from P is

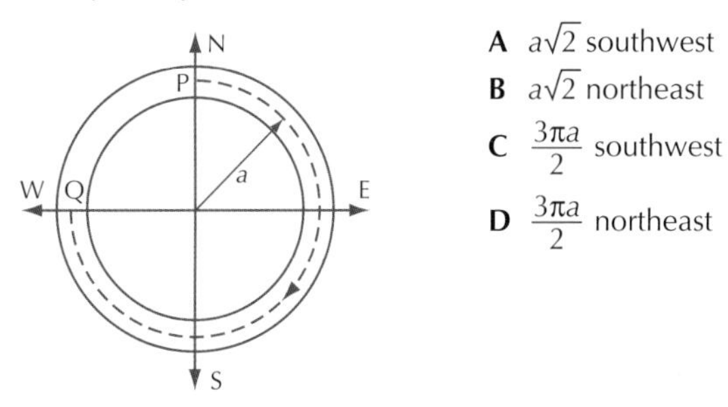

A $a\sqrt{2}$ southwest

B $a\sqrt{2}$ northeast

C $\frac{3\pi a}{2}$ southwest

D $\frac{3\pi a}{2}$ northeast

6 The frequency of oscillation f of a mass m suspended from a vertical spring is given by

$$f = \frac{1}{2\pi}\sqrt{\frac{k}{m}}$$

where k is the spring constant.

Which **one** of the following plots would produce a straight-line graph?

A f against m

B f^2 against $\frac{1}{m}$

C f against $\sqrt{m}$

D $\frac{1}{f}$ against m

7 Repeated measurements of a quantity can reduce the effects of

A both random and systematic errors

B only random errors

C only systematic errors

D neither random nor systematic errors

8 A vector $\boldsymbol{v}$ makes an angle θ with the x axis as shown.

As the angle θ increases from 0° to 90°, how do the x and y components of $\boldsymbol{v}$ vary?

	***x* component**	***y* component**
A	Increases	Increases
B	Increases	Decreases
C	Decreases	Increases
D	Decreases	Decreases

9 An object is rolled from rest down an inclined plane. The distance travelled by the object was measured at seven different times. A graph was then constructed of the distance travelled against the (time taken)^2 as shown below.

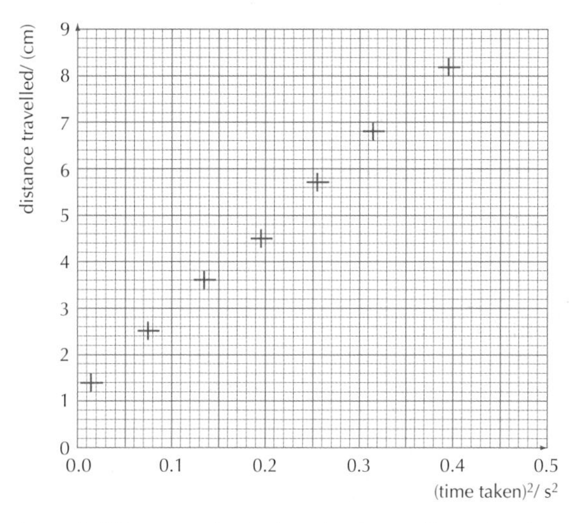

(a) **(i)** What quantity is given by the gradient of such a graph? [2]

(ii) Explain why the graph suggests that the collected data is valid but includes a **systematic error**. [2]

(iii) Do these results suggest that distance is proportional to (time taken)^2? Explain your answer. [2]

(iv) Making allowance for the systematic error, calculate the acceleration of the object. [2]

(b) The following graph shows that same data after the uncertainty ranges have been calculated and drawn as error bars.

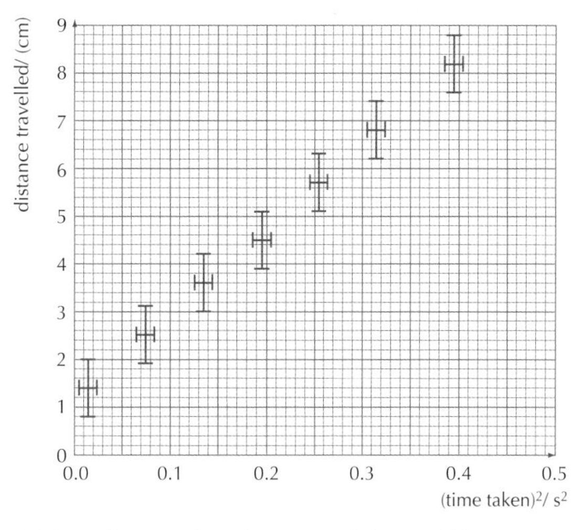

Add two lines to show the range of the possible acceptable values for the gradient of the graph. [2]

10 The lengths of the sides of a rectangular plate are measured, and the diagram shows the measured values with their uncertainties.

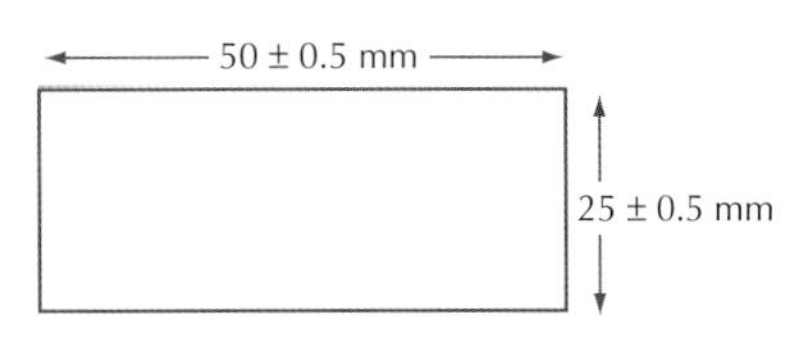

Which one of the following would be the best estimate of the percentage uncertainty in the calculated area of the plate?

A ± 0.02% **C** ± 3%

B ± 1% **D** ± 5%

11 A stone is dropped down a well and hits the water 2.0 s after it is released. Using the equation $d = \frac{1}{2}g\,t^2$ and taking $g = 9.81 \text{ m s}^{-2}$, a calculator yields a value for the depth d of the well as 19.62 m. If the time is measured to ±0.1 s then the best estimate of the absolute error in d is

A ±0.1 m **C** ±1.0 m

B ±0.2 m **D** ±2.0 m

12 In order to determine the density of a certain type of wood, the following measurements were made on a **cube** of the wood.

Mass = 493 g
Length of **each** side = 9.3 cm

The percentage uncertainty in the measurement of mass is ±0.5% and the percentage uncertainty in the measurement of length is ±1.0%.

The best estimate for the uncertainty in the density is

A ±0.5% **C** ±3.0%

B ±1.5% **D** ±3.5%

13 The graphs A to D below are plots of log y against log x in arbitrary units.

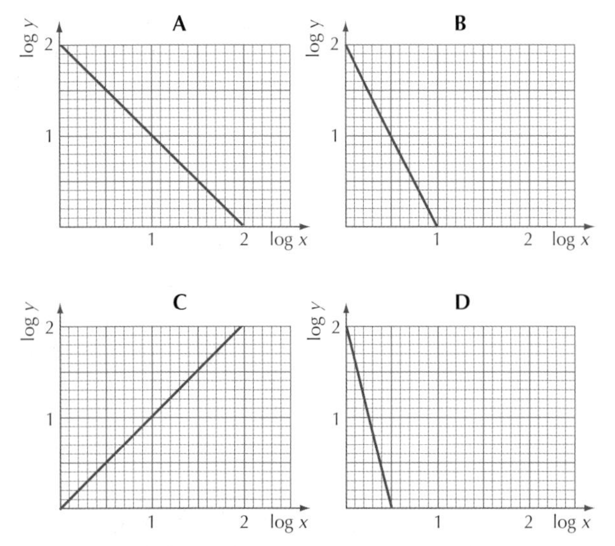

Which one of the graphs best represents the variation of y, the electrostatic potential due to a positive point charge, with x, the distance from the point charge?

14 This question is about finding the relationship between the forces between magnets and their separations.

In an experiment, two magnets were placed with their North-seeking poles facing one another. The force of repulsion, f, and the separation of the magnets, d, were measured and the results are shown in the table below.

Separation d/m	Force of repulsion f/N
0.04	4.00
0.05	1.98
0.07	0.74
0.09	0.32

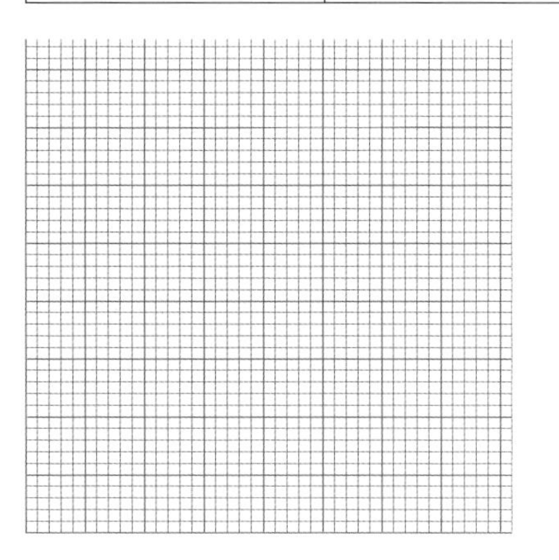

(a) Plot a graph of log (force) against log (distance). [3]

(b) The law relating the force to the separation is of the form

$f = kd^n$

(i) Use the graph to find the value of n. [2]

(ii) Calculate a value for k, giving its units. [3]

15 Astronauts wish to determine the gravitational acceleration on Planet X by dropping stones from an overhanging cliff. Using a steel tape measure they measure the height of the cliff as $s = 7.64 \text{ m} \pm 0.01 \text{ m}$. They then drop three similar stones from the cliff, timing each fall using a hand-held electronic stopwatch which displays readings to one-hundredth of a second. The recorded times for three drops are 2.46 s, 2.31 s and 2.40 s.

(a) Explain why the time readings vary by more than a tenth of a second, although the stopwatch gives readings to one hundredth of a second. [1]

(b) Obtain the average time t to fall, and write it in the form (value ± uncertainty), to the appropriate number of significant digits. [1]

(c) The astronauts then determine the gravitational acceleration a_g on the planet using the formula $a_g = \frac{2s}{t^2}$. Calculate a_g from the values of s and t, and determine the uncertainty in the calculated value. Express the result in the form a_g = (value ± uncertainty), to the appropriate number of significant digits. [3]

Kinematic concepts

DEFINITIONS

These technical terms should not be confused with their 'everyday' use. In particular one should note that

- Vector quantities always have a direction associated with them.
- Generally, velocity and speed are NOT the same thing. This is particularly important if the object is not going in a straight line.
- The units of acceleration come from its definition. (m s^{-1}) ÷ s = m s^{-2}.
- The definition of acceleration is precise. It is related to the change in **velocity** (not the same thing as the change in speed). Whenever the motion of an object changes, it is called acceleration. For this reason acceleration does not necessarily mean constantly increasing speed – it is possible to accelerate while at constant speed if the direction is changed.
- A deceleration means slowing down i.e. negative acceleration if velocity is positive.

	Symbol	Definition	Example	SI Unit	Vector or scalar?
Displacement	s	The distance moved in a particular direction	The displacement from London to Rome is 1.43×10^6 m southeast.	m	Vector
Velocity	v or u	The rate of change of displacement. $\text{velocity} = \frac{\text{change of displacement}}{\text{time taken}}$	The average velocity during a flight from London to Rome is 160 m s^{-1} southeast.	m s^{-1}	Vector
Speed	v or u	The rate of change of distance. $\text{speed} = \frac{\text{distance gone}}{\text{time taken}}$	The average speed during a flight from London to Rome is 160 m s^{-1}	m s^{-1}	Scalar
Acceleration	a	The rate of change of velocity $\text{acceleration} = \frac{\text{change of velocity}}{\text{time taken}}$	The average acceleration of a plane on the runway during take-off is 3.5 m s^{-2} in a forwards direction. This means that on average, its velocity changes every second by 3.5 m s^{-1}	m s^{-2}	Vector

INSTANTANEOUS VS AVERAGE

It should be noticed that the **average** value (over a period of time) is very different to the **instantaneous** value (at one particular time).

In the example below, the positions of a sprinter are shown at different times after the start of a race.

The average speed over the whole race is easy to work out. It is the total distance (100 m) divided by the total time (11.3 s) giving 8.8 m s^{-1}.

But during the race, her instantaneous speed was changing all the time. At the end of the first 2.0 seconds, she had travelled 10.04 m. This means that her average speed over the first 2.0 seconds was 5.02 m s^{-1}. During these first two seconds, her instantaneous speed was increasing – she was accelerating. If she started at rest (speed = 0.00 m s^{-1}) and her **average** speed (over the whole two seconds) was 5.02 m s^{-1} then her instantaneous speed at 2 seconds must be more than this. In fact the instantaneous speed for this sprinter was 9.23 m s^{-1}, but it would not be possible to work this out from the information given.

start | finish

d = 0.00 m, t = 0.0 s
d = 10.04 m, t = 2.0 s
d = 28.21 m, t = 4.0 s
d = 47.89 m, t = 6.0 s
d = 69.12 m, t = 8.0 s
d = 100.00 m, t = 11.3 s

FRAMES OF REFERENCE

If two things are moving in the same straight line but are travelling at different speeds, then we can work out their **relative velocities** by simple addition or subtraction as appropriate. For example, imagine two cars travelling along a straight road at different speeds.

If one car (travelling at 30 m s^{-1}) overtakes the other car (travelling at 25 m s^{-1}), then according to the driver of the slow car, the relative velocity of the fast car is + 5 m s^{-1}.

In technical terms what we are doing is moving from one **frame of reference** into another. The velocities of 25 m s^{-1} and 30 m s^{-1} were measured according to a stationary observer on the side of the road. We moved from this frame of reference into the driver's frame of reference.

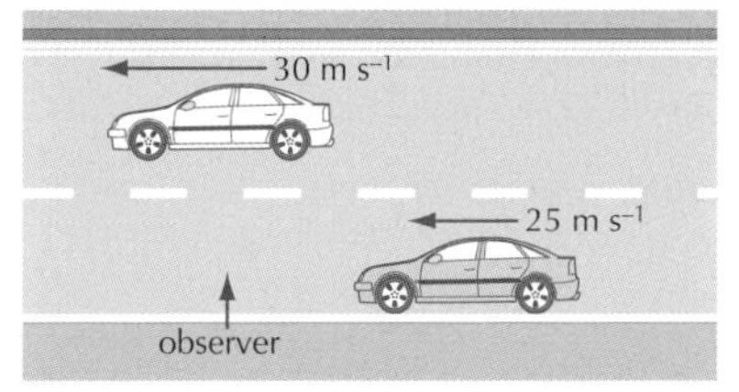

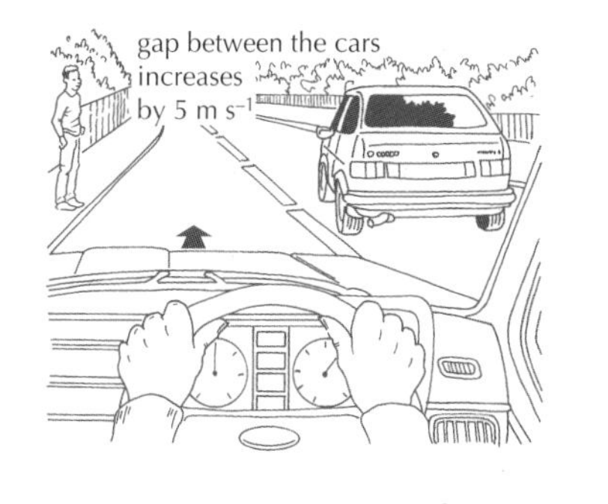

Above: one car overtaking another, as seen by an observer on the side of the road. Right: one car overtaking another, as seen by the driver of the slow car.

Graphical representation of motion

THE USE OF GRAPHS

Graphs are very useful for representing the changes that happen when an object is in motion. There are three possible graphs that can provide useful information

- displacement–time or distance–time graphs
- velocity–time or speed–time graphs
- acceleration–time graphs.

There are two common methods of determining particular physical quantities from these graphs. The particular physical quantity determined depends on what is being plotted on the graph.

1. Finding the gradient of the line.

To be a little more precise, one could find either the gradient of

- a straight-line section of the graph (this finds an average value), or
- the tangent to the graph at one point (this finds an instantaneous value).

2. Finding the area under the line.

To make things simple at the beginning, the graphs are normally introduced by considering objects that are just moving in one particular direction. If this is the case then there is not much difference between the scalar versions (distance or speed) and the vector versions (displacement or velocity) as the directions are clear from the situation. More complicated graphs can look at the component of a velocity in a particular direction.

If the object moves forward then backward (or up then down), we distinguish the two velocities by choosing which direction to call positive. It does not matter which direction we choose, but it should be clearly labelled on the graph.

Many examination candidates get the three types of graph muddled up. For example a speed–time graph might be interpreted as a distance–time graph or even an acceleration–time graph. Always look at the axes of a graph very carefully.

DISPLACEMENT–TIME GRAPHS

- The gradient of a displacement–time graph is the velocity
- The area under a displacement–time graph does not represent anything useful

Examples

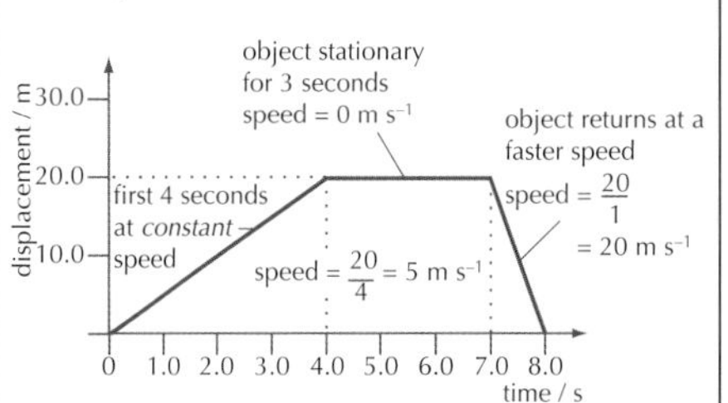

Object moves at constant speed, stops then returns.

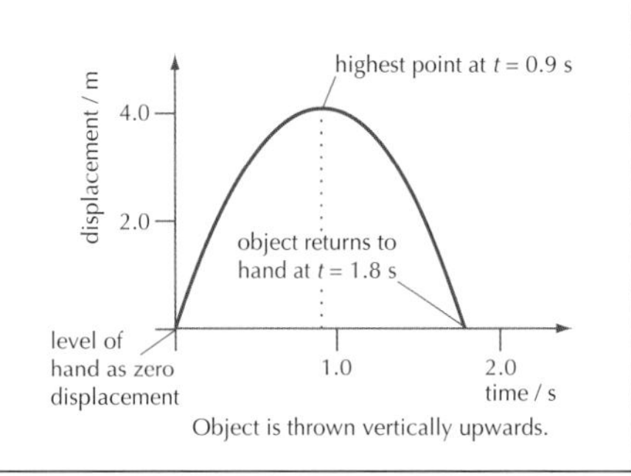

Object is thrown vertically upwards.

VELOCITY–TIME GRAPHS

- The gradient of a velocity–time graph is the acceleration
- The area under a velocity–time graph is the displacement

Examples

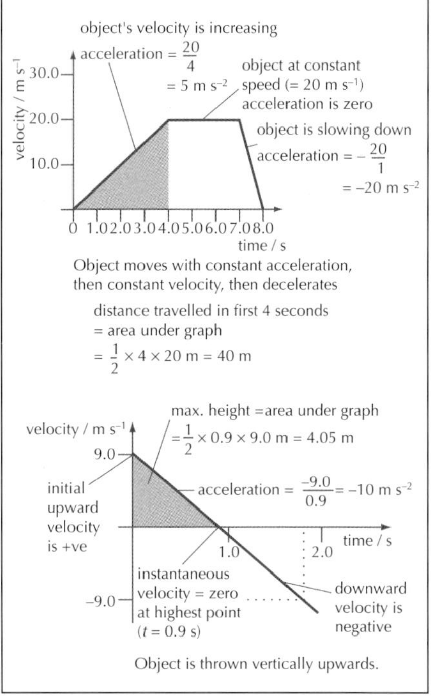

ACCELERATION–TIME GRAPHS

- The gradient of an acceleration–time graph is not often useful (it is actually the rate of change of acceleration).
- The area under an acceleration–time graph is the change in velocity

Examples

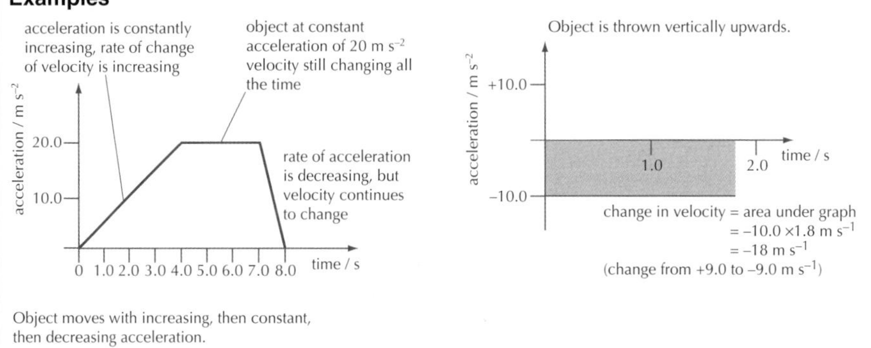

Object moves with increasing, then constant, then decreasing acceleration.

EXAMPLE OF EQUATION OF UNIFORM MOTION

A car accelerates uniformly from rest. After 8 s it has travelled 120 m.
Calculate: (i) its average acceleration (ii) its instantaneous speed after 8 s

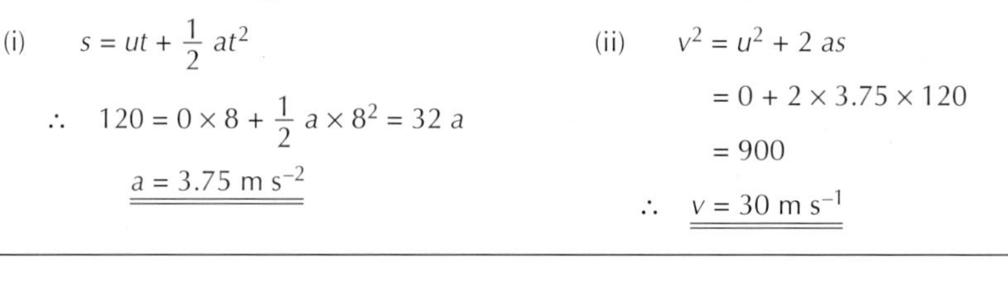

(i) $s = ut + \frac{1}{2}at^2$

$\therefore \quad 120 = 0 \times 8 + \frac{1}{2} a \times 8^2 = 32\,a$

$a = 3.75 \text{ m s}^{-2}$

(ii) $v^2 = u^2 + 2\,as$

$= 0 + 2 \times 3.75 \times 120$

$= 900$

$\therefore \quad v = 30 \text{ m s}^{-1}$

Uniformly accelerated motion

PRACTICAL CALCULATIONS

In order to determine how the velocity (or the acceleration) of an object varies in real situations, it is often necessary to record its motion. Possible laboratory methods include

Light gates
A light gate is a device that senses when an object cuts through a beam of light. The time for which the beam is broken is recorded. If the length of the object that breaks the beam is known, the average speed of the object through the gate can be calculated.

Alternatively, two light gates and a timer can be used to calculate the average velocity between the two gates. Several light gates and a computer can be joined together to make direct calculations of velocity or acceleration.

Strobe photography
A strobe light gives out very brief flashes of light at fixed time intervals. If a camera is pointed at an object and the only source of light is the strobe light, then the developed picture will have captured an object's motion.

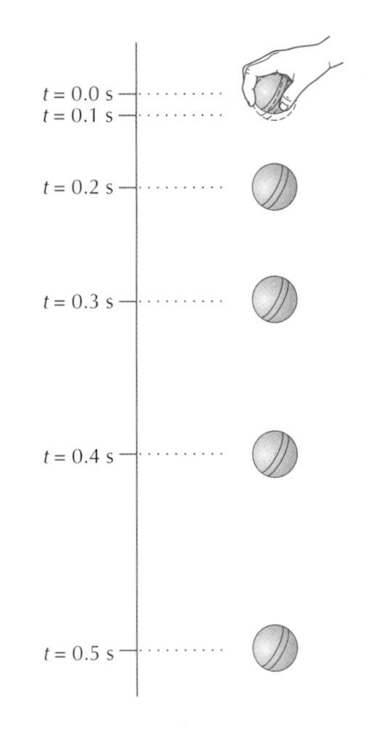

Ticker timer
A ticker timer can be arranged to make dots on a strip of paper at regular intervals of time (typically every fiftieth of a second). If the piece of paper is attached to an object, and the object is allowed to fall, the dots on the strip will have recorded the distance moved by the object in a known time.

EQUATIONS OF UNIFORM MOTION

These equations can only be used when the acceleration is constant – don't forget to check if this is the case!

The list of variables to be considered (and their symbols) is as follows

- u initial velocity
- v final velocity
- a acceleration (constant)
- t time taken
- s distance travelled

The following equations link these different quantities.

$$v = u + at$$

$$s = \left(\frac{u + v}{2}\right) t$$

$$v^2 = u^2 + 2as$$

$$s = ut + \frac{1}{2}at^2$$

$$s = vt - \frac{1}{2}at^2$$

The first equation is derived from the definition of acceleration. In terms of these symbols, this definition would be

$$a = \frac{(v - u)}{t}$$

This can be rearranged to give the first equation.

$$v = u + at \quad (1)$$

The second equation comes from the definition of average velocity.

$$\text{average velocity} = \frac{s}{t}$$

Since the velocity is changing uniformly we know that this average velocity must be given by

$$\text{average velocity} = \frac{(v + u)}{2}$$

or $$\frac{s}{t} = \frac{(u + v)}{2}$$

This can be rearranged to give

$$s = \frac{(u + v)t}{2} \quad (2)$$

The other equations of motion can be derived by using these two equations and substituting for one of the variables (see previous page for example).

FALLING OBJECTS

A very important example of uniformly accelerated motion is the vertical motion of an object in a uniform **gravitational field**. If we ignore the effects of air resistance, this is known as being in **free-fall**.

Taking down as positive, the graphs of the motion of any object in free-fall are

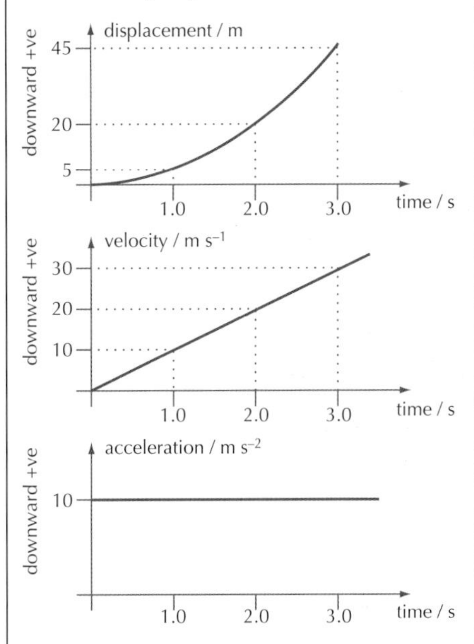

In the absence of air resistance, all falling objects have the SAME acceleration of free-fall, INDEPENDENT of their mass.

Air resistance will (eventually) affect the motion of all objects. Typically, the graphs of a falling object affected by air resistance become the shapes shown below

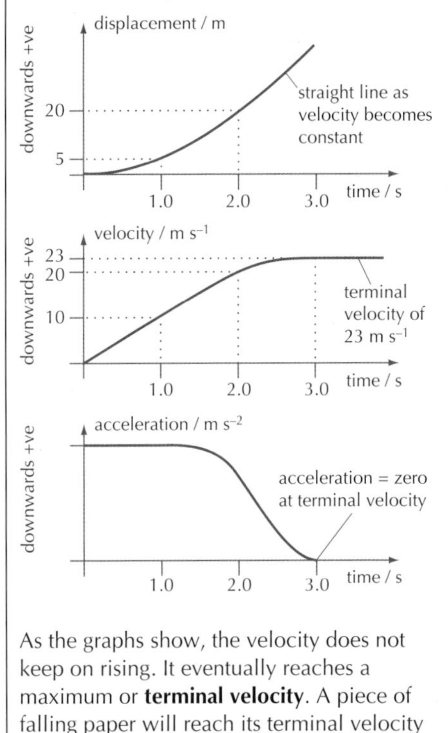

As the graphs show, the velocity does not keep on rising. It eventually reaches a maximum or **terminal velocity**. A piece of falling paper will reach its terminal velocity in a much shorter time than a falling book.

Forces and free-body diagrams

FORCES – WHAT THEY ARE AND WHAT THEY DO

In the examples below, a force (the kick) can cause deformation (the ball changes shape) or a change in motion (the ball gains a velocity). There are many different types of forces, but in general terms one can describe any force as 'the cause of a deformation or a velocity change'. The SI unit for the measurement of forces is the newton (N).

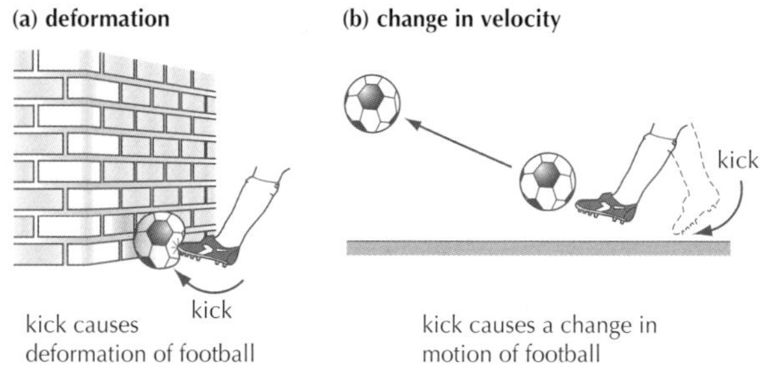

Effect of a force on a football

- A (resultant) force causes a CHANGE in velocity. If the (resultant) force is zero then the velocity is constant. Remember a change in velocity is called an acceleration, so we can say that **a force causes an acceleration**. A (resultant) force is NOT needed for a constant velocity (see page 17).
- The fact that a force can cause deformation is also important, but the deformation of the ball was, in fact, not caused by just one force – there was another one from the wall.
- One force can act on only one object. To be absolutely precise the description of a force should include
  - its magnitude
  - its direction
  - the object on which it acts (or the part of a large object)
  - the object that exerts the force
  - the nature of the force

A description of the force shown in the example would thus be 'a 50 N push at 20° to the horizontal acting ON the football FROM the boot'.

DIFFERENT TYPES OF FORCES

The following words all describe the forces (the pushes or pulls) that exist in nature

Gravitational force	**Normal reaction**	**Compression**
Electrostatic force	**Friction**	**Upthrust**
Magnetic force	**Tension**	**Lift**

One way of categorising these forces is whether they result from the contact between two surfaces or whether the force exists even if a distance separates the objects.

The origin of all these everyday forces is either gravitational or electromagnetic. The vast majority of everyday effects that we observe are due to electromagnetic forces.

MEASURING FORCES

The simplest experimental method for measuring the size of a force is to use the **extension** of a spring. When a spring is in tension it increases in length. The difference between the natural length and stretched length is called the extension of a spring.

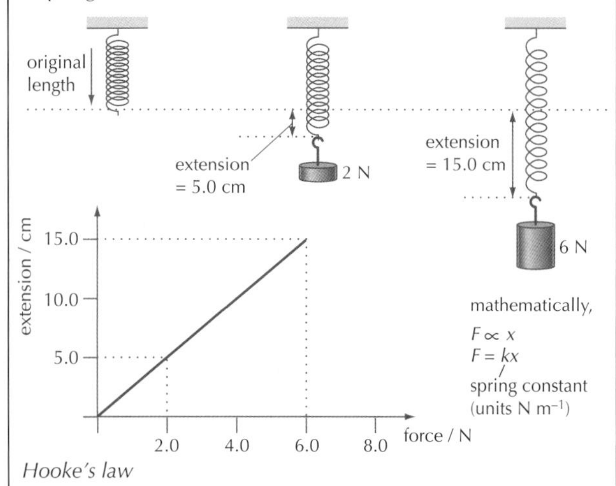

Hooke's law

Hooke's law states that up to the elastic limit, the extension, x, of a spring is proportional to the tension force, F. The constant of proportionality k is called the **spring constant**. The SI units for the spring constant are N m^{-1}. Thus by measuring the extension, we can calculate the force.

FORCES AS VECTORS

Since forces are vectors, vector mathematics must be used to find the resultant force from two or more other forces. A force can also be split into its components. See page 10 for more details.

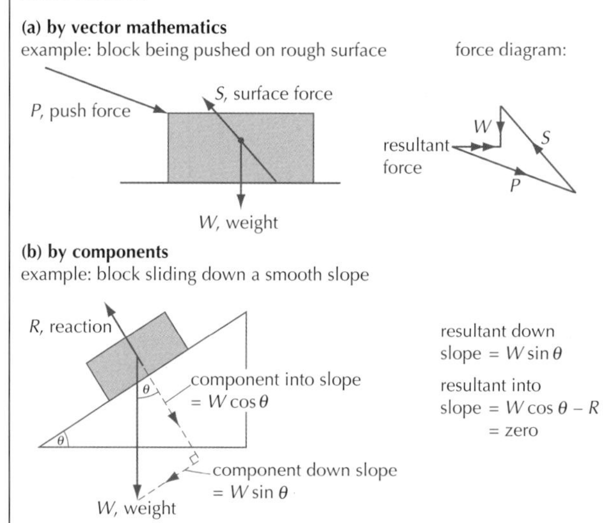

Vector addition

FREE-BODY DIAGRAMS

In a **free-body diagram**,

- one object (and ONLY one object) is chosen.
- all the forces on that object are shown and labelled

For example, if we considered the simple situation of a book resting on a table, we can construct free-body diagrams for either the book or the table.

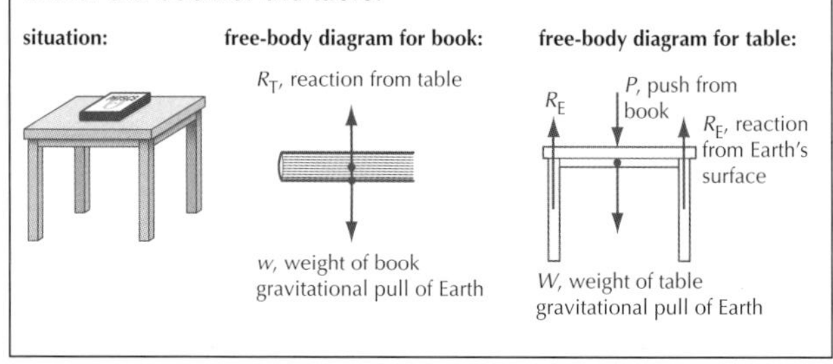

Newton's first law

NEWTON'S FIRST LAW

Newton's first law of motion states that 'an object continues in uniform motion in a straight line or at rest unless a resultant external force acts'. On first reading this can sound complicated but it does not really add anything to the description of a force given on page 16. All it says is that a resultant force causes acceleration. No resultant force means no acceleration – i.e 'uniform motion in a straight line'.

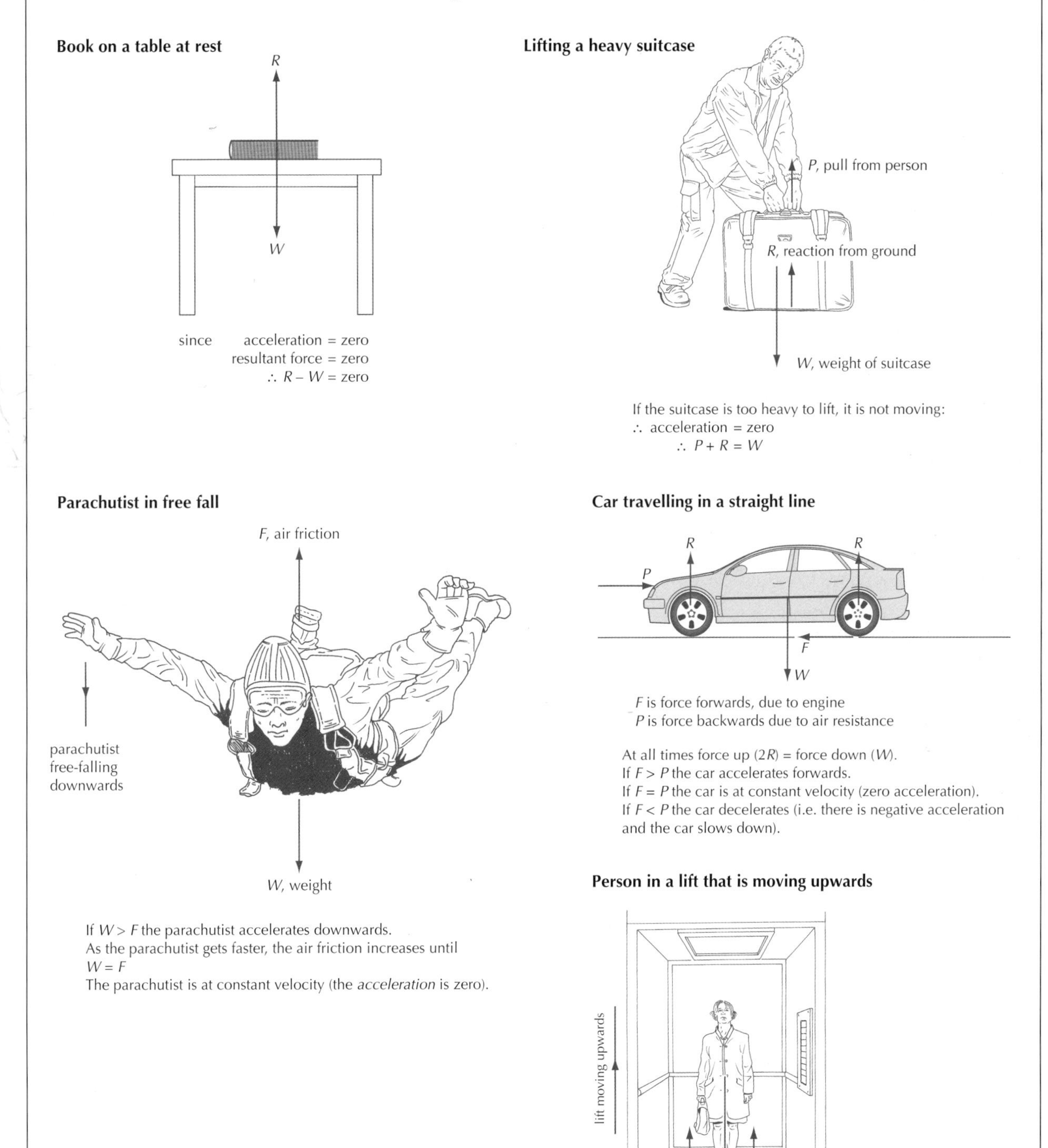

Book on a table at rest
R
W
since acceleration = zero
resultant force = zero
∴ R – W = zero
Lifting a heavy suitcase
P, pull from person
R, reaction from ground
W, weight of suitcase
If the suitcase is too heavy to lift, it is not moving:
∴ acceleration = zero
∴ P + R = W
Parachutist in free fall
F, air friction
parachutist free-falling downwards
W, weight
If W > F the parachutist accelerates downwards.
As the parachutist gets faster, the air friction increases until W = F
The parachutist is at constant velocity (the acceleration is zero).
Car travelling in a straight line
R
R
P
F
W
F is force forwards, due to engine
P is force backwards due to air resistance
At all times force up (2R) = force down (W).
If F > P the car accelerates forwards.
If F = P the car is at constant velocity (zero acceleration).
If F < P the car decelerates (i.e. there is negative acceleration and the car slows down).
Person in a lift that is moving upwards
lift moving upwards
R/2
R/2
W
The total force up from the floor of the lift = R.
The total force down due to gravity = W.
If R > W the person is accelerating upwards.
If R = W the person is at constant velocity (acceleration = zero).
If R < W the person is decelerating (acceleration is negative).

Equilibrium

EQUILIBRIUM

If the resultant force on an object is zero then it is said to be in **translational equilibrium** (or just in equilibrium). Mathematically this is expressed as follows

$$\Sigma F = zero$$

From Newton's first law, we know that the objects in the following situations must be in equilibrium.

1. An object that is constantly at rest.
2. An object that is moving with constant (uniform) velocity in a straight line.

Since forces are vector quantities, a zero resultant force means no force IN ANY DIRECTION.

For 2-dimensional problems it is sufficient to show that the forces balance in any two non-parallel directions. If this is the case then the object is in equilibrium.

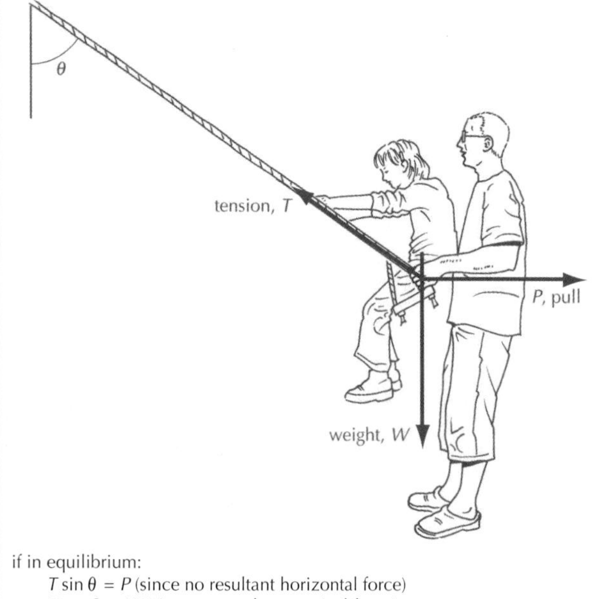

if in equilibrium:
$T \sin\theta = P$ (since no resultant horizontal force)
$T \cos\theta = W$ (since no resultant vertical force)

Translational equilibrium does NOT mean the same thing as being at rest. For example if the child in the previous example is allowed to swing back and forth, there are times when she is instantaneously at rest but not in equilibrium.

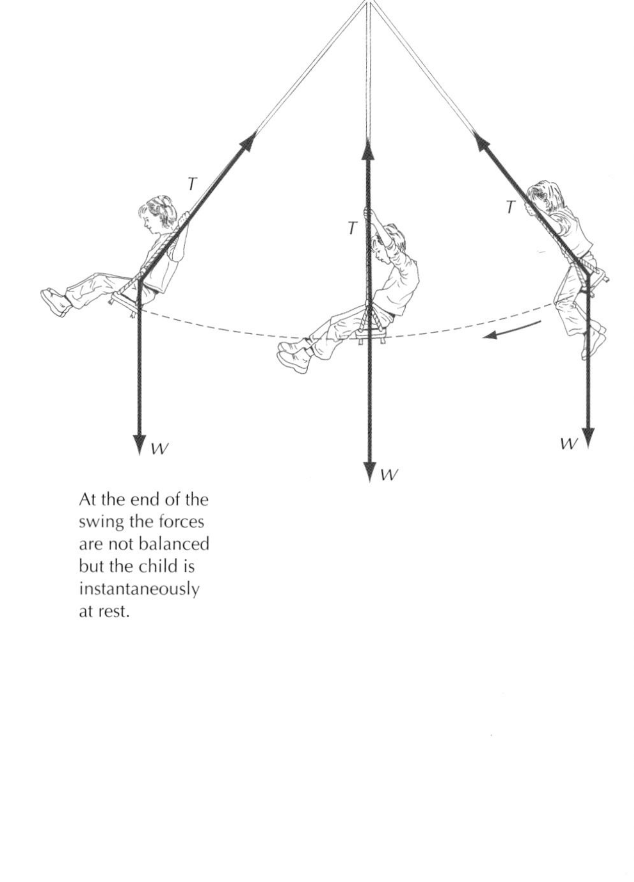

At the end of the swing the forces are not balanced but the child is instantaneously at rest.

DIFFERENT TYPES OF FORCES

Name of force	Description
Gravitational force	The force between objects as a results of their masses. This is sometimes referred to as the **weight** of the object but this term is, unfortunately, ambiguous – see page 21.
Electrostatic force	The force between objects as a result of their electric charges.
Magnetic force	The force between magnets and/or electric currents.
Normal reaction	The force between two surfaces that acts at right angles to the surfaces. If two surfaces are smooth then this is the only force that acts between them.
Friction	The force that opposes the relative motion of two surfaces and acts along the surfaces. **Air resistance** or **drag** can be thought of as a frictional force – technically this is known as **fluid friction**.
Tension	When a string (or a spring) is stretched, it has equal and opposite forces on its ends pulling outwards. The tension force is the force that the end of the string applies to another object.
Compression	When a rod is compressed (squashed), it has equal and opposite forces on its ends pushing inwards. The compression force is the force that the ends of the rod applies to another object. This is the opposite of the tension force.
Upthrust	This is the upward force that acts on an object when it is submerged in a fluid. It is the force that causes some objects to float in water.
Lift	This force can be exerted on an object when a fluid flows over it in an asymmetrical way. The shape of the wing of an aircraft causes the aerodynamic lift that enables the aircraft to fly.

Newton's second law

NEWTON'S SECOND LAW OF MOTION

Newton's first law states that a resultant force causes an acceleration. His second law provides a means of calculating the value of this acceleration. The best way of stating the second law is use the concept of the **momentum** of an object. This concept is explained on page 22.

A correct statement of Newton's second law using momentum would be 'the resultant force is proportional to the rate of change of momentum'. If we use SI units (and you always should) then the law is even easier to state – 'the resultant force is equal to the rate of change of momentum'. In symbols, this is expressed as follows

In SI units, $F = \frac{\Delta p}{\Delta t}$.

or, in full calculus notation, $F = \frac{dp}{dt}$

p is the symbol for the momentum of a body.

Until you have studied what this means this will not make much sense, but this version of the law is given here for completeness.

An equivalent (but more common) way of stating Newton's second law applies when we consider the action of a force on a single mass. If the amount of mass stays constant we can state the law as follows. 'The resultant force is proportional to the acceleration.' If we also use SI units then 'the resultant force is equal to the product of the mass and the acceleration'.

In symbols, in SI units,

$$F = m\ a$$

- F: resultant force measured in newtons
- m: mass measured in kilograms
- a: acceleration measured in m s^{-2}

Note:

- The '$F=ma$' version of the law only applies if we use SI units – for the equation to work the mass must be in **kilograms** rather than in grams.
- F is the resultant force. If there are several forces acting on an object (and this is usually true) then one needs to work out the resultant force before applying the law.
- This is an experimental law.
- There are no exceptions – Newton's laws apply throughout the Universe. (To be absolutely precise, Einstein's theory of relativity takes over at very large values of speed and mass. See page 173.)

The $F = ma$ version of the law can be used whenever the situation is simple – for example, a constant force acting on a constant mass giving a constant acceleration. If the situation is more difficult (e.g. a changing force or a changing mass) then one needs to use the $F = \frac{dp}{dt}$ version.

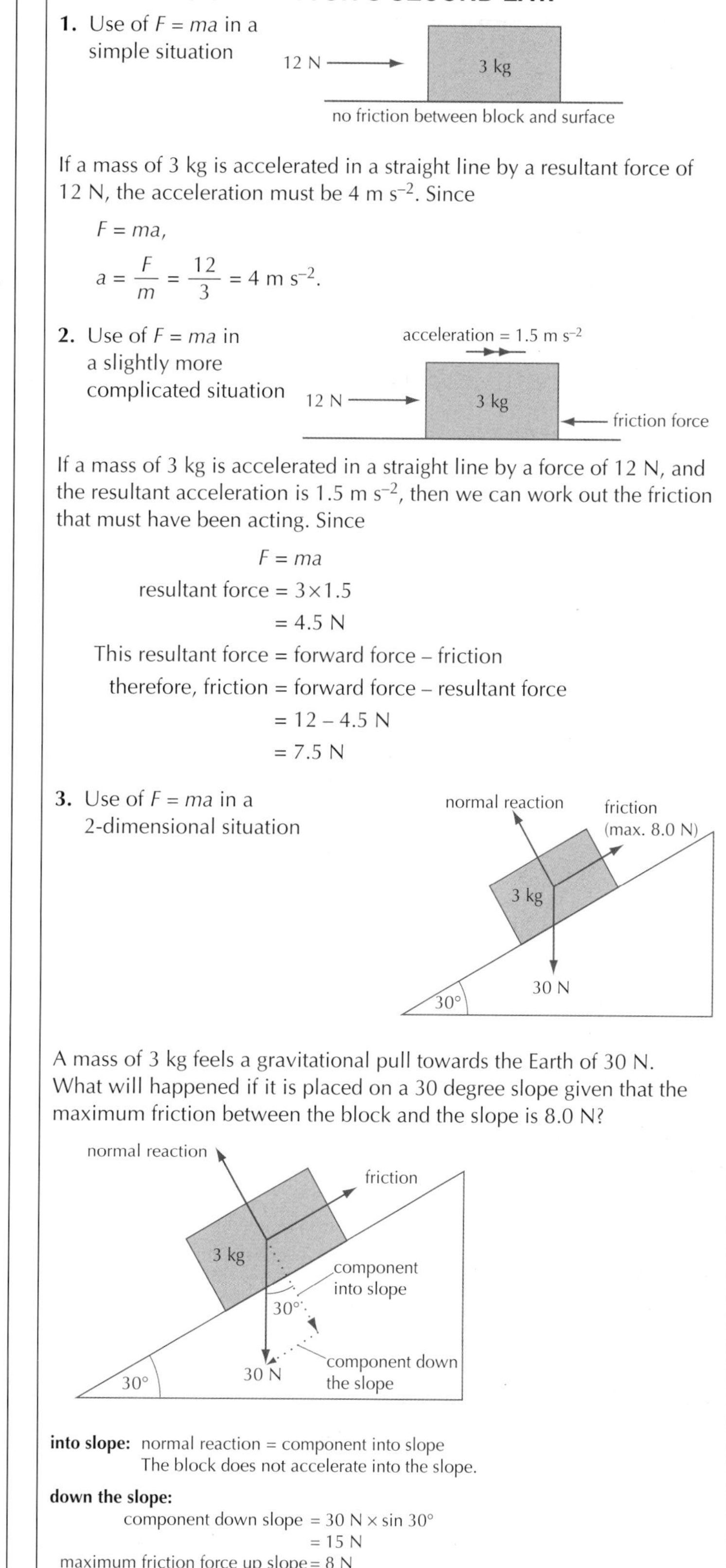

EXAMPLES OF NEWTON'S SECOND LAW

1. Use of $F = ma$ in a simple situation

If a mass of 3 kg is accelerated in a straight line by a resultant force of 12 N, the acceleration must be 4 m s^{-2}. Since

$F = ma$,

$a = \frac{F}{m} = \frac{12}{3} = 4$ m s^{-2}.

2. Use of $F = ma$ in a slightly more complicated situation

If a mass of 3 kg is accelerated in a straight line by a force of 12 N, and the resultant acceleration is 1.5 m s^{-2}, then we can work out the friction that must have been acting. Since

$$F = ma$$

resultant force = 3×1.5
= 4.5 N

This resultant force = forward force – friction

therefore, friction = forward force – resultant force
= 12 – 4.5 N
= 7.5 N

3. Use of $F = ma$ in a 2-dimensional situation

A mass of 3 kg feels a gravitational pull towards the Earth of 30 N. What will happened if it is placed on a 30 degree slope given that the maximum friction between the block and the slope is 8.0 N?

into slope: normal reaction = component into slope
The block does not accelerate into the slope.

down the slope:

component down slope = 30 N × sin 30°
= 15 N

maximum friction force up slope = 8 N

∴ resultant force down slope = 15 – 8
= 7 N

$F = ma$

∴ acceleration down slope = $\frac{F}{m}$

$= \frac{7}{3} = 2.3$ m s^{-2}

Newton's third law

STATEMENT OF THE LAW

Newton's second law is an experimental law that allows us to calculate the effect that a force has. Newton's third law highlights the fact that forces always come in pairs. It provides a way of checking to see if we have remembered all the forces involved.

It is very easy to state. 'When two bodies A and B interact, the force that A exerts on B is equal and opposite to the force that B exerts on A'. Another way of saying the same thing is that 'for every action on one object there is an equal but opposite reaction on another object'.

In symbols,

$$F_{AB} = -F_{BA}$$

Key points to notice include

- The two forces in the pair act on DIFFERENT objects – this means that equal and opposite forces that act on the same object are NOT Newton's third law pairs.
- Not only are the forces equal and opposite, but they must be of the same type. In other words, if the force that A exerts on B is a gravitational force, then the equal and opposite force exerted by B on A is also a gravitational force.

EXAMPLES OF THE LAW

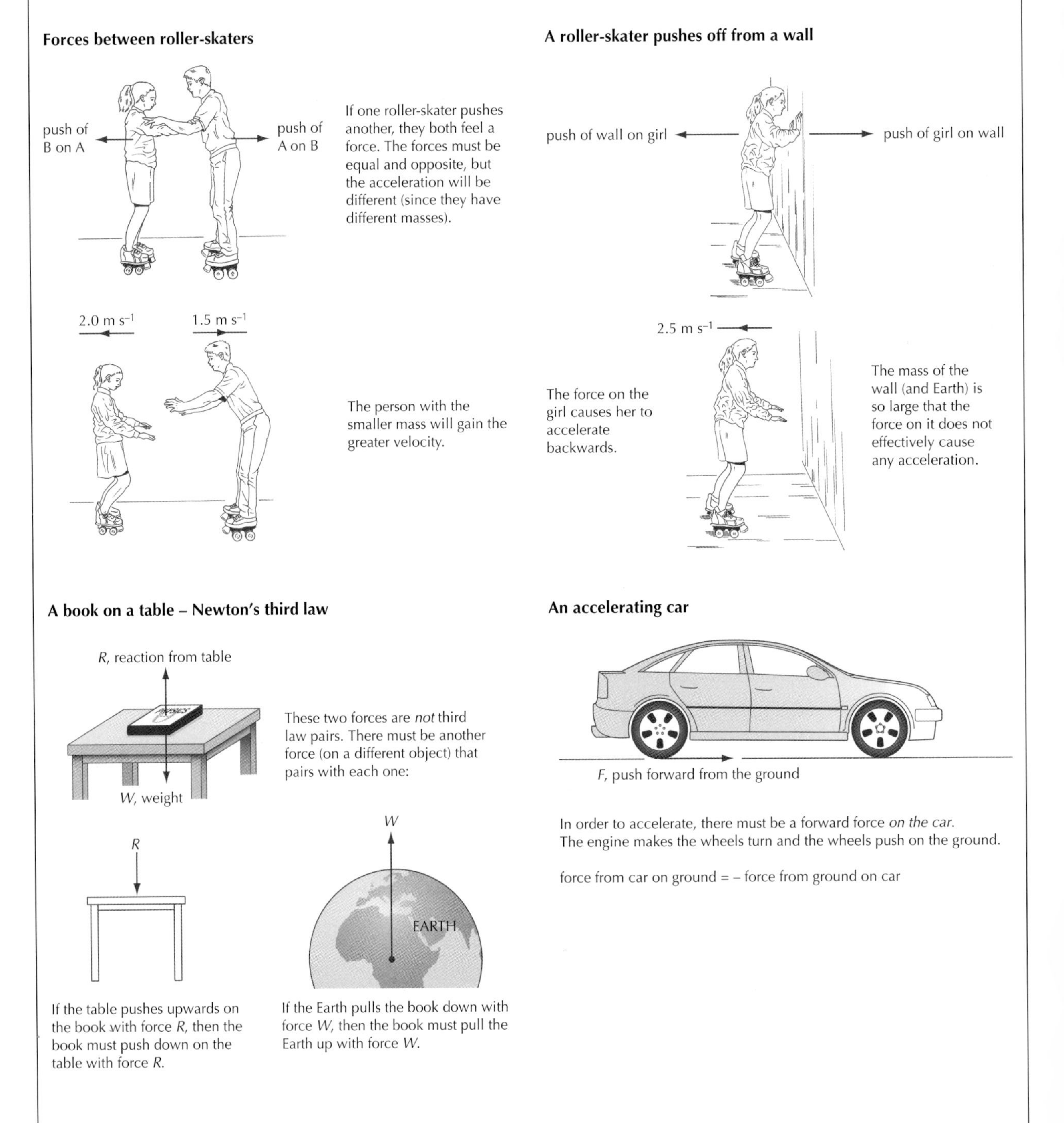

Forces between roller-skaters
push of B on A
push of A on B
If one roller-skater pushes another, they both feel a force. The forces must be equal and opposite, but the acceleration will be different (since they have different masses).
2.0 m s⁻¹
1.5 m s⁻¹
The person with the smaller mass will gain the greater velocity.
A roller-skater pushes off from a wall
push of wall on girl
push of girl on wall
2.5 m s⁻¹
The force on the girl causes her to accelerate backwards.
The mass of the wall (and Earth) is so large that the force on it does not effectively cause any acceleration.
A book on a table – Newton's third law
R, reaction from table
W, weight
These two forces are not third law pairs. There must be another force (on a different object) that pairs with each one:
R
W
EARTH
If the table pushes upwards on the book with force R, then the book must push down on the table with force R.
If the Earth pulls the book down with force W, then the book must pull the Earth up with force W.
An accelerating car
F, push forward from the ground
In order to accelerate, there must be a forward force on the car.
The engine makes the wheels turn and the wheels push on the ground.
force from car on ground = – force from ground on car

Mass and weight

WEIGHT

There are two different ways of defining mass – see page 172. Mass and **weight** are two very different things. Unfortunately their meanings have become muddled in everyday language. Mass is the amount of matter contained in an object (measured in kg) whereas the weight of an object is a force (measured in N).

If an object is taken to the moon, its mass would be the same, but its weight would be less (the gravitational forces on the moon are less than on the Earth). On the Earth the two terms are often muddled because they are proportional. People talk about wanting to gain or lose weight – what they are actually worried about is gaining or losing mass.

Double the mass means double the weight

To make things worse, the term 'weight' can be ambiguous even to physicists. Some people choose to define weight as the gravitational force on an object. Other people define it to be the reading on a supporting scale. Whichever definition you use, you weigh less at the top of a building compared with at the bottom – the pull of gravity is slightly less!

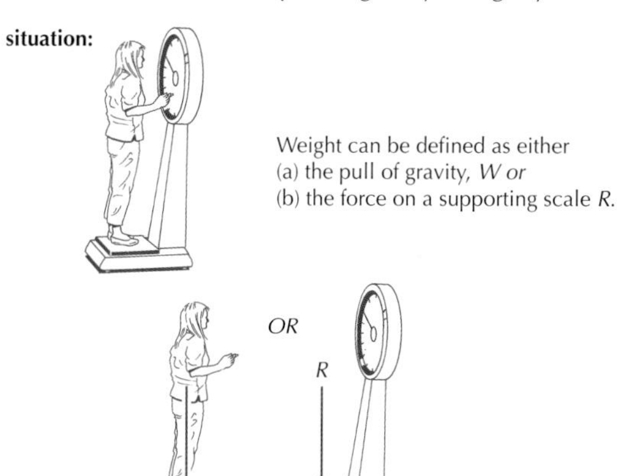

situation:
Weight can be defined as either
(a) the pull of gravity, W or
(b) the force on a supporting scale R.
OR
R
W

Two different definitions of 'weight'

Although these two definitions are the same if the object is in equilibrium, they are very different in non-equilibrium situations. For example, if both the object and the scale were put into a lift and the lift accelerated upwards then the definitions would give different values.

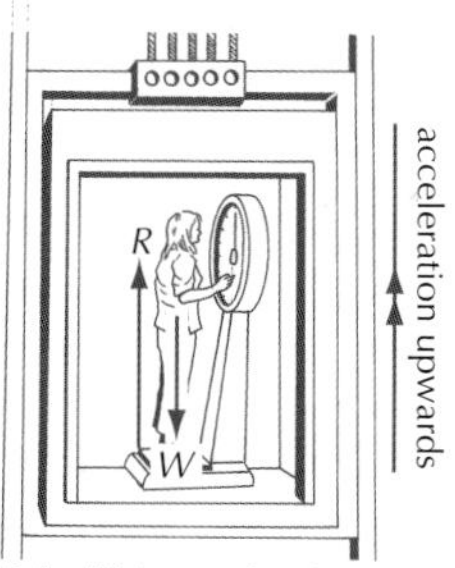

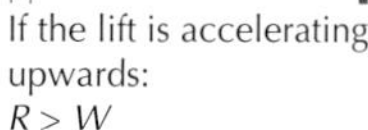

If the lift is accelerating upwards:
$R > W$

The safe thing to do is to avoid using the term weight if at all possible! Stick to the phrase 'gravitational force' and you cannot go wrong.

Gravitational force = $m\,g$

On the surface of the Earth, g is approximately 10 N kg^{-1}, whereas on the surface of the moon, $g \approx 1.6$ N kg^{-1}

Momentum

DEFINITIONS – LINEAR MOMENTUM AND IMPULSE

Linear momentum (always given the symbol $\boldsymbol{p}$) is defined as the product of mass and velocity.

Momentum = mass × velocity

$$\boldsymbol{p} = m\,\boldsymbol{v}$$

The SI units for momentum must be kg m s^{-1}. Alternative units of N s can also be used (see below). Since velocity is a vector, momentum must be a vector. In any situation, particularly if it happens quickly, the change of momentum Δp is called the **impulse** ($\Delta p = F\,\Delta t$).

USE OF MOMENTUM IN NEWTON'S SECOND LAW

Newton's second law states that the resultant force is proportional to the rate of change of momentum. Mathematically we can write this as

$$F = \frac{\textit{(final momentum – initial momentum)}}{\textit{time taken}} = \frac{\Delta p}{\Delta t}$$

Example 1

A jet of water leaves a hose and hits a wall where its velocity is brought to rest. If the hose cross-sectional area is 25 cm^2, the velocity of the water is 50 m s^{-1} and the density of the water is 1000 kg m^{-3}, what is the force acting on the wall?

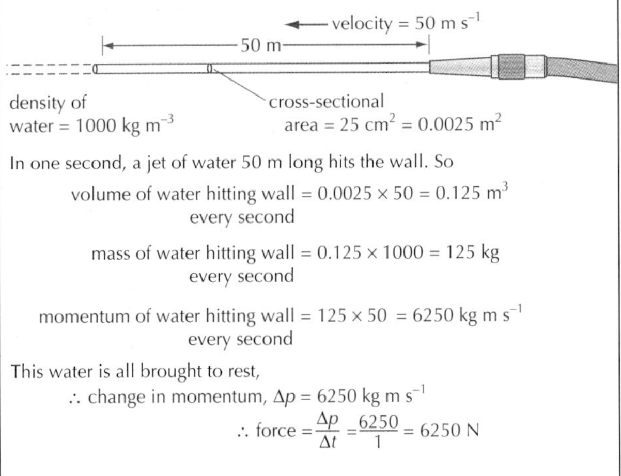

velocity = 50 m s⁻¹
50 m
density of water = 1000 kg m⁻³
cross-sectional area = 25 cm² = 0.0025 m²
In one second, a jet of water 50 m long hits the wall. So
volume of water hitting wall = 0.0025 × 50 = 0.125 m³ every second
mass of water hitting wall = 0.125 × 1000 = 125 kg every second
momentum of water hitting wall = 125 × 50 = 6250 kg m s⁻¹ every second
This water is all brought to rest,
∴ change in momentum, Δp = 6250 kg m s⁻¹
∴ force = Δp/Δt = 6250/1 = 6250 N

Example 2

The graph below shows the variation with time of the force on a football of mass 500g. Calculate the final velocity of the ball.

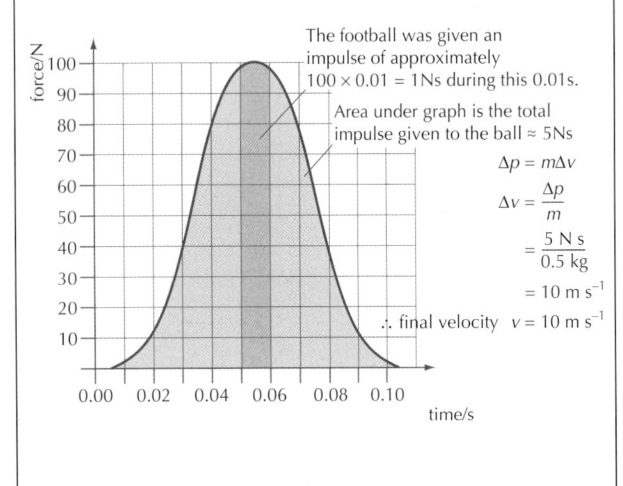

CONSERVATION OF MOMENTUM

The law of conservation of linear momentum states that 'the total linear momentum of a system of interacting particles remains constant **provided there is no resultant external force'**.

To see why, we start by imagining two isolated particles A and B that collide with one another.

- The force from A onto B, F_{AB} will cause B's momentum to change by a certain amount.
- If the time taken was Δt, then the momentum change (the impulse) given to B will be given by $\Delta p_B = F_{AB}\,\Delta t$
- By Newton's third law, the force from B onto A, F_{BA} will be equal and opposite to the force from A onto B, $F_{AB} = -F_{BA}$.
- Since the time of contact for A and B is the same, then the momentum change for A is equal and opposite to the momentum change for B, $\Delta p_A = -F_{AB}\,\Delta t$.
- This means that the total momentum (momentum of A plus the momentum of B) will remain the same. Total momentum is conserved.

This argument can be extended up to any number of interacting particles so long as the system of particles is still isolated. If this is the case, the momentum is still conserved.

ELASTIC AND INELASTIC COLLISIONS

The law of conservation of linear momentum is not enough to always predict the outcome after a collision (or an explosion). This depends on the nature of the colliding bodies. For example, a moving railway truck, m_A, velocity v, collides with an identical stationary truck m_B. Possible outcomes are:

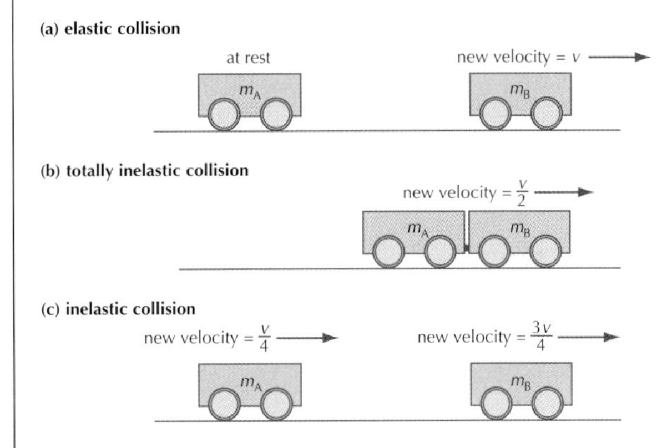

In (a), the trucks would have to have elastic bumpers. If this were the case then no mechanical energy at all would be lost in the collision. A collision in which no mechanical energy is lost is called an **elastic collision**. In reality, collisions between everyday objects always lose some energy – the only real example of elastic collisions is the collision between molecules. For an elastic collision, the relative velocity of approach always equals the relative velocity of separation.

In (b), the railway trucks stick together during the collision (the relative velocity of separation is zero). This collision is what is known as a **totally inelastic collision**. A large amount of mechanical energy is lost (as heat and sound), but the total momentum is still conserved.

In energy terms, (c) is somewhere between (a) and (b). Some energy is lost, but the railway trucks do not join together. This is an example of an **inelastic collision**. Once again the total momentum is conserved.

Work

WHEN IS WORK DONE?

Work is done when a force moves its point of application in the direction of the force. If the force moves at right angles to the direction of the force, then no work has been done.

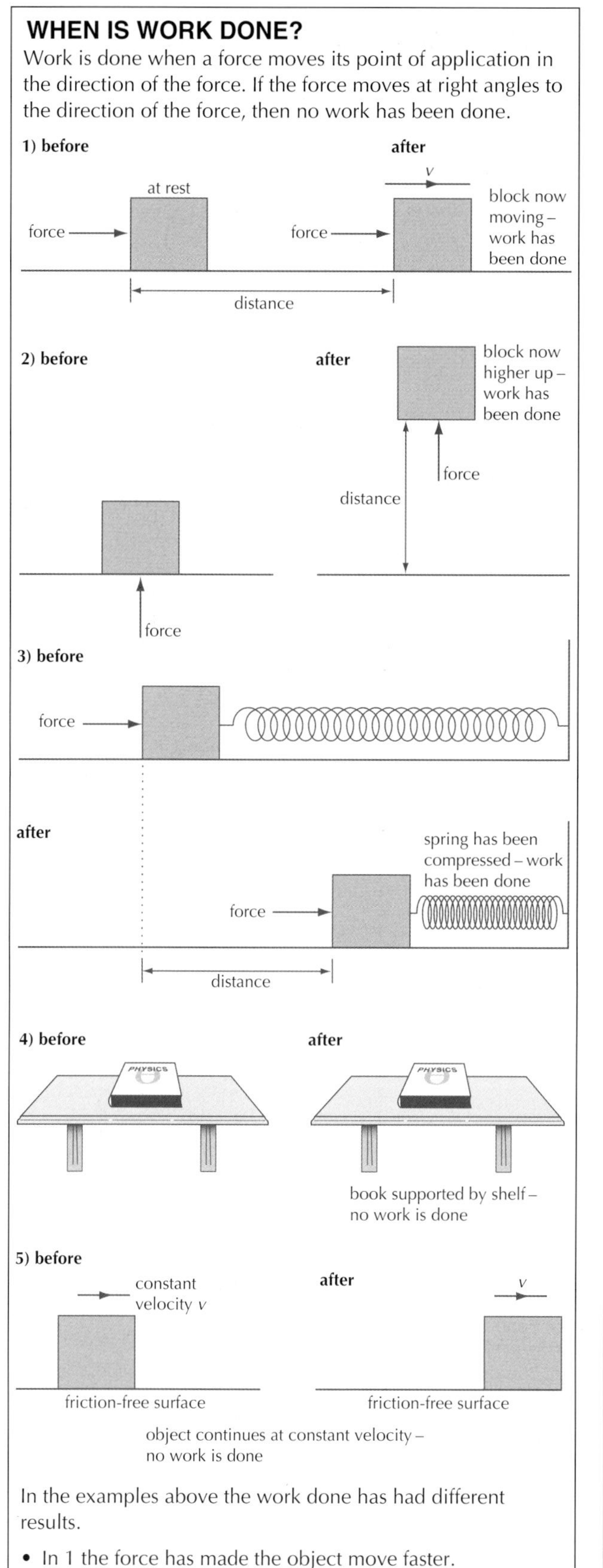

In the examples above the work done has had different results.

- In 1 the force has made the object move faster.
- In 2 the object has been lifted higher in the gravitational field.
- In 3 the spring has been compressed.
- In 4 and 5, NO work is done. Note that even though the object is moving in the last example, there is no force moving along its direction of action so no work is done.

DEFINITION OF WORK

Work is a scalar quantity. Its definition is as follows.

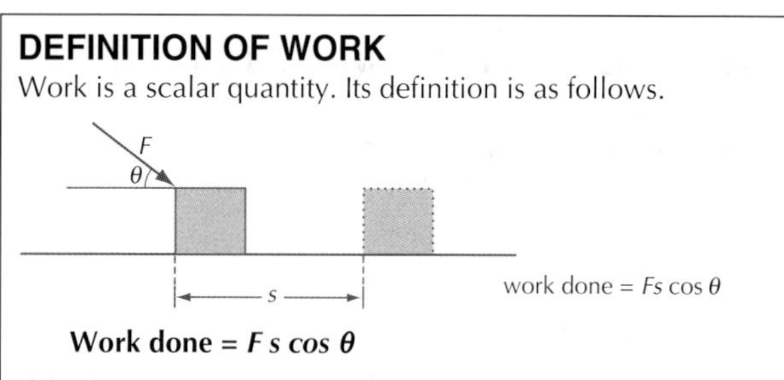

Work done = $F\,s\cos\theta$

If the force and the displacement are in the same direction, this can be simplified to

'Work done = force×distance'

From this definition, the SI units for work done are N m. We define a new unit called the joule: 1 J = 1 N m.

EXAMPLES

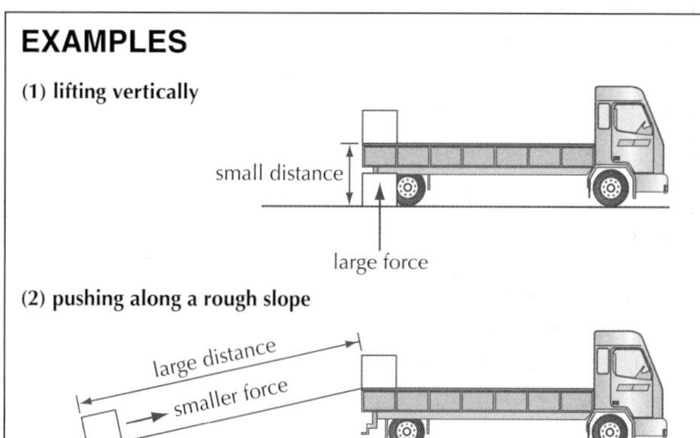

The task in the second case would be easier to perform (it involves less force) but overall it takes more work since work has to be done to overcome friction. In each case, the useful work is the same.

If the force doing work is not constant (for example, when a spring is compressed), then graphical techniques can be used.

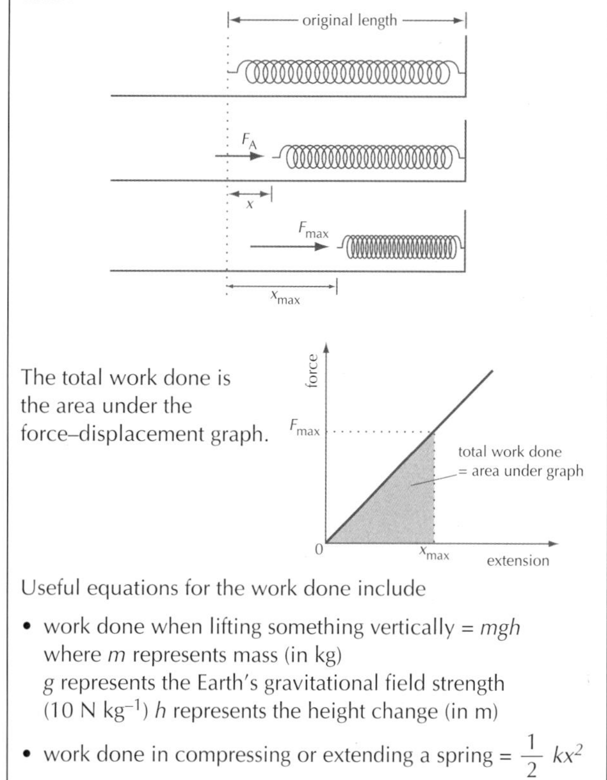

Useful equations for the work done include

- work done when lifting something vertically = mgh
  where m represents mass (in kg)
  g represents the Earth's gravitational field strength (10 N kg^{-1}) h represents the height change (in m)
- work done in compressing or extending a spring = $\frac{1}{2}kx^2$

Energy and power

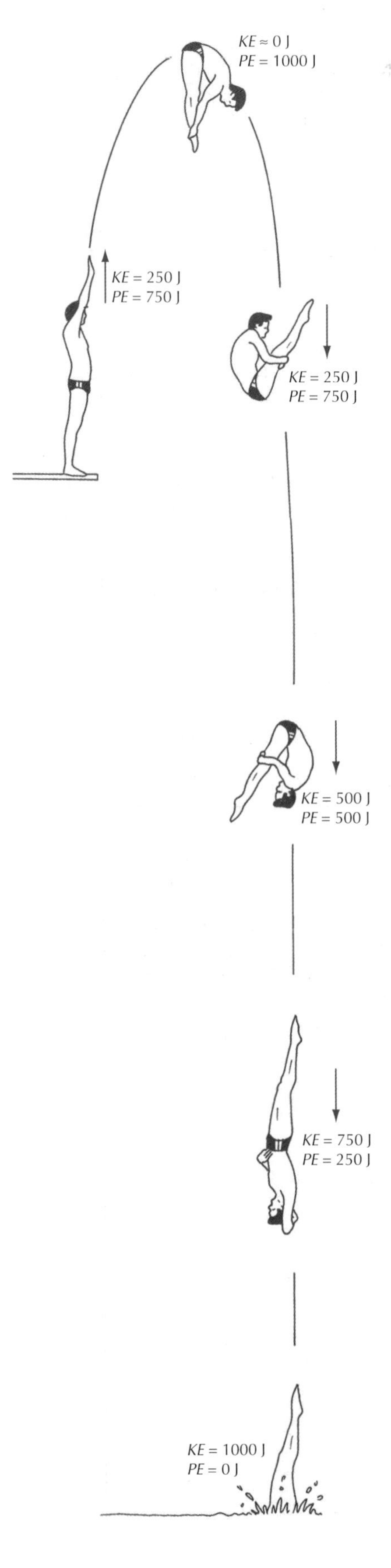

CONCEPTS OF ENERGY AND WORK

Energy and work are linked together. When you do work on an object, it gains energy and you lose energy. **The amount of energy transferred is equal to the work done**. Energy is a measure of the amount of work done. This means that the units of energy must be the same as the units of work – joules.

ENERGY TRANSFORMATIONS – CONSERVATION OF ENERGY

In any situation, we must be able to account for the changes in energy. If it is 'lost' by one object, it must be gained by another. This is known as the **principle of conservation of energy**. There are several ways of stating this principle:

- Overall the total energy of any closed system must be constant.
- Energy is neither created or destroyed, it just changes form.
- There is no change in the total energy in the Universe.

ENERGY TYPES

Kinetic energy	**Electrostatic potential**	**Chemical energy**	**Radiant energy**
Gravitational potential	**Thermal energy**	**Nuclear energy**	**Solar energy**
Elastic potential energy	**Electrical energy**	**Internal energy**	**Light energy**

Equations for the first three types of energy are given below.

Kinetic energy = $\frac{1}{2}mv^2$ where m is the mass (in kg), v is the velocity (in m s^{-1})

Gravitational potential energy = mgh where m represents mass (in kg), g represents the Earth's gravitational field (10 N kg^{-1}), h represents the height change (in m)

Elastic potential energy = $\frac{1}{2}kx^2$ where k is the spring constant (in N m^{-1}), x is the extension (in m)

EFFICIENCY AND POWER

1. Power

Power is defined as the RATE at which energy is transferred. This is the same as the rate at which work is done.

$$\text{Power} = \frac{\text{energy transferred}}{\text{time taken}} = \frac{\text{work done}}{\text{time taken}}$$

The SI unit for power is the joule per second (J s^{-1}). Another unit for power is defined – the watt (W). 1 W = 1 J s^{-1}.

If something is moving at a constant velocity v against a constant frictional force f, the power P needed is $P = f\,v$

2. Efficiency

Depending on the situation, we can categorise the energy transferred (work done) as useful or not. In a light bulb, the useful energy would light energy, the 'wasted' energy would be thermal energy (and non-visible forms of radiant energy).

We define efficiency as the ratio of useful energy to the total energy transferred. Possible forms of the equation include:

$$\text{Efficiency} = \frac{\text{useful work OUT}}{\text{total energy transformed}}$$

$$\text{Efficiency} = \frac{\text{useful energy OUT}}{\text{total energy IN}}$$

$$\text{Efficiency} = \frac{\text{useful power OUT}}{\text{total power IN}}$$

Since this is a ratio it does not have any units. Often it is expressed as a percentage.

EXAMPLE

A grasshopper (mass 8g) uses its hindlegs to push for 0.1s and as a result jumps 1.8m high. Calculate (i) its take off speed, (ii) the power developed.

(i) P.E. gained = mgh

K.E at start = $\frac{1}{2}mv^2$

$\frac{1}{2}mv^2 = mgh$ (conservation of energy)

$v = \sqrt{2gh} = \sqrt{2 \times 10 \times 1.8} = 6\text{ ms}^{-1}$

(ii) Power = $\frac{mgh}{t}$

$= \frac{0.008 \times 10 \times 1.8}{0.1}$

≈ 1.4 W

Uniform circular motion

MECHANICS OF CIRCULAR MOTION

The phrase 'uniform circular motion' is used to describe an object that is going around a circle at constant speed. Most of the time this also means that the circle is horizontal. An example of uniform circular motion would be the motion of a small mass on the end of a string as shown below.

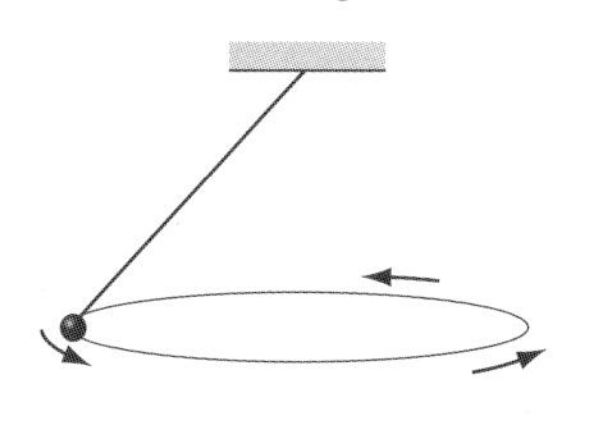

Example of uniform circular motion

It is important to remember that even though the speed of the object is constant, its direction is changing all the time.

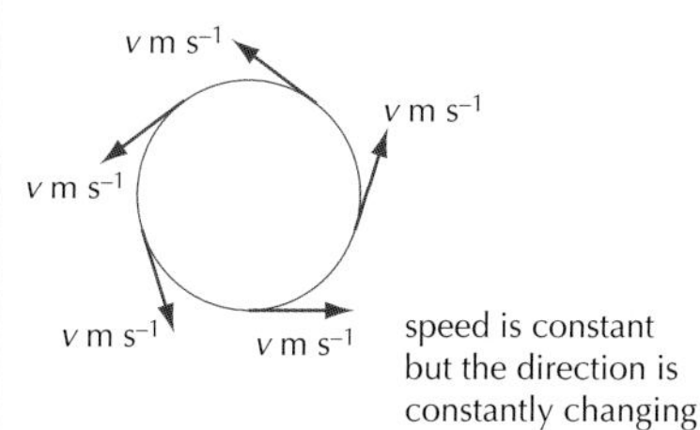

Circular motion. The direction of motion is changing all the time.

This constantly changing direction means that the velocity of the object is constantly changing. The word 'acceleration' is used whenever an object's velocity changes. This means that an object in uniform circular motion MUST be accelerating even if the speed is constant.

The acceleration of a particle travelling in circular motion is called the **centripetal acceleration**. The force needed to cause the centripetal acceleration is called the **centripetal force**.

MATHEMATICS OF CIRCULAR MOTION

The diagram below allows us to work out the direction of the centripetal acceleration – which must also be the direction of the centripetal force. This direction is constantly changing.

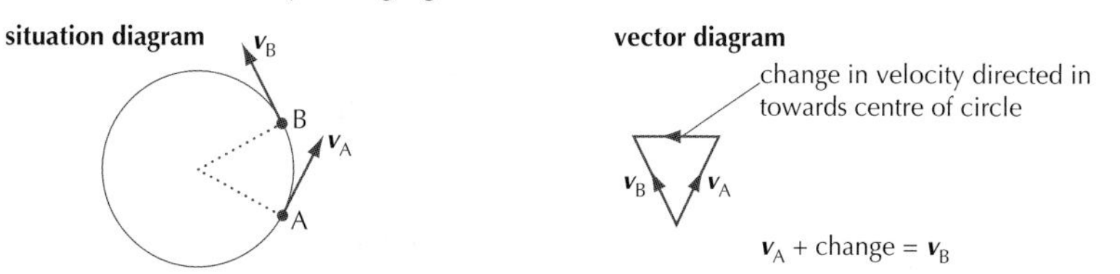

The object is shown moving between two points A and B on a horizontal circle. Its velocity has changed from v_A to v_B. The magnitude of velocity is always the same, but the direction has changed. Since velocities are vector quantities we need to use vector mathematics to work out the average change in velocity. This vector diagram is also shown above.

In this example, the direction of the average change in velocity is towards the centre of the circle. This is always the case and thus true for the instantaneous acceleration. For a mass m moving at a speed v in uniform circular motion of radius r,

Centripetal acceleration $a_{centripetal} = \frac{v^2}{r}$ [In towards the centre of the circle]

A force must have caused this acceleration. The value of the force is worked out using Newton's second law:

Centripetal force (CPF) $f_{centripetal} = m\, a_{centripetal}$

$= \frac{m\, v^2}{r}$ [In towards the centre of the circle]

For example, if a car of mass 1500 kg is travelling at a constant speed of 20 m s^{-1} around a circular track of radius 50 m, the resultant force that must be acting on it works out to be

$$F = \frac{1500\,(20)^2}{50} = 12\,000 \text{ N}$$

It is really important to understand that centripetal force is NOT a new force that starts acting on something when it goes in a circle. It is a way of working out what the total force must have been. This total force must result from all the other forces on the object. See the examples below for more details.

One final point to note is that the centripetal force does NOT do any work. (Work done = force×distance **in the direction of the force**.)

EXAMPLES

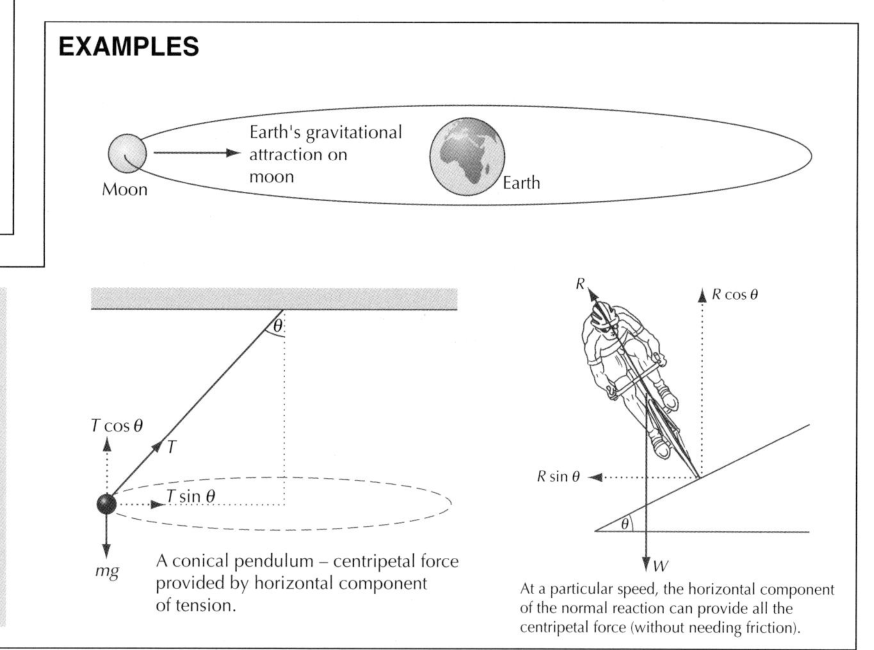

F
friction forces between tyres and road

IB QUESTIONS – MECHANICS

1 Two identical objects A and B fall from rest from different heights. If B takes twice as long as A to reach the ground, what is the ratio of the heights from which A and B fell? Neglect air resistance.

A $1:\sqrt{2}$ **B** 1:2 **C** 1:4 **D** 1:8

2 A trolley is given an initial push along a horizontal floor to get it moving. The trolley then travels forward along the floor, gradually slowing. What is true of the horizontal force(s) on the trolley while it is slowing?

A There is a forward force and a backward force, but the forward force is larger.

B There is a forward force and a backward force, but the backward force is larger.

C There is only a forward force, which diminishes with time.

D There is only a backward force.

3 A mass is suspended by cord from a ring which is attached by two further cords to the ceiling and the wall as shown. The cord from the ceiling makes an angle of less than 45° with the vertical as shown. The tensions in the three cords are labelled R, S and T in the diagram.

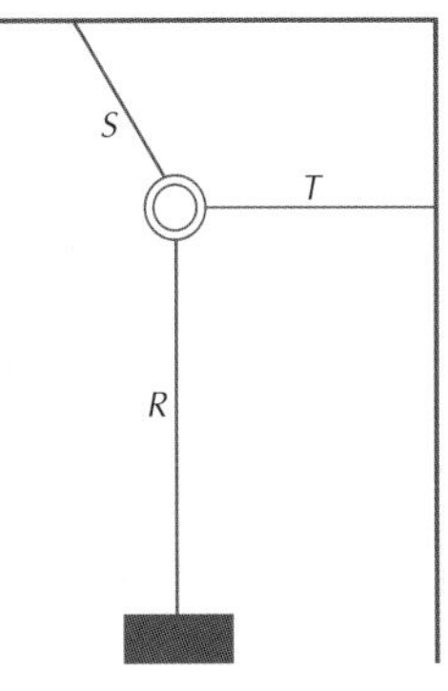

How do the tensions R, S and T in the three cords compare in magnitude?

A $R > T > S$ **B** $S > R > T$

C $R = S = T$ **D** $R = S > T$

4 In any collision between two objects, what is true about the total momentum and the total kinetic energy of the system of two objects?

	Total momentum	Total kinetic energy
A	always stays the same	always stays the same
B	always stays the same	can change
C	can change	always stays the same
D	can change	can change

5 A ball is dropped on to a hard surface and makes several bounces before coming to rest. Which one of the graphs below best represents how the velocity of the ball varies with time?

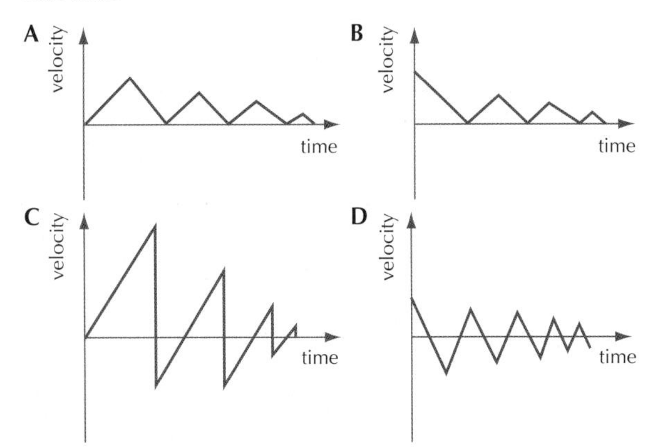

6 A car and a truck are both travelling at the speed limit of 60 km h^{-1} but in opposite directions as shown. The truck has **twice** the mass of the car.

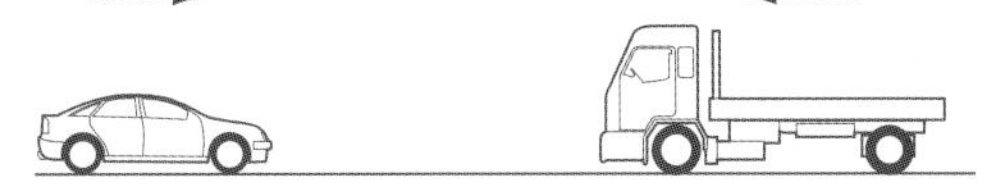

The vehicles collide head-on and become entangled together.

(a) During the collision, how does the force exerted by the car on the truck compare with the force exerted by the truck on the car? Explain. [2]

(b) In what direction will the entangled vehicles move after collision, or will they be stationary? Support your answer, referring to a physics principle. [2]

(c) Determine the speed (in km h^{-1}) of the combined wreck immediately after the collision. [3]

(d) How does the acceleration of the car compare with the acceleration of the truck during the collision? Explain. [2]

(e) Both the car and truck drivers are wearing seat belts. Which driver is likely to be more severely jolted in the collision? Explain. [2]

(f) The total kinetic energy of the system decreases as a result of the collision. Is the principle of conservation of energy violated? Explain. [1]

7 A car travels at a steady speed v in a circular path of radius r on a circular track banked at an angle θ, as shown in the plan and side views in the diagram.

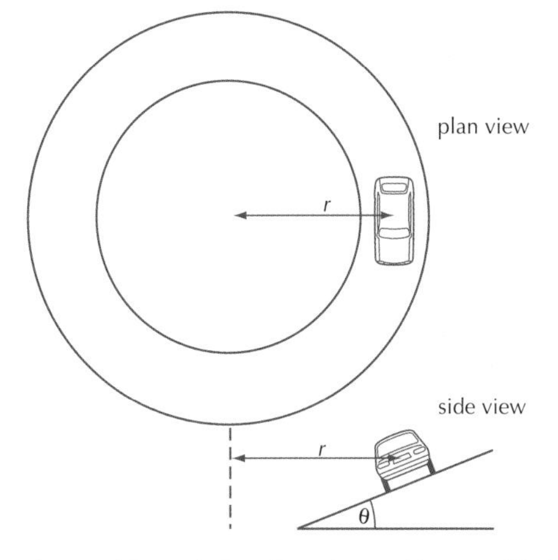

The car's speed is such that there is no sideways frictional force between the tyres and the track.

(a) Does the car have an acceleration? Explain why or why not. If you say yes, state its direction. [2]

(b) The car on the track is represented by a block in the figure below, moving perpendicular to the page. Draw a force diagram, showing and labelling all the forces acting on the moving car. [2]

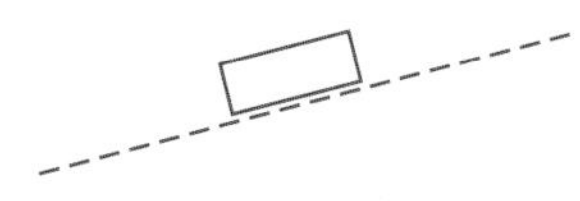

(c) The track is banked at an angle of 17° and the circular path of the car has radius 30 m. Calculate the speed at which the car must travel in order that there be no sideways frictional force between the tyres and the track. Show all working. [4]

Thermal concepts

TEMPERATURE AND HEAT FLOW

Hot and cold are just labels that identify the direction in which thermal energy (otherwise known as heat) will be naturally transferred when two objects are placed in thermal contact. This leads to the concept of the 'hotness' of an object. The direction of the natural flow of thermal energy between two objects is determined by the 'hotness' of each object. Thermal energy naturally flows from hot to cold.

The temperature of an object is a measure of how hot it is. In other words, if two objects are placed in thermal contact, then the temperature difference between the two objects will determine the direction of the natural transfer of thermal energy. Thermal energy is naturally transferred 'down' the temperature difference – from high temperature to low temperature. Eventually, the two objects would be expected to reach the same temperature. When this happens, they are said to be in **thermal equilibrium**.

Heat is not a substance that flows from one object to another. What has happened is that thermal energy has been transferred. Thermal energy (heat) refers to the non-mechanical transfer of energy between a system and its surroundings.

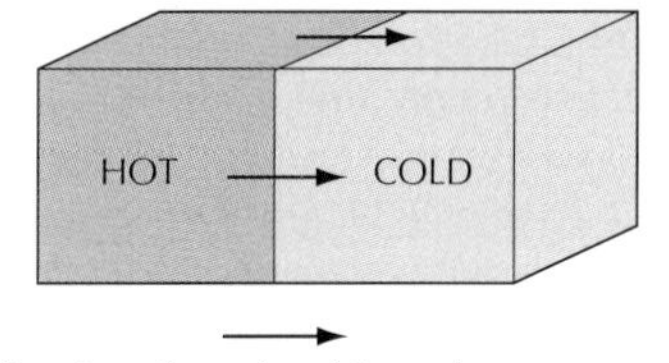

direction of transfer of thermal energy

KELVIN AND CELSIUS

Most of the time, there are only two sensible temperature scales to chose between – the Kelvin scale and the Celsius scale.

In order to use them, you do not need to understand the details of how either of these scales has been defined, but you do need to know the relation between them. Most everyday thermometers are marked with the Celsius scale – temperature is quoted in degrees Celsius (°C).

There is an easy relationship between a temperature T as measured on the Kelvin scale and the corresponding temperature t as measured on the Celsius scale. The approximate relationship is

T (K) = t (°C) + 273

This means that the 'size' of the units used on each scale is identical, but they have different zero points.

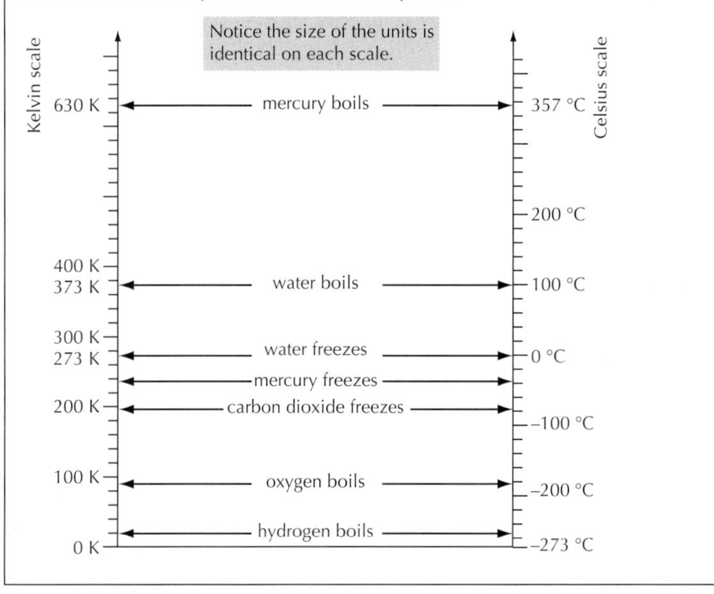

DEFINITIONS

The concepts of the **mole**, **molar mass** and the **Avogadro constant** are all introduced so as to be able to relate the mass of a gas (an easily measurable quantity) to the number of molecules that are present in the gas.

Ideal gas — An ideal gas is one that follows the gas laws for all values of P, V and T (see pages 31 and 87).

Mole — The mole is the basic SI unit for 'amount of substance'. One mole of any substance is equal to the amount of that substance that contains the same number of atoms as 0.012 kg of carbon–12 (^{12}C). When writing the unit it is (slightly) shortened to the mol.

Avogadro constant — This is the number of atoms in 0.012 kg of carbon–12 (^{12}C). It is 6.02×10^{23}.

Molar mass — The mass of one mole of a substance is called the molar mass. A simple rule applies. If an element has a certain mass number, A, then the molar mass will be A grams.

EXAMPLE

How many atoms are there in 8 g of helium (mass number 4)?

$n = \frac{8}{4} = 2$ moles

number of atoms $= 2 \times 6.02 \times 10^{23}$
$= 1.2 \times 10^{24}$

Heat and internal energy

MICROSCOPIC VS MACROSCOPIC

When analysing something physical, we have a choice.

- The **macroscopic** point of view considers the system as a whole and sees how it interacts with its surroundings.
- The **microscopic** point of view looks inside the system to see how its component parts interact with each other.

So far we have looked at the temperature of a system in a macroscopic way, but all objects are made up of **atoms** and **molecules**.

According to **kinetic theory** these particles are constantly in random motion – hence the name. See below for more details. Although atoms and molecules are different things (a molecule is a combination of atoms), the difference is not important at this stage. The particles can be thought of as little 'points' of mass with velocities that are continually changing.

INTERNAL ENERGY

If the temperature of an object changes then it must have gained (or lost) energy. From the microscopic point of view, the molecules must have gained (or lost) this energy.

The two possible forms are kinetic energy and potential energy.

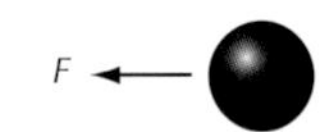
speed in a random direction
v
∴ molecule has KE
F
equilibrium position
resultant force back towards equilibrium position due to neighbouring molecules ∴ molecule has PE

- The molecules have kinetic energy because they are moving. To be absolutely precise, a molecule can have either translational kinetic energy (the whole molecule is moving in a certain direction) or rotational kinetic energy (the molecule is rotating about one or more axes).
- The molecules have potential energy because of the **intermolecular** forces. If we imagine pulling two molecules further apart, this would require work against the intermolecular forces.

The total energy that the molecules possess (kinetic plus potential) is called the **internal energy** of a substance. Whenever we heat a substance, we increase its internal energy.

Temperature is a measure of the average kinetic energy of the molecules in a substance.

If two substances have the same temperature, then their molecules have the same average kinetic energy.

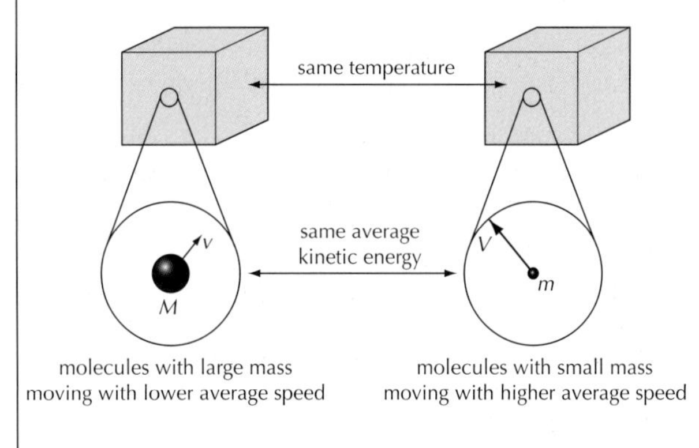
same temperature
same average kinetic energy
v
M
V
m
molecules with large mass moving with lower average speed
molecules with small mass moving with higher average speed

KINETIC THEORY

Molecules are arranged in different ways depending on the **phase** of the substance (i.e. solid, liquid or gas).

Solids

Macroscopically, solids have a fixed volume and a fixed shape. This is because the molecules are held in position by bonds. However the bonds are not absolutely rigid. The molecules vibrate around a mean (average) position. The higher the temperature, the greater the vibrations.

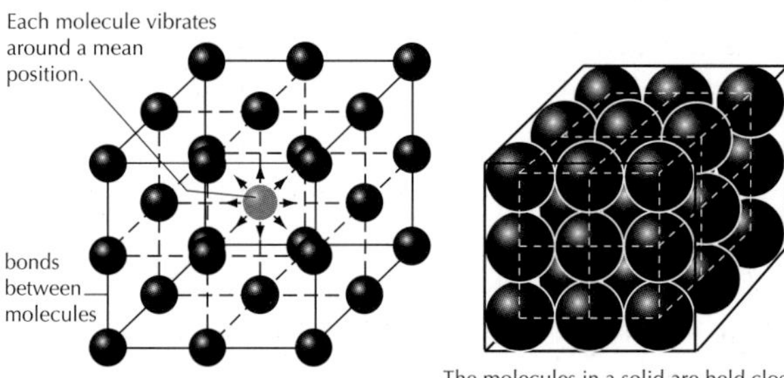
Each molecule vibrates around a mean position.
bonds between molecules

The molecules in a solid are held close together by the intermolecular bonds.

Liquids

A liquid also has a fixed volume but its shape can change. The molecules are also vibrating, but they are not completely fixed in position. There are still strong forces between the molecules. This keeps the molecules close to one another, but they are free to move around each other.

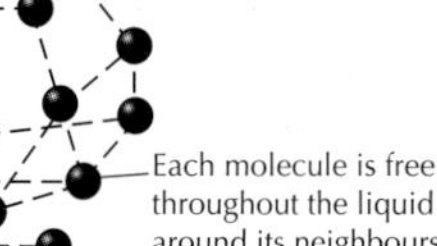
Bonds between neighbouring molecules; these can be made and broken, allowing a molecule to move.
Each molecule is free to move throughout the liquid by moving around its neighbours.

Gases

A gas will always expand to fill the container in which it is put. The molecules are not fixed in position, and any forces between the molecules are very weak. This means that the molecules are essentially independent of one another, but they do occasionally collide. More detail is given on page 31.

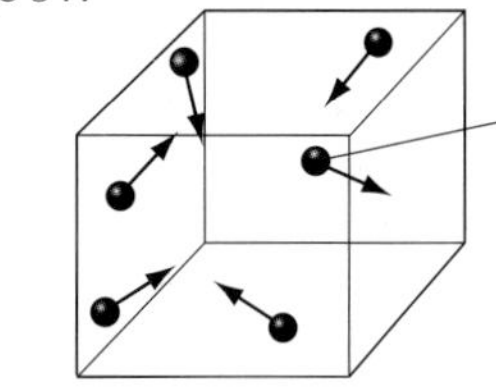
molecules in random motion
no fixed bonds between molecules so they are free to move

HEAT AND WORK

Many people have confused ideas about heat and work. In answers to examination questions it is very common to read, for example, that 'heat rises' – when what is meant is that the transfer of thermal energy is upwards.

- When a force moves through a distance, we say that work is done. Work is the energy that has been transmitted from one system to another from the macroscopic point of view.
- When work is done on a microscopic level (i.e. on individual molecules), we say that heating has taken place. Heat is the energy that has been transmitted. It can either increase the kinetic energy of the molecules or their potential energy or, of course, both.

In both cases energy is being transferred.

Specific heat capacity

DEFINITIONS & MICROSCOPIC EXPLANATION

In theory, if an object could be heated up with no energy loss, then the increase in temperature ΔT depends on three things:

- the energy given to the object Q,
- the mass, m, and,
- the substance from which the object is made.

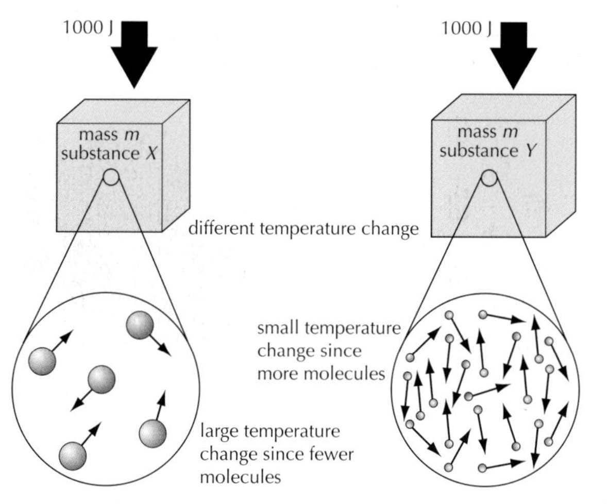

Two different blocks with the same mass and same energy input will have a different temperature change.

We define the **thermal capacity** C of an object as the energy required to raise its temperature by 1 K. Different objects (even different samples of the same substance) will have different values of heat capacity. **Specific heat capacity** is the energy required to raise a unit mass of a substance by 1 K. 'Specific' here just means 'per unit mass'.

In symbols,

Thermal capacity $C = \frac{Q}{\Delta T}$ (J K^{-1} or J °C^{-1})

Specific heat capacity $c = \frac{Q}{(m\,\Delta T)}$ (J kg^{-1} K^{-1} or J kg^{-1} °C^{-1})

Note

- A particular gas can have many different values of specific heat capacity – it depends on the conditions used – see page 88.
- These equations refer to the **temperature difference** resulting from the addition of a certain amount of energy. In other words, it generally takes the same amount of energy to raise the temperature of an object from 25 °C to 35 °C as it does for the same object to go from 402 °C to 412 °C. This is only true so long as energy is not lost from the object.
- If an object is raised above room temperature, it starts to lose energy. The hotter it becomes, the greater the rate at which it loses energy.

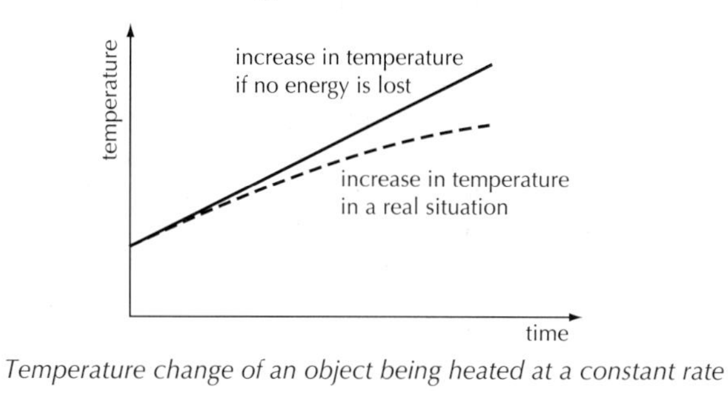

Temperature change of an object being heated at a constant rate

METHODS OF MEASURING HEAT CAPACITIES AND SPECIFIC HEAT CAPACITIES

The are two basic ways to measure heat capacity.

1. Electrical method

The experiment would be set up as below:

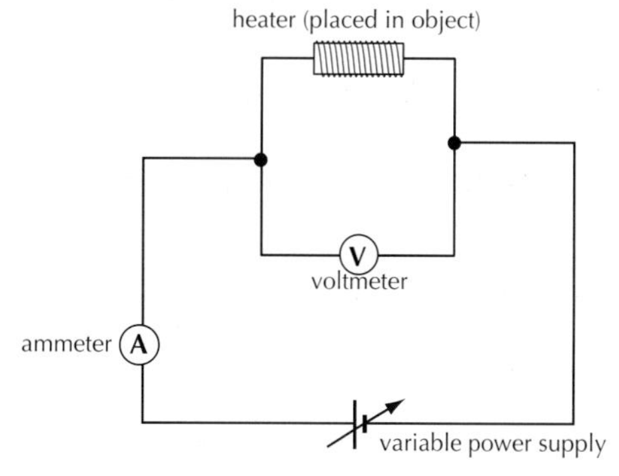

- the specific heat capacity $c = \frac{I\,t\,V}{m\,(T_2 - T_1)}$.

Sources of experimental error

- the loss of thermal energy from the apparatus.
- container for the substance and the heater will also be warmed up.
- it will take some time for the energy to be shared uniformly through the substance.

2. Method of mixtures

The known specific heat capacity of one substance can be used to find the specific heat capacity of another substance.

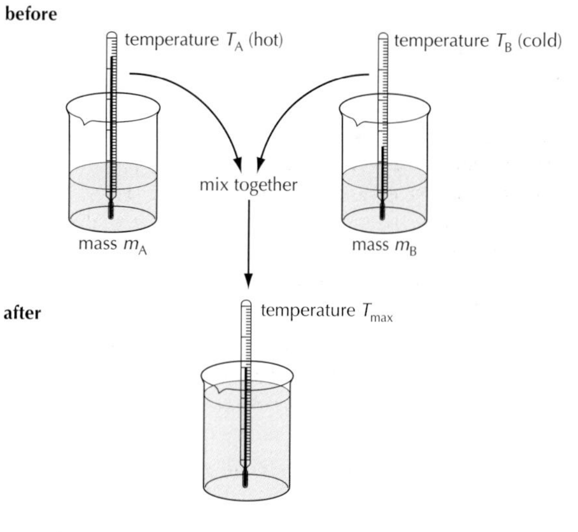

Procedure:

- measure the masses of the liquids m_A and m_B.
- measure the two starting temperatures T_A and T_B.
- mix the two liquids together.
- record the maximum temperature of the mixture T_{max}.

If no energy is lost from the system then,

energy lost by hot substance cooling down = energy gained by cold substance heating up

$$m_A c_A (T_A - T_{max}) = m_B c_B (T_{max} - T_B)$$

Again, the main source of experimental error is the loss of thermal energy from the apparatus – particularly while the liquids are being transferred. The changes of temperature of the container also need to be taken into consideration for a more accurate result.

Phases (states) of matter and latent heat

DEFINITIONS AND MICROSCOPIC VIEW

When a substance changes phase, the temperature remains constant even though thermal energy is still being transferred.

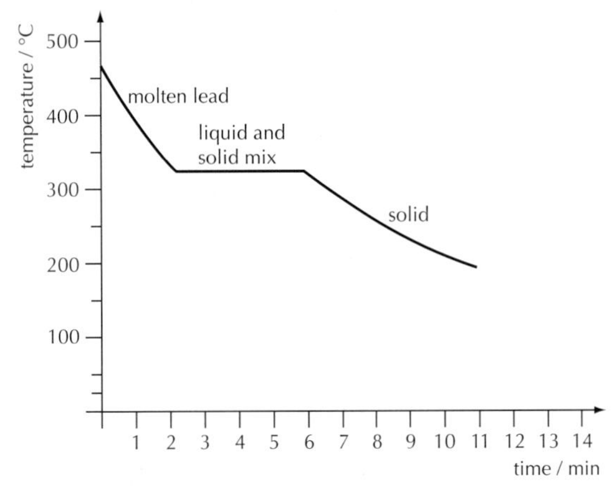

Cooling curve for molten lead (idealized)

The amount of energy associated with the phase change is called the **latent heat**. The technical term for the change of phase from solid to liquid is **fusion** and the term for the change from liquid to gas is **vaporization**.

The energy given to the molecules does not increase their kinetic energy so it must be increasing their potential energy. Intermolecular bonds are being broken and this takes energy. When the substance freezes bonds are created and this process releases energy.

It is a very common mistake to think that the molecules must speed up during a phase change. The molecules in water vapour at 100 °C must be moving with the same average speed as the molecules in liquid water at 100 °C.

The **specific latent heat** of a substance is defined as the amount of energy per unit mass absorbed or released during a change of phase.

In symbols,

Specific latent heat $L = \frac{Q}{m}$ (J kg^{-1}.)

EVAPORATION

Evaporation takes place at the surface of liquids. If a liquid is below its boiling point, on average the molecules do not have sufficient energy to leave the surface.

The result of the overall process is for the faster moving molecules to escape the liquid. This means that it is the slower moving ones that are left behind – in other words the temperature of the liquid falls as a result of evaporation. Evaporation causes cooling.

The rate at which evaporation takes place depends on

- the surface area of the liquid (increased area means increased evaporation rate).
- the temperature of the liquid (increased temperature means increased evaporation rate).
- the pressure (or moisture content) of the air above the liquid (increased pressure means decreased evaporation rate).
- any draught that exists above the liquid (increased draught means increased evaporation rate).

METHODS OF MEASURING

The two possible methods for measuring latent heats shown below are very similar in principle to the methods for measuring specific heat capacities (see previous page)

1. A method for measuring the specific latent heat of vaporisation of water

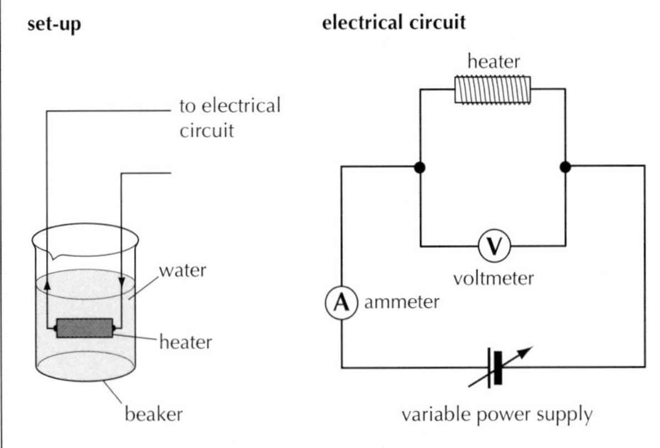

A method for measuring the latent heat of vaporization.

The amount of thermal energy provided to water at its boiling point is calculated using electrical energy = $I\,t\,V$. The mass vaporised needs to be recorded.

- The specific latent heat $L = \frac{I\,t\,V}{(m_1 - m_2)}$.

Sources of experimental error

- Loss of thermal energy from the apparatus.
- Some water vapour will be lost before and after timing.

2. A method for measuring the specific latent heat of fusion of water

Providing we know the specific heat capacity of water, we can calculate the specific latent heat of fusion for water. In the example below ice (at 0 °C) is added to warm water and the temperature of the resulting mix is measured.

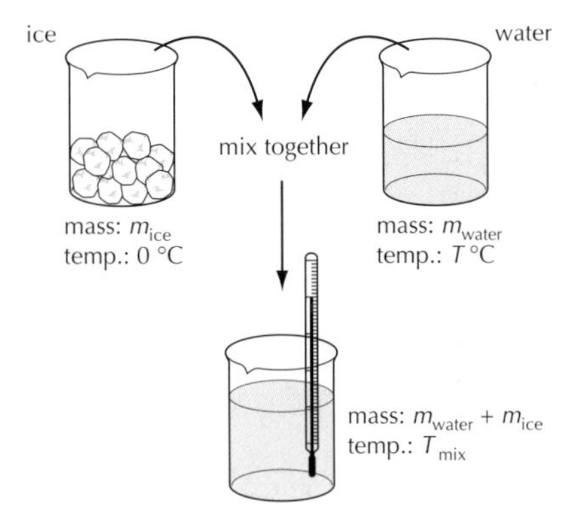

A method for measuring the latent heat of fusion.

If no energy is lost from the system then,

energy lost by water cooling down = energy gained by ice

$$m_{water}\,c_{water}\,(T_{water} - T_{mix}) = m_{ice}\,L_{fusion} + m_{ice}\,c_{water}\,T_{mix}$$

Sources of experimental error

- Loss (or gain) of thermal energy from the apparatus.
- If the ice had not started at exactly zero, then there would be an additional term in the equation in order to account for the energy needed to warm the ice up to 0 °C.
- Water clinging to the ice before the transfer.

Molecular model of an ideal gas

KINETIC MODEL OF AN IDEAL GAS

Assumptions:

- Newton's laws apply to molecular behaviour
- there are no intermolecular forces
- the molecules are treated as points
- the molecules are in random motion
- the collisions between the molecules are elastic (no energy is lost)
- there is no time spent in these collisions.

The pressure of a gas is explained as follows:

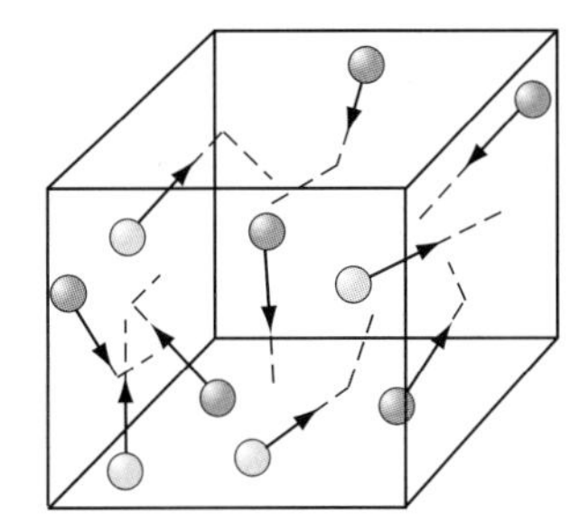

The pressure of a gas is a result of collisions between the molecules and the walls of the container.

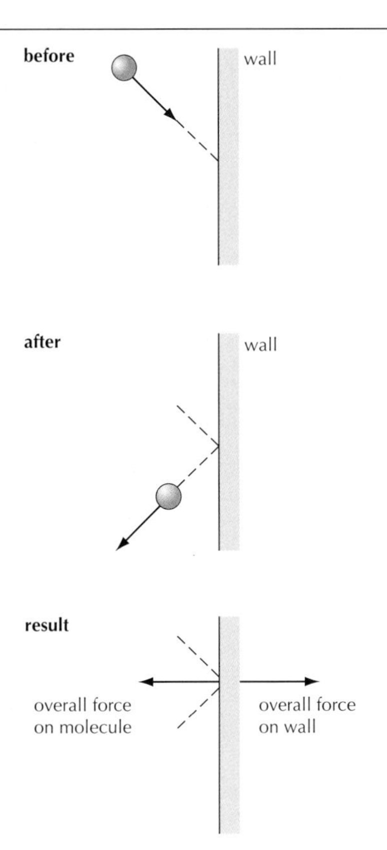

A single molecule hitting the walls of the container.

- When a molecule bounces off the walls of a container its momentum changes (due to the change in direction – momentum is a vector).
- There must have been a force on the molecule from the wall (Newton II).
- There must have been an equal and opposite force on the wall from the molecule (Newton III).
- Each time there is a collision between a molecule and the wall, a force is exerted on the wall.
- The average of all the microscopic forces on the wall over a period of time means that there is effectively a constant force on the wall from the gas.
- This force per unit area of the wall is what we call pressure.

$$P = \frac{F}{A}$$

Since the temperature of a gas is a measure of the average kinetic energy of the molecules, as we lower the temperature of a gas the molecules will move slower. At absolute zero, we imagine the molecules to have zero kinetic energy. We cannot go any lower because we cannot reduce their kinetic energy any further!

PRESSURE LAW

Macroscopically, at a constant volume the pressure of a gas is proportional to its temperature in kelvin (see page 87). Microscopically this can be analysed as follows

- If the temperature of a gas goes up, the molecules have more average kinetic energy – they are moving faster on average.
- Fast moving molecules will have a greater change of momentum when they hit the walls of the container.
- Thus the microscopic force from each molecule will be greater.
- The molecules are moving faster so they hit the walls more often.
- For both these reasons, the total force on the wall goes up.
- Thus the pressure goes up.

low temperature high temperature

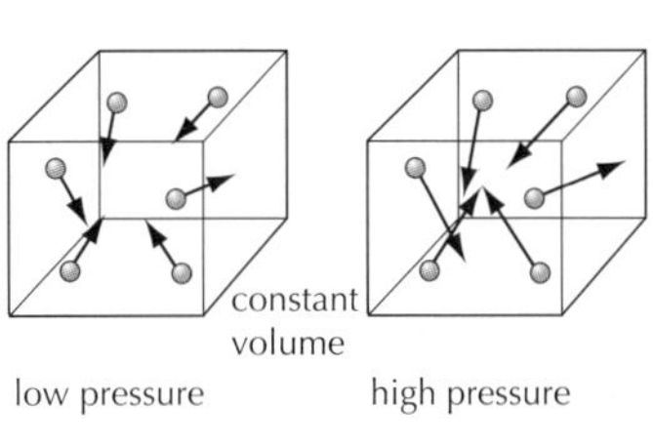

low pressure high pressure

Microscopic justification of the pressure law

CHARLES'S LAW

Macroscopically, at a constant pressure, the volume of a gas is proportional to its temperature in kelvin (see page 87). Microscopically this can be analysed as follows

- A higher temperature means faster moving molecules (see above).
- Faster moving molecules hit the walls with a greater microscopic force (see above).
- If the volume of the gas increases, then the rate at which these collisions take place on a unit area of the wall must goes down.
- The average force on a unit area of the wall can thus be the same.
- Thus the pressure remains the same.

low temperature high temperature

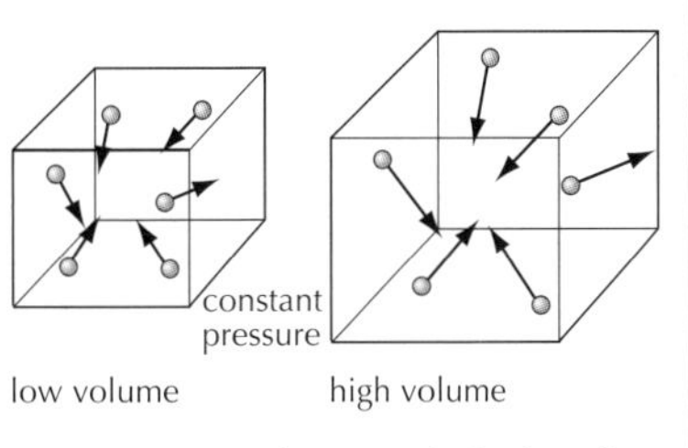

low volume high volume

Microscopic justification of Charles's law

BOYLE'S LAW

Macroscopically, at a constant temperature, the pressure of a gas is inversely proportional to its volume (see page 87). Microscopically this can be seen to be correct.

- The constant temperature of gas means that the molecules have a constant average speed.
- The microscopic force that each molecule exerts on the wall will remain constant.
- Increasing the volume of the container decreases the rate with which the molecules hit the wall – average total force decreases.
- If the average total force decreases the pressure decreases.

high pressure low pressure

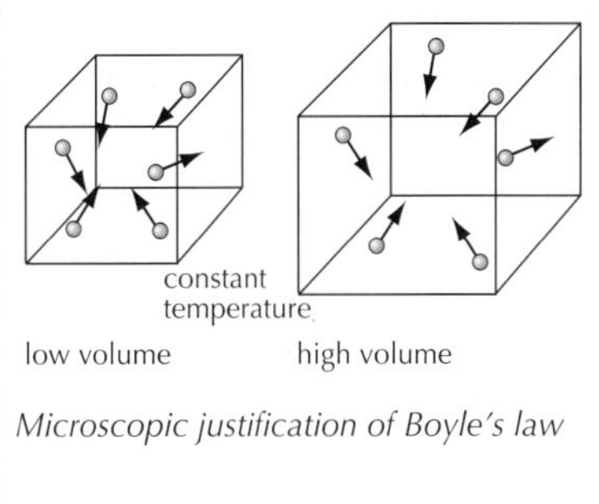

low volume high volume

Microscopic justification of Boyle's law

IB QUESTIONS – THERMAL PHYSICS

The following information relates to questions 1 and 2 below.

A substance is heated at a constant rate of energy transfer. A graph of its temperature against time is shown below.

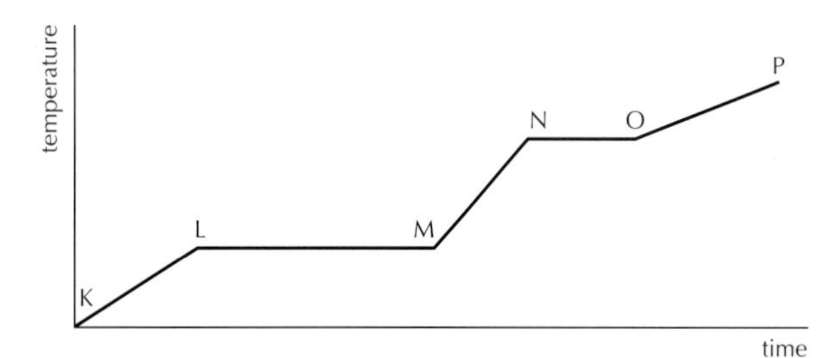

1 Which regions of the graph correspond to the substance existing in a mixture of two phases?

A KL, MN and OP **C** All regions

B LM and NO **D** No regions

2 In which region of the graph is the specific heat capacity of the substance greatest?

A KL **C** MN

B LM **D** OP

3 When the volume of a gas is isothermally compressed to a smaller volume, the pressure exerted by the gas on the container walls increases. The best microscopic explanation for this pressure increase is that at the smaller volume

A the individual gas molecules are compressed

B the gas molecules repel each other more strongly

C the average velocity of gas molecules hitting the wall is greater

D the frequency of collisions with gas molecules with the walls is greater

4 A lead bullet is fired into an iron plate, where it deforms and stops. As a result, the temperature of the lead increases by an amount ΔT. For an identical bullet hitting the plate with twice the speed, what is the best estimate of the temperature increase?

A ΔT **C** $2\,\Delta T$

B $\sqrt{2}\,\Delta T$ **D** $4\,\Delta T$

5 In winter, in some countries, the water in a swimming pool needs to be heated.

(a) Estimate the cost of heating the water in a typical swimming pool from 5 °C to a suitable temperature for swimming. You may choose to consider any reasonable size of pool.

Clearly show any estimated values. The following information will be useful:

Specific heat capacity of water 4186 J kg^{-1} K^{-1}
Density of water 1000 kg m^{-3}
Cost per kW h of electrical energy $0.10

(i) Estimated values [4]
(ii) Calculations [7]

(b) An electrical heater for swimming pools has the following information written on its side:

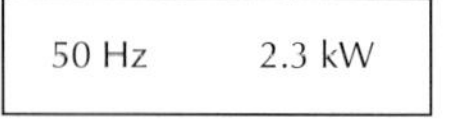

(i) Estimate how many days it would take this heater to heat the water in the swimming pool. [4]
(ii) Suggest two reasons why this can only be an approximation. [2]

(c) Overnight the water in the swimming pool cools. The temperature loss depends on the conditions during the night. Two possible factors affecting the temperature loss are listed below.

a CLOUDY night or a CLEAR night

a DAMP night or a DRY night.

(i) For each of these factors underline which of the extremes given above would cause the smallest temperature loss, and explain your reasoning. [6]
(ii) Suggest one other factor that might affect the overnight temperature loss and explain its effect. [2]

6 This question is about determining the specific latent heat of fusion of ice.

A student determines the specific latent heat of fusion of ice at home. She takes some ice from the freezer, measures its mass and mixes it with a known mass of water in an insulating jug. She stirs until all the ice has melted and measures the final temperature of the mixture. She also measured the temperature in the freezer and the initial temperature of the water.

She records her measurements as follows:

Mass of ice used	m_i	0.12 kg
Initial temperature of ice	T_i	–12 °C
Initial mass of water	m_w	0.40 kg
Initial temperature of water	T_w	22 °C
Final temperature of mixture	T_f	15 °C

The specific heat capacities of water and ice are $c_w = 4.2$ kJ kg^{-1} $°C^{-1}$ and $c_i = 2.1$ kJ kg^{-1} $°C^{-1}$

(a) Set up the appropriate equation, representing energy transfers during the process of coming to thermal equilibrium, that will enable her to solve for the specific latent heat L_i of ice. Insert values into the equation from the data above, **but do not solve the equation**. [5]

(b) Explain the physical meaning of each *energy transfer term* in your equation (but not each symbol). [4]

(c) State an assumption you have made about the experiment, in setting up your equation in **(a)**. [1]

(d) Why should she take the temperature of the mixture *immediately* after all the ice has melted? [1]

(e) Explain from the microscopic point of view, in terms of molecular behaviour, why the temperature of the ice does not increase while it is melting. [4]

Simple harmonic motion (SHM)

DEFINITIONS

Many systems involve vibrations or oscillations; an object continually moves to-and-fro about a fixed average point (the **mean position**) retracing the same path through space taking a fixed time between repeats. Oscillations involve the interchange of energy between kinetic and potential.

	Kinetic energy	Potential energy store
Mass moving between two horizontal springs	Moving mass	Elastic potential energy in the springs
Mass moving on a vertical spring	Moving mass	Elastic potential energy in the springs and gravitational potential energy
Simple pendulum	Moving pendulum bob	Gravitational potential energy of bob
Buoy bouncing up and down in a water	Moving buoy	Gravitational PE of buoy and water
A oscillating ruler as a result of one end being displaced while the other is fixed	Moving sections of the ruler	Elastic PE of the bent ruler

	Definition
Displacement, x	The instantaneous distance (SI measurement: m) of the moving object from its mean position (in a specified direction)
Amplitude, A	The maximum displacement (SI measurement: m) from the mean position
Frequency, f	The number of oscillations completed per unit time. The SI measurement is the number of cycles per second or Hertz (Hz).
Period, T	The time taken (SI measurement: s) for one complete oscillation. $T = \frac{1}{f}$
Phase difference, ϕ	This is a measure of how "in step" different particles are. If moving together they are **in phase**. ϕ is measured in either degrees (°) or radians (rad). 360° or 2π rad is one complete cycle so 180° or π rad is completely **out of phase** by half a cycle. A phase difference of 90° or $\pi/2$ rad is a quarter of a cycle.

object oscillates between extremes

displacement, x

amplitude, A

mean position

SIMPLE HARMONIC MOTION (SHM)

Simple harmonic motion is defined as the motion that takes place when the acceleration, a, of object is always directed towards, and is proportional to, its displacement from a fixed point. This acceleration is caused by a **restoring force** that must always be pointed towards the mean position and also proportional to the displacement from the mean position.

$$F \propto -x \text{ or } F = -(\text{constant}) \times x$$

Since F = ma

$$a \propto -x \text{ or } a = -(\text{constant}) \times x$$

The negative sign signifies that the acceleration is always pointing back towards the mean position.

The constant of proportionality between acceleration and displacement is often identified as the square of a constant ω which is referred to as the **angular frequency**.

$$a = -\omega^2 x$$

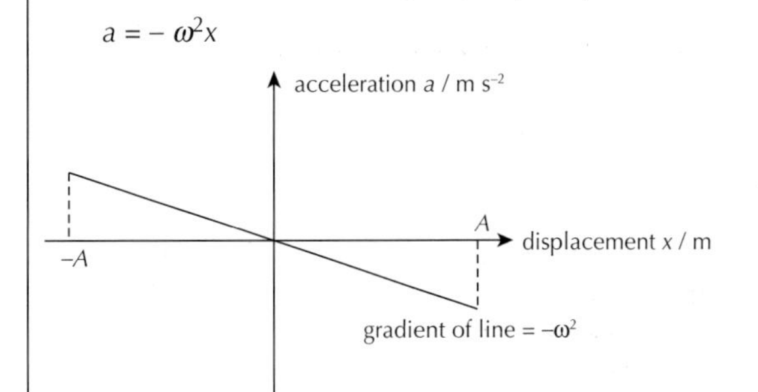

Points to note about SHM:

- The time period T does not depend on the amplitude A.
- Not all oscillations are SHM, but there are many everyday examples of natural SHM oscillations.

EXAMPLE OF SHM: MASS BETWEEN TWO SPRINGS

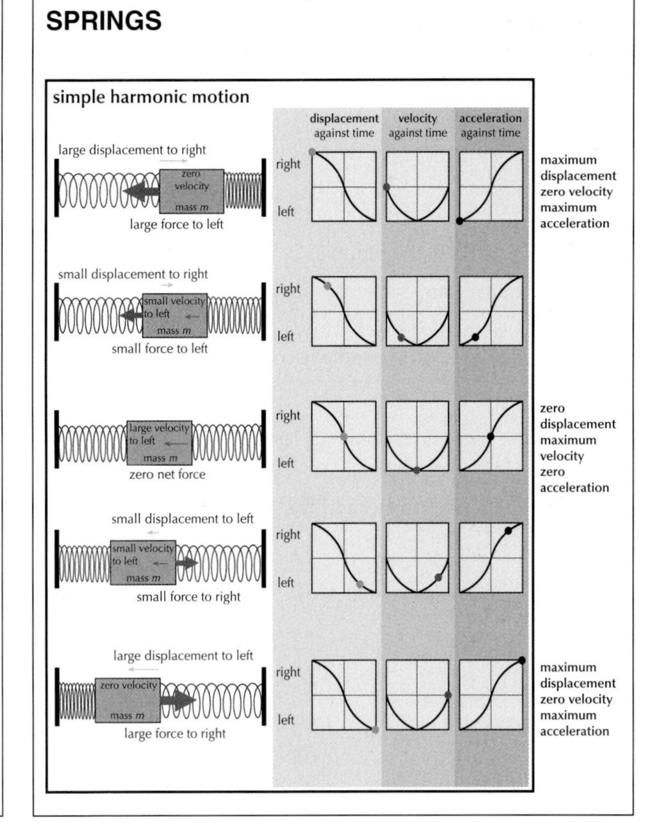

Kinematics of simple harmonic motion

IDENTIFICATION OF SHM

In order to analyse a situation to decide if SHM is taking place, the following procedure should be followed.

- Identify all the forces acting on an object when it is displaced an arbitrary distance x from its rest position using a free body diagram.
- Calculate the resultant force using Newton's second law. If this force is proportional to the displacement and always points back towards the mean position (i.e. $F \propto -x$) then the motion of the object must be SHM.
- Once SHM has been identified, the equation of motion must be in the following form:

$$a = -\left(\frac{\text{restoring force per unit displacement, k}}{\text{oscillating mass, m}}\right) \times x$$

- This identifies the angular frequency ω as $\omega^2 = \left(\frac{k}{m}\right)$ or

$\omega = \sqrt{\left(\frac{k}{m}\right)}$. Identification of ω allows quantitative equations to be applied.

EXAMPLE

A 600 g mass is attached to a light spring with spring constant 30 N m^{-1}.

(a) Show that the mass does SHM.

(b) Calculate the frequency of its oscillation.

(a) Weight of mass = mg = 6.0 N

Additional displacement x down means that resultant force on mass = $k\,x$ upwards. Since $F \propto -x$, the mass will oscillate with SHM.

(b) Since SHM, $T = 2\pi\sqrt{\left(\frac{m}{k}\right)} = 2\pi\sqrt{\left(\frac{0.6}{30}\right)} = 0.889s$

$$f = \frac{1}{T} = \frac{1}{0.889} = 1.1\text{ Hz}$$

ACCELERATION, VELOCITY AND DISPLACEMENT DURING SHM

The variation with time of the acceleration, a, velocity, v and displacement, x of an object doing SHM depends on the angular frequency ω as

$$x = A\sin(\omega t)$$

$$v = A\omega\cos(\omega t)$$

$$a = -A\omega^2\sin(\omega t)$$

Note that these equations assume that the time measurement starts when the object is in the centre and moving at its maximum positive velocity. The first two equations can be rearranged to produce the following relationship:

$$v = \pm\,\omega\sqrt{\left(A^2 - x^2\right)}$$

A is the amplitude of the oscillation measured in m

t is the time taken measured in s

ω is the angular frequency measured in rad s^{-1}

ωt is an angle that increases with time measured in radians. A full oscillation is completed when $(\omega t) = 2\pi$ rad.

The angular frequency is related to the time period T by the following equation.

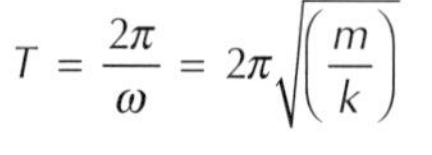

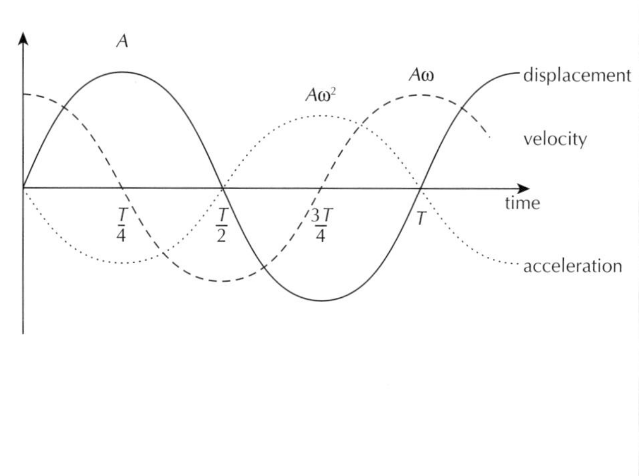

Energy changes during simple harmonic motion

During SHM, energy is interchanged between KE and PE. Providing there are no resistive forces which dissipate this energy, the total energy must remain constant. The oscillation is said to be **undamped**.

The kinetic energy can be calculated from

$$E_k = \frac{1}{2}mv^2 = \frac{1}{2}m\omega^2\left(A^2 - x^2\right)$$

The potential energy can be calculated from

$$E_p = \frac{1}{2}m\omega^2 x^2$$

The total energy is

$$E_t = E_k + E_p = \frac{1}{2}m\omega^2\left(A^2 - x^2\right) + \frac{1}{2}m\omega^2 x^2 = \frac{1}{2}m\omega^2 A^2$$

Energy in SHM is proportional to:

- the mass m
- the (amplitude)^2
- the (frequency)^2

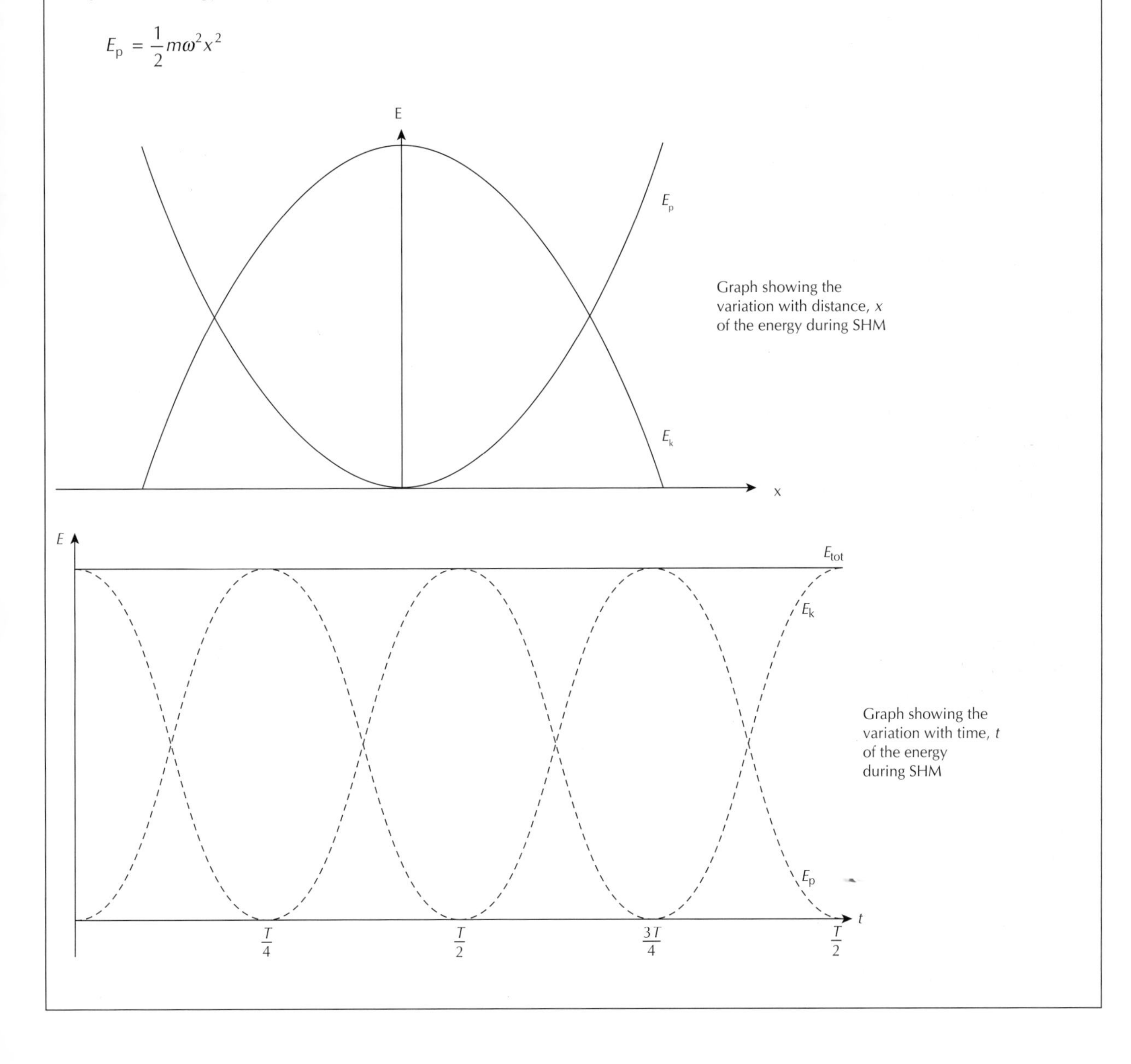

Graph showing the variation with distance, x of the energy during SHM

Graph showing the variation with time, t of the energy during SHM

Forced oscillations and resonance (1)

DAMPING

Damping involves a frictional force that is always in the opposite direction to the direction of motion of the oscillating particle. As the particle oscillates, it does work against this resistive (or dissipative) force and so the particle loses energy. As the total energy of the particle is proportional to the (amplitude)^2 of the SHM, the amplitude decreases exponentially with time.

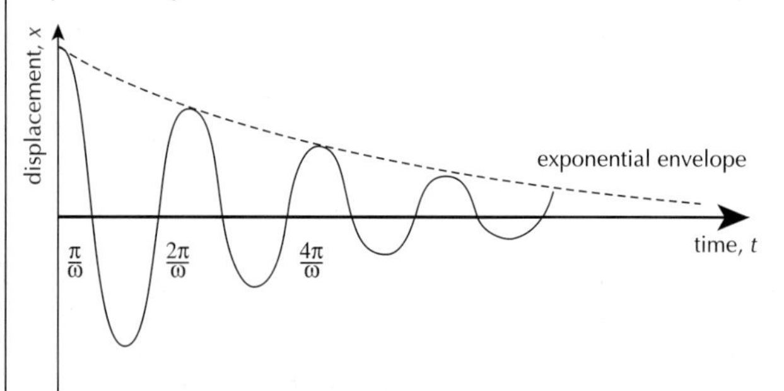

The above example shows the effect of **light damping** (the system is said to be **under damped**) where the resistive force is small so a small fraction of the total energy is removed each cycle. The time period of the oscillations is not affected and the oscillations continue for a significant number of cycles. The time taken for the oscillations to "die out" can be long.

Heavy damping or **over damping** involves large resistive forces (e.g. the SHM taking place in a viscous liquid) and can completely prevent the oscillations from taking place. The time taken for the particle to return to zero displacement can again be long.

Critical damping involves an intermediate value for resistive force such that the time taken for the particle to return to zero displacement is a minimum. Effectively there is no "overshoot". Examples of critically damped systems include electric meters with moving pointers and door closing mechanisms.

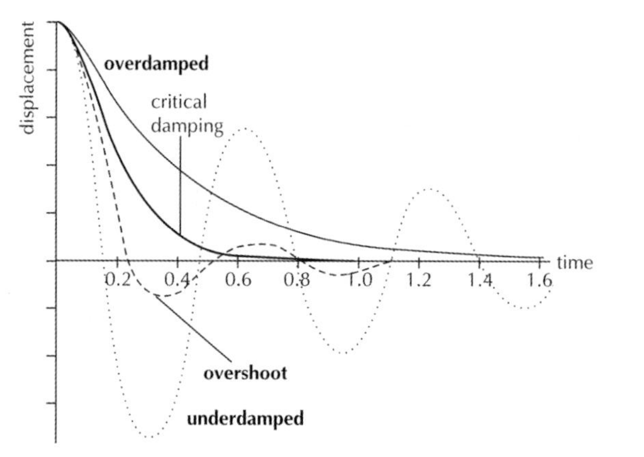

RESONANCE

If the system is temporarily displaced from its equilibrium position, the system will oscillate as a result. This oscillation will be at the **natural frequency of vibration** of the system. For example, if you tap the rim of a wine glass with knife, it will oscillate and you can hear a note for a short while. Complex systems tend to have many possible modes of vibration each with its own natural frequency.

It is also possible to force a system to oscillate at any frequency that we choose by subjecting it to a changing force that varies with the chosen frequency. This periodic driving force must be provided from outside the system. When this **driving frequency** is first applied, a combination of natural and forced oscillations take place which produces complex **transient** oscillations. Once the amplitude of the transient oscillations "die down", a steady condition is achieved in which:

- The system oscillates at the driving frequency.
- The amplitude of the forced oscillations is fixed. Each cycle energy is dissipated as a result of damping and the driving force does work on the system. The overall result is that the energy of the system remains constant.

- The amplitude of the forced oscillations depends on:
  - the comparative values of the natural frequency and the driving frequency
  - the amount of damping present in the system.

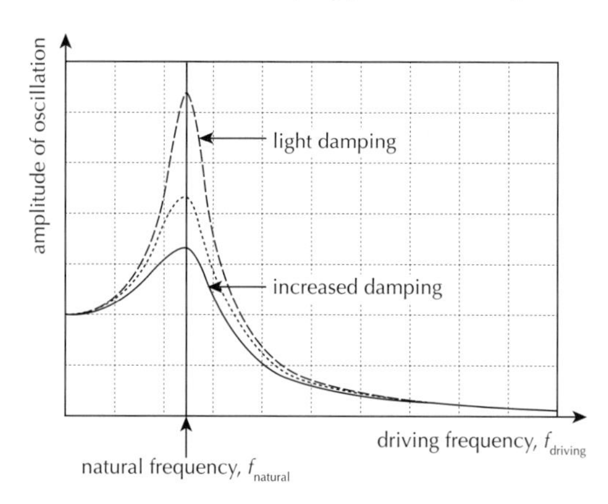

Resonance occurs when a system is subject to an oscillating force at exactly the same frequency as the natural frequency of oscillation of the system.

Resonance (2)

EXAMPLES OF RESONANCE

	Comment
Vibrations in machinery	When in operation, the moving parts of machinery provide regular driving forces on the other sections of the machinery. If the driving frequency is equal to the natural frequency, the amplitdue of a particular vibration may get dangerously high. e.g. at a particular engine speed a truck's rear view mirror can be seen to vibrate.
Quartz oscillators	A quartz crystal feels a force if placed in an electric field. When the field is removed, the crystal will oscillate. Appropriate electronics are added to generate an oscillating voltage from the mechanical movements of the crystal and this is used to drive the crystal at its own natural frequency. These devices provide accurate clocks for microprocessor systems.
Microwave generator	Microwave ovens produce electromagnetic waves at a known frequency. The changing electric field is a driving force that causes all charges to oscillate. The driving frequency of the microwaves provides energy, which means that water molecules in particular are provided with kinetic energy – i.e. the temperature is increased.
Radio receivers	Electrical circuits can be designed (using capacitors, resistors and inductors) that have their own natural frequency of electrical oscillations. The free charges (electrons) in an aerial will feel a driving force as a result of the frequency of the radio waves that it receives. Adjusting the components of the connected circuit allows its natural frequency to be adjusted to equal the driving frequency provided by a particular radio station. When the driving frequency equals the circuit's natural frequency, the electrical oscillations will increase in amplitude and the chosen radio station's signal will dominate the other stations.
Musical instruments	Many musical instruments produce their sounds by arranging for column of air or a string to be driven at its natural frequency which causes the amplitude of the oscillations to increase.
Greenhouse effect	The natural frequency of oscillation of the molecules of greenhouse gases is in the infrared region. Radiation emitted from the Earth can be readily absorbed by the greenhouse gases in the atmosphere. See page 76 for more details.

Travelling waves

INTRODUCTION – RAYS AND WAVE FRONTS

Light, sound and ripples on the surface of a pond are all examples of wave motion

- they all transfer energy from one place to another.
- they do so without a net motion of the medium through which they travel.
- they all involve oscillations (vibrations) of one sort or another. The oscillations are SHM.

A **continuous wave** involves a succession of individual oscillations. A **wave pulse** involves just one oscillation. Two important categories of wave are **transverse** and **longitudinal** (see below). The table gives some examples.

	Example of energy transfer
Water ripples (Transverse)	A floating object gains an 'up and down' motion.
Sound waves (Longitudinal)	The sound received at an ear makes the eardrum vibrate.
Light wave (Transverse)	The back of the eye (the retina) is stimulated when light is received.
Earthquake waves (Both T and L)	Buildings collapse during an earthquake.
Waves along a stretched rope (Transverse)	A 'sideways pulse' will travel down a rope that is held taut between two people.
Compression waves down a spring (Longitudinal)	A compression pulse will travel down a spring that is is held taught between two people.

The following pages analyse some of the properties that are common to all waves.

TRANSVERSE WAVES

Suppose a stone is thrown into a pond. Waves spread out as shown below.

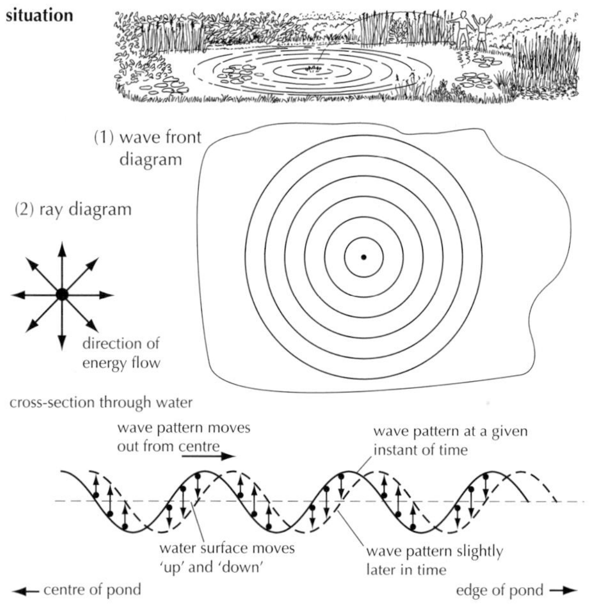

The top of the wave is known as the **crest**, whereas the bottom of the wave is known as the **trough**.

Note that there are several aspects to this wave that can be studied. These aspects are important to all waves.

- The movement of the wave pattern. The **wave fronts** highlight the parts of the wave that are moving together.
- The direction of energy transfer. The **rays** highlight the direction of energy transfer.
- The oscillations of the medium.

It should be noted that the rays are at right angles to the wave fronts in the above diagrams. This is always the case.

This wave is an example of a transverse wave because the oscillations are **at right angles** to the direction of energy transfer.

LONGITUDINAL WAVES

Sound is a longitudinal wave. This is because the oscillations are **parallel** to the direction of energy transfer.

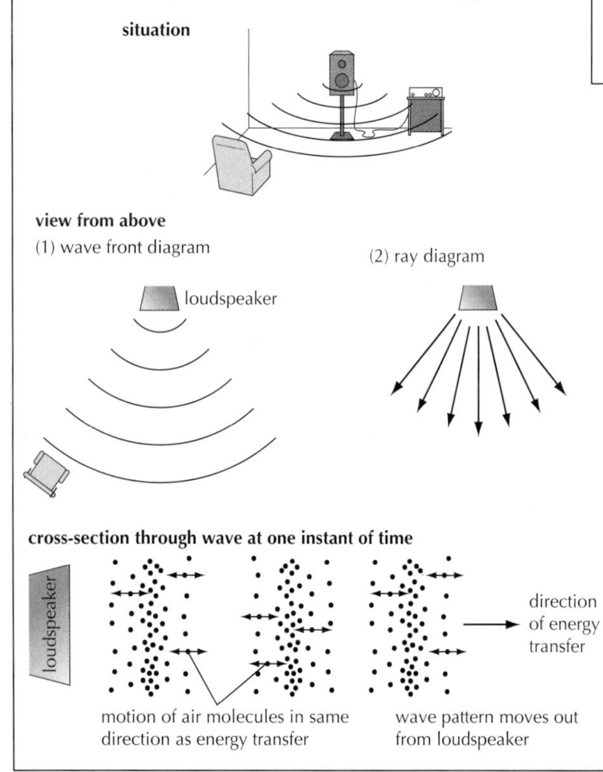

A point on the wave where everything is 'bunched together' (high pressure) is known as a **compression**. A point where everything is 'far apart' (low pressure) is known as a **rarefaction**.

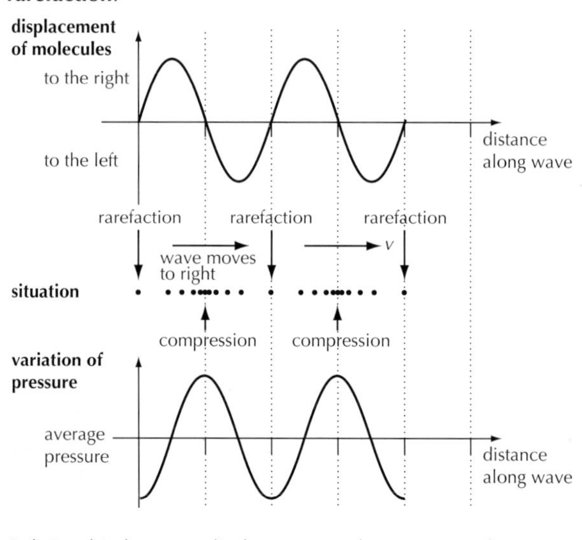

Relationship between displacement and pressure graphs

Wave characteristics

DEFINITIONS

There are some useful terms that need to be defined in order to analyse wave motion in more detail. The table below attempts to explain these terms and they are also shown on the graphs.

Because the graphs seem to be identical, you need to look at the axes of the graphs carefully.

- The displacement – time graph represents the oscillations for one point on the wave. All the other points on the wave will oscillate in a similar manner, but they will not start their oscillations at exactly the same time.
- The displacement – position graph represents a 'snap shot' of all the points along the wave at one instant of time. At a later time, the wave will have moved on but it will retain the same shape.
- The graphs can be used to represent longitudinal AND transverse waves because the y-axis records only the value of the displacement. It does NOT specify the direction of this displacement. So, if this displacement were parallel to the direction of the wave energy, the wave would be a longitudinal wave. If this displacement were at right angles to the direction of the wave energy, the wave would be a longitudinal wave.

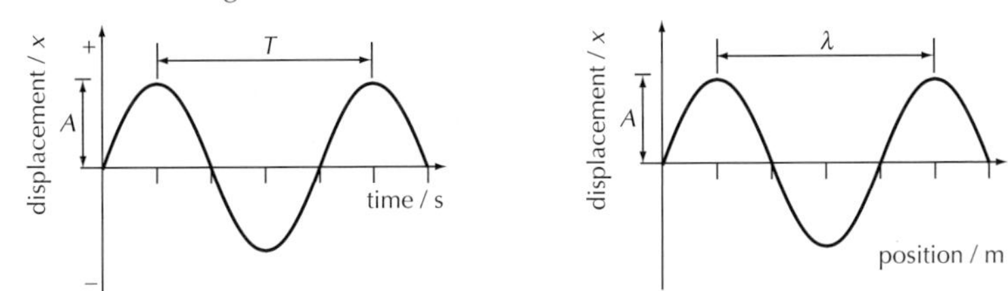

Term	Symbol	Definition
Displacement	x	This measures the change that has taken place as a result of a wave passing a particular point. Zero displacement refers to the mean (or average) position. For mechanical waves the displacement is the distance (in metres) that the particle moves from its undisturbed position.
Amplitude	A	This is the maximum displacement from the mean position. If the wave does not lose any of its energy its amplitude is constant.
Period	T	This is the time taken (in seconds) for one complete oscillation. It is the time taken for one complete wave to pass any given point.
Frequency	f	This is the number of oscillations that take place in one second. The unit used is the Hertz (Hz). A frequency of 50 Hz means that 50 cycles are completed every second.
Wavelength	λ	This is the shortest distance (in metres) along the wave between two points that are **in phase** with one another. 'In phase' means that the two points are moving exactly in step with one another. For example, the distance from one crest to the next crest on a water ripple or the distance from one compression to the next one on a sound wave.
Wave speed	v	This is the speed (in m s^{-1}) at which the wave fronts pass a stationary observer.
Intensity	I	The intensity of a wave is the power per unit area that is received by the observer. The unit is W m^{-2}. the intensity of a wave is proportional to the square of its amplitude: $I \propto A^2$.

The period and the frequency of any wave are inversely related. For example, if the frequency of a wave is 100 Hz, then its period must be exactly $\frac{1}{100}$ of a second. In symbols,

$$T = \frac{1}{f}$$

WAVE EQUATIONS

There is a very simple relationship that links wave speed, wavelength and frequency. It applies to all waves.

The time taken for one complete oscillation is the period of the wave, T.

In this time, the wave pattern will have moved on by one wavelength, λ.

This means that the speed of the wave must be given by

$$v = \frac{\text{distance}}{\text{time}} = \frac{\lambda}{T}.$$

Since $\frac{1}{T} = f$

$$v = f\lambda$$

In words,

velocity = frequency × wavelength

EXAMPLE

A stone is thrown onto a still water surface and creates a wave. A small floating cork 1.0 m away from the impact point has the following displacement–time graph (time is measured from the instant the stone hits the water):

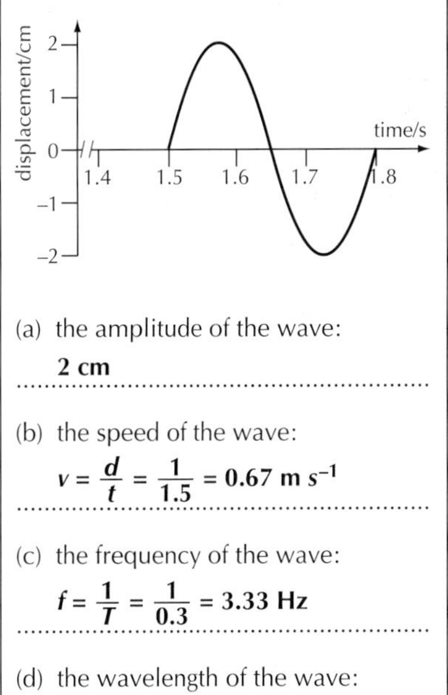

(a) the amplitude of the wave:

2 cm

(b) the speed of the wave:

$v = \frac{d}{t} = \frac{1}{1.5} = 0.67 \text{ m s}^{-1}$

(c) the frequency of the wave:

$f = \frac{1}{T} = \frac{1}{0.3} = 3.33 \text{ Hz}$

(d) the wavelength of the wave:

$\lambda = \frac{v}{f} = \frac{0.666}{3.33} = 0.2 \text{ m}$

Electromagnetic spectrum

ELECTROMAGNETIC WAVES

Visible light is one part of a much larger spectrum of similar waves that are all electromagnetic.

Charges that are accelerating generate electromagnetic fields. If an electric charge oscillates, it will produce a varying electric and magnetic field at right angles to one another.

These oscillating fields propagate (move) as a transverse wave through space. Since no physical matter is involved in this propagation, they can travel through a vacuum. The speed of this wave can be calculated from basic electric and magnetic constants and it is the same for all electromagnetic waves, 3.0×10^8 m s^{-1}. See page 155.

Although all electromagnetic waves are identical in their nature, they have very different properties. This is because of the huge range of frequencies (and thus energies) involved in the electromagnetic spectrum.

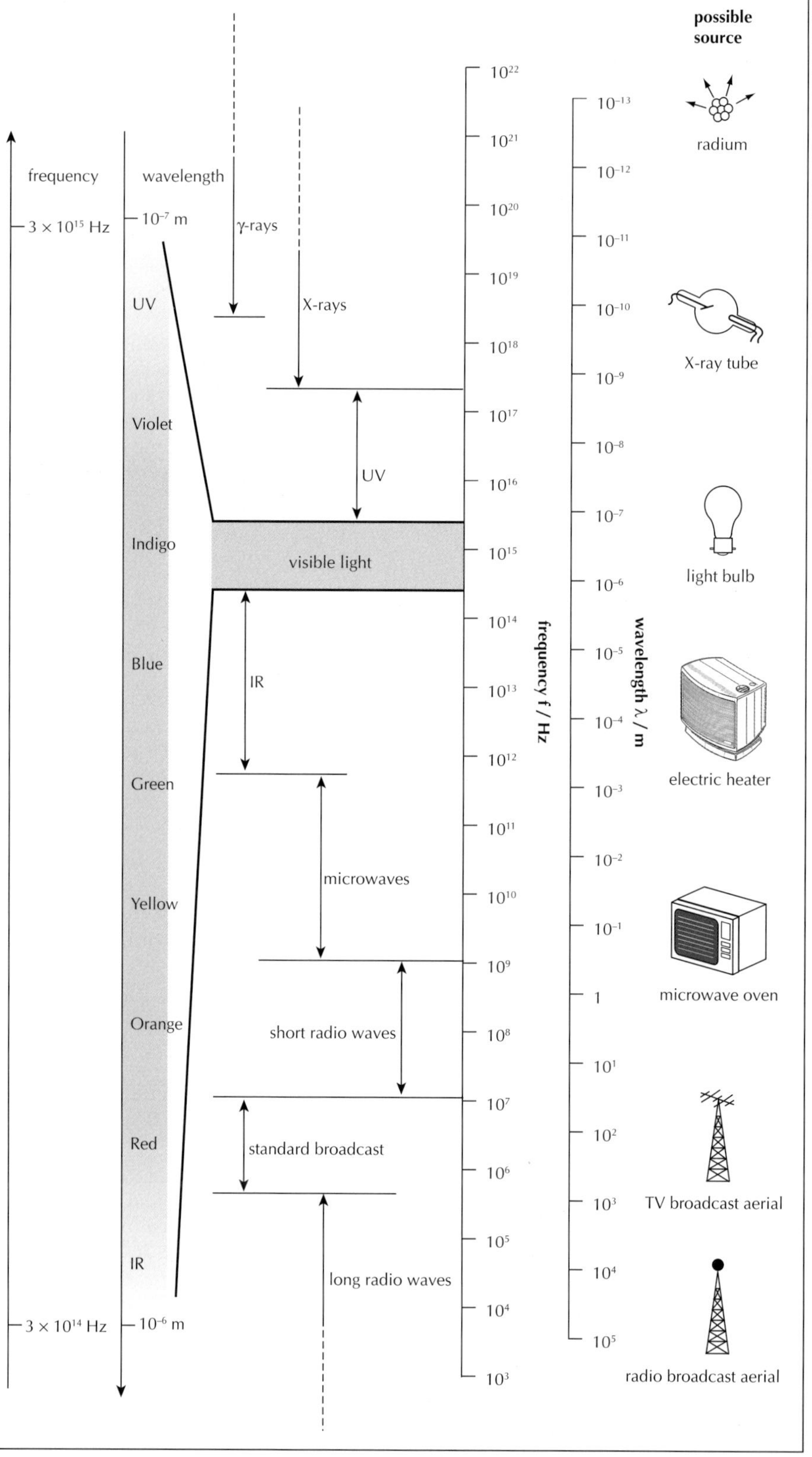

Wave properties (1)

REFLECTION & TRANSMISSION

In general, when any wave meets the boundary between two different media it is partially reflected and partially transmitted.

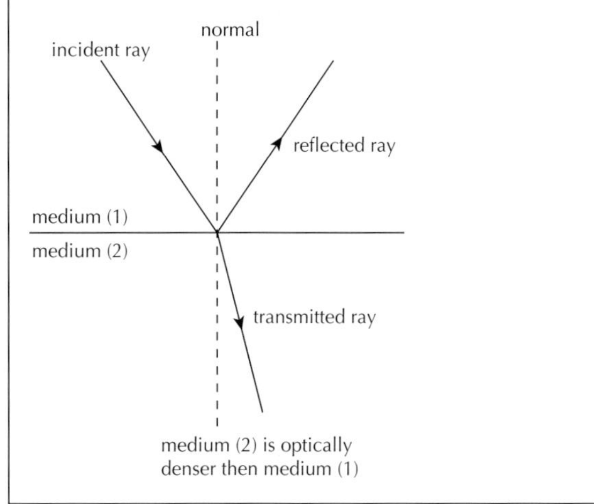

REFLECTION OF TWO-DIMENSIONAL PLANE WAVES

The diagram below shows what happens when plane waves are reflected at a boundary. When working with rays, by convention we always measure the angles between the rays and the **normal**. The normal is a construction line that is drawn at right angles to the surface.

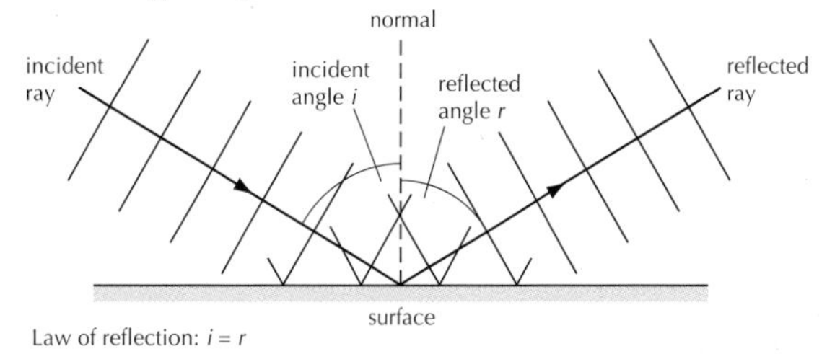

DIFFRACTION

When waves pass through apertures they tend to spread out. Waves also spread around obstacles. This wave property is called **diffraction**.

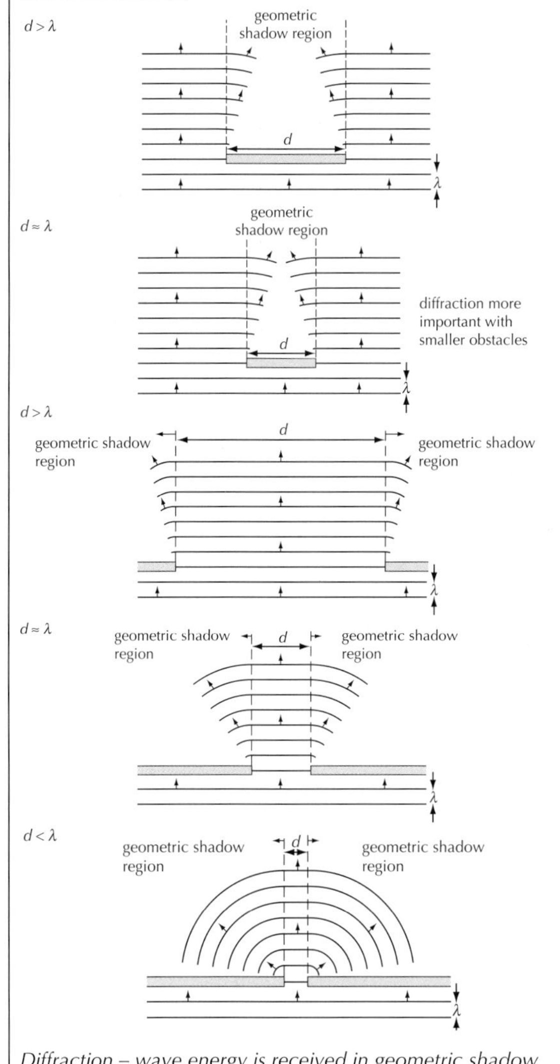

Diffraction – wave energy is received in geometric shadow region d = width of obstacle/gap

REFRACTION OF PLANE WAVES

If plane waves are incident at an angle on the boundary between two media, the transmitted wave will change direction – it has been **refracted**. The reason for this change in direction is the change in speed of the wave.

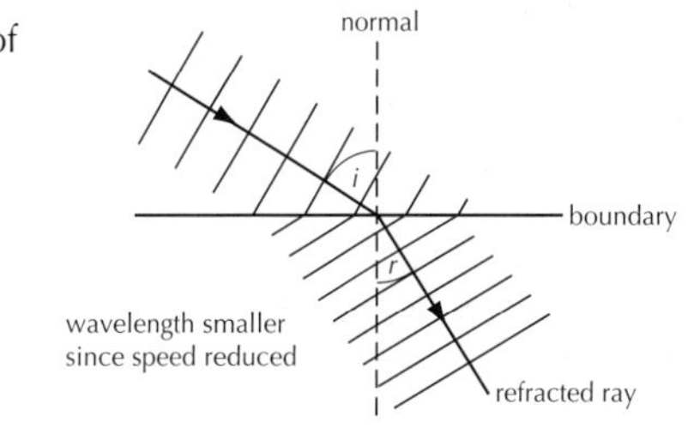

Snell's law (an experimental law of refraction) states that the ratio $\frac{\sin i}{\sin r}$ = constant, for a given frequency.

The ratio is equal to the ratio of the speeds in the different media

$$\frac{\sin \theta_1}{\sin \theta_2} = \frac{V_1}{V_2}$$

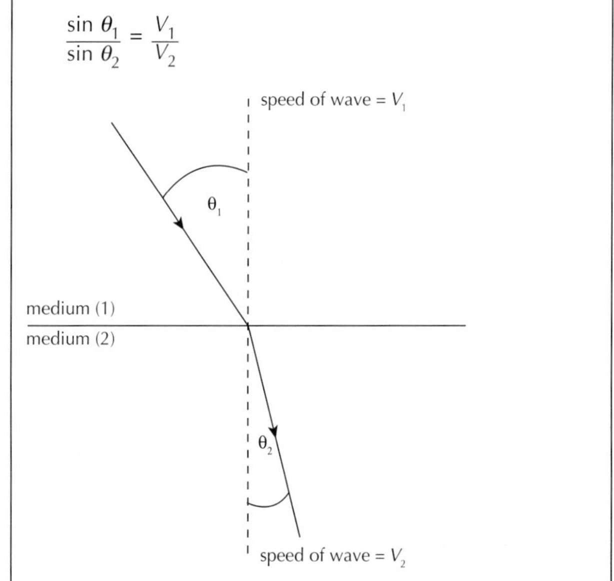

The are some important points to note from these diagrams.

- diffraction becomes relatively more important when the wavelength is large in comparison to the size of the aperture (or the object).
- the wavelength needs to be of the same order of magnitude as the aperture for diffraction to be noticeable.

Wave properties (2)

INTERFERENCE

When two waves of the same type meet, they **interfere** and we can work out the resulting wave using the **principle of superposition**. The overall disturbance at any point and at any time where the waves meet is the vector sum of the disturbances that would have been produced by each of the individual waves. This is shown below.

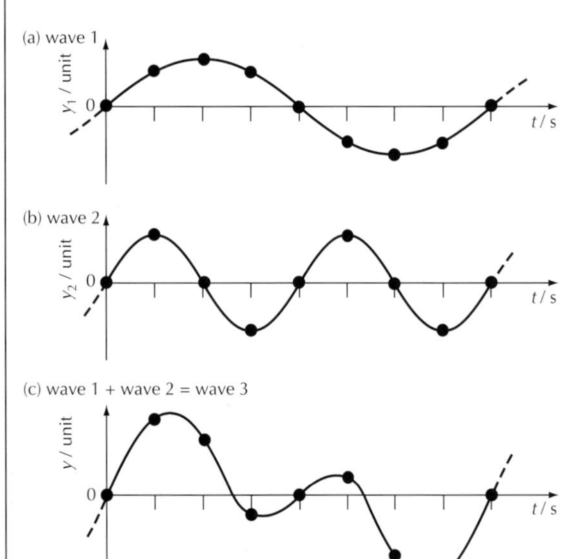

Wave superposition

If the waves have the same amplitude and the same frequency then the interference **at a particular point** can be **constructive** or **destructive**.

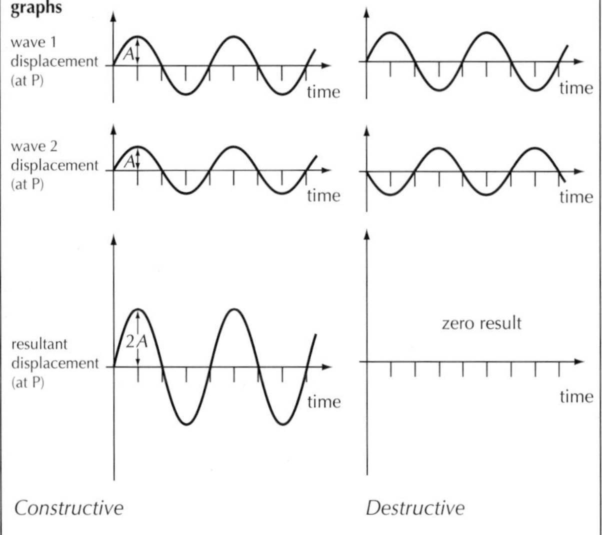

Constructive *Destructive*

TECHNICAL LANGUAGE

Constructive interference takes place when the two waves are 'in step' with one another – they are said to be **in phase**. There is a zero **phase difference** between them. Destructive interference takes place when the waves are exactly 'out of step' – they are said to be **out of phase**. There are several different ways of saying this. One could say that the phase difference is equal to 'half a cycle' or '180 degrees' or 'π radians'.

Interference can take place if there are two possible routes for a ray to travel from source to observer. If the path difference between the two rays is a whole number of wavelengths, then constructive interference will take place.

path difference = $n\,\lambda \rightarrow$ constructive

path difference = $(n + \frac{1}{2})\,\lambda \rightarrow$ destructive

$n = 0, 1, 2, 3 \ldots$

For constructive or destructive interference to take place, the sources of the waves must be phase linked or **coherent**.

EXAMPLES OF DIFFRACTION

Diffraction provides the reason why we can hear something even if we can not see it.

If you look at a distant street light at night and then squint your eyes the light spreads sideways – this is as a result of diffraction taking place around your eyelashes! (Needless to say, this explanation is a simplification.)

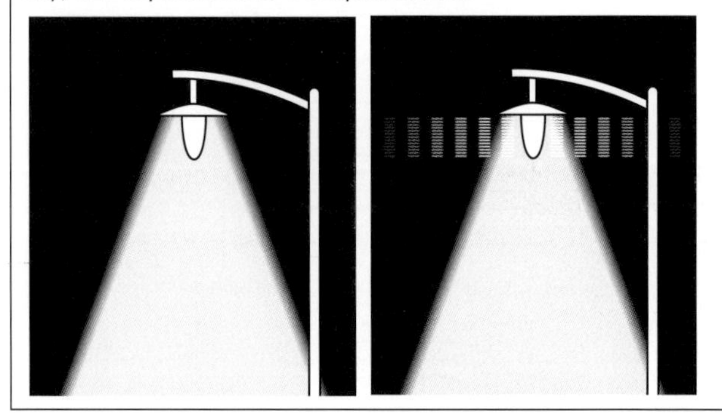

EXAMPLES OF INTERFERENCE

Water waves

A ripple tank can be used to view the interference of water waves. Regions of large-amplitude waves are constructive interference. Regions of still water are destructive interference.

Sound

It is possible to analyse any noise in terms of the component frequencies that make it up. A computer can then generate exactly the same frequencies but of different phase. This 'antisound' will interfere with the original sound. An observer in a particular position in space could have the overall noise level reduced if the waves superimposed destructively at that position.

Light

The colours seen on the surface of a soap bubble are a result of constructive and destructive interference of two light rays. One ray is reflected off the outer surface of the bubble whereas the other is reflected off the inner surface.

Nature of reflection

TYPES OF REFLECTION

When a single ray of light strikes a smooth mirror it produces a single reflected ray. This type of 'perfect' reflection is very different to the reflection that takes place from an uneven surface such as the walls of a room. In this situation, a single incident ray is generally scattered in all directions. This is an example of a **diffuse** reflection.

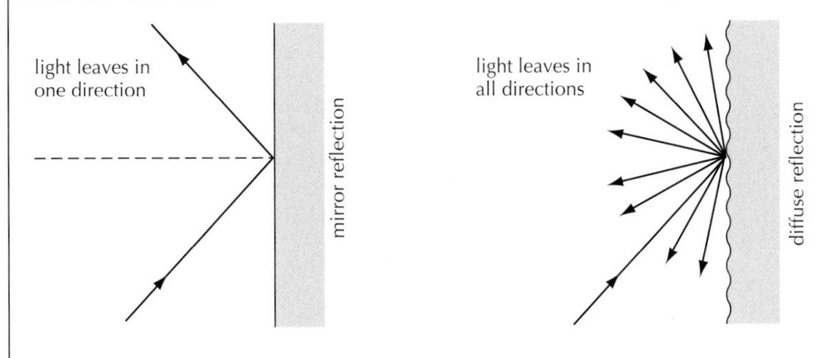

light leaves in one direction
mirror reflection
light leaves in all directions
diffuse reflection

We see objects by receiving light that has come from them. Most objects do not give out light by themselves so we cannot see them in the dark. Objects become visible with a source of light (e.g. the Sun or a light bulb) because diffuse reflections have taken place that scatter light from the source towards our eyes.

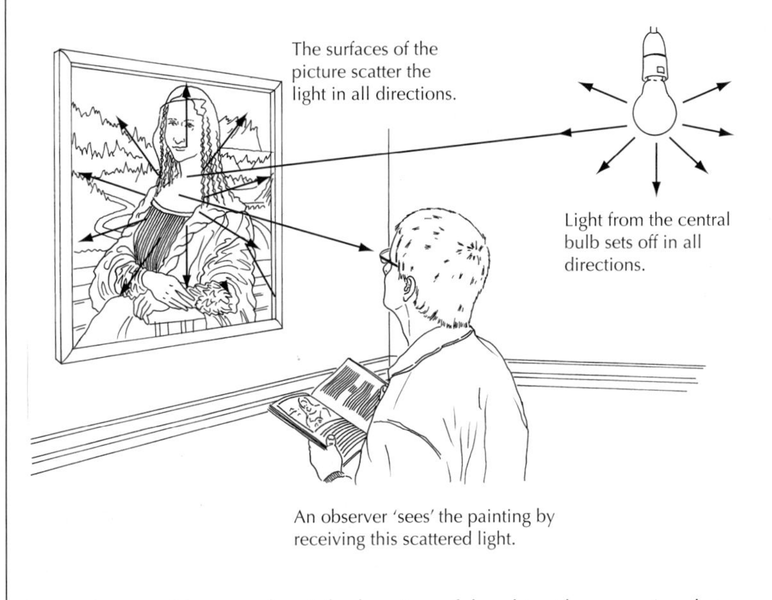

The surfaces of the picture scatter the light in all directions.
Light from the central bulb sets off in all directions.
An observer 'sees' the painting by receiving this scattered light.

Our brains are able to work out the location of the object by assuming that rays travel in straight lines.

LAW OF REFLECTION

The location and nature of optical images can be worked out using **ray diagrams** and the principles of **geometric optics**. A ray is a line showing the direction in which light energy is propagated. The ray must always be at right angles to the wave front. The study of geometric optics ignores the wave and particle nature of light.

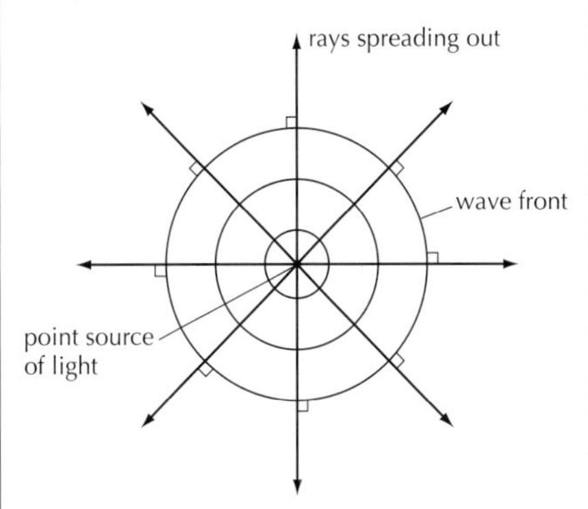

rays spreading out
wave front
point source of light

When a mirror reflection takes place, the direction of the reflected ray can be predicted using the laws of reflection. In order to specify the ray directions involved, it is usual to measure all angles with respect to an imaginary construction line called the **normal**. For example, the incident angle is always taken as the angle between the incident ray and the normal. The normal to a surface is the line at right angles to the surface as shown below.

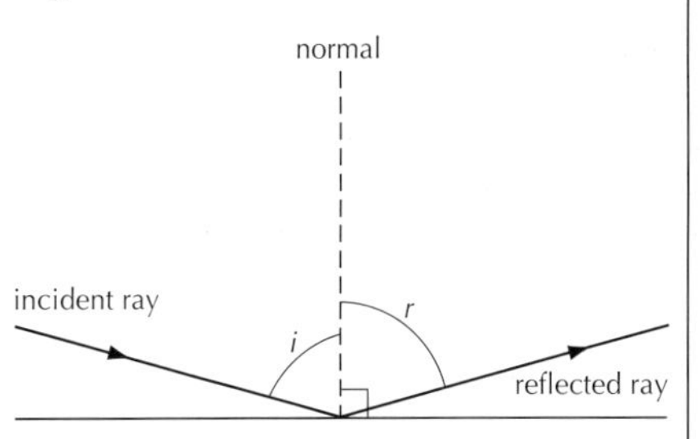

normal
incident ray
i
r
reflected ray

The laws of reflection are that:

- the incident angle is equal to the reflected angle.
- the incident ray, the reflected ray and the normal all lie in the same plane (as shown in the diagram).

The second statement is only included in order to be precise and is often omitted. It should be obvious that a ray arriving at a mirror (such as the one represented above) is not suddenly reflected in an odd direction (e.g. out of the plane of the page).

Snell's law and refractive index

REFRACTIVE INDEX AND SNELL'S LAW.

Refraction takes place at the boundary between two media. In general, a wave that crosses the boundary will undergo a change of direction. The reason for this change in direction is the change in wave speed that has taken place.

As with reflection, the ray directions are always specified by considering the angles between the ray **and the normal.** If a ray travels into an optically denser medium (e.g. from air into water), then the ray of light is refracted **towards** the normal. If the ray travels into an optically less dense medium then the ray of light is refracted **away from** the normal.

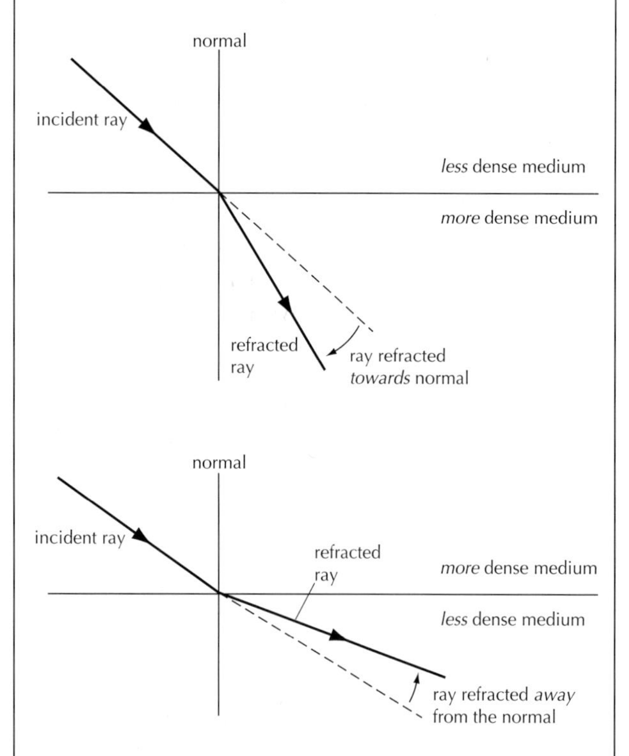

Snell's law allows us to work out the angles involved. When a ray is refracted between two different media,

the ratio $\frac{\sin(\text{angle of incidence})}{\sin(\text{angle of refraction})}$ is a constant.

The constant is called the refractive index n between the two media. As seen on page 41 this ratio is equal to the ratio of the speeds of the waves in the two media.

$$\frac{\sin i}{\sin r} = n$$

If the refractive index for a particular substance is given as a particular number and the other medium is not mentioned then you can assume that the other medium is air (or to be absolutely correct, a vacuum). Another way of expressing this is to say that the refractive index of air can be taken to be 1.0.

For example the refractive index for a type of glass might be given as

$$n_{glass} = 1.34$$

This means that a ray entering the glass from air with an incident angle of 40° would have a refracted angle given by

$$\sin r = \frac{\sin 40°}{1.34} = 0.4797$$

$$\therefore r = 28.7°$$

EXAMPLES

1. Parallel-sided block

A ray will always leave a parallel-sided block travelling in a parallel direction to the one with which it entered the block. The overall effect of the block has been to move the ray sideways. An example of this is shown below.

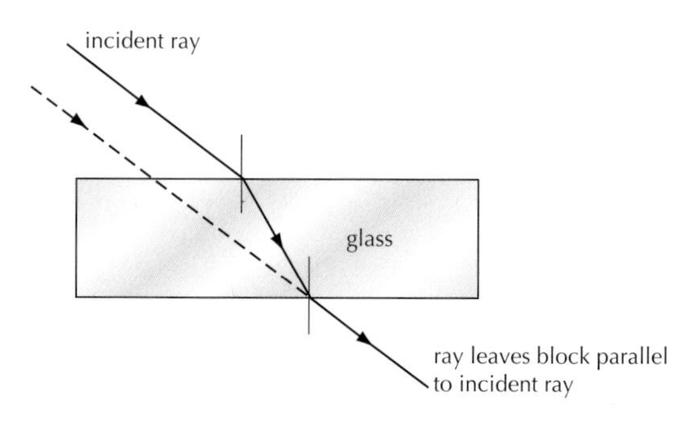

2. Ray travelling between two media

If a ray goes between two different media, the two individual refractive indices can be used to calculate the overall refraction using the following equation

$$n_1 \sin \theta_1 = n_2 \sin \theta_2$$

n_1 refractive index of medium 1
θ_1 angle in medium 1
n_2 refractive index of medium 2
θ_2 angle in medium 2

Suppose a ray of light is shone into a fish tank that contains water. The refraction that takes place would be calculated as shown below:

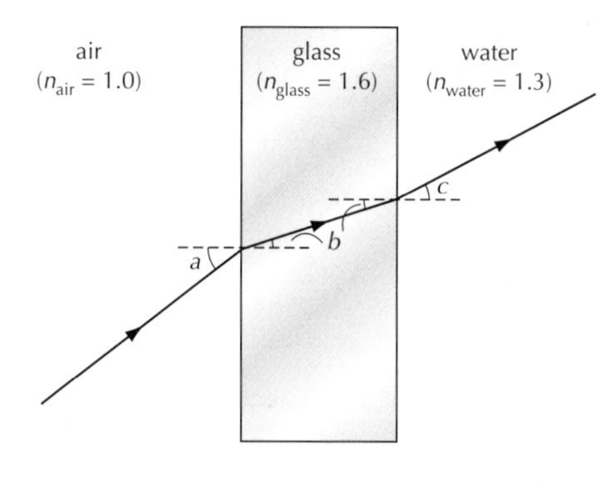

1st refraction:

$$n_{glass} = \frac{\sin a}{\sin b}$$

2nd refraction:

$$n_{glass} \times \sin b = n_{water} \times \sin c$$

$$\frac{n_{glass}}{n_{water}} = \frac{\sin c}{\sin b}$$

Overall the refraction is from incident angle a to refracted angle c.

i.e. $$n_{overall} = \frac{\sin a}{\sin c}$$

$$= n_{water}$$

IB QUESTIONS – OSCILLATIONS AND WAVES

1 A surfer is out beyond the breaking surf in a deep-water region where the ocean waves are sinusoidal in shape. The crests are 20 m apart and the surfer rises a vertical distance of 4.0 m from wave **trough** to **crest**, in a time of 2.0 s. What is the speed of the waves?

A 1.0 m s^{-1} **B** 2.0 m s^{-1} **C** 5.0 m s^{-1} **D** 10.0 m s^{-1}

2 Radio waves of wavelength 30 cm have a frequency of about

A 10 MHz **B** 90 MHz **C** 1000 MHz **D** 9000 MHz

3 Light passes through three different media, being refracted at each interface as shown.

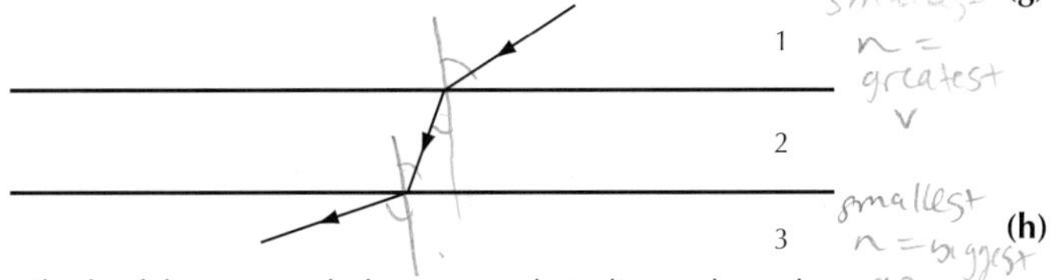

Which of the options below correctly indicates how the speed of light compares in the three media?

A $v_1 > v_2 > v_3$
B $v_3 > v_2 > v_1$
C $v_3 > v_1 > v_2$
D $v_2 > v_1 > v_3$

4 In order to measure the speed of sound in water a loudspeaker and microphone are set up to float in the middle of a swimming pool as shown in **Figure A** below. The microphone output is recorded directly by an oscilloscope and the recording electronics. The water depth is measured as 0.85 m.

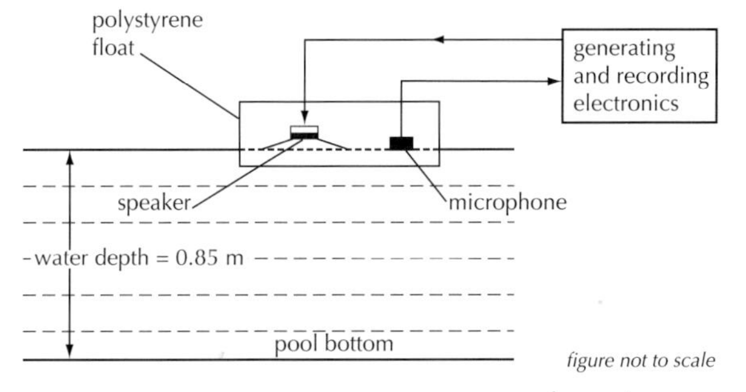

Figure A

The experiment proceeds as follows: At some time, call it $t = 0$, a pulse of sound is generated in the speaker. **At the same time** the recording equipment is triggered to start recording. The burst of sound is of 1.0 ms duration and travels to the bottom of the pool and is reflected, back to the top, where it is detected by the microphone.

A typical recorded signal is shown in **Figure B** below.

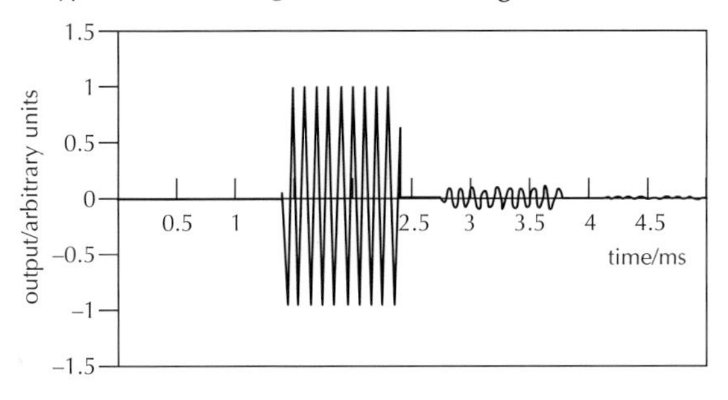

Figure B

(a) Describe the features of the whole recorded signal. (At least **three** different points should be noted.) [3]

(b) From the recorded data, determine the speed of sound in the water. [4]

(c) Why is it best to do this experiment in the middle of the pool and not near the sides? [1]

(d) What is the frequency of the sound wave in the 1.0 m s^{-1} sound pulse? [2]

(e) What is the wavelength of the sound in water at the frequency used? [1]

(f) If the frequency of the sound wave in the pulse was changed, what effect would this have on the measured speed of sound? [1]

(g) A more precise value for the speed of sound in water could be obtained by using a number of different water depths, particularly the greater depths in the diving pool. Why would greater depths give more precise values? [2]

(h) Consider whether a similar pulse timing technique could be used to measure the speed of light by replacing the speaker with a light source and the microphone with a light detector. Suggest at least **two** difficulties that would be experienced in trying to carry out such an experiment. [3]

5 Which one of the following graphs best represents the variation with displacement, x of the acceleration, a of an object undergoing simple harmonic motion?

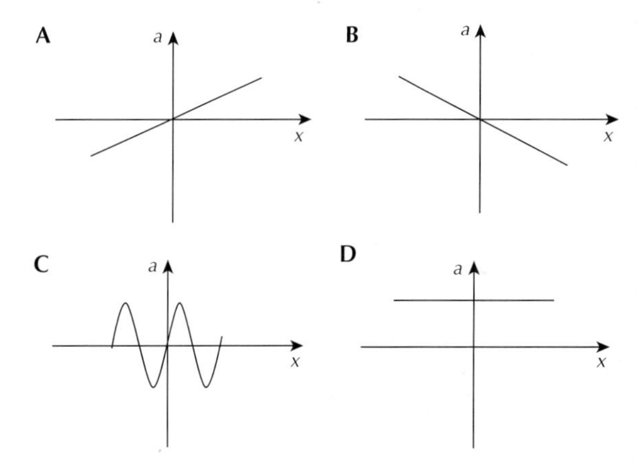

6 A 200 g mass oscillates at the end of a spring with an amplitude of 3.2 cm. It takes 13.8 s to complete 10 oscillations. Calculate:

(a) the angular frequency ω of the oscillations [2]

(b) the value of the spring constant [3]

(c) the maximum speed of the mass [2]

(d) the maximum acceleration of the mass. [2]

Electric potential energy and electric potential difference

ENERGY DIFFERENCE IN AN ELECTRIC FIELD

When placed in an electric field, a charge feels a force. This means that if it moves around in an electric field work will be done. As a result, the charge will either gain or lose electric potential energy. Electric potential energy is the energy that a charge has as a result of its position in an electric field. This is the same idea as a mass in a gravitational field. If we lift a mass up, its gravitational potential energy increases. If the mass falls, its gravitational potential energy decreases. In the example below a positive charge is moved from position *A* to position *B*. This results in an increase in electric potential energy. Since the field is uniform, the force is constant. This makes it is very easy to calculate the work done.

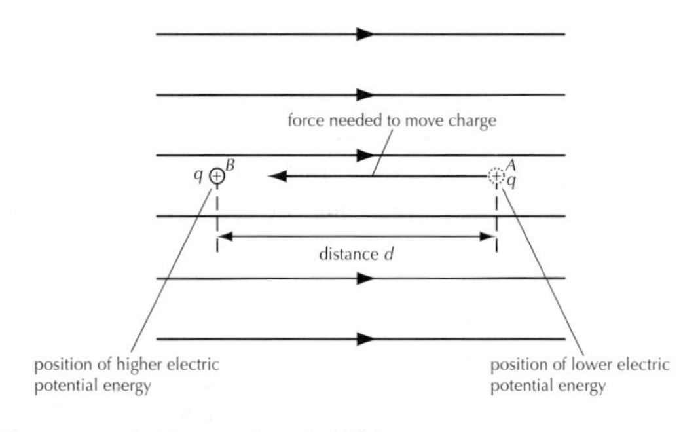

Charge moving in an electric field

Change in electric potential energy = force × distance

$$= E\,q \times d$$

See page 53 for a definition of electric field, *E*.

In the example above the electric potential energy at B is greater than the electric potential energy at A. We would have to put in this amount of work to push the charge from A to B. If we let go of the charge at B it would be pushed by the electric field. This push would accelerate it so that the loss in electrical potential energy would be the same as the gain in kinetic energy.

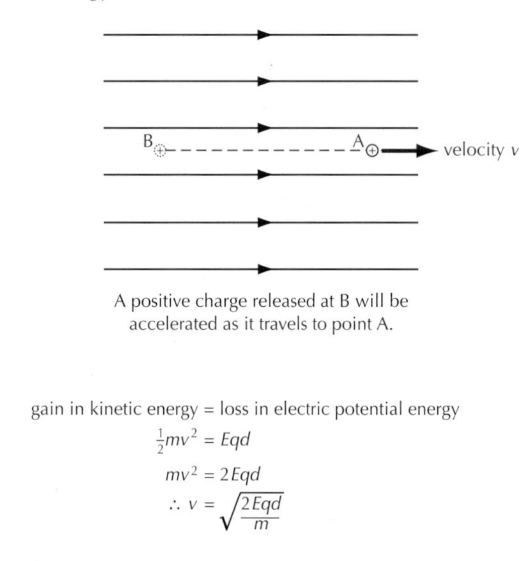

A positive charge released at B will be accelerated as it travels to point A.

gain in kinetic energy = loss in electric potential energy

$$\tfrac{1}{2}mv^2 = Eqd$$

$$mv^2 = 2Eqd$$

$$\therefore v = \sqrt{\frac{2Eqd}{m}}$$

ELECTRIC POTENTIAL DIFFERENCE

In the example left, the actual energy difference between *A* and *B* depended on the charge that was moved. If we doubled the charge we would double the energy difference. The quantity that remains fixed between *A* and *B* is the energy difference **per unit charge**. This is called the **potential difference,** or **p.d.**, between the points.

$$\text{Potential difference between two points} = \frac{\text{energy difference}}{\text{per unit charge moved}}$$

$$= \frac{\text{energy difference}}{\text{charge}}$$

The basic unit for potential difference is the joule/coulomb, J C^{-1}. A very important point to note is that for a given electric field, the potential difference between any two points is a single fixed scalar quantity. The work done between these two points does not depend on the path taken by the test charge. A technical way of saying this is 'the electric field is **conservative**'.

UNITS

The smallest amount of negative charge available is the charge on an electron; the smallest amount of positive charge is the charge on a proton. In everyday situations this unit is far too small so we use the **coulomb, C**. One coulomb of negative charge is the charge carried by a total of 6.25×10^{18} electrons.

From its definition, the units of potential difference (p.d.) are J C^{-1}. This is given a new name, the volt, V. Thus:

1 volt = 1 J C^{-1}

Voltage and potential difference are different words for the same thing. Potential difference is probably the better name to use as it reminds you that it is measuring the difference between two points.

When working at the atomic scale, the joule is far too big to use for a unit for energy. The everyday unit used by physicists for this situation is the electronvolt. As could be guessed from its name, the electronvolt is simply the energy that would be gained by an electron moving through a potential difference of 1 volt.

Since energy gained = p.d. × charge

$$1 \text{ electronvolt} = 1 \text{ volt} \times 1.6 \times 10^{-19}\text{ C}$$

$$= 1.6 \times 10^{-19}\text{ J}$$

The normal SI prefixes also apply so one can measure energies in kiloelectronvolts (keV) or megaelectronvolts (MeV). The latter unit is very common in particle physics.

Example

Calculate the speed of an electron accelerated in a vacuum by a p.d. of 1000V.

KE of electron $= V \times e = 1000 \times 1.6 \times 10^{-19}$

$$= 1.6 \times 10^{-16}\text{ J}$$

$$\frac{1}{2}mv^2 = 1.6 \times 10^{-16}\text{ J}$$

$$v = 1.87 \times 10^{7}\text{ ms}^{-1}$$

Electric current

ELECTRICAL CONDUCTION IN A METAL

Whenever charges move we say that a **current** is flowing. A current is the name for moving charges and the path that they follow is called the **circuit**. Without a complete circuit, a current cannot be maintained for any length of time.

Current flows THROUGH an object when there is a potential difference ACROSS the object. A battery (or power supply) is the device that creates the potential difference.

By convention, currents are always represented as the flow of positive charge. Thus **conventional current**, as it is known, flows from positive to negative. Although currents can flow in solids, liquids and gases, in most everyday electrical circuits the currents flow through wires. In this case the things that actually move are the negative electrons – the **conduction electrons**. The direction in which they move is opposite to the direction of the representation of conventional current. As they move the interactions between the conduction electrons and the lattice ions means that work needs to be done. Therefore, when a current flows, the metal heats up. The speed of the electrons due to the current is called their **drift velocity**.

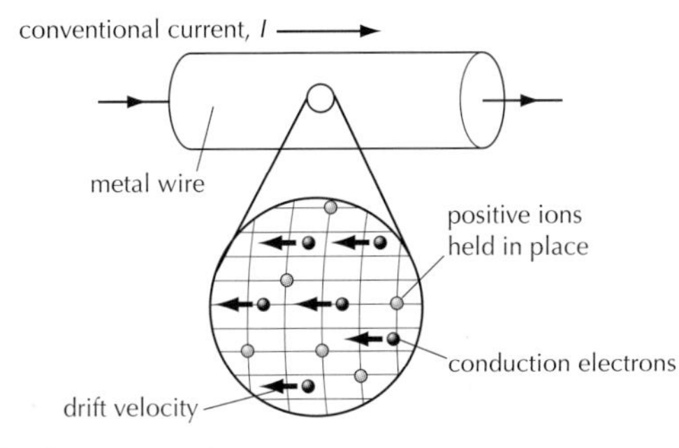

Electrical conduction in a metal

CURRENT

Current is defined as the **rate of flow of electrical charge**. It is always given the symbol, I. Mathematically the definition for current is expressed as follows:

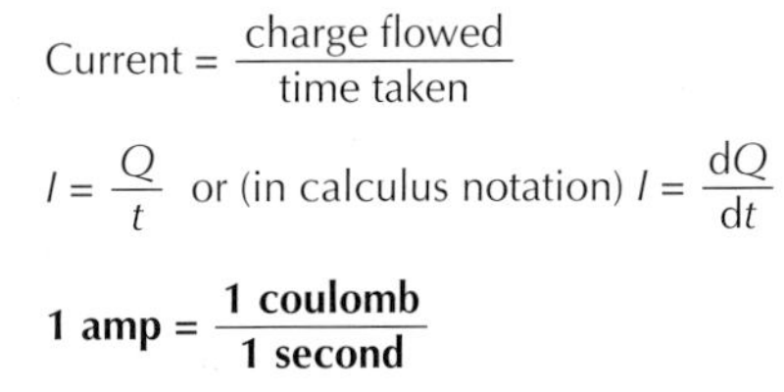

$$\text{Current} = \frac{\text{charge flowed}}{\text{time taken}}$$

$$I = \frac{Q}{t} \quad \text{or (in calculus notation)} \quad I = \frac{dQ}{dt}$$

$$\mathbf{1\ amp} = \frac{\mathbf{1\ coulomb}}{\mathbf{1\ second}}$$

If a current flows in just one direction it is known as a **direct current**. A current that constantly changes direction (first one way then the other) is known as an **alternating current** or a.c.

RESISTANCE

Resistance is the mathematical ratio between potential difference and current. If something has a high resistance, it means that you would need a large potential difference across it in order to get a current to flow.

$$\text{Resistance} = \frac{\text{potential difference}}{\text{current}}$$

In symbols, $R = \frac{V}{I}$

We define a new unit, the ohm, Ω, to be equal to one volt per amp.

1 ohm = 1 V A^{-1}

OHM'S LAW – OHMIC AND NON-OHMIC BEHAVIOUR

The graphs below show how the current varies with potential difference for some typical devices.

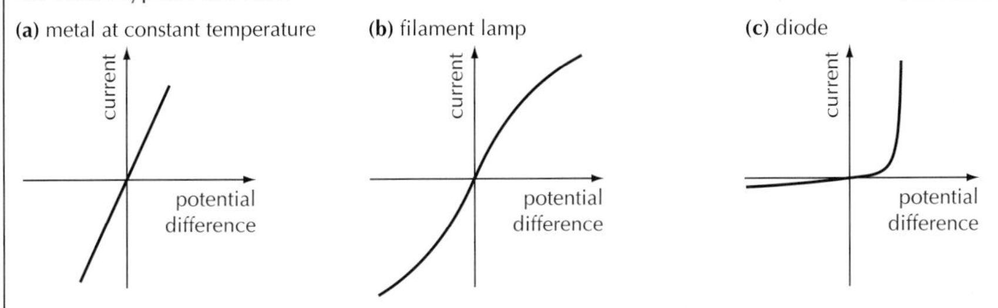

If current and potential difference are proportional (like the metal at constant temperature) the device is said to be **ohmic**. Devices where current and potential difference are not proportional (like the filament lamp or the diode) are said to be **non-ohmic**.

Ohm's law states that the current flowing through a piece of metal is proportional to the potential difference across it providing the temperature remains constant.

In symbols,

$V \propto I$ [if temperature is constant]

A device with constant resistance (in other words an ohmic device) is called a **resistor**.

POWER DISSIPATION

$$\text{Since potential difference} = \frac{\text{energy difference}}{\text{charge flowed}}$$

$$\text{And current} = \frac{\text{charge flowed}}{\text{time taken}}$$

This means that potential difference × current

$$= \frac{\text{(energy difference)}}{\text{(charge flowed)}} \times \frac{\text{(charge flowed)}}{\text{(time taken)}}$$

$$= \frac{\text{energy difference}}{\text{time}}$$

This energy difference per time is the power dissipated by the resistor. All this energy is going into heating up the resistor. In symbols:

$P = V \times I$

Sometimes it is more useful to use this equation in a slightly different form, e.g.

$P = V \times I$ but $V = I \times R$ so

$P = (I \times R) \times I$

$P = I^2 R$

Similarly $P = \frac{V^2}{R}$

EXAMPLE

A 1.2 kW electric kettle is plugged into the 250 V mains supply. Calculate

(i) the current drawn

(ii) its resistance

(i) $I = \frac{1200}{250} = 4.8$ A

(ii) $R = \frac{250}{4.8} = 52\ \Omega$

Electric circuits

CIRCUITS

An electric circuit can contain many different devices or **components**. The mathematical relationship $V = IR$ can be applied to any component or groups of components in a circuit.

When analysing a circuit it is important to look at the circuit as a whole. The power supply is the device that is providing the energy, but it is the whole circuit that determines what current flows through the circuit.

RESISTORS IN SERIES

A **series circuit** has components connected one after another in a continuous chain. The current must be the same everywhere in the circuit since charge is conserved. The total potential difference is shared among the components.

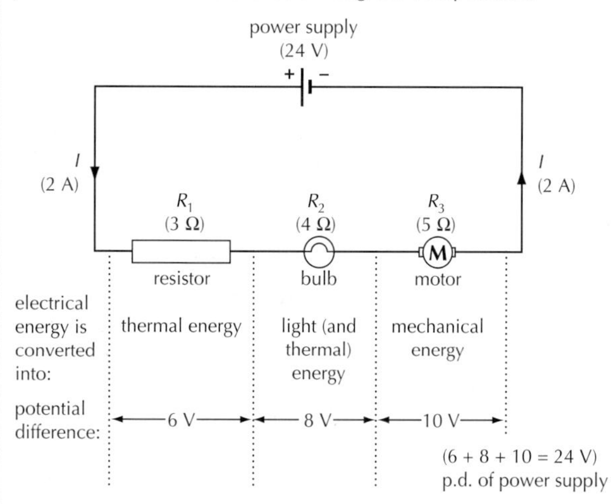

Example of a series circuit

We can work out what share they take by looking at each component in turn, e.g.

The potential difference across the resistor = $I \times R_1$

The potential difference across the bulb = $I \times R_2$

$$\boldsymbol{R_{total} = R_1 + R_2 + R_3}$$

This always applies to a series circuit. Note that $V = IR$ correctly calculates the potential difference across each individual component as well as calculating it across the total.

RESISTORS IN PARALLEL

A **parallel circuit** branches and allows the charges more than one possible route around the circuit.

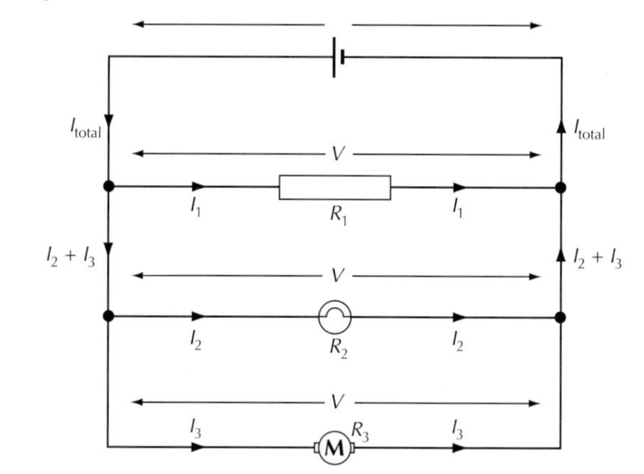

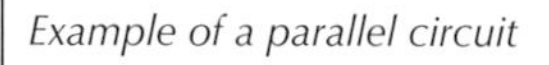

Example of a parallel circuit

Since the power supply fixes the potential difference, each component has the same potential difference across it. The total current is just the addition of the currents in each branch.

$$I_{total} = I_1 + I_2 + I_3$$

$$= \frac{V}{R_1} + \frac{V}{R_2} + \frac{V}{R_3}$$

$$\frac{1}{R_{total}} = \frac{1}{R_1} + \frac{1}{R_2} + \frac{1}{R_3}$$

ELECTRICAL METERS

A current-measuring meter is called an **ammeter**. It should be connected in series at the point where the current needs to be measured. A perfect ammeter would have zero resistance.

A meter that measures potential difference is called a **voltmeter**. It should be placed in parallel with the component or components being considered. A perfect voltmeter has infinite resistance.

ELECTROMOTIVE FORCE AND INTERNAL RESISTANCE

When a 6V battery is connected in a circuit some energy will be used up inside the battery itself. In other words, the battery has some **internal resistance**. The TOTAL energy difference per unit charge around the circuit is still 6 volts, but some of this energy is used up inside the battery. The energy difference per unit charge from one terminal of the battery to the other is less than the total made available by the chemical reaction in the battery.

For historical reasons, the TOTAL energy difference per unit charge around a circuit is called the **electromotive force (e.m.f.)**. However, remember that it is not a force (measured in newtons) but an energy difference per charge (measured in volts).

In practical terms, e.m.f. is exactly the same as potential difference if no current flows.

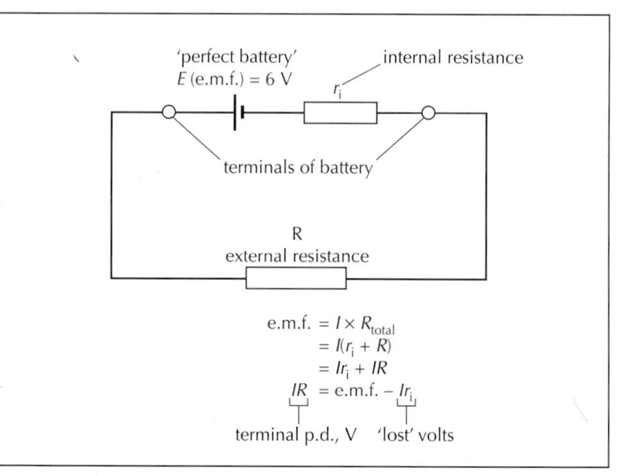

Potential divider circuits and sensors

POTENTIAL DIVIDER CIRCUIT

The problem below is an example of a circuit involving a **potential divider**. It is so called because the two resistors 'divide up' the potential difference of the battery. You can calculate the 'share' taken by one resistor using from the ratio of the resistances but this approach does not work unless the voltmeter's resistance is also considered. An ammeter's internal resistance also needs to be considered. One of the most common mistakes when solving problems involving electrical circuits is to assume the current or potential difference remains constant after a change to the circuit. After a change, the only way to ensure your calculations are correct is to start again.

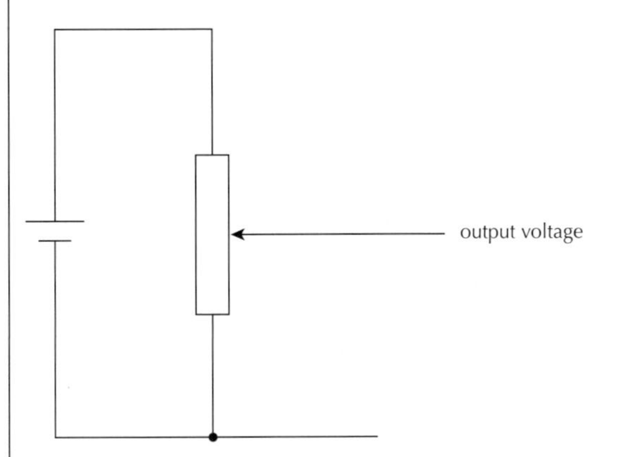
output voltage

A variable potential divider (a **potentiometer**) is often the best way to produce a variable power supply. When designing the potential divider, the smallest resistor that is going to be connected needs to be taken into account: the potentiometer's resistance should be significantly smaller.

EXAMPLE

In the circuit below the voltmeter has a resistance of 20 kΩ. Calculate:

(a) the p.d. across the 20 kΩ resistor with the switch open

(b) the reading on the voltmeter with the switch closed

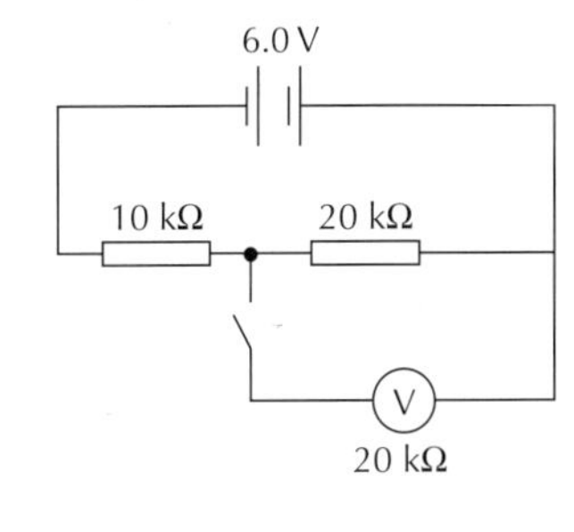
6.0 V
10 kΩ
20 kΩ
V
20 kΩ

(a) $\text{p.d.} = \dfrac{20}{(20 + 10)} \times 6.0 = 4.0\ \text{V}$

(b) resistance of 20 kΩ resistor and voltmeter combination, R, given by:

$$\frac{1}{R} = \frac{1}{20} + \frac{1}{20}\ \text{k}\Omega^{-1}$$

$$\therefore R = 10\ \text{k}\Omega$$

$$\therefore \text{p.d.} = \frac{10}{(10 + 10)} \times 6.0 = 3.0\ \text{V}$$

SENSORS

A **light dependent resistor** (**LDR**), is a device whose resistance depends on the amount of light shining on its surface. An increase in light causes a decrease in resistance.

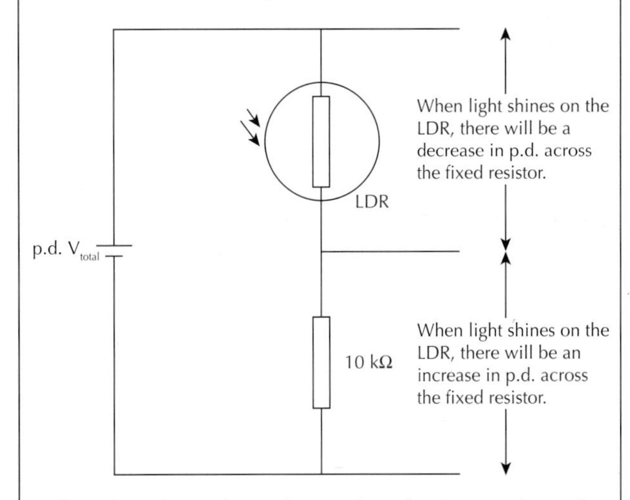
When light shines on the LDR, there will be a decrease in p.d. across the fixed resistor.
LDR
p.d. V total
10 kΩ
When light shines on the LDR, there will be an increase in p.d. across the fixed resistor.

A **thermistor** is a resistor whose value of resistance depends on its temperature. Most are semi-conducting devices that have a **negative temperature coefficient** (**NTC**). This means that an increase in temperature causes a decrease in resistance. Both of these devices can be used in potential divider circuits to create sensor circuits. The output potential difference of a **sensor circuit** depends on an external factor.

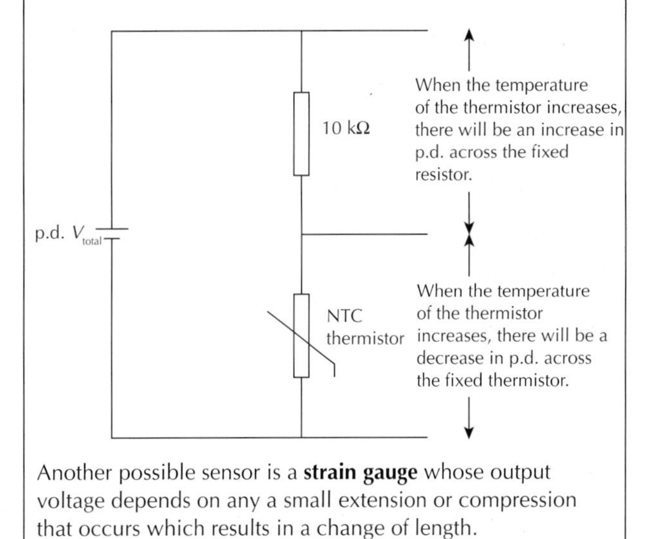
10 kΩ
When the temperature of the thermistor increases, there will be an increase in p.d. across the fixed resistor.
p.d. V total
NTC thermistor
When the temperature of the thermistor increases, there will be a decrease in p.d. across the fixed thermistor.

Another possible sensor is a **strain gauge** whose output voltage depends on any a small extension or compression that occurs which results in a change of length.

Resistivity

RESISTIVITY

The resistivity, ρ, of a material is defined in terms of its resistance, R, its length l and its cross-sectional area A.

$$R = \rho \frac{l}{A}$$

The units of resistivity must be ohm metres (Ω m). Note that this is the ohm multiplied by the metre, not "ohms per meter".

Example

The resistivity of copper is 3.3×10^{-7} Ω m, the resistance of a 100 m length of wire of cross-sectional area 1.0 mm² is:

$$R = 3.3 \times 10^{-7} \times \frac{100}{10^{-4}} \approx 0.3\Omega$$

IB QUESTIONS – ELECTRIC CURRENTS

1 A helium nucleus $^4_2$He and a proton $^1_1$H are both accelerated from rest through the same potential difference. The ratio of the kinetic energy of the helium nucleus to that of the proton will be

A 1:2 **B** 1:$\sqrt{2}$ **C** $\sqrt{2}$:1 **D** 2:1

2 A circuit consists of a battery and three resistors as shown below. The currents at different parts of the circuit are labelled. Which of the following gives a correct relationship between currents?

A $I_2 = I_3$ **B** $I_1 = I_2$ **C** $4I_3 = I_2$ **D** $4I_2 = I_3$

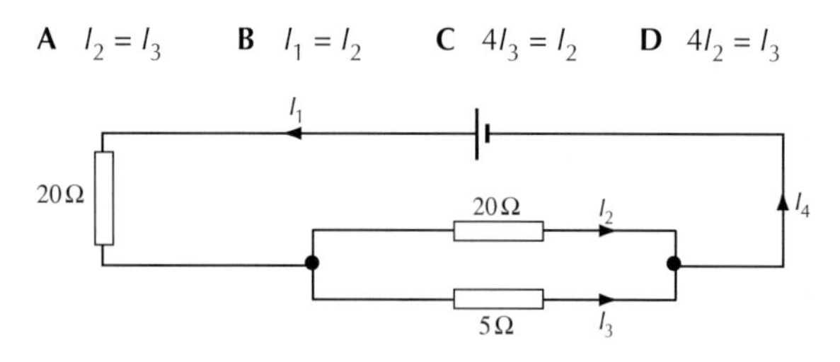

3 The identical lamps in the circuit below each have a resistance R at their rated voltage of 6 V. The lamps are to be run in parallel with each other using a 12 V source and a series resistor Z in the circuit as shown.

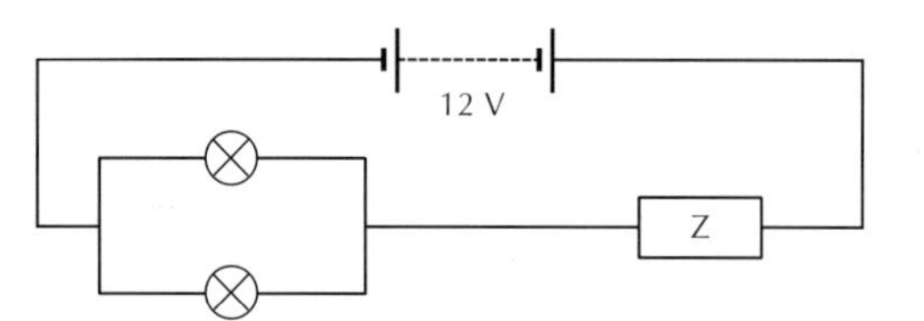

If the lamps are to operate at their rated voltage, what must be the resistance of Z?

A $\frac{R}{2}$ **B** R **C** $2R$ **D** zero

4 The element of an electric heater has a resistance R when in operation. What is the resistance of the element of a second heater which has twice the power output at the same voltage?

A $\frac{R}{2}$ **B** R **C** $2R$ **D** $4R$

5 In the following circuit, the voltmeter has an internal resistance of 100 kΩ.

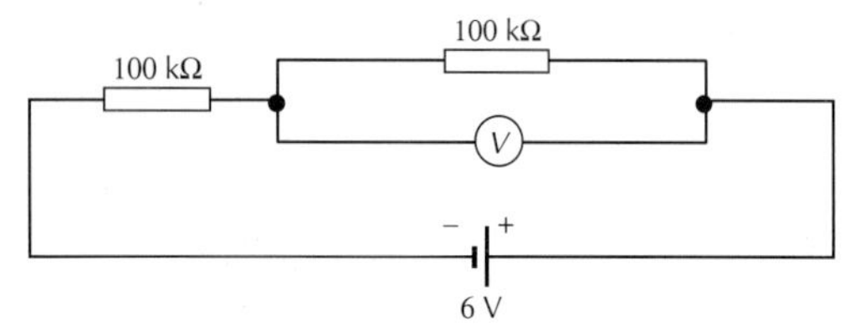

The reading on the voltmeter will be

A 1 V **B** 2 V **C** 3 V **D** 6 V

6 This question involves physical reasoning and calculations for electric circuits.

Light bulbs are marked with the rating 10 V; 3 W. Suppose you connect three of the bulbs in series with a switch and a 30 V battery as shown in **Figure 1** below. Switch **S** is initially open.

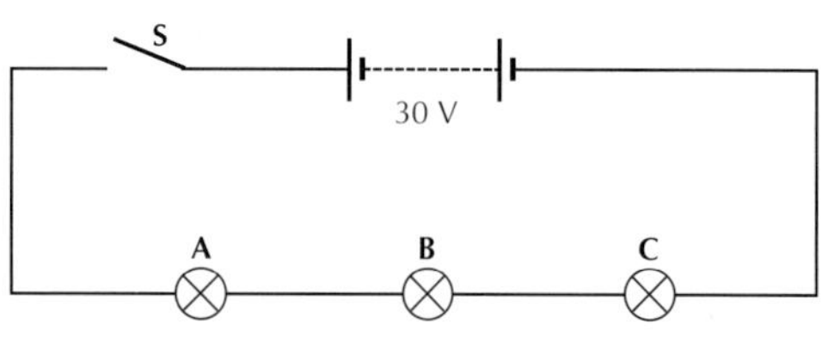

Figure 1

(a) A student tells you that after switch **S** is closed, bulb **C** will light up first, because electrons from the negative terminal of the battery will reach it first, and then go on to light bulbs **B** and **A** in succession. Is this prediction and reasoning correct? How would you reply? [2]

(b) State how the brightness of the three bulbs in the circuit will compare with each other. [1]

(c) The student now connects a fourth bulb **D** across bulb **B** as shown in **Figure 2** below.

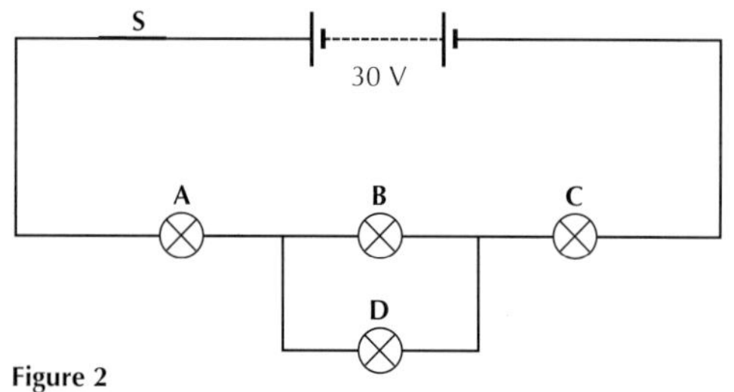

Figure 2

When she connects **D**, what will happen to the brightnesses of bulbs **A, B** and **C**? Explain your reasoning. [3]

(d) Assuming that the resistance of the bulbs remains constant, calculate the power output of bulb **B** in the modified circuit in **Figure 2**. [3]

Gravitational force and field

NEWTON'S LAW OF UNIVERSAL GRAVITATION

If you trip over, you will fall down towards the ground.

Newton's theory of **universal gravitation** explains what is going on. It is called 'universal' gravitation because at the core of this theory is the statement that every mass in the Universe attracts all the other masses in the Universe. The value of the attraction between two **point** masses is given by an equation.

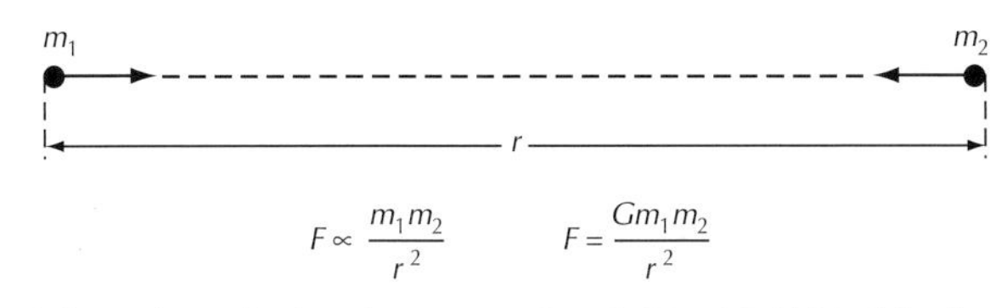

$$F \propto \frac{m_1 m_2}{r^2} \qquad F = \frac{G m_1 m_2}{r^2}$$

Universal gravitational constant $G = 6.67 \times 10^{-11}$ Nm^2 kg^{-2}

The following points should be noticed:

- the law only deals with point masses.
- technically speaking, the masses in this equation are gravitational masses (as opposed to inertial masses – see page 172 for more details).
- there is a force acting on each of the masses. These forces are EQUAL and OPPOSITE (even if the masses are not equal).
- the forces are always attractive.
- gravitation forces act between ALL objects in the Universe. The forces only become significant if one (or both) of the objects involved are massive, but they are there nonetheless.

The interaction between two spherical masses turns out to be the same as if the masses were concentrated at the centres of the spheres.

GRAVITATIONAL FIELD STRENGTH

The table below should be compared with the one on page 54.

	Gravitational field strength
Symbol	g
Caused by...	Masses
Affects...	Masses
One type of...	Mass
Simple force rule:	All masses attract

The gravitational field is therefore defined as the force per unit mass. $g = \frac{F}{m}$ $\quad m$ = test mass

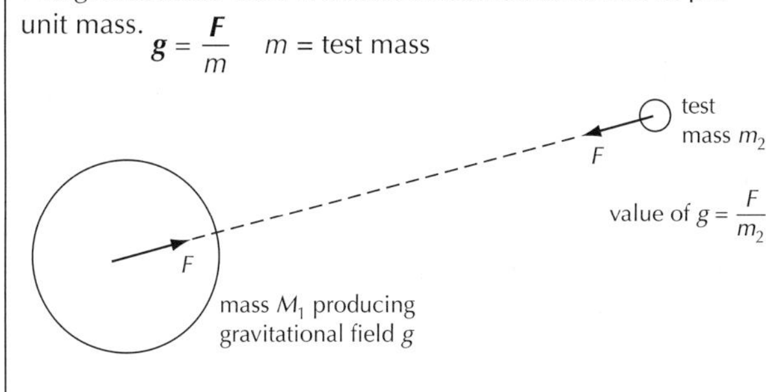

The SI units for g are N kg^{-1}. These are the same as m s^{-2}. Field strength is a vector quantity and can be represented by the use of field lines.

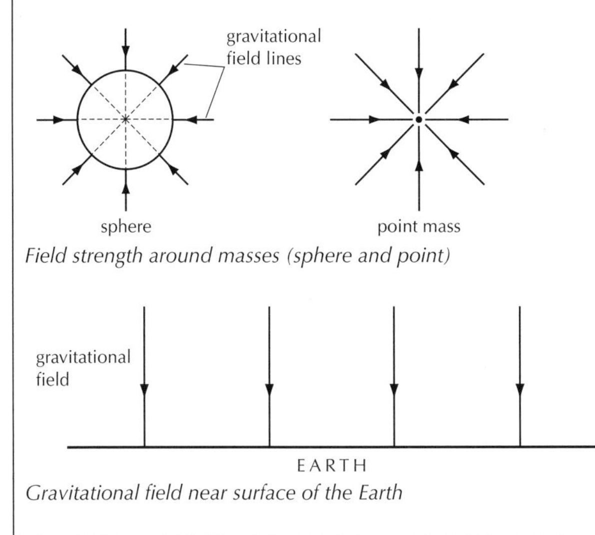

Field strength around masses (sphere and point)

Gravitational field near surface of the Earth

In the example on the left the numerical value for the gravitational field can be calculated using Newton's law:

$$g = \frac{GM}{r^2}$$

The gravitational field strength at the surface of a planet must be the same as the acceleration due to gravity on the surface.

Field strength is defined to be $\frac{\text{force}}{\text{mass}}$.

Acceleration $= \frac{\text{force}}{\text{mass}}$ (from $F = ma$)

For the Earth

$M = 6.0 \times 10^{24}$ kg

$r = 6.4 \times 10^6$ m

$$g = \frac{6.67 \times 10^{-11} \times 6.0 \times 10^{24}}{(6.4 \times 10^6)^2} = 9.8 \text{ ms}^{-2}$$

EXAMPLE

In order to calculate the overall gravitational field strength at any point we must use vector addition. The overall gravitational field strength at any point between the Earth and the Moon must be a result of both pulls.

There will be a single point somewhere between the Earth and the Moon where the total gravitational field due to these two masses is zero. Up to this point the overall pull is back to the Earth, after this point the overall pull is towards the Moon.

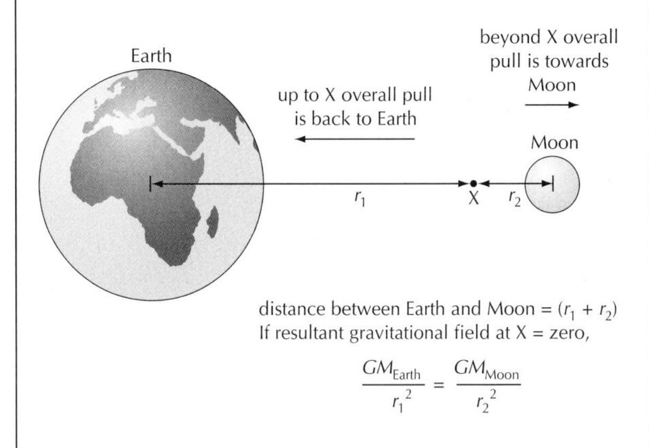

distance between Earth and Moon = $(r_1 + r_2)$

If resultant gravitational field at X = zero,

$$\frac{GM_{Earth}}{r_1^2} = \frac{GM_{Moon}}{r_2^2}$$

Electric charge

CONSERVATION OF CHARGE

Two types of charge exist – positive and negative. Equal amounts of positive and negative charge cancel each other. Matter that contains no charge, or matter that contains equal amounts of positive and negative charge, is said to be electrically **neutral**.

Charges are known to exist because of the forces that exist between all charges, called the **electrostatic force**: like charges repel, unlike charges attract.

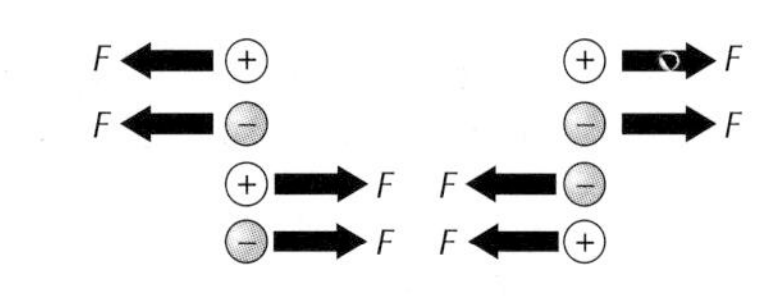

Like charges repel, unlike charges attract.

A very important experimental observation is that charge is always conserved.

Charged objects can be created by friction. In this process electrons are physically moved from one object to another – in order for the charge to remain on the object, it normally needs to be an insulator.

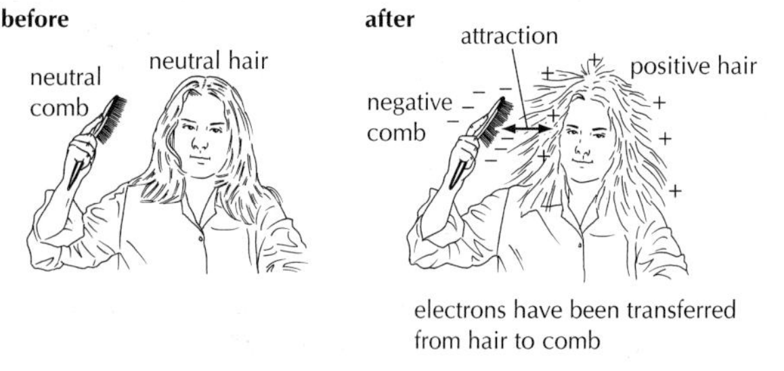

electrons have been transferred from hair to comb

CONDUCTORS AND INSULATORS

A material that allows the flow of charge through it is called an electrical **conductor**. If charge cannot flow through a material it is called an electrical **insulator**. In solid conductors the flow of charge is always as a result of the flow of electrons from atom to atom.

Electrical conductors	Electrical insulators
all metals	plastics
e.g. copper	e.g. polythene
aluminium	nylon
brass	acetate
graphite	rubber
	dry wood
	glass
	ceramics

Electric force and field

COULOMB'S LAW

The diagram shows the force between two point charges that are far away from the influence of any other charges.

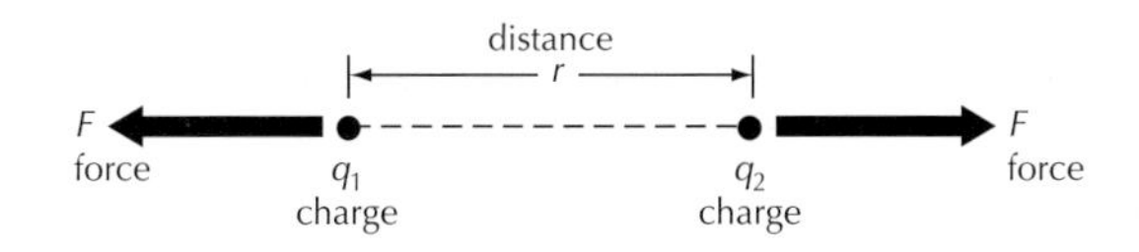

The directions of the forces are along the line joining the charges. If they are like charges, the forces are away from each other – they repel. If they are unlike charges, the forces are towards each other – they attract.

Each charge must feel a force of the same size as the force on the other one

Experimentally, the force is proportional to the size of both charges and inversely proportional to the square of the distance between the charges.

$$F = \frac{k\,q_1\,q_2}{r^2}$$

This is known as Coulomb's law and the constant k is called the Coulomb constant. In fact, the law is often quoted in a slightly different form using a different constant for the medium called the permittivity, ε.

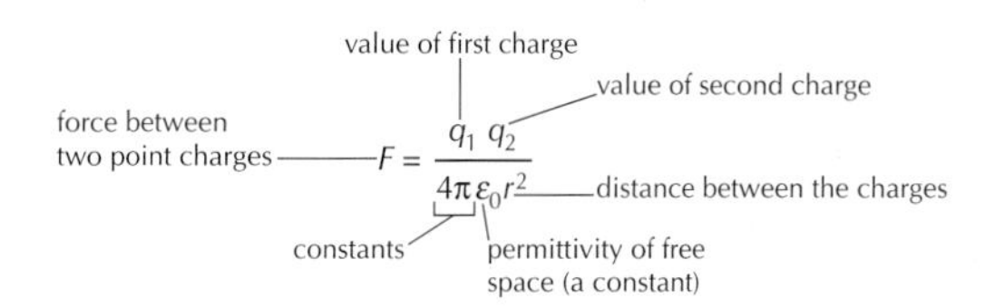

If there are two or more charges near another charge, the overall force can be worked out using vector addition.

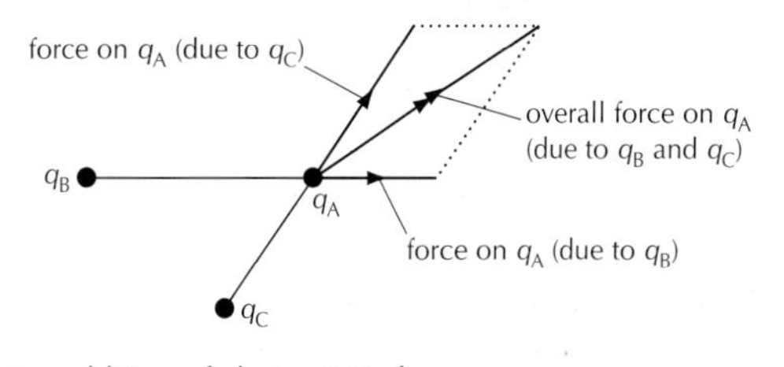

Vector addition of electrostatic forces

ELECTRIC FIELDS – DEFINITION

A charge, or combination of charges, is said to produce an **electric field** around it. If we place a **test charge** at any point in the field, the value of the force that it feels at any point will depend on the value of the test charge only.

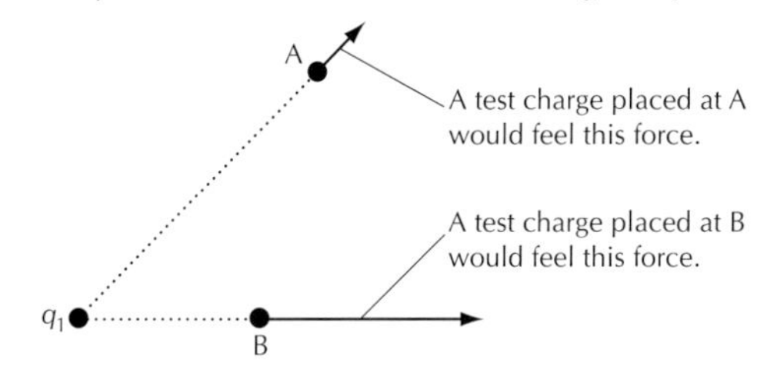

A test charge would feel a different force at different points around a charge q_1.

In practical situations, the test charge needs to be small so that it doesn't disturb the charge or charges that are being considered.

The definition of electric field, E, is

$$E = \frac{F}{q_2} = \text{force per positive unit test charge.}$$

Coulomb's law can be used to relate the electric field around a point charge to the charge producing the field.

$$E = \frac{q_1}{4\pi\varepsilon_0 r^2}$$

When using these equations you have to be very careful

- not to muddle up the charge producing the field and the charge sitting in the field (and thus feeling a force).
- not to use the mathematical equation for the field around a point charge for other situations (e.g. parallel plates).

REPRESENTATION OF ELECTRIC FIELDS

This is done using **field lines**.

At any point in a field

- the direction of field is represented by the direction of the field lines closest to that point.
- the magnitude of the field is represented by the number of field lines passing near that point.

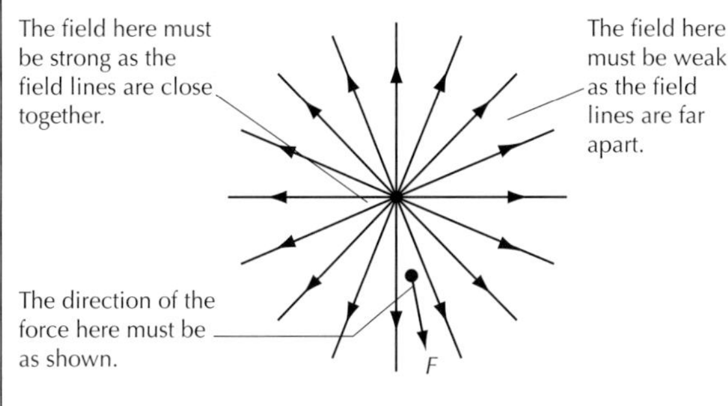

Field around a positive point charge

The resultant electric field at any position due to a collection of point charges is shown to the right.

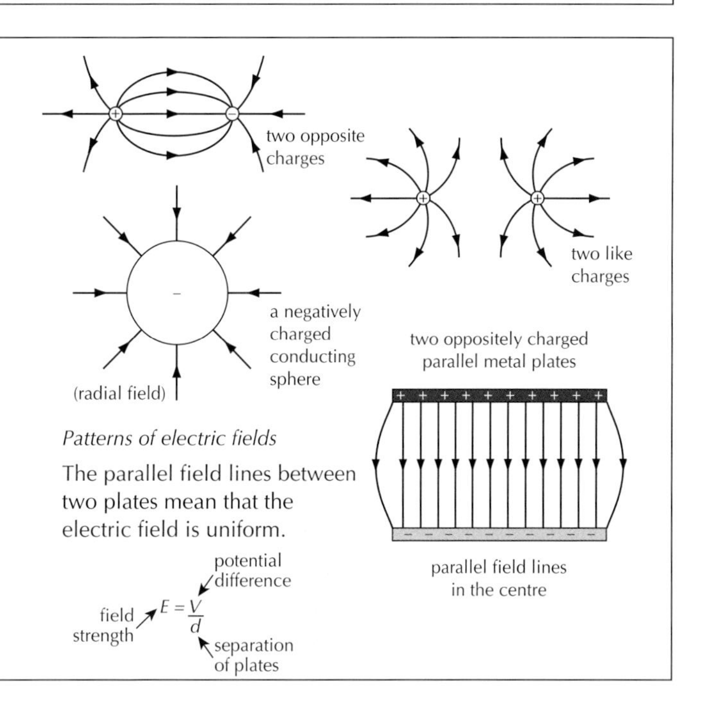

Patterns of electric fields

The parallel field lines between two plates mean that the electric field is uniform.

$$E = \frac{V}{d}$$

(field strength E; potential difference V; separation of plates d)

Magnetic force and fields

MAGNETIC FIELD LINES

There are many similarities between the magnetic force and the electrostatic force. In fact, both forces have been shown to be two aspects of one force – the electromagnetic interaction (see page 199). It is, however, much easier to consider them as completely separate forces to avoid confusion.

Page 53 introduced the idea of electric fields. A similar concept is used for magnetic fields. A table of the comparisons between these two fields is shown below.

	Electric field	Magnetic field
Symbol	E	B
Caused by ...	Charges	Magnets (or electric currents)
Affects ...	Charges	Magnets (or electric currents)
Two types of ...	Charge: positive and negative	Pole: North and South
Simple force rule:	Like charges repel, Unlike charges attract	Like poles repel, Unlike poles attract

In order to help visualise a magnetic field we, once again, use the concept of field lines. This time the field lines are lines of magnetic field – also called **flux** lines. If a 'test' magnetic North pole is place in a magnetic field, it will feel a force.

- The direction of the force is shown by the direction of the field lines.
- The strength of the force is shown by how close the lines are to one another.

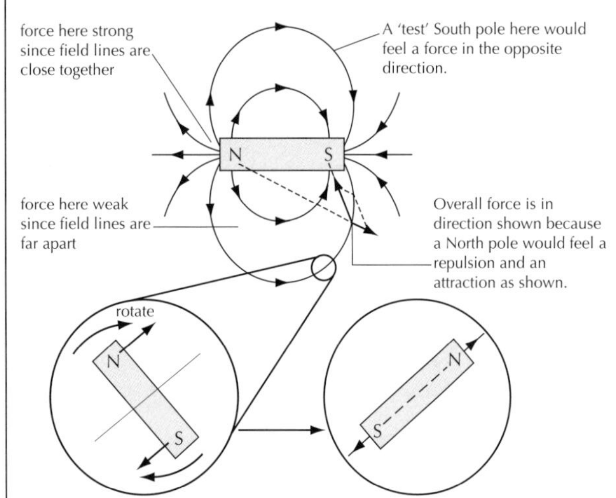
force here strong since field lines are close together
A 'test' South pole here would feel a force in the opposite direction.
N
S
force here weak since field lines are far apart
Overall force is in direction shown because a North pole would feel a repulsion and an attraction as shown.
rotate
N
S
N
S

A small magnet placed in the field would rotate until lined up with the field lines. This is how a compass works. Small pieces of iron (iron filings) will also line up with the field lines – they will be induced to become little magnets.

Field pattern of an isolated bar magnet

Despite all the similarities between electric fields and magnetic fields, it should be remembered that they are very different. For example:

- A magnet does not feel a force when placed in an electric field.
- A positive charge does not feel a force when placed stationary in a magnetic field.
- Isolated charges exist whereas isolated poles do not.

The Earth itself has a magnetic field. It turns out to be similar to that of a bar magnet with a magnetic South pole near the geographic North Pole as shown above right.

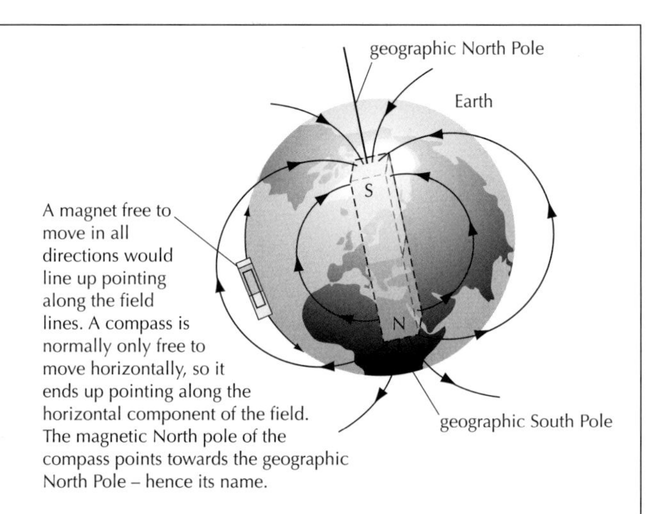
geographic North Pole
Earth
S
N
A magnet free to move in all directions would line up pointing along the field lines. A compass is normally only free to move horizontally, so it ends up pointing along the horizontal component of the field. The magnetic North pole of the compass points towards the geographic North Pole – hence its name.
geographic South Pole

An electric current can also cause a magnetic field. The mathematical value of the magnetic fields produced in this way is given on page 56. The field patterns due to different currents can be seen in the diagrams below.

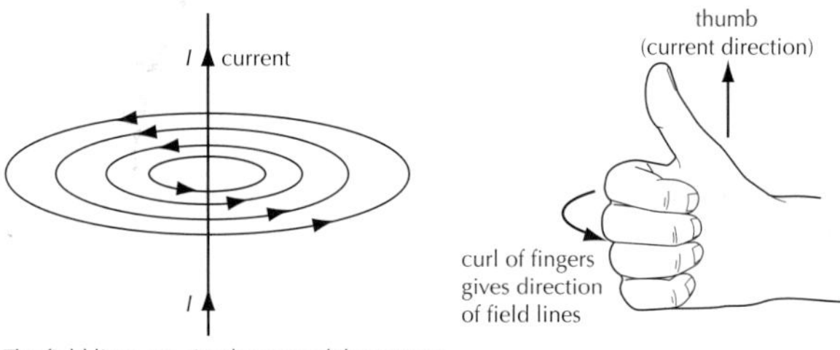
I current
I
thumb (current direction)
curl of fingers gives direction of field lines

The field lines are circular around the current.

The direction of the field lines can be remembered with the right hand grip rule. If the thumb of the right hand is arranged to point along the direction of a current, the way the fingers of the right hand naturally curl will give the direction of the field lines.

Field pattern of a straight wire carrying current

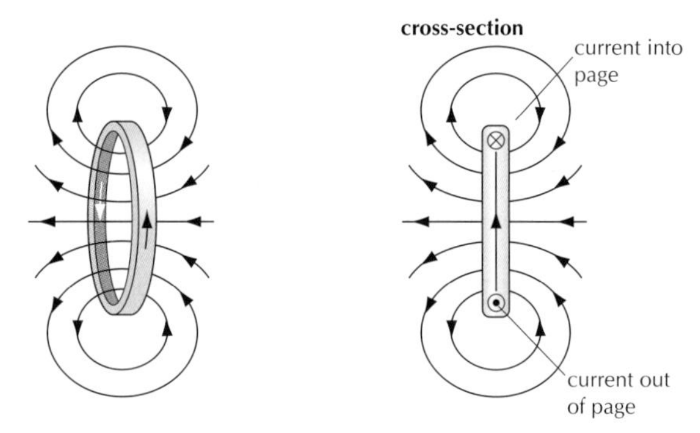
cross-section
current into page
current out of page

Field pattern of a flat circular coil

A long current-carrying coil is called a solenoid.

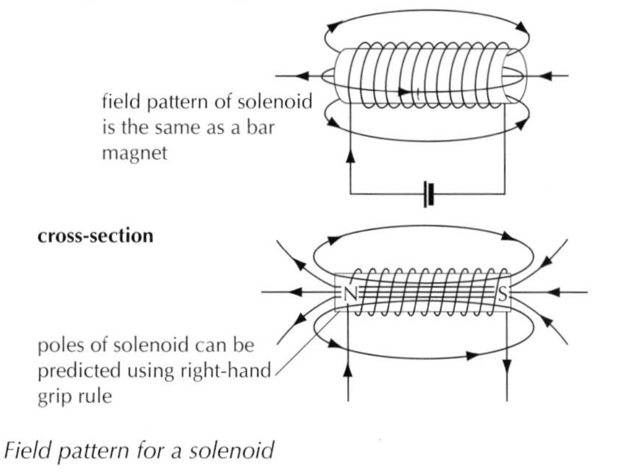
field pattern of solenoid is the same as a bar magnet
cross-section
N
S
poles of solenoid can be predicted using right-hand grip rule

Field pattern for a solenoid

Magnetic forces

MAGNETIC FORCE ON A CURRENT

When a current-carrying wire is placed in a magnetic field the magnetic interaction between the two results in a force. This is known as the **motor effect**. The direction of this force is at right angles to the plane that contains the field and the current as shown below.

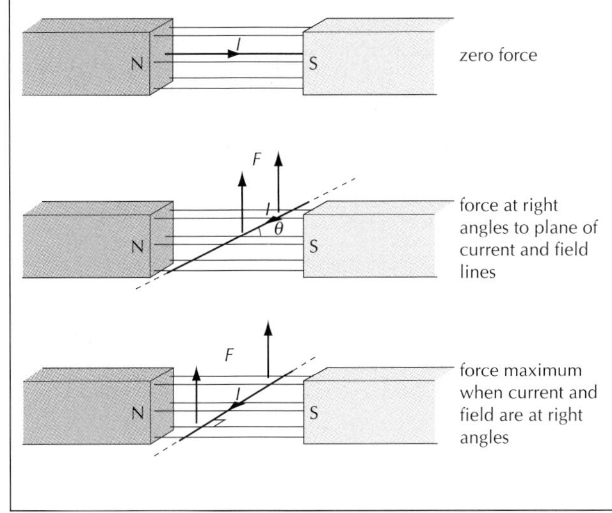

Motor effect force on current

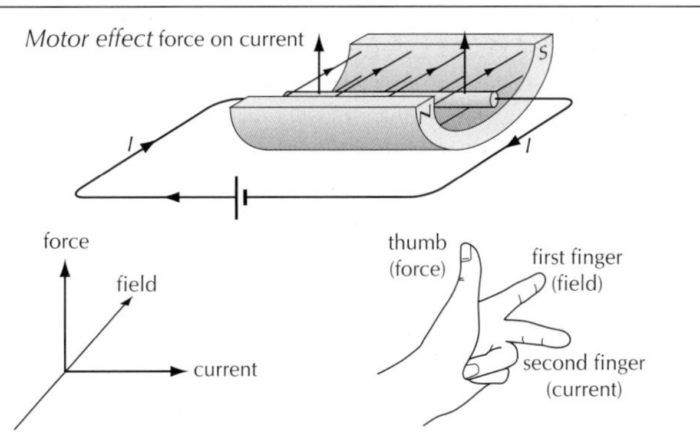

Fleming's left-hand rule

Experiments show that the force is proportional to:

- the magnitude of the magnetic field, B.
- the magnitude of the current, I.
- the length of the current, L, that is in the magnetic field.
- the sine of the angle, θ, between the field and current.

The magnetic field strength, B is defined as follows:

$$\boldsymbol{F = BIL \sin \theta} \quad \text{or}$$

$$\boldsymbol{B = \frac{F}{IL \sin \theta}}$$

A new unit, the tesla, is introduced. 1 T is defined to be equal to 1 N A^{-1} m^{-1}. Another possible unit for magnetic field strength is Wb m^{-2}. Another possible term is magnetic flux density.

MAGNETIC FORCE ON A MOVING CHARGE

A single charge moving through a magnetic field also feels a force in exactly the same way that a current feels a force.

In this case the force on a moving charge is proportional to:

- the magnitude of the magnetic field, B.
- the magnitude of the charge, q.
- the velocity of the charge, v.
- the sine of the angle, θ, between the velocity of the charge and the field.

We can use these relationships to give an alternative definition of the magnetic field strength, B. This definition is exactly equivalent to the previous definition.

$$F = Bqv \sin \theta \text{ or } B = \frac{F}{qv \sin \theta}$$

Since the force on a moving charge is always at right angles to the velocity of the charge the resultant motion can be circular. An example of this would be when an electron enters a region where the magnetic field is at right angles to its velocity as shown below.

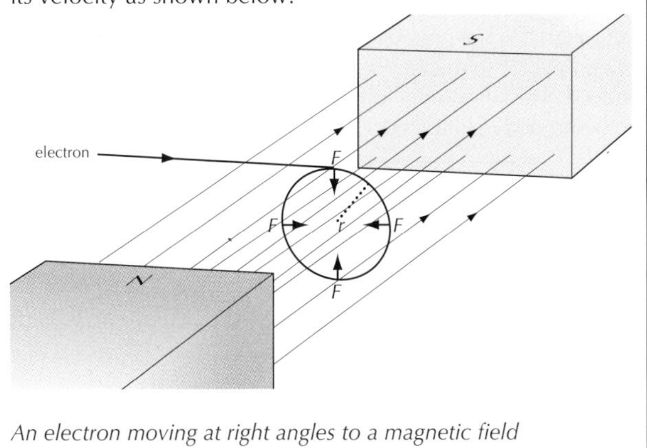

An electron moving at right angles to a magnetic field

Examples of the magnetic field due to currents

The formula used on this page do not need to be remembered.

STRAIGHT WIRE

The field pattern around a long straight wire shows that as one moves away from the wire, the strength of the field gets weaker. Experimentally the field is proportional to

- the value of the current, I.
- the inverse of the distance away from the wire, r. If the distance away is doubled, the magnetic field will halve.

The field also depends on the medium around the wire. These factors are summarised in the equation:

$$B = \frac{\mu I}{2\pi r}$$

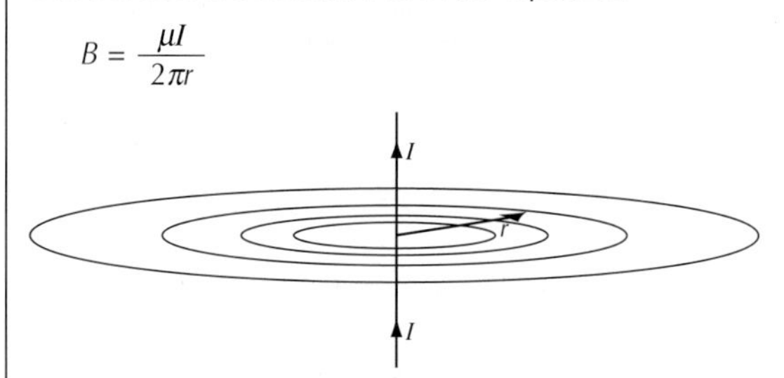

Magnetic field of a straight current

The constant μ is called the permeability and changes if the medium around the wire changes. Most of the time we consider the field around a wire when there is nothing there – so we use the value for the permeability of a vacuum, μ_0. There is almost no difference between the permeability of air and the permeability of a vacuum. There are many possible units for this constant, but it is common to use N A^{-2} or T m A^{-1}.

Permeability and permittivity are related constants. In other words, if you know one constant you can calculate the other. In the SI system of units, the permeability of a vacuum is defined to have a value of exactly $4\pi \times 10^{-7}$ N A^{-2}. See the definition of the ampere (right) for more detail.

MAGNETIC FIELD IN A SOLENOID

The magnetic field of a solenoid is very similar to the magnetic field of a bar magnet. As shown by the parallel field lines, the magnetic field inside the solenoid is constant. It might seem surprising that the field does not vary at all inside the solenoid, but this can be experimentally verified near the centre of a long solenoid. It does tend to decrease near the ends of the solenoid as shown in the graph below.

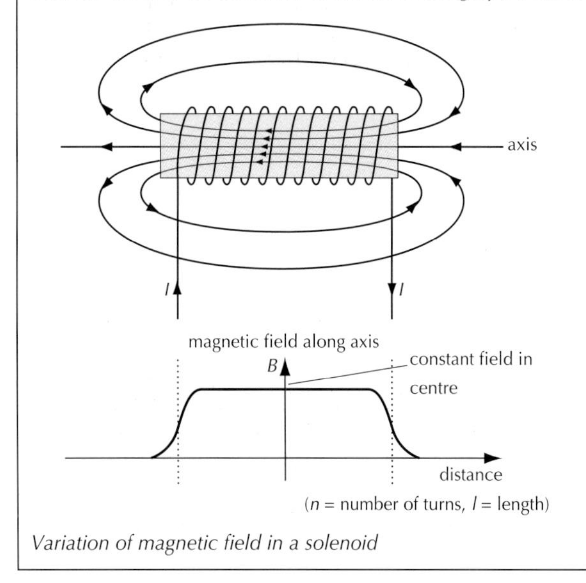

Variation of magnetic field in a solenoid

The mathematical equation for this constant field at the centre of a long solenoid is

$$B = \mu \left(\frac{n}{l}\right) I_l$$

Thus the field only depends on:

- the current, I
- the number of turns per unit length, $\frac{n}{l}$
- the nature of the solenoid core, μ

It is independent of the cross-sectional area of the solenoid.

TWO PARALLEL WIRES – DEFINITION OF THE AMPERE

Two parallel current-carrying wires provide a good example of the concepts of magnetic field and magnetic force. Because there is a current flowing down the wire, each wire is producing a magnetic field. The other wire is in this field so it feels a force. The forces on the wires are an example of a Newton's third law pair of forces.

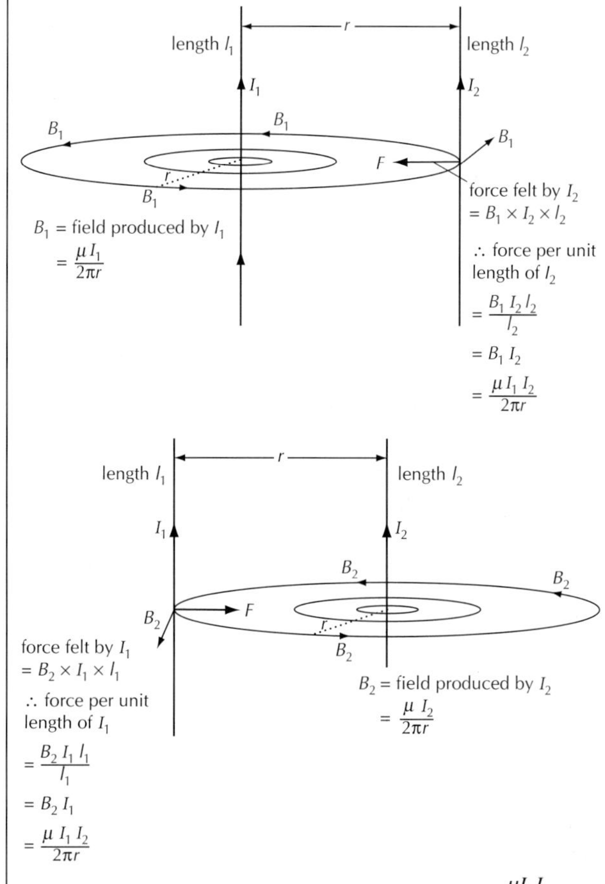

Magnitude of force per unit length on either wire $= \frac{\mu I_1 I_2}{2\pi r}$

This equation is experimentally used to define the ampere. The coulomb is then defined to be one ampere second. If we imagine two infinitely long wires carrying a current of one amp separated by a distance of one metre, the equation would predict the force per unit length to be 2×10^{-7} N. Although it is not possible to have infinitely long wires, an experimental set-up can be arranged with very long wires indeed. This allows the forces to be measured and ammeters to be properly calibrated.

IB QUESTIONS – FIELDS AND FORCES

1 Which **one** of the field patterns below could be produced by two point charges?

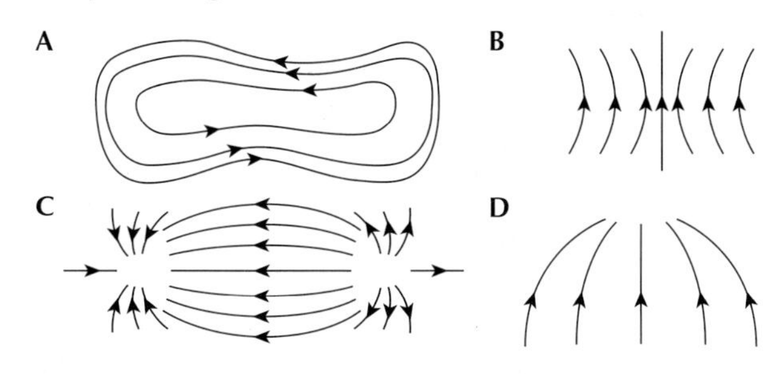

2 Two long, vertical wires **X** and **Y** carry currents in the same direction and pass through a horizontal sheet of card.

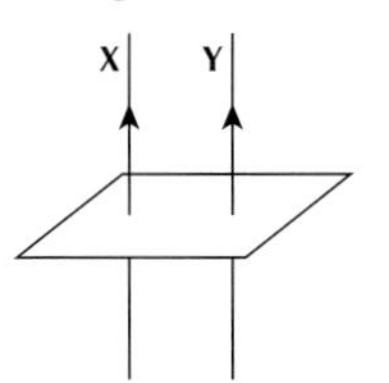

Iron filings are scattered on the card. Which **one** of the following diagrams best shows the pattern formed by the iron filings? (The dots show where the wires **X** and **Y** enter the card.)

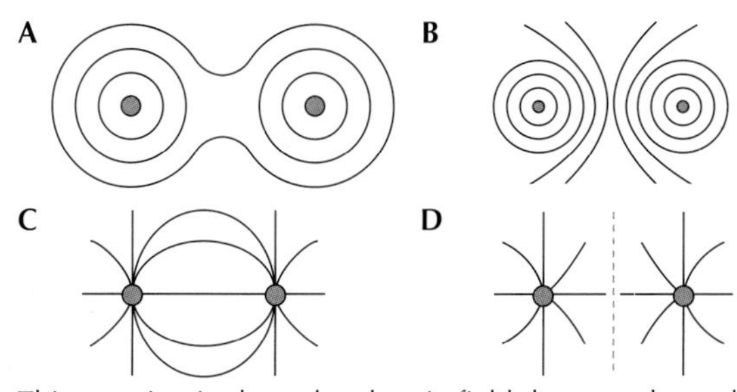

3 This question is about the electric field due to a charged sphere and the motion of electrons in that field.

The diagram below shows an isolated metal sphere in a vacuum that carries a negative electric charge of 9.0 nC.

–

(a) On the diagram draw arrows to represent the electric field pattern due to the charged sphere. [3]

(b) The electric field strength at the surface of the sphere and at points outside the sphere can be determined by assuming that the sphere acts as though a point charge of magnitude 9.0 nC is situated at its centre. The radius of the sphere is 4.5×10^{-2} m. Deduce that the magnitude of the field strength at the surface of the sphere is 4.0×10^{4} Vm^{-1}. [1]

An electron is initially at rest on the surface of the sphere.

(c) (i) Describe the path followed by the electron as it leaves the surface of the sphere. [1]

(ii) Calculate the initial acceleration of the electron. [3]

(iii) State and explain whether the acceleration of the electron remains constant, increases or decreases as it moves away from the sphere. [2]

(iv) At a certain point P, the speed of the electron is 6.0×10^{6} ms^{-1}. Determine the potential difference between the point P and the surface of the sphere. [2]

4 A current-carrying solenoid is placed with its axis pointing east–west as shown below. A small compass is situated near one end of the solenoid.

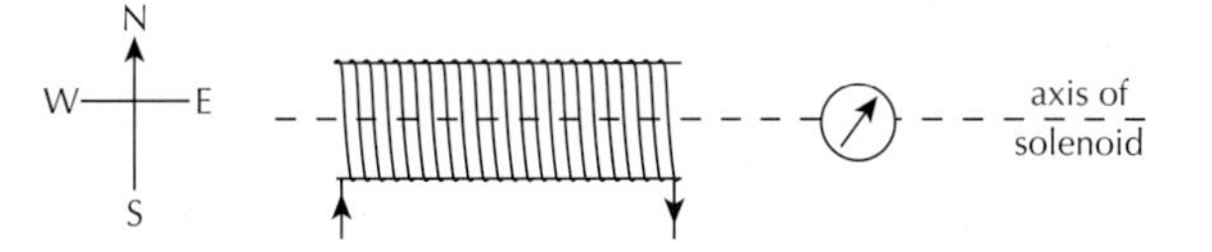

The axis of the needle of the compass is approximately 45° to the axis of the solenoid. The current in the solenoid is then doubled. Which of the following diagrams best shows the new position of the compass needle?

A W– – – (→) – – –E **B** W– – – (↗) – – –E

C W– – – (↑) – – –E **D** W– – – (↗) – – –E

5 This question is about the force between current-carrying wires.

Figure 1 shows two long, parallel vertical wires each carrying equal currents in the same direction. The wires pass through a horizontal sheet of card.

Figure 2 shows a plan view of the wires looking down onto the card.

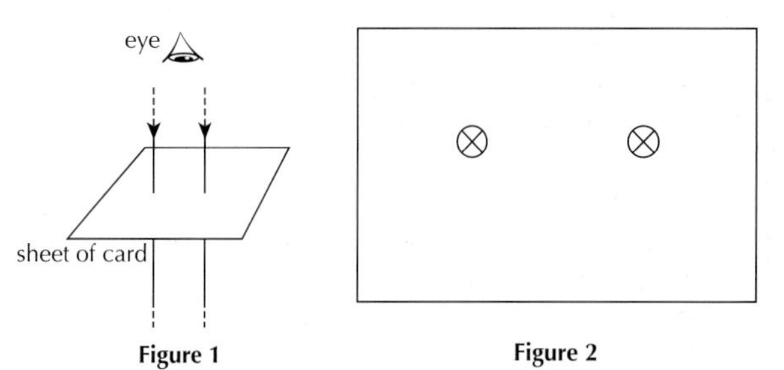

Figure 1 **Figure 2**

(a) (i) Draw on Figure 1 the direction of the force acting on each wire. [1]

(ii) Draw on Figure 2 the magnetic field pattern due to the currents in the wire. [3]

(b) The card is removed and one of the two wires is free to move. Describe and explain the changes in the velocity and in acceleration of the moveable wire. [3]

6 The acceleration of free fall of a small sphere of mass 5.0×10^{-3} kg when close to the surface of Jupiter is 25 ms^{-2}. The gravitational field strength at the surface of Jupiter is

A 2.0×10^{-4} N Kg^{-1}

B 1.3×10^{-1} N Kg^{-1}

C 25 N Kg^{-1}

D 5.0×10^{3} N Kg^{-1}

7 The Earth is distance R_M from the Moon and distance R_S from the Sun. The ratio

$$\frac{\text{gravitational field strength at the Earth due to the Moon}}{\text{gravitational field strength at the Earth due to the Sun}}$$

is proportional to which of the following?

A $\frac{R_M^2}{R_S^2}$ **B** $\frac{R_M}{R_S}$ **C** $\frac{R_S^2}{R_M^2}$ **D** $\frac{R_S}{R_M}$

Atomic structure (1)

INTRODUCTION

All matter that surrounds us, living or otherwise, is made up of different combinations of atoms. There are only a hundred, or so, different types of atoms present in nature. Atoms of a single type form an element. Each of these elements has a name and a chemical symbol e.g. hydrogen, the simplest of all the elements, has the chemical symbol H. Oxygen has the chemical symbol O. The combination of two hydrogen atoms with one oxygen atom is called a water molecule – H_2O. The full list of elements is shown in a periodic table. Atoms consist of a combination of three things: protons, neutrons and electrons.

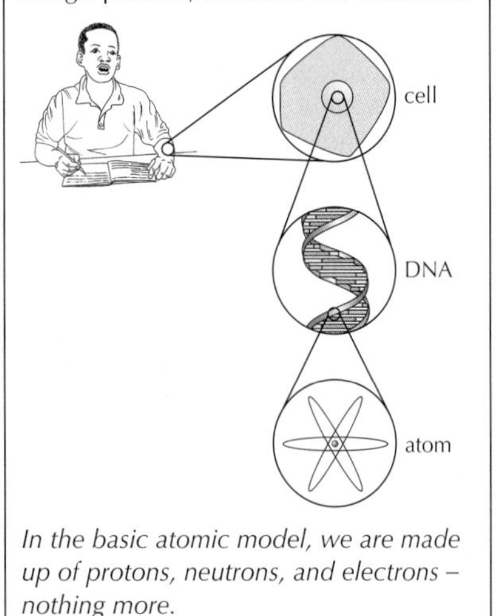

In the basic atomic model, we are made up of protons, neutrons, and electrons – nothing more.

ATOMIC MODEL

The basic atomic model, known as the nuclear model, was developed during the last century and describes a very small central nucleus surrounded by electrons arranged in different energy levels. The nucleus itself contains protons and neutrons (collectively called **nucleons**). All of the positive charge and almost all the mass of the atom is in the nucleus. The electrons provide only a tiny bit of the mass but all of the negative charge. Overall an atom is neutral. The vast majority of the volume is nothing at all – a vacuum. The nuclear model of the atom seems so strange that there must be good evidence to support it.

	Protons	Neutrons	Electrons
Relative mass	1	1	Negligible
Charge	+ 1	Neutral	– 1

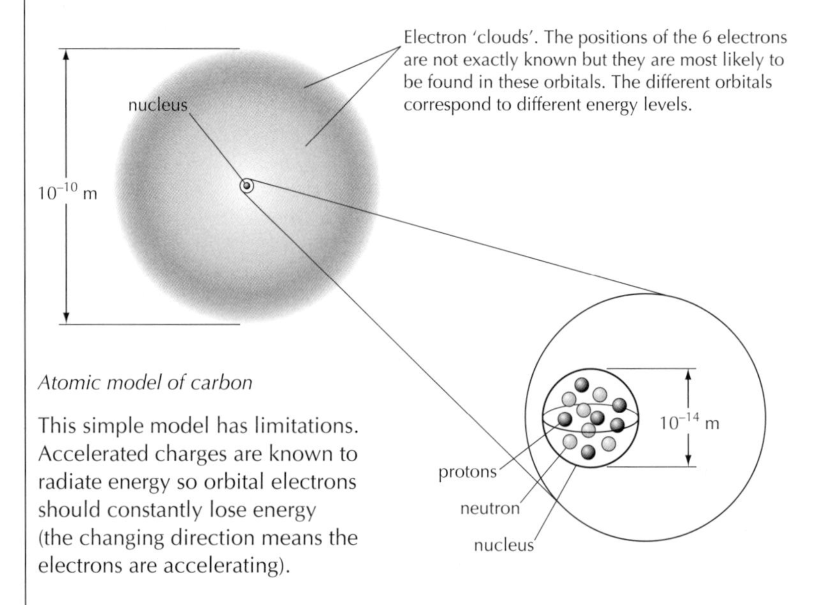

Atomic model of carbon

This simple model has limitations. Accelerated charges are known to radiate energy so orbital electrons should constantly lose energy (the changing direction means the electrons are accelerating).

EVIDENCE

One of the most convincing pieces of evidence for the nuclear model of the atom comes from the Geiger–Marsden experiment. Positive alpha particles were "fired" at a thin gold leaf. The relative size and velocity of the alpha particles meant that most of them were expected to travel straight through the gold leaf. The idea behind this experiment was to see if there was any detectable structure within the gold atoms. The amazing discovery was that some of the alpha particles were deflected through huge angles. The mathematics of the experiment showed that numbers being deflected at any given angle agreed with an inverse square law of repulsion from the nucleus. Evidence for electron energy levels comes from emission and absorption spectra. The existence of isotopes provides evidence for neutrons.

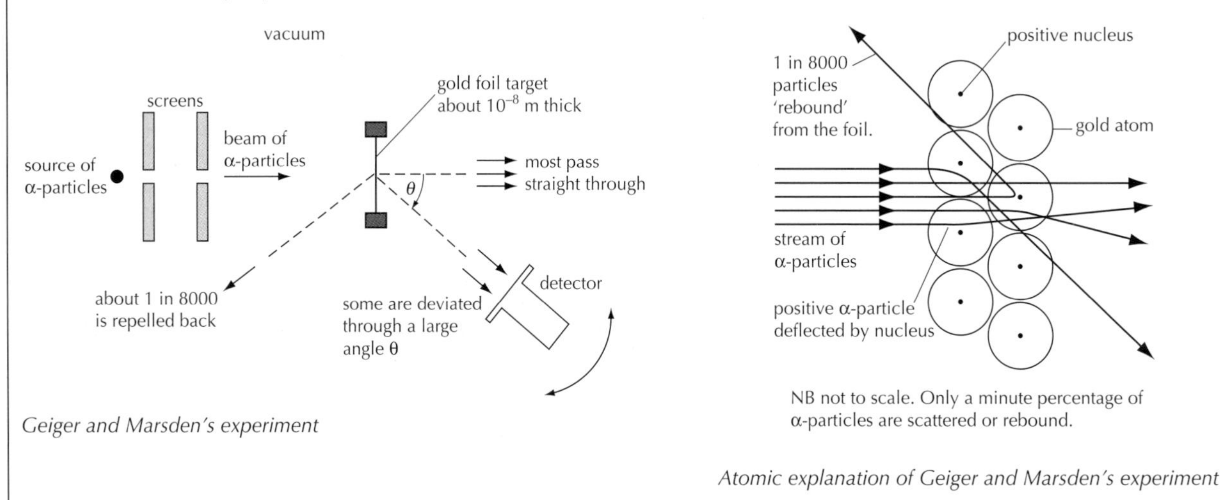

Geiger and Marsden's experiment

Atomic explanation of Geiger and Marsden's experiment

Atomic structure (2) – emission and absorption spectra

EMISSION SPECTRA AND ABSORPTION SPECTRA

When an element is given enough energy it emits light. This light can be analysed by splitting it into its various colours (or frequencies) using a prism or a diffraction grating. If all possible frequencies of light were present, this would be called a **continuous spectrum**. The light an element emits, its **emission spectrum**, is not continuous, but contains only a few characteristic colours. The frequencies emitted are particular to the element in question. For example, the yellow-orange light from a street lamp is often a sign that the element sodium is present in the lamp. Exactly the same particular frequencies are **absent** if a continuous spectrum of light is shone through an element when it is in gaseous form. This is called an **absorption** spectrum.

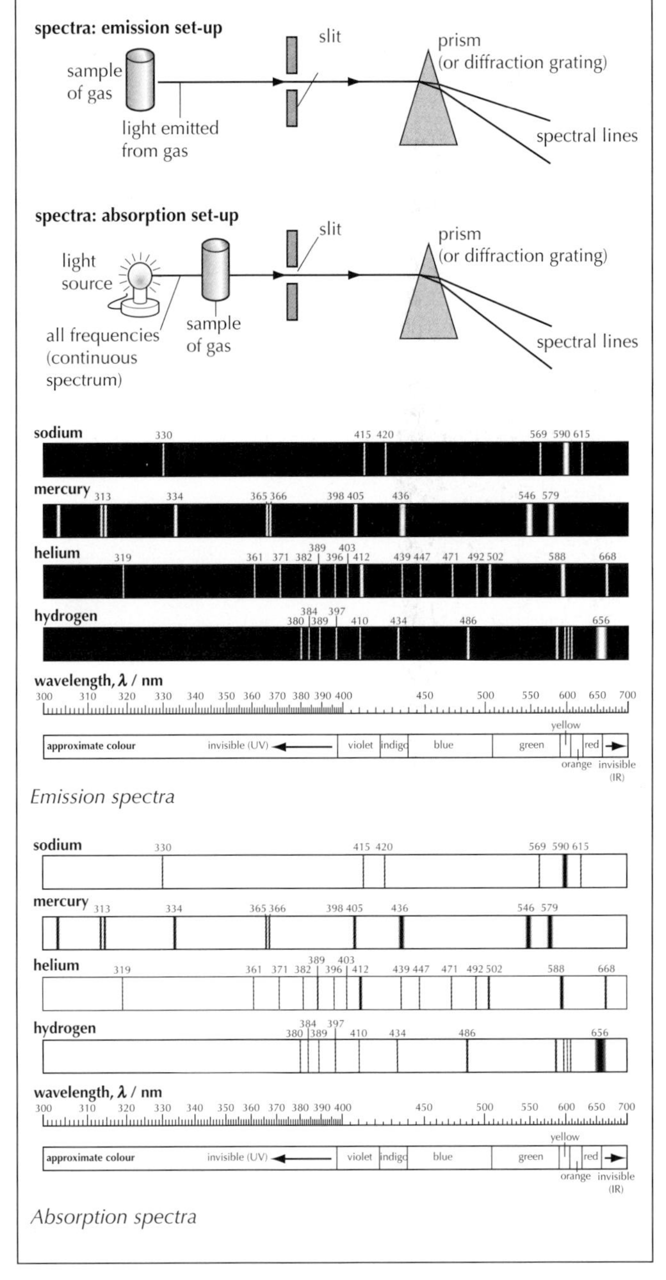

Emission spectra

Absorption spectra

EXPLANATION OF ATOMIC SPECTRA

In an atom, electrons are bound to the nucleus. This means that they cannot "escape" without the input of energy. If enough energy is put in, an electron can leave the atom. If this happens, the atom is now positive overall and is said to be ionised. Electrons can only occupy given energy levels – the energy of the electron is said to be **quantized**. These energy levels are fixed for particular elements and correspond to "allowed" obitals. The reason why only these energies are "allowed" forms a significant part of quantum theory (see HL topic 13).

When an electron moves between energy levels it must emit or absorb energy. The energy emitted or absorbed corresponds to the difference between the two allowed energy levels. This energy is emitted or absorbed as "packets" of light called photons (for more information, see page 104). A higher energy photon corresponds to a higher frequency (shorter wavelength) of light.

The energy of a photon is given by

$$E = h\,f$$

energy in joules (E); frequency of light in Hz (f); Planck's constant (h) 6.63×10^{-34} Js

Thus the frequency of the light, emitted or absorbed, is fixed by the energy difference between the levels. Since the energy levels are unique to a given element, this means that the emission (and the absorption) spectrum will also be unique.

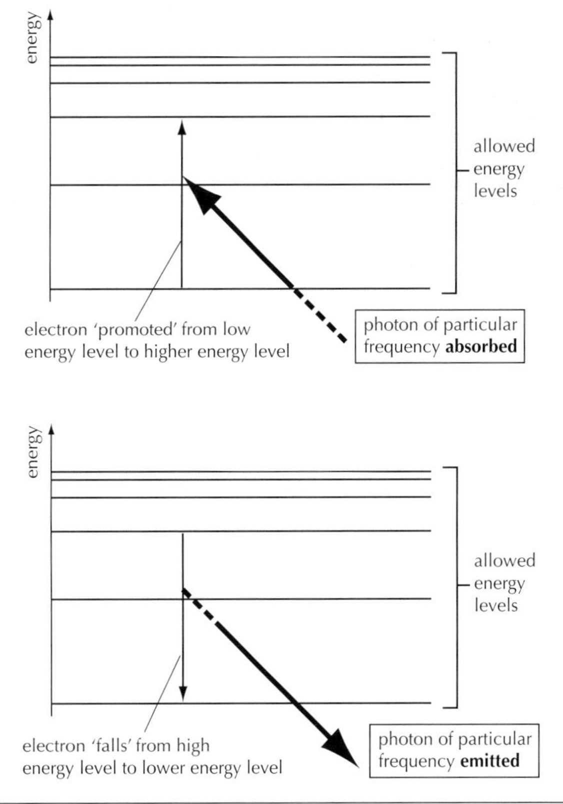

Nuclear structure

ISOTOPES

When a chemical reaction takes place, it involves the outer electrons of the atoms concerned. Different elements have different chemical properties because the arrangement of outer electrons varies from element to element. The chemical properties of a particular element are fixed by the amount of positive charge that exists in the nucleus – in other words, the number of protons. In general, different nuclear structures will imply different chemical properties. A **nuclide** is the name given to a particular species of atom (one whose nucleus contains a specified number of protons and a specified number of neutrons). Some nuclides are the same element – they have the same chemical properties and contain the same number of protons. These nuclides are called **isotopes** – they contain the same number of protons but different numbers of neutrons.

NOTATION

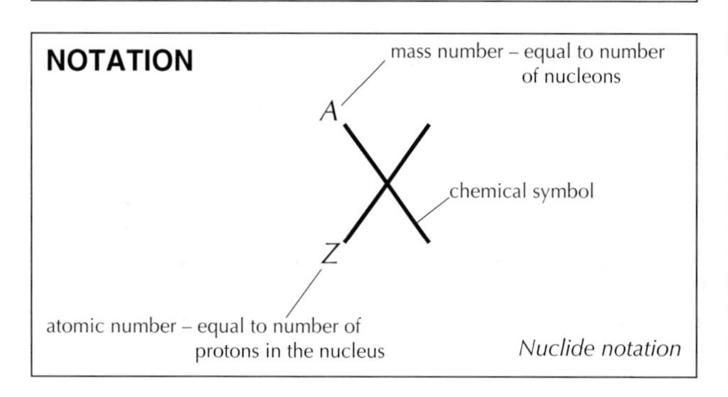

Nuclide notation

EXAMPLES

	Notation	Description	Comment
1	${}^{12}_{6}C$	carbon-12	isotope of **2**
2	${}^{13}_{6}C$	carbon-13	isotope of **1**
3	${}^{238}_{92}U$	uranium-238	
4	${}^{198}_{78}Pt$	platinum-198	same mass number as **5**
5	${}^{198}_{80}Hg$	mercury-198	same mass number as **4**

Each element has a unique chemical symbol and its own atomic number. *No.1* and *No.2* are examples of two isotopes, whereas *No.4* and *No.5* are not.

In general, when physicists use this notation they are concerned with the nucleus rather than the whole atom. Chemists use the same notation but tend to include the overall charge on the atom. Thus ${}^{12}_{6}C$ can represent the carbon nucleus to a physicist or the carbon atom to a chemist depending on the context. If the charge is present the situation becomes unambiguous. ${}^{35}_{17}Cl^-$ must refer to a chlorine ion – an atom that has gained one extra electron.

STRONG NUCLEAR FORCE

The protons in a nucleus are all positive. Since like charges repel, they must be repelling one another all the time. This means there must be another force keeping the nucleus together. Without it the nucleus would "fly apart". We know a few things about this force.

- It must be strong. If the proton repulsions are calculated it is clear that the gravitational attraction between the nucleons is far too small to be able to keep the nucleus together.
- It must be very short-ranged as we do not observe this force anywhere other than inside the nucleus.
- It is likely to involve the neutrons as well. Small nuclei tend to have equal numbers of protons and neutrons. Large nuclei need proportionately more neutrons in order to keep the nucleus together.

The name given to this force is the **strong nuclear force**.

NUCLEAR STABILITY

Many atomic nuclei are unstable. The stability of a particular nuclide depends greatly on the numbers of neutrons present. The graph below shows the stable nuclides that exist.

- For small nuclei, the number of neutrons tends to equal the number of protons.
- For large nuclei there are more neutrons than protons.
- Nuclides above the band of stability have "too many neutrons" and will tend to decay with either alpha or beta decay (see page 61).
- Nuclides below the band of stability have "too few neutrons" and will tend to emit positrons (see page 64).

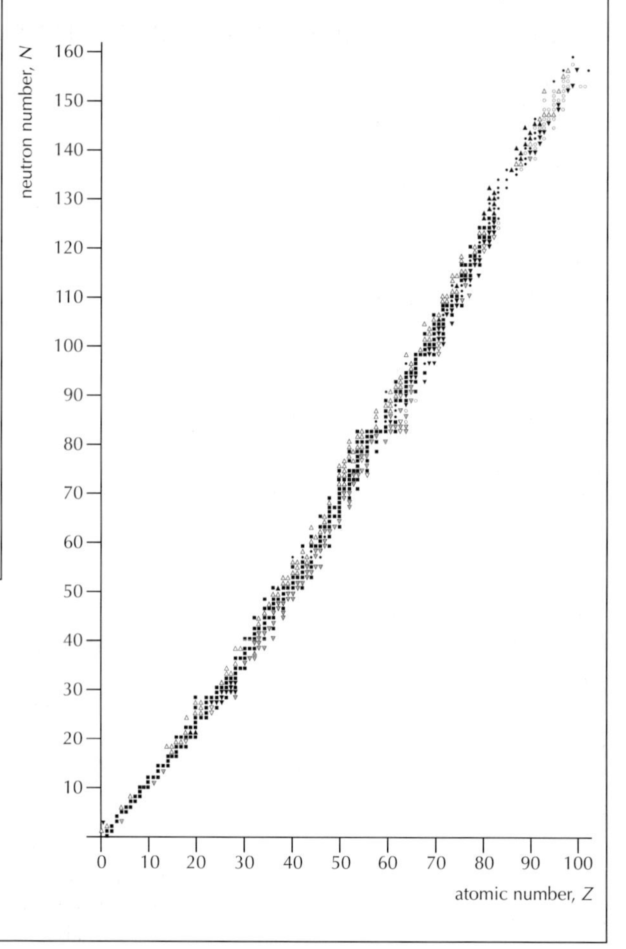

Key

N number of neutrons

Z number of protons

- ■ naturally occurring stable nuclide
- • naturally occurring α-emitting nuclide
- ○ artificially produced α-emitting nuclide
- ▲ naturally occurring β^--emitting nuclide
- △ artificially produced β^+-emitting nuclide
- ▽ artificially produced β^--emitting nuclide
- ▽ artificially produced electron-capturing nuclide
- ▼ artificial nuclide decaying by spontaneous fissio

Radioactivity

IONISING PROPERTIES

Many atomic nuclei are unstable. The process by which they decay is called **radioactive decay**. Every decay involves the emission of one of three different possible radiations from the nucleus: alpha (α), beta (β) or gamma (γ) (see also page 64).

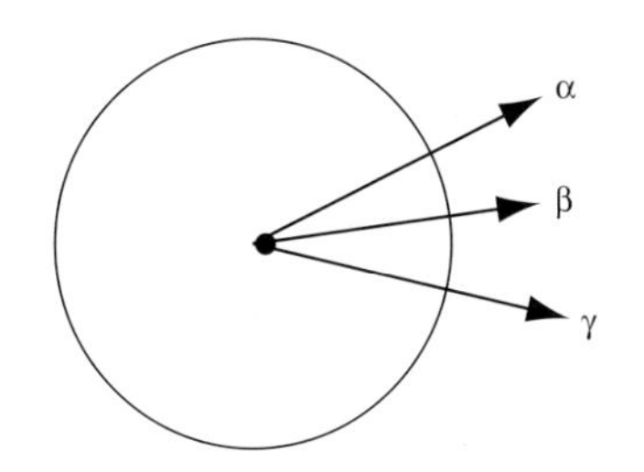

Alpha, beta and gamma all come from the nucleus

All three radiations are ionising. This means that as they go through a substance, collisions occur which cause electrons to be removed from atoms. Atoms that have lost or gained electrons are called ions. This ionising property allows the radiations to be detected. It also explains their dangerous nature. When ionisations occur in biologically important molecules, such as DNA, mutations can occur.

EFFECTS OF RADIATION

At the molecular level, an ionisation could cause damage directly to a biologically important molecule such as DNA or RNA. This could cause it to cease functioning. Alternatively, an ionisation in the surrounding medium is enough to interfere with the complex chemical reactions (called **metabolic pathways**) taking place.

Molecular damage can result in a disruption to the functions that are taking place within the cells that make up the organism. As well as potentially causing the cell to die, this could just prevent cells from dividing and multiplying. On top of this, it could be the cause of the transformation of the cell into a malignant form.

As all body tissues are built up of cells, damage to these can result in damage to the body systems that have been affected. The non-functioning of these systems can result in death for the animal. If malignant cells continue to grow then this is called **cancer**.

RADIATION SAFETY

There is no such thing as a safe dose of ionising radiation. Any hospital procedures that result in a patient receiving an extra dose (for example having an X-ray scan) should be justifiable in terms of the information received or the benefit it gives.

There are three main ways of protecting oneself from too large a dose. These can be summarised as follows:

- **Run away!**
  The simplest method of reducing the dose received is to increase the distance between you and the source. Only electromagnetic radiation can travel large distances and this follows an inverse square relationship with distance.
- **Don't waste time!**
  If you have to receive a dose, then it is important to keep the time of this exposure to a minimum.
- **If you can't run away, hide behind something!**
  Shielding can always be used to reduce the dose received. Lead-lined aprons can also be used to limit the exposure for both patient and operator.

PROPERTIES OF ALPHA, BETA AND GAMMA RADIATIONS

Property	Alpha, α	Beta, β	Gamma, γ
Effect on photographic film	Yes	Yes	Yes
Approximate number of ion pairs produced in air	10^4 per mm travelled	10^2 per mm travelled	1 per mm travelled
Typical material needed to absorb it	10^{-2} mm aluminium; piece of paper	A few mm aluminium	10 cm lead
Penetration ability	Low	Medium	High
Typical path length in air	A few cm	Less than one m	Effectively infinite
Deflection by E and B fields	Behaves like a positive charge	Behaves like a negative charge	Not deflected
Speed	About 10^7 m s^{-1}	About 10^8 m s^{-1}, very variable	3×10^8 m s^{-1}

NATURE OF ALPHA, BETA AND GAMMA DECAY

When a nucleus decays the mass numbers and the atomic numbers must balance on each side of the nuclear equation.

- Alpha particles are helium nuclei, ${}^4_2\alpha$ or ${}^4_2He^{2+}$. In alpha decay, a "chunk" of the nucleus is emitted. The portion that remains will be a different nuclide.

$${}^A_ZX \rightarrow {}^{(A-4)}_{(Z-2)}Y + {}^4_2\alpha$$

e.g. $${}^{241}_{95}Am \rightarrow {}^{237}_{93}Np + {}^4_2\alpha$$

The atomic numbers and the mass numbers balance on each side of the equation.

(95 = 93 + 2 and 241 = 237 +4)

- Beta particles are electrons, ${}^{\;0}_{-1}\beta$ or ${}^{\;0}_{-1}e^-$, emitted **from the nucleus**. The explanation is that the electron is formed when a neutron decays. At the same time, another particle is emitted called an antineutrino.

$${}^1_0n \rightarrow {}^1_1p + {}^{\;0}_{-1}\beta + \bar{\upsilon}$$

Since an antineutrino has no charge and virtually no mass it does not affect the equation and so is sometimes ignored. See page 110 for more details.

$${}^A_ZX \rightarrow {}^{\;\;\;A}_{(Z+1)}Y + {}^{\;0}_{-1}\beta + \bar{\upsilon}$$

e.g. $${}^{90}_{38}Sr \rightarrow {}^{90}_{39}Y + {}^{\;0}_{-1}\beta + \bar{\upsilon}$$

- Gamma rays are unlike the other two radiations in that they are part of the electromagnetic spectrum. After their emission, the nucleus has less energy but its mass number and its atomic number have not changed. It is said to have changed from an **excited state** to a lower energy state.

$${}^A_ZX^* \rightarrow {}^A_ZX + {}^0_0\gamma$$

Excited state — Lower energy state

Half-life

RANDOM DECAY

Radioactive decay is a **random** process and is not affected by external conditions. For example, increasing the temperature of a sample of radioactive material does not affect the rate of decay. This means that is there no way of knowing whether or not a particular nucleus is going to decay within a certain period of time. All we know is the *chances* of a decay happening in that time.

Although the process is random, the large numbers of atoms involved allows us to make some accurate predictions. If we start with a given number of atoms then we can expect a certain number to decay in the next minute. If there were more atoms in the sample, we would expect the number decaying to be larger. On average the rate of decay of a sample is proportional to the number of atoms in the sample. This proportionality means that radioactive decay is an **exponential** process. The number of atoms of a certain element, N, decreases exponentially over time. Mathematically this is expressed as:

$$\frac{dN}{dt} \propto -N$$

HALF-LIFE

There is a temptation to think that every quantity that decreases with time is an exponential decrease, but exponential curves have a particular mathematical property. In the graph shown below, the time taken for half the number of nuclides to decay is always the same, whatever starting value we choose. This allows us to express the chances of decay happening in a property called the **half-life,** $T_{1/2}$. The half-life of a nuclide is the time taken for half the number of nuclides present in a sample to decay. An equivalent statement is that the half-life is the time taken for the rate of decay of a particular sample of nuclides to halve. A substance with a large half-life takes a long time to decay. A substance with a short half-life will decay quickly. Half-lives can vary from fractions of a second to millions of years.

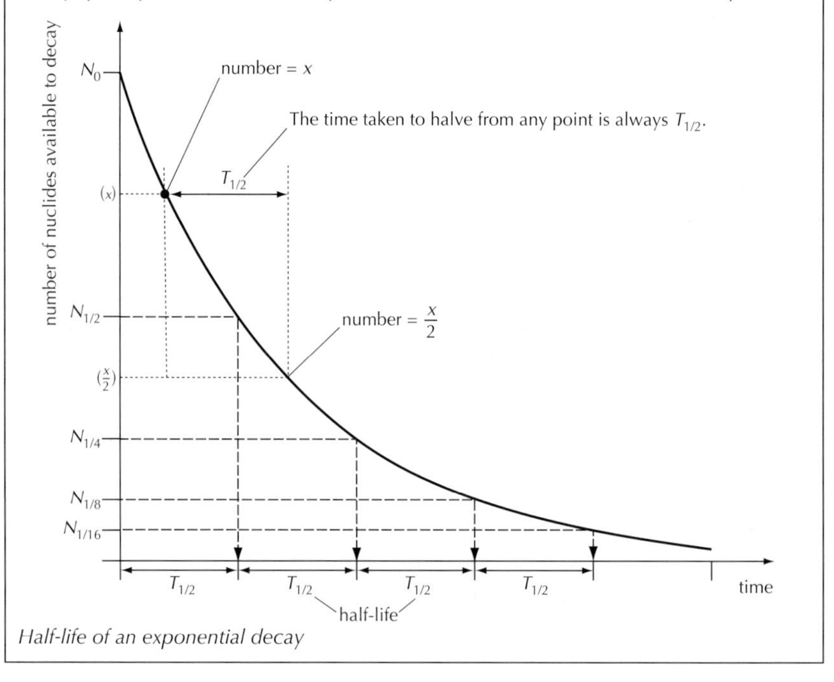

Half-life of an exponential decay

EXAMPLE

In simple situations, working out how much radioactive material remains is a matter of applying the half-life property several times. A common mistake is to think that if the half-life of a radioactive material is 3 days then it will all decay in six days. In reality, after six days (two half-lives) a "half of a half" will remain i.e. a quarter.

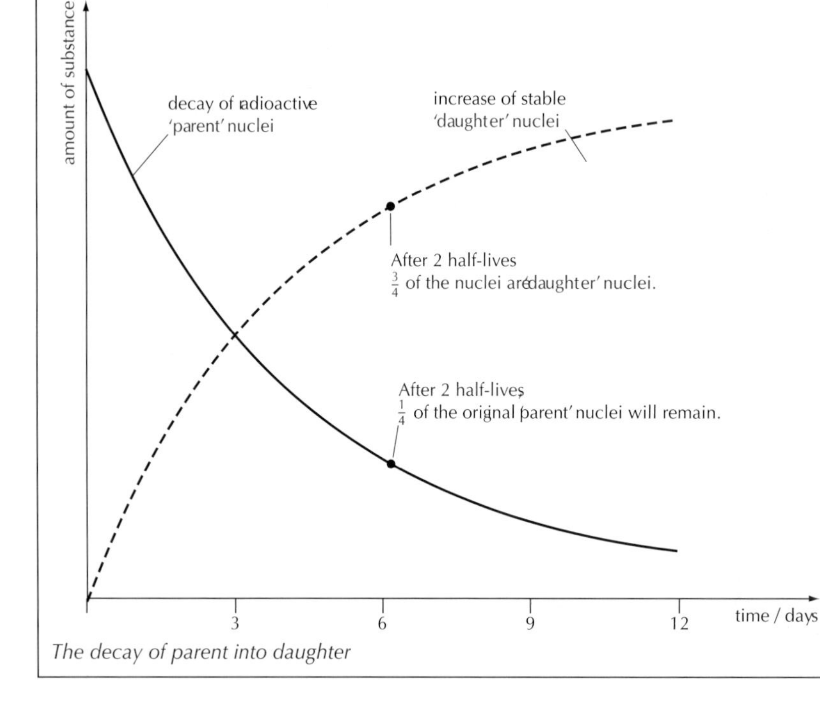

The decay of parent into daughter

e.g. The half-life of $^{14}_{6}C$ is 5570 years. Approximately how long is needed before less than 1% of a sample of $^{14}_{6}C$ remains?

Time	Fraction left
$T_{1/2}$	50%
$2T_{1/2}$	25%
$3T_{1/2}$	12.5%
$4T_{1/2}$	~ 6.3%
$5T_{1/2}$	~ 3.1%
$6T_{1/2}$	~ 1.6%
$7T_{1/2}$	~ 0.8%

6 half lives = 33 420 years

7 half lives = 38 990 years

∴ approximately 37 000 years needed

Nuclear reactions

ARTIFICIAL TRANSMUTATIONS

There is nothing that we can do to change the likelihood of a certain radioactive decay happening, but under certain conditions we can make nuclear reactions happen. This can be done by bombarding a nucleus with a nucleon, an alpha particle or another small nucleus. Such reactions are called **artificial transmutations**. In general, the target nucleus first "captures" the incoming object and then an emission takes place. The first ever artificial transmutation was carried out by Rutherford in 1919. Nitrogen was bombarded by alpha particles and the presence of oxygen was detected spectroscopically.

$${}^{4}_{2}\mathrm{He}^{2+} + {}^{14}_{7}\mathrm{N} \rightarrow {}^{17}_{8}\mathrm{O} + {}^{1}_{1}\mathrm{p}$$

The mass numbers (4 + 14 = 17 +1) and the atomic numbers (2 + 7 = 8 +1) on both sides of the equation must balance.

UNIFIED MASS UNITS

The individual masses involved in nuclear reactions are tiny. In order to compare atomic masses physicists often use unified mass units, u. These are defined in terms of the most common isotope of carbon, carbon-12. There are 12 nucleons in the carbon-12 atom (6 protons and 6 neutons) and one unified mass unit is defined as exactly one twelfth the mass of a carbon-12 atom. Essentially, the mass of a proton and the mass of a neutron are both 1 u as shown in the table below.

$$1\ \mathrm{u} = \frac{1}{12}\ \text{mass of a (carbon-12) atom} = 1.66\times10^{-27}\ \mathrm{kg}$$

mass* of 1 proton = 1.007 276 u

mass* of 1 neutron = 1.008 665 u

mass* of 1 electron = 0.000 549 u

** = Technically these are all "rest masses" – see Relativity option*

MASS DEFECT AND BINDING ENERGY

The table above shows the masses of neutrons and protons. It should be obvious that if we add together the masses of 6 protons, 6 neutrons and 6 electrons we will get a number bigger than 12 u, the mass of a carbon-12 atom. What has gone wrong? The answer becomes clear when we investigate what keeps the nucleus bound together.

The difference between the mass of a nucleus and the masses of its component nucleons is called the **mass defect**. If one imagined assembling a nucleus, the protons and neutrons would initially need to be brought together. Doing this takes work because the protons repel one another. Creating the bonds between the protons and neutrons releases a greater amount of energy than the work done in bringing them together. This energy released must come from somewhere. The answer lies in Einstein's famous mass–energy equivalence relationship.

$$E = mc^2$$

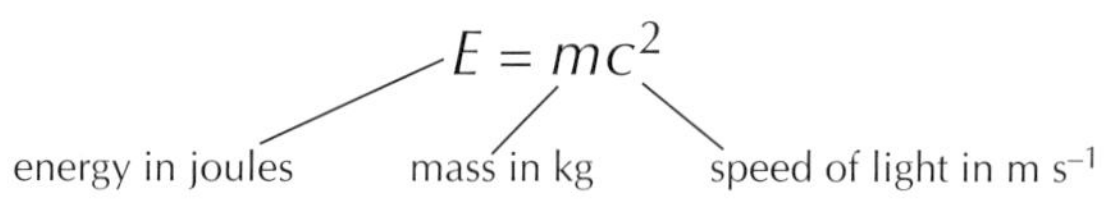

In Einstein's equation, mass is another form of energy and it is possible to convert mass directly into energy and vice versa. The **binding energy** is the amount of energy that is released when a nucleus is assembled from its component nucleons. It comes from a decrease in mass. The binding energy would also be the energy that needs to be added in order to separate a nucleus into its individual nucleons. The mass defect is thus a measure of the binding energy.

UNITS

Using Einstein's equation, 1 kg of mass is equivalent to 9×10^{16} J of energy. This is a huge amount of energy. At the atomic scale other units of energy tend to be more useful. The electronvolt (see topic 5), or more usually, the Megaelectronvolt are often used.

$1\ \mathrm{eV} = 1.6\times10^{-19}\ \mathrm{J}$

$1\ \mathrm{MeV} = 1.6\times10^{-13}\ \mathrm{J}$

1 u of mass converts into 931.5 MeV

Since mass and energy are equivalent it is sometimes useful to work in units that avoid having to do repeated multiplications by the (speed of light)^2. A new possible unit for mass is thus MeV c^{-2}. It works like this:

If 1 MeV c^{-2} worth of mass is converted you get 1 MeV worth of energy.

WORKED EXAMPLES

Question:
How much energy would be released if 14 g of carbon-14 decayed as shown in the equation below?

$${}^{14}_{6}\mathrm{C} \rightarrow {}^{14}_{7}\mathrm{N} + {}^{0}_{-1}\beta$$

Answer:
Information given
atomic mass of carbon-14 = 14.003242 u;
atomic mass of nitrogen-14 = 14.003074 u;
mass of electron = 0.000549 u

mass of left-hand side = nuclear mass of ${}^{14}_{6}\mathrm{C}$
= 14.003242 – 6(0.000549) u
= 13.999948 u

nuclear mass of ${}^{14}_{7}\mathrm{N}$ = 14.003074 – 7(0.000549) u
= 13.999231 u

mass of right-hand side = 13.999231 + 0.000549 u
= 13.999780 u

mass difference = LHS – RHS
= 0.000168 u

energy released per decay = 0.000168 × 931.5 MeV
= 0.156492 MeV

14g of C-14 is 1 mol

∴ Total number of decays = $N_A = 6.022 \times 10^{23}$

∴ Total energy release = $6.022 \times 10^{23} \times 0.156492$ MeV
= 9.424×10^{22} MeV
= 15142 J
≈15 kJ

Fission, fusion and antimatter

FISSION

Fission is the name given to the nuclear reaction whereby large nuclei are induced to break up into smaller nuclei and release energy in the process. It is the reaction that is used in nuclear reactors and atomic bombs. A typical single reaction might involve bombarding a uranium nucleus with a neutron. This can cause the uranium nucleus to break up into two smaller nuclei. A typical reaction might be:

$$ {}^{1}_{0}n + {}^{235}_{92}U \rightarrow {}^{141}_{56}Ba + {}^{92}_{36}Kr + 3\,{}^{1}_{0}n + \text{energy} $$

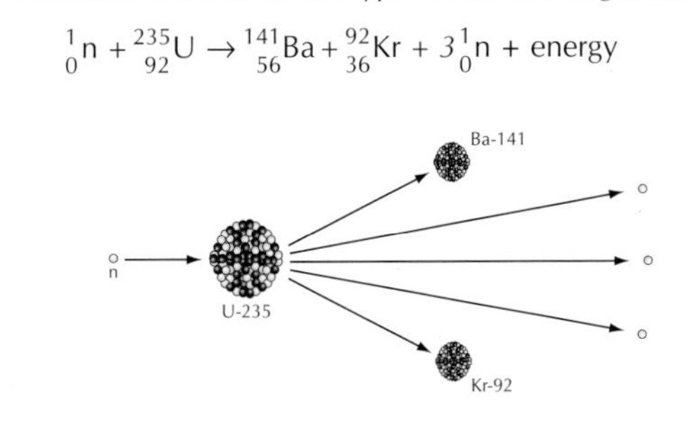

A fission reaction

Since the one original neutron causing the reaction has resulted in the production of three neutrons, there is the possibility of a **chain reaction** occurring. It is technically quite difficult to get the neutrons to lose enough energy to go on and initiate further reactions, but it is achievable.

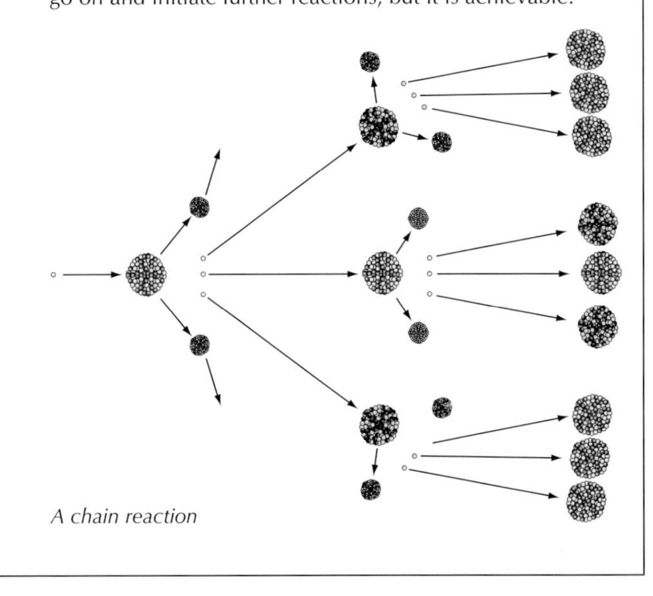

A chain reaction

FUSION

Fusion is the name given to the nuclear reaction whereby small nuclei are induced to join together into larger nuclei and release energy in the process. It is the reaction that "fuels" all stars including the Sun. A typical reaction that is taking place in the Sun is the fusion of two different isotopes of hydrogen to produce helium.

$$ {}^{2}_{1}H + {}^{3}_{1}H \rightarrow {}^{4}_{2}He + {}^{1}_{0}n + \text{energy} $$

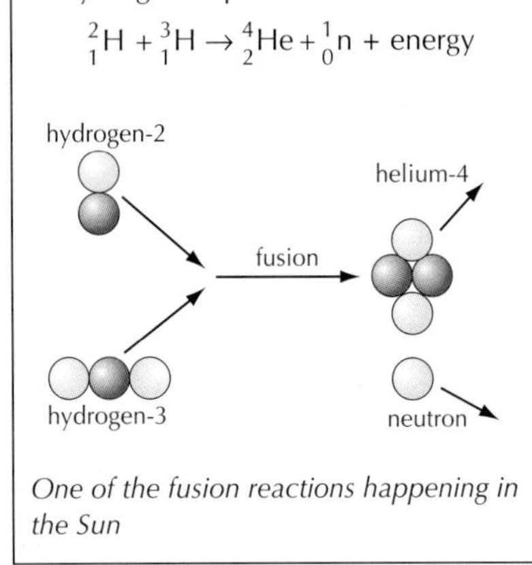

One of the fusion reactions happening in the Sun

BINDING ENERGY PER NUCLEON

Whenever a nuclear reaction (fission or fusion) releases energy, the products of the reaction are in a lower energy state than the reactants. Mass loss must be the source of this energy. In order to compare the energy states of different nuclei, physicists calculate the binding energy per nucleon. This is the total binding energy for the nucleus divided by the total number of nucleons. The nucleus with the largest binding energy per nucleon is iron-56, ${}^{56}_{26}Fe$.

A reaction is energetically feasible if the products of the reaction have a greater binding energy per nucleon when compared with the reactants.

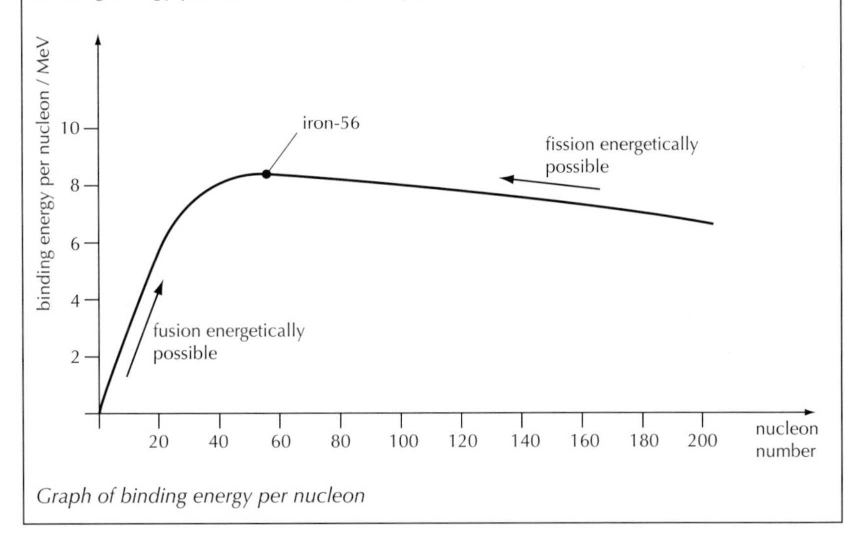

Graph of binding energy per nucleon

ANTIMATTER

The nuclear model given in the previous pages is somewhat simplified. One important thing that has not been mentioned is the existence of antimatter. Every form of matter has its equivalent form of antimatter. If matter and antimatter came together they would annihilate each other. Not surprisingly, antimatter is rare but it does exist. For example, another form of radioactive decay that can take place is beta plus or positron decay. In this decay a proton decays into a neutron, and the antimatter version of an electron, a positron, is emitted.

$$ {}^{1}_{1}p \rightarrow {}^{1}_{0}n + {}^{0}_{+1}\beta^{+} + \upsilon $$

$$ {}^{19}_{10}Ne \rightarrow {}^{19}_{9}F + {}^{0}_{+1}\beta^{+} + \upsilon $$

The positron, β^{+}, emission is accompanied by a neutrino.

The antineutrino is the antimatter form of the neutrino.

For more details see the Higher level material in Chapter 13.

IB QUESTIONS – ATOMIC AND NUCLEAR PHYSICS

1 A sample of radioactive material contains the element Ra 226. The half-life of Ra 226 can be defined as the time it takes for

A the mass of the sample to fall to half its original value.

B half the number of atoms of Ra 226 in the sample to decay.

C half the number of atoms in the sample to decay.

D the volume of the sample to fall to half its original value.

2 Oxygen-15 decays to nitrogen-15 with a half-life of approximately 2 minutes. A pure sample of oxygen-15, with a mass of 100 g, is placed in an airtight container. After 4 minutes, the masses of oxygen and nitrogen in the container will be

	Mass of oxygen	Mass of nitrogen
A	0 g	100 g
B	25 g	25 g
C	50 g	50 g
D	25 g	75 g

3 A radioactive nuclide ${}_{Z}X$ undergoes a sequence of radioactive decays to form a new nuclide ${}_{Z+2}Y$. The sequence of emitted radiations could be

A β, β **B** α, β, β **C** α, α **D** α, β, γ

4 In the Rutherford scattering experiment, a stream of α particles is fired at a thin gold foil. Most of the α particles

A are scattered randomly.

B rebound.

C are scattered uniformly.

D go through the foil.

5 A piece of radioactive material now has about 1/16 of its previous activity. If the half-life is 4 hours the difference in time between measurements is approximately

A 8 hours. **B** 16 hours. **C** 32 hours. **D** 60 hours.

6 The nuclide ${}^{14}_{6}C$ undergoes radioactive beta-minus decay. The resulting daughter nuclide is

A ${}^{14}_{6}C$ **B** ${}^{10}_{6}Be$ **C** ${}^{14}_{7}N$ **D** ${}^{14}_{5}B$

7 Although there are protons in close proximity to each other in a nucleus of an atom, the nucleus does not blow apart by electrostatic Coulomb repulsion because

A there are an equal number of electrons in the nucleus which neutralise the protons.

B the Coulomb force does not operate in a nucleus.

C the neutrons in the nucleus shield the protons from each other.

D there is a strong nuclear force which counteracts the repulsive Coulomb force.

8 **(a)** Two properties of the isotope of uranium, ${}^{238}_{92}U$ are:

(i) it decays radioactively (to ${}^{234}_{90}Th$)

(ii) it reacts chemically (e.g. with fluorine to form UF_6).

What features of the structure of uranium atoms are responsible for these two widely different properties? [2]

(b) A beam of deuterons (deuterium nuclei, ${}^{2}_{1}H$) are accelerated through a potential difference and are then incident on a magnesium target (${}^{26}_{12}Mg$). A nuclear reaction occurs resulting in the production of a sodium nucleus and an alpha particle.

(i) Write a balanced nuclear equation for this reaction. [2]

(ii) Explain why it is necessary to give the deuterons a certain minimum kinetic energy before they can react with the magnesium nuclei. [2]

9 *Radioactive carbon dating*

The carbon in trees is mostly carbon-12, which is stable, but there is also a small proportion of carbon-14, which is radioactive. When a tree is cut down, the carbon-14 present in the wood at that time decays with a half-life of 5800 years.

(a) Carbon-14 decays by beta-minus emission to nitrogen-14. Write the equation for this decay. [2]

(b) For an old wooden bowl from an archaeological site, the average count-rate of beta particles detected per kg of carbon is 13 counts per minute. The corresponding count rate from newly cut wood is 52 counts per minute.

(i) Explain why the beta activity from the bowl diminishes with time, even though the probability of decay of any individual carbon-14 nucleus is constant. [3]

(ii) Calculate the approximate age of the wooden bowl. [3]

10 This question is about a nuclear fission reactor for providing electrical power.

In a nuclear reactor, power is to be generated by the fission of uranium-235. The absorption of a neutron by ${}^{235}U$ results in the splitting of the nucleus into two smaller nuclei plus a number of neutrons and the release of energy. The splitting can occur in many ways; for example

$$n + {}^{235}_{92}U \rightarrow {}^{90}_{38}Sr + {}^{143}_{54}Xe + \text{neutrons} + \text{energy}$$

(a) *The nuclear fission reaction*

(i) How many neutrons are produced in this reaction? [1]

(ii) Explain why the release of several neutrons in each reaction is crucial for the operation of a fission reactor. [2]

(iii) The sum of the rest masses of the uranium plus neutron before the reaction is 0.22 u greater than the sum of the rest masses of the fission products. What becomes of this 'missing mass'? [1]

(iv) Show that the energy released in the above fission reaction is about 200 MeV. [2]

(b) *A nuclear fission power station*

(i) Suppose a nuclear fission power station generates electrical power at 550 MW. Estimate the minimum number of fission reactions occurring each second in the reactor, stating any assumption you have made about efficiency. [4]

Energy degradation and power generation

ENERGY CONVERSIONS

The production of electrical power around the world is achieved using a variety of different systems, often starting with the release of thermal energy from a fuel. In principle, thermal energy can be completely converted to work in a single process, but the continuous conversion of this energy into work implies the use of machines that are continuously repeating their actions in a fixed cycle. Any cyclical process must involve the transfer of some energy from the system to the surroundings that is no longer available to perform useful work. This unavailable energy is known as **degraded energy**, in accordance with the principle of the second law of thermodynamics (see page 89).

Energy conversions are represented using **Sankey diagrams**. An arrow (drawn from left to right) represents the energy changes taking place. The width of the arrow represents the power or energy involved at a given stage. Degraded energy is shown with an arrow up or down.

ELECTRICAL POWER PRODUCTION

In all electrical power stations the process is essentially the same. A fuel is used to release thermal energy. This thermal energy is used to boil water to make steam. The steam is used to turn turbines and the motion of the turbines is used to generate electrical energy. Transformers alter the potential difference (see page 101).

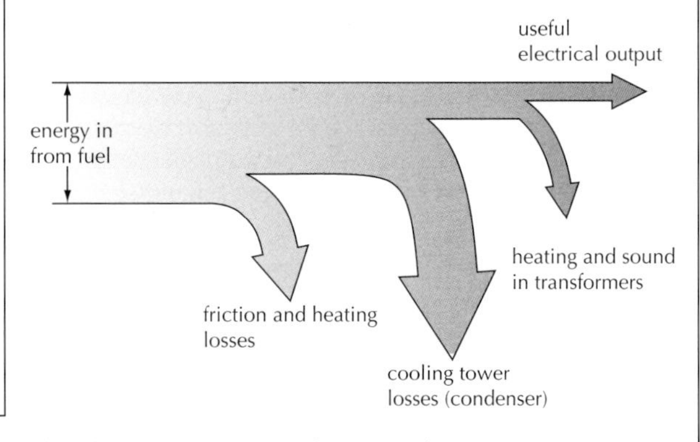

Sankey diagram representing the energy flow in a typical power station

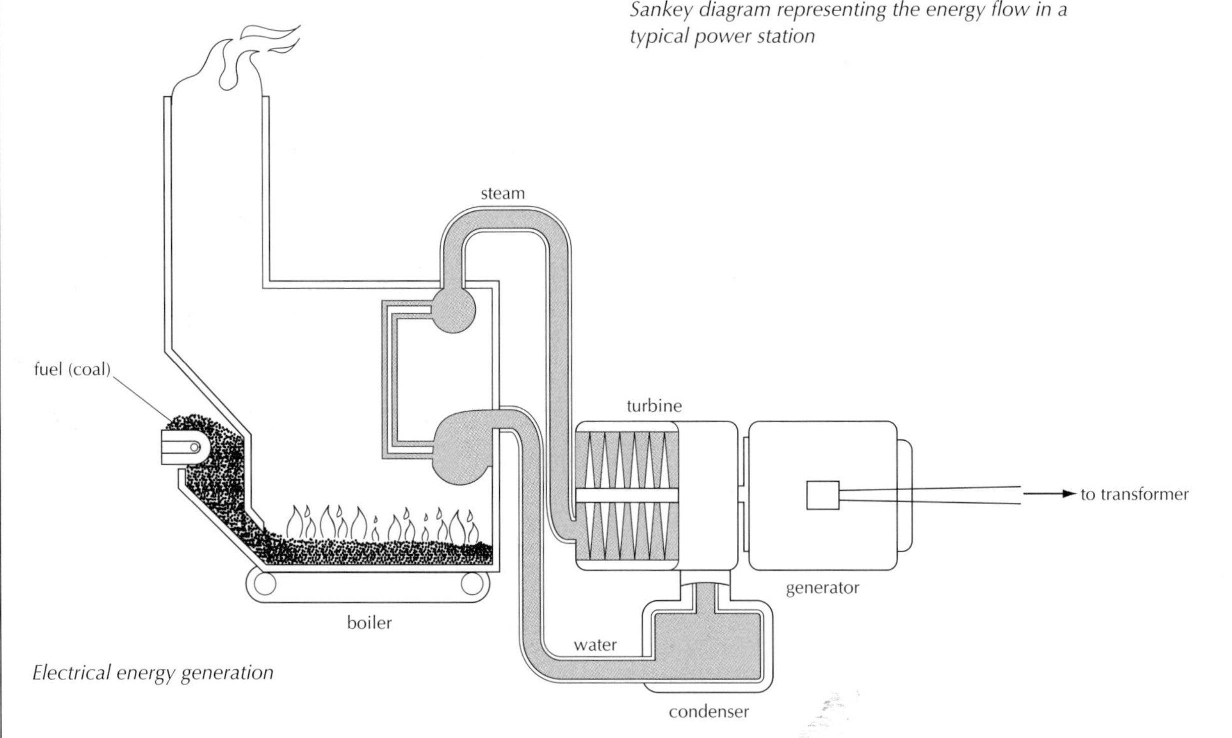

Electrical energy generation

World energy sources

RENEWABLE / NON-RENEWABLE ENERGY SOURCES

The law of conservation of energy states that energy is neither created or destroyed, it just changes form. As far as human societies are concerned, if we wish to use devices that require the input of energy, we need to identify sources of energy. **Renewable** sources of energy are those that cannot be used up, whereas **non-renewable** sources of energy can be used up and eventually run out.

Renewable sources	Non-renewable sources
hydroelectric	coal
photovoltaic cells	oil
active solar heaters	natural gas
wind	nuclear
biofuels	

Sometimes the sources are hard to classify so care needs to be taken when deciding whether a source is renewable or not. One point that sometimes worries students is that the Sun will eventually run out as a source of energy for the Earth, so no source is perfectly renewable! This is true, but all of these sources are considered from the point of view of life on Earth. When the Sun runs out, then so will life on Earth. Other things to keep in mind include:

- Nuclear sources (both fission and fusion) consume a material as their source so they must be non-renewable. On the other hand, the supply available can make the source **effectively** renewable.
- It is possible for a fuel to be managed in a renewable or a non-renewable way. For example, if trees are cut down as a source of wood to burn then this is clearly non-renewable. It is, however, possible to replant trees at the same rate as they are cut down. If this is properly managed, it could be a renewable source of energy.

Of course these possible sources must have got their energy from somewhere in the first place. Most of the energy used by humans can be traced back to energy radiated from the Sun, but not quite all of it. Possible sources are:

- the Sun's radiated energy
- gravitational energy of the Sun and the Moon
- nuclear energy stored within atoms
- the Earth's internal heat energy

Although you might think that there are other sources of energy, the above list is complete. Many everyday sources of energy (such as coal or oil) can be shown to have derived their energy from the Sun's radiated energy. On the industrial scale, electrical energy needs to be generated from another source. When you plug anything electrical into the mains electricity you have to pay the electricity-generating company for the energy you use. In order to provide you with this energy, the company must be using one (or more) of the original list of sources.

Energy density

Energy density provides a useful comparison between fuels and is defined as the energy liberated per unit mass of fuel consumed. Energy density is measured in J kg^{-1}.

$$\text{energy density} = \frac{\text{energy release from fuel}}{\text{mass of fuel consumed}}$$

Fuel choice can be particularly influenced by energy density when the fuel needs to be transported: the greater the mass of fuel that needs to be transported, the greater the cost.

COMPARISON OF WORLD ENERGY SOURCES

Fuel	Renewable?	CO_2 emission	Energy density (MJ kg^{-1}) (values vary depending on type)
Coal	No	Yes	22–33
Oil	No	Yes	42
Gas	No	Yes	54
Nuclear (Uranium)	No	No	90 000 000
Waste	No	Yes	10
Solar	Yes	No	n/a
Wind	Yes	No	n/a
Hydro - water stored in dams	Yes	No	n/a
Tidal	Yes	No	n/a
Pumped storage	n/a	No	n/a
Wave	Yes	No	n/a
Geothermal	Yes	No	n/a
Biofuels e.g. ethanol	Some types	Yes	30

RELATIVE PROPORTIONS OF GLOBAL USE OF DIFFERENT ENERGY SOURCES

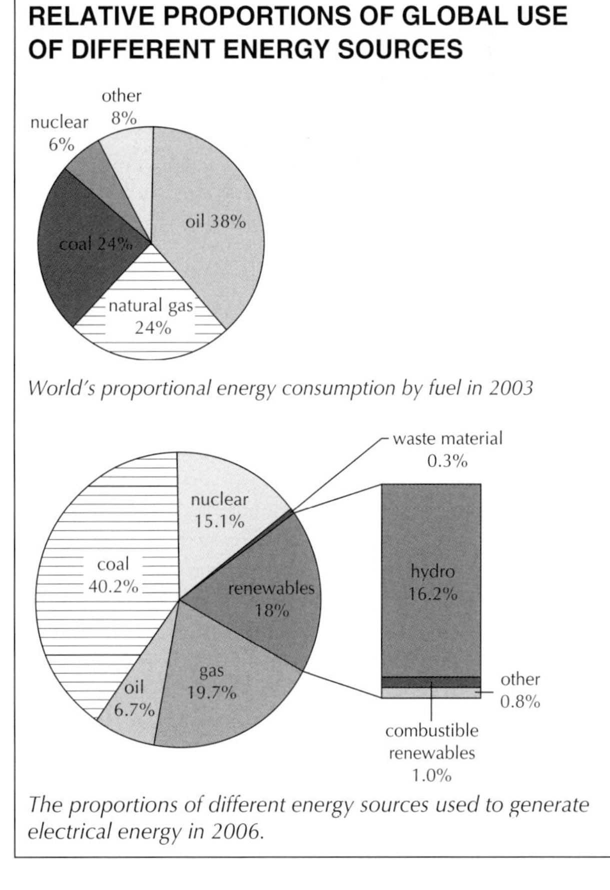

World's proportional energy consumption by fuel in 2003

The proportions of different energy sources used to generate electrical energy in 2006.

Fossil fuel power production

ORIGIN OF FOSSIL FUEL

Coal, oil and natural gas are known as **fossil fuels**. These fuels have been produced over a time scale that involves tens or hundreds of millions of years from accumulations of dead matter. This matter has been converted into fossil fuels by exposure to the very high temperatures and pressure that exist beneath the Earth's surface.

Coal is formed from the dead plant matter that used to grow in swamps. Layer upon layer of decaying matter decomposed. As it was buried by more plant matter and other substances, the material became more compressed. Over the geological time scale this turned into coal.

Oil is formed in a similar manner from the remains of microscopic marine life. The compression took place under the sea. Natural gas, as well as occurring in underground pockets, can be obtained as a by-product during the production of oil. It is also possible to manufacture gas from coal.

ENERGY TRANSFORMATIONS

Fossil fuel power stations release energy in fuel by burning it. The thermal energy is then used to convert water into steam that once again can be used to turn turbines. Since all fossil fuels were originally living matter, the original source of this energy was the Sun. For example, millions of years ago energy radiated from the Sun was converted (by photosynthesis) into living plant matter. Some of this matter has eventually been converted into coal.

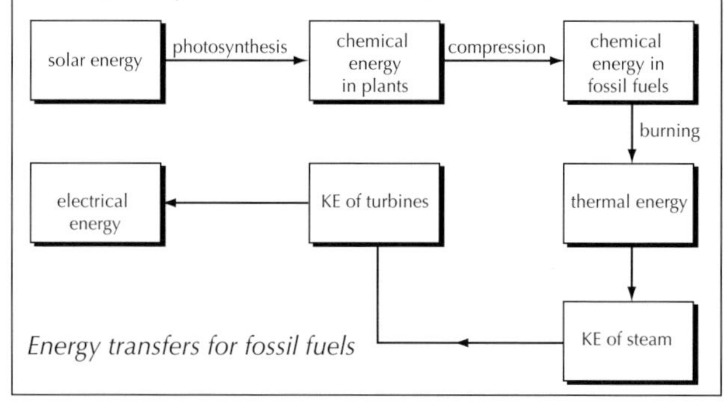

Energy transfers for fossil fuels

HISTORICAL AND GEOGRAPHICAL FACTORS

The **industrial revolution** took place in Western Europe in the late 18th and early 19th centuries in which large-scale manufacturing industries were developed and the factory as a place of work was introduced. Machines were designed and built which replaced traditional manual labour. The industrial growth of particular towns and regions in the United Kingdom started a process that spread throughout the world.

Three particular inventions were particularly important.

- Machines that allowed materials for textiles to be manufactured (e.g. the *Spinning Jenny* or *Arkwright's frame*). The establishment of cotton mills, for example, in Manchester (United Kingdom) is seen as a key event at the beginning of the industrial revolution.
- The design and improvement of the steam engine (for example by Thomas Newcomen and James Watt) meant that factories could be built that did not rely on the flow of water to provide their energy requirements.
- The development of the iron smelting industries meant that iron and steel could be produced more cheaply and allow for the growth industries based around the manufacture of iron and steel.

As the industrial revolution spread, the rate of energy usage greatly increased and industry tended to develop near to existing deposits of fossil fuels. Once factories were established, people seeking work would tend to migrate towards the cities. In addition, infrastructure was created to allow coal and other fossil fuels to be easily transported as the higher rates of energy usage demanded the use of fuels with a high energy density. This encouraged the growth of industries located near the raw materials.

EXAMPLE

Use the data on this page and the previous page to calculate the typical rate (in tonnes per hour) at which coal must be supplied to a 500 MW coal fired power station

Answer

Electrical power supply $= 500 \text{ MW} = 5 \times 10^8 \text{ J s}^{-1}$

Power released from fuel $= 5 \times 10^8 / \text{efficiency}$
$= 5 \times 10^8 / 0.35$
$= 1.43 \times 10^9 \text{ J s}^{-1}$

Rate of consumption of coal $= 1.43 \times 10^9 / 3.3 \times 10^7 \text{ kg s}^{-1}$
$= 43.3 \text{ kg s}^{-1}$
$= 43.3 \times 60 \times 60 \text{ kg hr}^{-1}$
$= 1.56 \times 10^5 \text{ kg hr}^{-1}$
$\approx 160 \text{ tonnes hr}^{-1}$

EFFICIENCY OF FOSSIL FUEL POWER STATIONS

The efficiency of different power stations depends on the design. At the time of publishing, the following figures apply.

Fossil fuel	Typical efficiency	Current maximum efficiency
Coal	35%	42%
Natural gas	45%	52%
Oil	38%	45%

Note that thermodynamic considerations limit the maximum achievable efficiency (see page 90).

ADVANTAGES AND DISADVANTAGES

Advantages

- Very high 'energy density' – a great deal of energy is released from a small mass of fossil fuel
- Fossil fuels are relatively easy to transport
- Still cheap when compared to other sources of energy
- Can be built anywhere with good transport links and water availability
- Can be used directly in the home to provide heating

Disadvantages

- Combustion products can produce pollution, notably acid rain
- Combustion products contain 'greenhouse' gases
- Extraction of fossil fuels can damage the environment
- Nonrenewable
- Coal-fired power stations need large amounts of fuel.

Nuclear power (1)

PRINCIPLES OF ENERGY PRODUCTION

Many nuclear power stations use uranium–235 as the 'fuel'. This fuel is not burned – the release of energy is achieved using a fission reaction. An overview of this process is described on page 64. In each individual reaction, an incoming neutron causes a uranium nucleus to split apart. The fragments are moving fast. In other words the temperature is very high. Among the fragments are more neutrons. If these neutrons go on to initiate further reactions then a chain reaction is created.

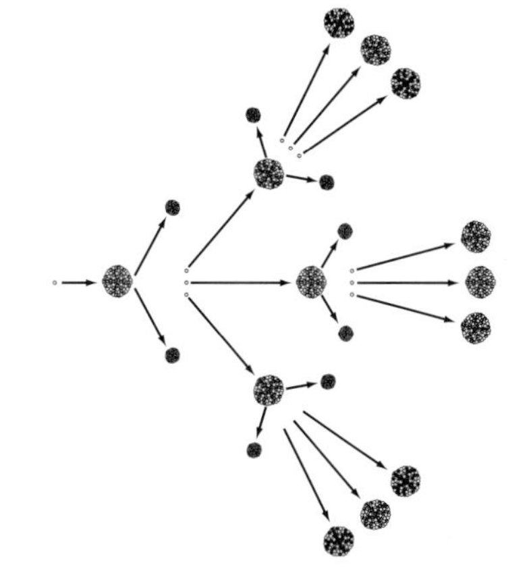

The design of a nuclear reactor needs to ensure that, on average, only one neutron from each reaction goes on to initiate a further reaction. If more reactions took place then the number of reactions would increase all that time and the chain reaction would run out of control. If fewer reactions took place, then the number of reactions would be decreasing and the fission process would soon stop.

The chance that a given neutron goes on to cause a fission reaction depends on several factors. Two important ones are:

- the number of potential nuclei 'in the way'.
- the speed (or the energy) of the neutrons.

As a general trend, as the size of a block of fuel increases so do the chances of a neutron causing a further reaction (before it is lost from the surface of the block). As the fuel is assembled together a stage is reached when a chain reaction can occur. This happens when a so-called **critical mass** of fuel has been assembled. The exact value of the critical mass depends on the exact nature of the fuel being used and the shape of the assembly.

There are particular neutrons energies that make them more likely to cause nuclear fission. In general, the neutrons created by the fission process are moving too fast to make reactions likely. Before they can cause further reactions the neutrons have to be slowed down.

MODERATOR, CONTROL RODS AND HEAT EXCHANGER

Three important components in the design of all nuclear reactors are the **moderator**, the **control rods** and the **heat exchanger**.

- Collisions between the neutrons and the nuclei of the moderator slow them down and allow further reactions to take place.
- The control rods are movable rods that readily absorb neutrons. They can be introduced or removed from the reaction chamber in order to control the chain reaction.
- The heat exchanger allows the nuclear reactions to occur in a place that is sealed off from the rest of the environment. The reactions increase the temperature in the core. This thermal energy is transferred to water and the steam that is produced turns the turbines.

A general design for one type of nuclear reactor (PWR or Pressurized Water Reactor) is shown here. It uses water as the moderator and as a coolant.

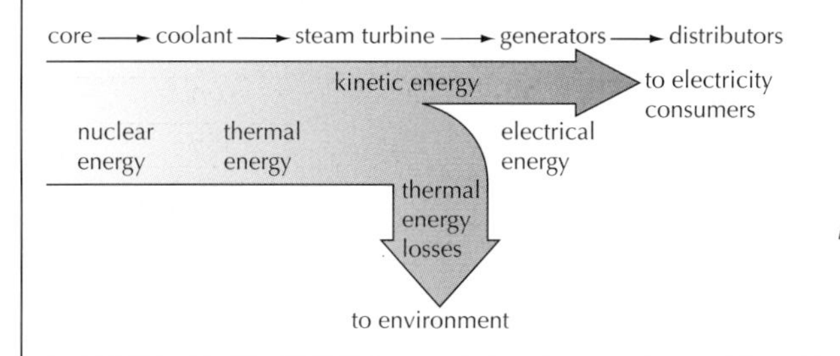

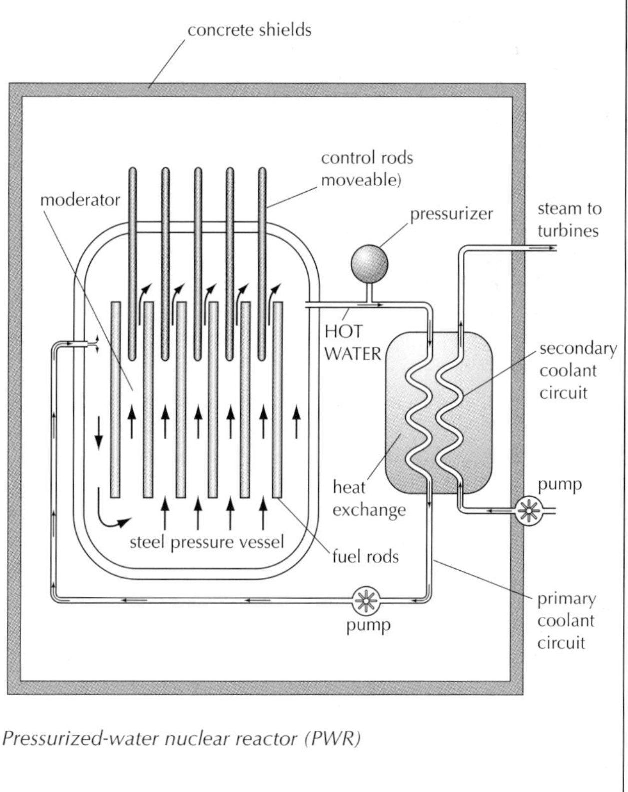

Pressurized-water nuclear reactor (PWR)

ADVANTAGES AND DISADVANTAGES

Advantages

- Extremely high 'energy density' – a great deal of energy is released from a very small mass of uranium
- Reserves of uranium large compared to oil

Disadvantages

- Process produces radioactive nuclear waste that is currently just stored
- Larger possible risk if anything should go wrong
- Nonrenewable (but should last a long time)

Nuclear power (2)

ENRICHMENT AND REPROCESSING

Naturally occurring uranium contains less than 1% of uranium-235. Enrichment is the process by which this percentage composition is increased to make nuclear fission more likely.

In addition to uranium-235, plutonium-239 is also capable of sustaining fission reactions. This nuclide is formed as a by-product of a conventional nuclear reactor. A uranium-238 nucleus can capture fast moving neutrons to form uranium-239. This undergoes β-decay to neptunium-239 which undergoes further β-decay to plutonium-239:

$$^{238}_{92}U + ^{1}_{0}n \rightarrow ^{239}_{92}U$$

$$^{239}_{92}U \rightarrow ^{239}_{93}Np + ^{0}_{-1}\beta + \bar{\upsilon}$$

$$^{239}_{93}Np \rightarrow ^{239}_{94}Pu + ^{0}_{-1}\beta + \bar{\upsilon}$$

Reprocessing involves treating used fuel waste from nuclear reactors to recover uranium and plutonium and to deal with other waste products. A **fast breeder reactor** is one design that utilizes plutonium-239.

HEALTH, SAFETY AND RISK

Issues associated with the use of nuclear power stations for generation of electrical energy include:

- If the control rods were all removed, the reaction would rapidly increase its rate of production. Completely uncontrolled nuclear fission would cause an explosion and **thermal meltdown** of the core. The radioactive material in the reactor could be distributed around the surrounding area causing many fatalities. Some argue that the terrible scale of such a disaster means that the use of nuclear energy is a risk not worth taking. Nuclear power stations could be targets for terrorist attack.
- The reaction produces radioactive nuclear waste. While much of this waste is of a low level risk and will radioactively decay within decades, a significant amount of material is produced which will remain dangerously radioactive for millions of years. The current solution is to bury this waste in geologically secure sites.
- The uranium fuel is mined from underground and any mining operation involves significant risk. The ore is also radioactive so extra precautions are necessary to protect the workers involved in uranium mines.
- The transportation of the uranium from the mine to a power station and of the waste from the nuclear power station to the reprocessing plant needs to be secure and safe.
- By-products of the civilian use of nuclear power can be used to produce nuclear weapons.

NUCLEAR WEAPONS

A nuclear power station involves controlled nuclear fission whereas an uncontrolled nuclear fission produces the huge amount of energy released in nuclear weapons. Weapons have been designed using both uranium and plutonium as the fuel. Issues associated with nuclear weapons include:

- Moral issues associated with any weapon of aggression that is associated with warfare. Nuclear weapons have such destructive capability that since the Second World War the threat of their deployment has been used as a deterrent to prevent non-nuclear aggressive acts against the possessors of nuclear capability.
- The unimaginable consequences of a nuclear war has forced many countries to agree to non-proliferation treaties, which attempt to limit nuclear power technologies to a small number of nations.
- A by-product of the peaceful use of uranium for energy production is the creation of plutonium-239 which could be used for the production of nuclear weapons. Is it right for the small number of countries that already have nuclear capability to prevent other countries from acquiring that knowledge?

FUSION REACTORS

Fusion reactors offer the theoretical potential of significant power generation without many of the problems associated with current nuclear fission reactors. The fuel used, hydrogen, is in plentiful supply and the reaction (if it could be sustained) would not produce significant amounts of radioactive waste.

The reaction is the same as takes place in the Sun (as outlined on page 64) and requires creating temperatures high enough to ionize atomic hydrogen into a plasma state (this is the 'fourth state of matter', in which electrons and protons are not bound in atoms but move independently). Currently the principal design challenges are associated with maintaining and confining the plasma at sufficiently high temperature and density for fusion to take place.

Solar power

ENERGY TRANSFORMATIONS (TWO TYPES)

There are two ways of harnessing the radiated energy that arrives at the Earth's surface from the Sun.

A **photovoltaic cell** (otherwise known as a solar cell or photocell) converts a portion of the radiated energy directly into a potential difference ('voltage'). It uses a piece of semiconductor to do this. Unfortunately, a typical photovoltaic cell produces a very small voltage and it is not able to provide much current. They are used to run electrical devices that do not require a great deal of energy. Using them in series would generate higher voltages and several in parallel can provide a higher current.

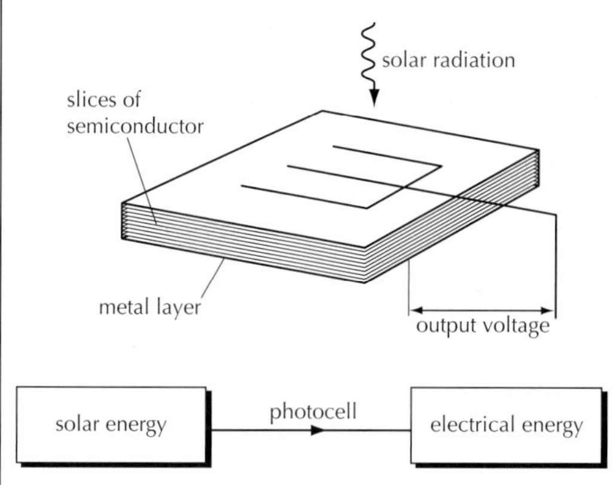

solar energy → photocell → electrical energy

An **active solar heater** (otherwise known as a solar panel) is designed to capture as much thermal energy as possible. The hot water that it typically produces can be used domestically and would save on the use of electrical energy.

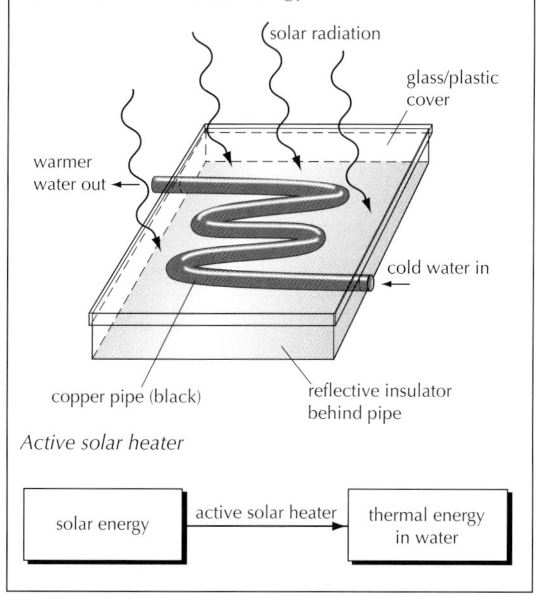

Active solar heater

solar energy → active solar heater → thermal energy in water

SOLAR CONSTANT

The amount of power that arrives from the Sun is measured by the solar constant. It is properly defined as the amount of solar energy that falls per second on an area of 1 m^2 above the Earth's atmosphere that is at right angles to the Sun's rays. Its average value is about 1400 W m^{-2}.

This is not the same as the power that arrives on 1 m^2 of the Earth's surface. Scattering and absorption in the atmosphere means that often less than half of this arrives at the Earth's surface. The amount that arrives depends greatly on the weather conditions.

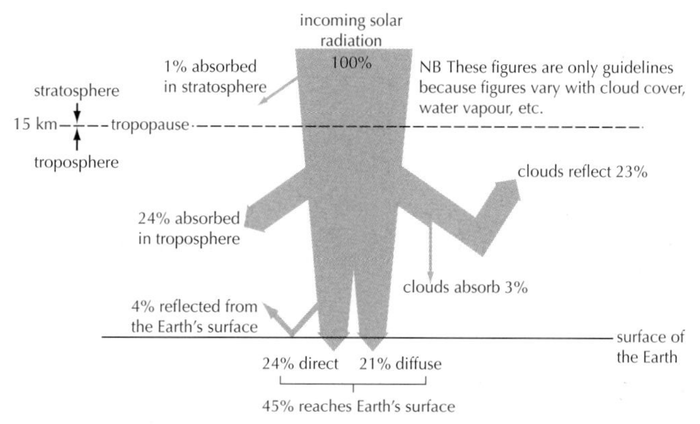

Fate of incoming radiation

Different parts of the Earth's surface (regions at different latitudes) will receive different amounts of solar radiation. The amount received will also vary with the seasons since this will affect how spread out the rays have become.

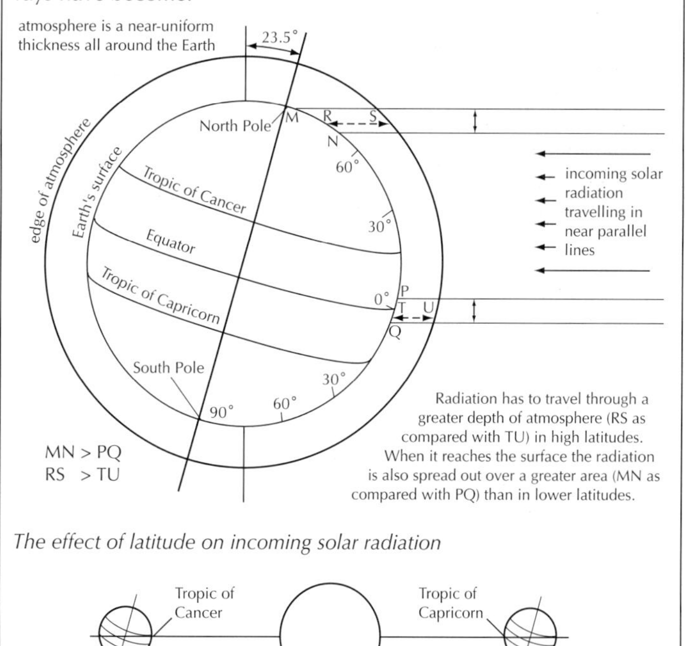

The effect of latitude on incoming solar radiation

Tropic of Cancer — Summer; Tropic of Capricorn — Winter

The Earth's orbit and the seasons

ADVANTAGES AND DISADVANTAGES

Advantages

- Very 'clean' production – no harmful chemical by-products
- Renewable source of energy
- Source of energy is free

Disadvantages

- Can only be utilized during the day
- Source of energy is unreliable – could be a cloudy day
- Low energy density – a very large area would be needed for a significant amount of energy

Solar radiation

INVERSE SQUARE LAW OF RADIATION

As the distance of an observer from a point source of light of an observer increases, the power received by the observer will decrease as the energy spreads out over a larger area. A doubling of distance will result in the reduction of the power received to a quarter of the original value.

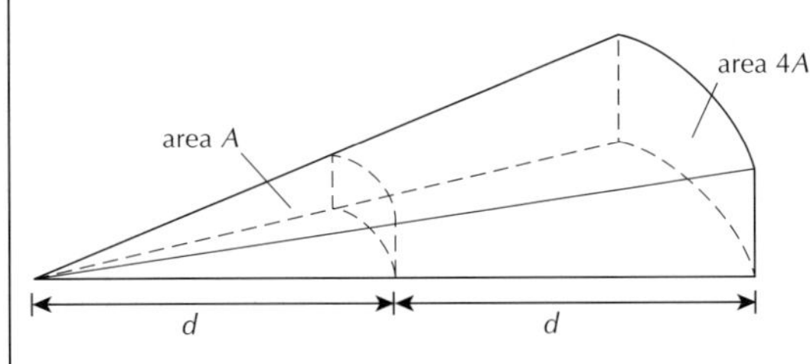

The surface area A of a sphere of radius r is calculated using:

$$A = 4\pi r^2$$

If the point source radiates a total power P in all directions, then the power received per unit area (the **intensity** I) at a distance r away from the point source is:

$$I = \frac{P}{4\pi r^2}$$

For a given area of receiver, the intensity of the received radiation is inversely proportional to the square of the distance from the point source to the receiver. This is known as the **inverse square law**.

ALBEDO

Some of the radiation received by a planet is reflected straight back into space. The fraction that is reflected back is called the **albedo,** α.

The Earth's albedo varies daily and is dependent on season (cloud formations) and latitude. Oceans have a low value but snow has a high value. The global annual mean albedo is 0.3 (30%) on Earth.

Wave power

Wave power is not the same as tidal power; the aim is to use the kinetic energy of waves to generate electrical energy. One successful technique is the **oscillating water column** (OWC) but there are a variety of different possible techniques currently being developed including:

- Pelarmis. This is a floating snake-like object. As the waves move across the sea, different sections of the pelarmis rise and fall. As the different sections move with respect to one another this relative motion can be used to generate electrical energy.
- The vertical motion of buoys due to passing waves can be used to generate electrical energy.
- Waves can be used to turn turbines.

The OWC is a device built on land. In-coming waves force air in and out of a turbine which generates electrical energy. The particular design of the turbine (**Wells turbine**) means that it generates electrical energy whatever the direction of flow of the air.

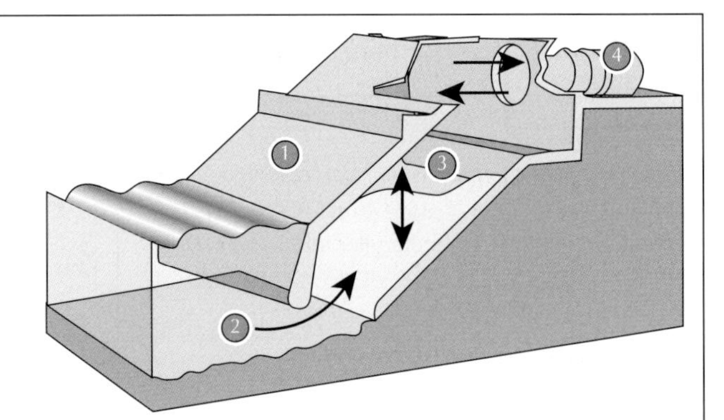

1. Wave capture chamber set into rock face
2. Tidal power forces water into chamber
3. Air alternately compressed and decompressed by "oscillating water column"
4. Rushes of air drive the Wells turbine, generating electrical power

MATHEMATICS

We model water waves as square waves to simplify the mathematics. Consider a square wave of amplitude A, speed v and wavelength λ as shown below:

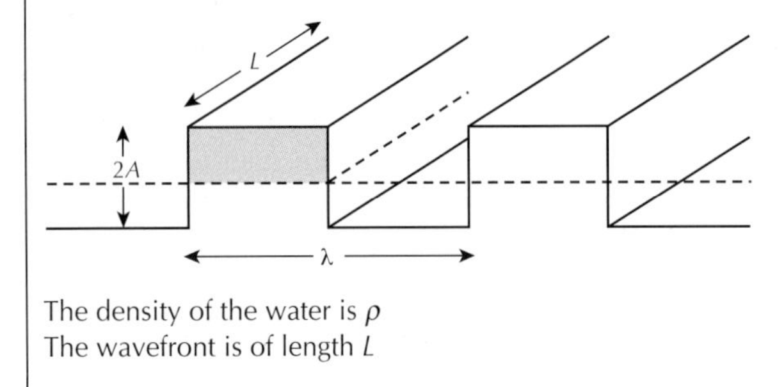

The density of the water is ρ
The wavefront is of length L

If the shaded top part of the wave is moved down (losing gravitational PE) into the trough, the sea would be flat.

This volume of water in the shaded top part $= A \times \frac{\lambda}{2} \times L$

The mass of water in shaded top part $= A \times \frac{\lambda}{2} \times L \times \rho$

Loss of PE of this water $= m\,g\,h = A\left(\frac{\lambda}{2}\right) L\,\rho\,g\,A$

Number of waves passing a point in unit time, $f = \frac{v}{\lambda}$

Loss of PE per unit time $= A^2 \left(\frac{\lambda}{2}\right) L\,\rho\,g \times \left(\frac{v}{\lambda}\right)$

Maximum power available $= \left(\frac{1}{2}\right) A^2\,L\,\rho\,g\,v$

Maximum power available per unit length $= \left(\frac{1}{2}\right) A^2\,\rho\,g\,v$

Hydroelectric power

ENERGY TRANSFORMATIONS

The source of energy in a hydroelectric power station is the gravitational potential energy of water. If water is allowed to move downhill, the flowing water can be used to generate electrical energy.

The water can gain its gravitational potential energy in several ways.

- As part of the 'water cycle', water can fall as rain. It can be stored in large reservoirs as high up as is feasible.
- Tidal power schemes trap water at high tides and release it during a low tide.
- Water can be pumped from a low reservoir to a high reservoir. Although the energy used to do this pumping must be more than the energy regained when the water flows back down hill, this 'pumped storage' system provides one of the few large-scale methods of storing energy.

ADVANTAGES AND DISADVANTAGES

Advantages

- Very 'clean' production – no harmful chemical by-products
- Renewable source of energy
- Source of energy is free

Disadvantages

- Can only be utilized in particular areas
- Construction of dam will involve land being buried under water

Wind power

ENERGY TRANSFORMATIONS

There is a great deal of kinetic energy involved in the winds that blow around the Earth. The original source of this energy is, of course, the Sun. Different parts of the atmosphere are heated to different temperatures. The temperature differences cause pressure differences, due to hot air rising or cold air sinking, and thus air flows as a result.

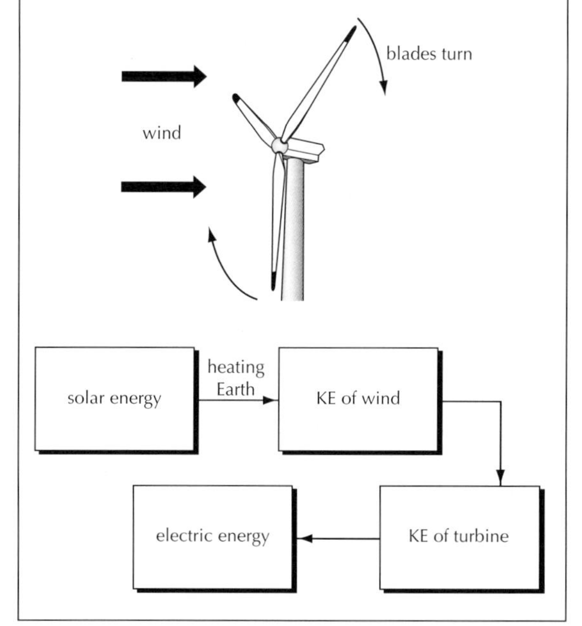

MATHEMATICS

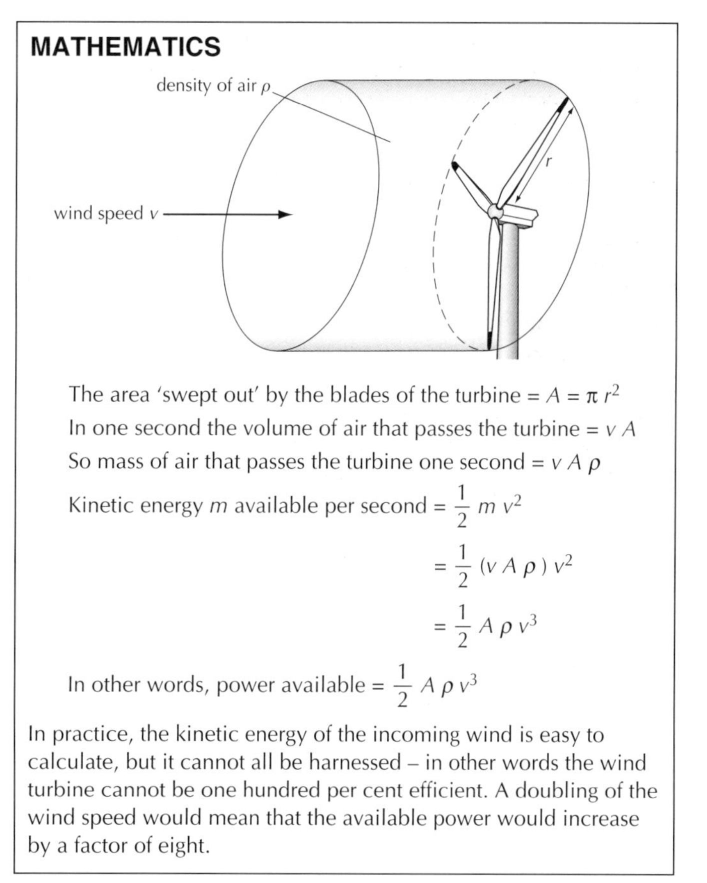

The area 'swept out' by the blades of the turbine = $A = \pi r^2$

In one second the volume of air that passes the turbine = $v\,A$

So mass of air that passes the turbine one second = $v\,A\,\rho$

Kinetic energy m available per second $= \frac{1}{2} m\, v^2$

$$= \frac{1}{2} (v\,A\,\rho)\, v^2$$

$$= \frac{1}{2} A\,\rho\, v^3$$

In other words, power available $= \frac{1}{2} A\,\rho\, v^3$

In practice, the kinetic energy of the incoming wind is easy to calculate, but it cannot all be harnessed – in other words the wind turbine cannot be one hundred per cent efficient. A doubling of the wind speed would mean that the available power would increase by a factor of eight.

ADVANTAGES AND DISADVANTAGES

Advantages

- Very 'clean' production – no harmful chemical by-products
- Renewable source of energy
- Source of energy is free

Disadvantages

- Source of energy is unreliable – could be a day without wind
- Low energy density – a very large area would need be covered for a significant amount of energy
- Some consider large wind generators to spoil the countryside
- Can be noisy
- Best positions for wind generators are often far from centres of population

Wien's law and the Stefan–Boltzmann law (1)

BLACK-BODY RADIATION

In general, the radiation given out from a hot object depends on many things. It is possible to come up with a theoretical model for the 'perfect' emitter of radiation. The 'perfect' emitter will also be a perfect absorber of radiation – a black object absorbs all of the light energy falling on it. For this reason the radiation from a theoretical 'perfect' emitter is known as **black-body radiation**.

Black-body radiation does not depend on the nature of the emitting surface, but it does depend upon its temperature. At any given temperature there will be a range of different wavelengths (and hence frequencies) of radiation that are emitted. Some wavelengths will be more intense than others. This variation is shown in the graphs below.

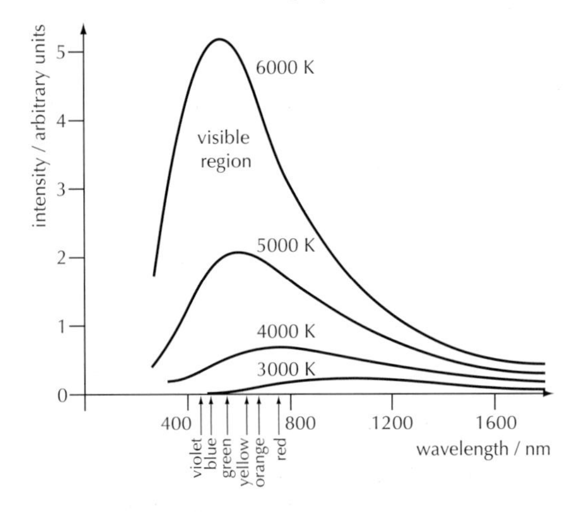

To be absolutely precise, it is not correct to label the y-axis on the above graph as the intensity, but this is often done. It is actually something that could be called the intensity function. This is defined so that the area under the graph (between two wavelengths) gives the intensity emitted in that wavelength range. The total area under the graph is thus a measure of the total power radiated.

Although stars and planets are not perfect emitters, their radiation spectrum is approximately the same as black-body radiation.

WIEN'S LAW

Wien's displacement law relates the wavelength at which the intensity of the radiation is a maximum λ_{max} to the temperature of the black body T. This states that

$$\lambda_{max}\, T = \text{constant}$$

The value of the constant can be found by experiment. It is 2.9×10^{-3} m K. It should be noted that in order to use this constant, the wavelength should be substituted into the equation in metres and the temperature in kelvin.

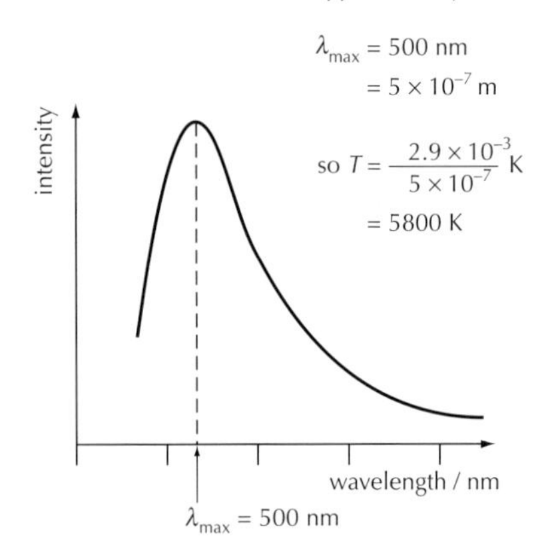

We can analyse light from a star and calculate a value for its surface temperature. This will be much less than the temperature in the core. Hot stars will give out all frequencies of visible light and so will tend to appear white in colour. Cooler stars might well only give out the higher wavelengths (lower frequencies) of visible light – they will appear red. Radiation emitted from planets will peak in the Infra-red.

STEFAN – BOLTZMANN LAW

The Stefan – Boltzmann law links the **total** power radiated by a black body (per unit area) to the temperature of the black-body. The important relationship is that

Total power radiated $\propto T^4$

In symbols we have,

Total power radiated = $\sigma A\, T^4$

Where

σ is a constant called the Stefan – Boltzmann constant. $\sigma = 5.67 \times 10^{-8}$ W m^{-2} K^{-4}

A is the surface area of the emitter (in m^2)

T is the absolute temperature of the emitter (in kelvin)

The radius of the Sun = 6.96×10^8 m.

Surface area $= 4\pi r^2$
$= 6.09 \times 10^{18}$ m^2

If temperature = 5800 K

then total power radiated = $\sigma A T^4$
$= 5.67 \times 10^{-8} \times 6.09 \times 10^{18} \times (5800^4)$
$= 3.9 \times 10^{26}$ W

The radius of the star r is linked to its surface area, A using the equation $A = 4\pi r^2$.

Wien's law and the Stefan–Boltzmann law (2)

BLACK BODY RADIATION AND STEFAN–BOLTZMANN

Matter is not involved in the transfer of thermal energy by radiation. All objects (that have a temperature above zero kelvin) radiate **electromagnetic waves**. If you hold your hand up to a fire to 'feel the heat', your hands are receiving the radiation.

For most everyday objects this radiation is in the **infra-red** part of the **electromagnetic spectrum**. For more details of the electromagnetic spectrum – see page 40.

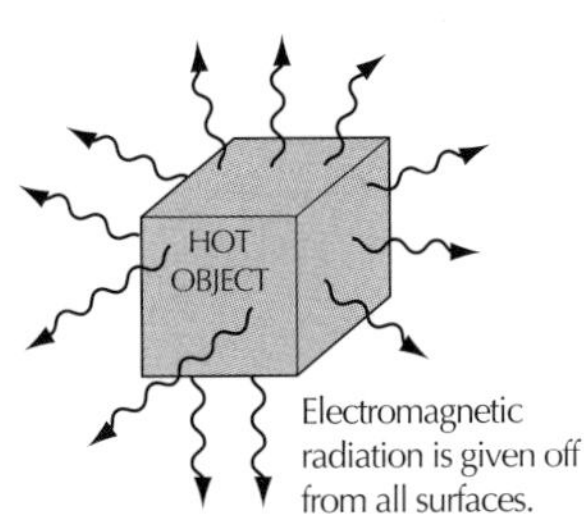

Electromagnetic radiation is given off from all surfaces.

Points to note:

- an object at room temperature absorbs and radiates energy. If it is at constant temperature (and not changing state) then the rates are the same.
- a surface that is a good radiator is also a good absorber.
- surfaces that are light in colour and smooth (shiny) are poor radiators (and poor absorbers).
- surfaces that are dark and rough are good radiators (and good absorbers).
- if the temperature of an object is increased then the frequency of the radiation increases. The total rate at which energy is radiated will also increase.
- radiation can travel through a vacuum (space).

Examples

- The Sun warms the Earth's surface by radiation.
- Clothes in summer tend to be white – so as not to absorb the radiation from the Sun.

EQUILIBRIUM

If the temperature of a planet is constant, then the power being absorbed by the planet must equal the rate at which energy is being radiated into space. The planet is in **thermal equilibrium**. If it absorbs more energy than it radiates, then the temperature must go up and if the rate of loss of energy is greater than its rate of absorption then its temperature must go down.

In order to estimate the power absorbed or emitted, the following concepts are useful.

Emissivity

The Earth and its atmosphere are not a perfect black body. Emissivity, ε, is defined as the ratio of power radiated per unit area by an object to the power radiated per unit area by a black body at the same temperature. It is a ratio and so has no units.

$$\varepsilon = \frac{\text{power radiated by object per unit area}}{\text{power radiated per unit area by black body at same temperature}}$$

Surface heat capacity C_s

Surface heat capacity is the energy required to raise the temperature of unit area of a planet's surface by one degree, and is measured in J m^{-2} K^{-1}.

$$C_s = \frac{\text{energy}}{\text{temperature change of surface} \times \text{area of surface}}$$

A planet of radius r (measured in m) receives a power per unit area from its sun, P (measured in W m^{-2}). Its albedo is α (a ratio without units).

Total power absorbed by the planet = $P(1 - \alpha)\pi r^2$

The Stefan–Boltzmann law and the concept of emissivity mean that:

Total power radiated from the surface of the planet $= \varepsilon \sigma 4\pi r^2 T^4$

$$\therefore P(1 - \alpha)\pi r^2 = \varepsilon\sigma 4\pi r^2 T^4$$

$$\therefore T = \sqrt[4]{\frac{P(1-\alpha)}{4\varepsilon\sigma}}$$

If the incoming radiation power and outgoing radiation power are not equal, then the change of the planet's temperature in a given period of time must be:

$$\text{Temperature change} = \frac{(\text{incoming radiation intensity} - \text{outgoing radiation intensity}) \times \text{time}}{C_s}$$

This simple energy balance model allows the temperature of a planet to be predicted from the incoming radiation power but contains many simplifications.

- It treats the whole planet as one single body and ignores any interactions taking place in, for example, the atmosphere and the oceans.
- It ignores any processes that involve **feedback** (positive and/or negative) in which the result of a change is a further change of one of the constants involved in the calculation (e.g. emissivity and/or albedo). For example an increase in a planet's temperature may result in the melting of some surface ice, which would cause a decrease in the average

The greenhouse effect

PHYSICAL PROCESSES

Short wavelength radiation is received from the sun and causes the surface of the Earth to warm up. The Earth will emit infra-red radiation (longer wavelengths than the radiation coming from the sun because the Earth is cooler than the sun). Some of this infra-red radiation is absorbed by gases in the atmosphere and re-radiated in all directions.

This is known as the **greenhouse effect** and the gases in the atmosphere that absorb infra-red radiation are called **greenhouse gases**. The net effect is that the upper atmosphere and the surface of the Earth are warmed. The name is potentially confusing, as real greenhouses are warm as a result of a different mechanism.

The temperature of the Earth's surface will be constant if the rate at which it radiates energy equals the rate at which it absorbs energy. The greenhouse effect is a natural process and without it the temperature of the Earth would be much lower; the average temperature of the moon is more than 30°C colder than the Earth.

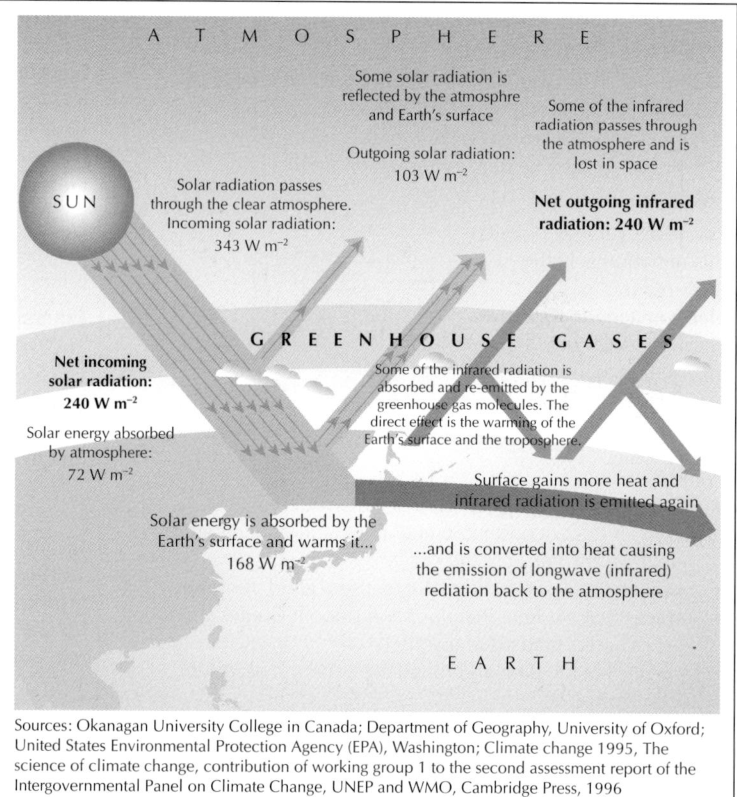

Sources: Okanagan University College in Canada; Department of Geography, University of Oxford; United States Environmental Protection Agency (EPA), Washington; Climate change 1995, The science of climate change, contribution of working group 1 to the second assessment report of the Intergovernmental Panel on Climate Change, UNEP and WMO, Cambridge Press, 1996

GREENHOUSE GASES

The main greenhouse gases are naturally occurring but the balance in the atmosphere can be altered as a result of their release due to industry and technology. They are:

- **Methane,** CH_4. This is the principal component of natural gas and the product of decay, decomposition or fermentation. Livestock and plants produce significant amounts of methane.
- **Water,** H_2O. The small amounts of water vapour in the upper atmosphere (as opposed to clouds which are condensed water vapour) have a significant effect. The average water vapour levels in the atmosphere do not appear to alter greatly as a result of industry, but local levels can vary.
- **Carbon dioxide,** CO_2. Combustion releases carbon dioxide into the atmosphere which can significantly increase the greenhouse effect. Overall, plants (providing they are growing) remove carbon dioxide from the atmosphere during photosynthesis. This is known as **carbon fixation**.
- **Nitrous oxide,** NO_2. Livestock and industries (e.g. the production of Nylon) are major sources of nitrous oxide. Its effect is significant as it can remain in the upper atmosphere for long periods.

In addition the following gases also contribute to the greenhouse effect:

- **Ozone,** O_3. The **ozone layer** is an important region of the atmosphere that absorbs high energy UV photons which would otherwise be harmful to living organisms. Ozone also adds to the greenhouse effect.
- **Chlorofluorocarbons** (CFCs). Used as refrigerants, propellants, and cleaning solvents. The also have the effect of depleting the ozone layer.

Each of these gases absorbs infrared radiation as a result of resonance (see page 36). The natural frequency of oscillation of the bonds within the molecules of the gas is in the infrared region. If the driving frequency (from the radiation emitted from the Earth) is equal to the natural frequency of the molecule, resonance will occur. The amplitude of the molecules' vibrations increases and the temperature will increase. The absorption will take place at specific frequencies depending on the molecular bonds.

Absorption spectra for major natural greenhouse gases in the Earth's atmosphere

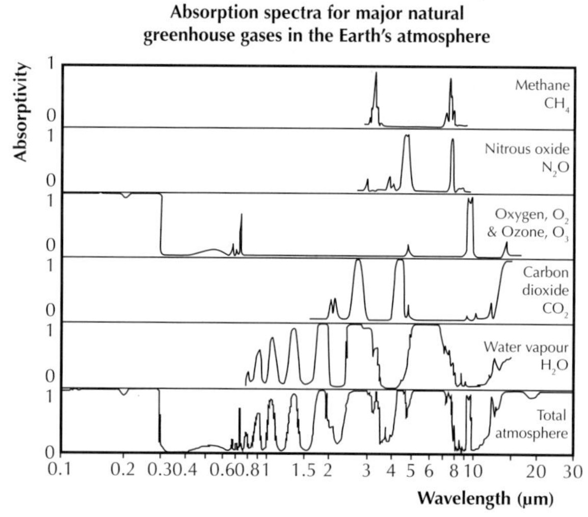

[After J.N. Howard, 1959: *Proc. I.R.E 47*, 1459: and R.M. Goody and G.D. Robinson, 1951: *Quart. J. Roy Meteorol. Soc. 77*, 153]

Global warming (1)

POSSIBLE CAUSES OF GLOBAL WARMING

Records show that the mean temperature of the Earth has been increasing in recent years.

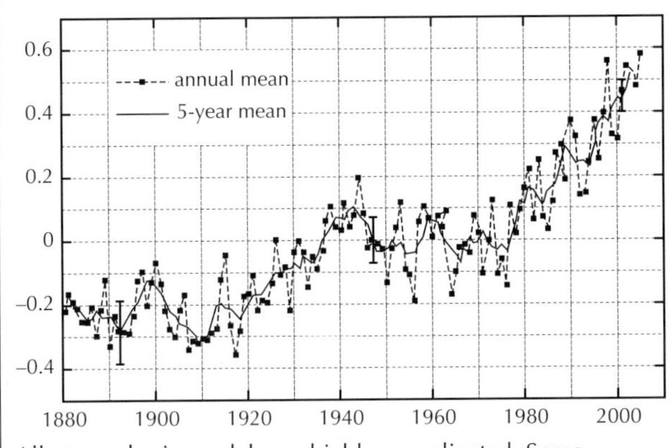

All atmospheric models are highly complicated. Some possible suggestions for this increase include:

- Changes in the composition of greenhouse gases in the atmosphere.
- Changes in the intensity of the radiation emitted by the sun linked to, for example, increased solar flare activity
- Cyclical changes in the Earth's orbit and volcanic activity.

The first suggestion could be caused by natural effects or could be caused by human activities (e.g. the increased burning of fossil fuels). An **enhanced greenhouse effect** is an increase in the greenhouse effect caused by human activities.

In 2007, the IPCC (see page 78) report stated that "Most of the observed increase in globally averaged temperature since the mid-20th century is very likely due to the observed increase in anthropogenic [human-caused] greenhouse gas concentrations".

Although it is still being debated, the generally accepted view is that that the increased combustion of fossil fuels has released extra carbon dioxide into the atmosphere, which has enhanced the greenhouse effect.

EVIDENCE FOR GLOBAL WARMING

One piece of evidence that links global warming to increased levels of greenhouse gases comes from ice core data. The ice core has been drilled in the Russian Antarctic base at Vostok. Each year's new snow fall adds another layer to the ice.

Isotopic analysis allows the temperature to be estimated and air bubbles trapped in the ice cores can be used to measure the atmospheric concentrations of greenhouse gases. The record provides data from over 400 000 years ago to the present. The variations of temperature and carbon dioxide are very closely correlated.

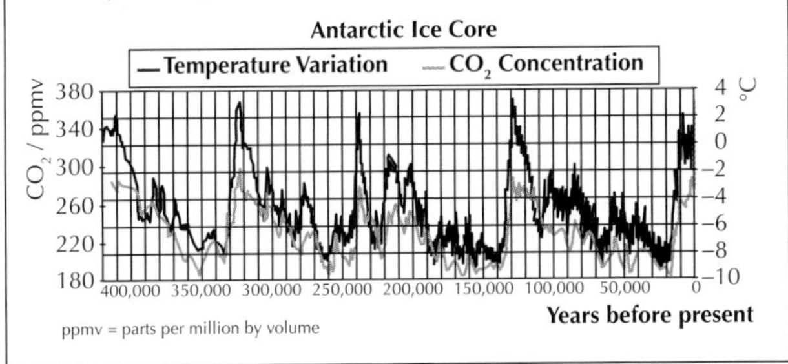

MECHANISMS

Predicting the future effects of global warming involves a great deal of uncertainty, as the interactions between different systems in the Earth and its atmosphere are extremely complex.

There are many mechanisms that may increase the rate of global warming.

- Global warming reduces ice/snow cover, which in turn reduces the albedo. This will result in an increase in the overall rate of heat absorption.
- Temperature increase reduces the solubility of CO_2 in the sea and thus increases atmospheric concentrations.
- Continued global warming will increase both evaporation and the atmosphere's ability to hold water vapour. Water vapour is a greenhouse gas.
- Regions with frozen subsoil exist (called tundra) that support simple vegetation. An increase in temperature may cause a significant release of trapped CO_2.
- Not only does deforestation result in the release of further CO_2 into the atmosphere, the reduction in number of trees reduces carbon fixation.

The first four mechanisms are examples of processes whereby a small initial temperature increase has gone on to cause a further increase in temperature. This process is known as **positive feedback**. Some people have suggested that the current temperature increases may be 'corrected' by a process which involves negative feedback, and temperatures may fall in the future.

Global warming (2)

PREDICTIONS OF THE EFFECTS OF GLOBAL WARMING

Most models of the effects of global warming suggest significant climate change and a rise in mean sea level but there is a great deal of uncertainty associated with these predictions. For example, most substances expand on heating but water's expansion is anomalous.

The **coefficient of volume expansion** records the fractional change in volume per degree change in temperature:

$$\gamma = \frac{\Delta V}{V_0 \Delta\theta}$$

ΔV is the increase in volume (measured in m^3)

$\Delta\theta$ is the increase in temperature (measured in K or °C)

V_0 is the original volume (measured in m^3)

γ is the coefficient of volume expansion (measured in K^{-1} or $°C^{-1}$)

Between 0°C and 4°C, the coefficient of volume expansion for water is negative. This means that if the temperature of water is increased within the range 0°C to 4°C this will cause a decrease in volume. When ice that is floating on seawater melts, the overall water level is predicted to initially decrease. The Arctic ice at the North Pole is floating.

Ice that is on land, however, is not displacing water and when it melts it will increase the sea level. Glaciers and snow on mountains are land-based and much of the Antarctic ice at the South Pole is attached to the land mass beneath.

Predictions identified in the 2007 IPCC report include:

- A temperature rise between 1.8 and 4 °C by the end of the century, with higher values possible.
- Sea levels are projected to rise between 18 and 59 cm by the end of the century with an additional 10 to 20 cm possible as a result of continued ice melting. Rises of 10 cm could cause flooding in large parts of Southeast Asia. Much of the world's population is concentrated in coastal cities.
- It is likely that hot extremes, heat waves, and heavy rains will continue to become more frequent. Strong hurricanes, droughts, wildfires, and other natural disasters may become commonplace in many parts of the world. The growth of deserts may also cause food shortages.
- Glaciers around the world could melt, adding to the increase in sea levels and creating water shortages in regions dependent on runoff for fresh water.
- More than a million species face extinction from disappearing habitat, changing ecosystems, and acidifying oceans.

POSSIBLE SOLUTIONS TO REDUCE THE ENHANCED GREENHOUSE EFFECT

A reduction in the amount of fossil fuels consumed would lead to a reduction in the emissions of greenhouse gases.

Possible approaches include:

- Advances in technology could be utilised to ensure
  - o greater efficiency of power production
  - o decarbonising exhaust gases from power plants (carbon dioxide capture and storage)
  - o fusion reactors are made operational.
- Reduction of energy requirements by
  - o improving thermal insulation in homes and dwellings
  - o reducing journeys and using more energy efficient methods of transport e.g. hybrid vehicles
  - o use of combined heating and power systems (CHP).
- Replacing the use of coal and oil
  - o with renewable energy sources and/or nuclear power to eliminate emissions
  - o with natural gas to reduce emissions
- Planting new trees and ensuring existing forests are maintained.

One country acting on its own cannot solve the problems of global warming. There are many international efforts aimed at reducing the enhanced greenhouse effect.

Intergovernmental Panel on Climate Change (IPCC). The World Meteorological Organization (WMO) and the United Nations Environment Programme (UNEP) established the IPCC in 1988. Hundreds of governmental scientific representatives from more than a hundred countries regularly assess the up to date evidence from international research into global warming and human induced climate change.

Kyoto Protocol. This is an amendment to United Nations Framework Convention on Climate Change. By signing the treaty, countries agree to work towards achieving a stipulated reduction in greenhouse gas emissions. Although over 160 countries have ratified the protocol, some significant industrialised nations have not signed, including the United States and Australia. Some other countries such as India and China, which have ratified the protocol, are not currently required to reduce their carbon emissions.

Asia-Pacific Partnership on Clean Development and Climate (APPCDC). Six countries that represent approximately 50% of the world's energy use 'have agreed to work together and with private sector partners to meet goals for energy security, national air pollution reduction, and climate change in ways that promote sustainable economic growth and poverty reduction'. The founding partners are Australia, China, India, Japan, Republic of Korea, and the United States.

IB QUESTIONS – ENERGY, POWER AND CLIMATE CHANGE

1 A wind generator converts wind energy into electric energy. The source of this wind energy can be traced back to solar energy arriving at the Earth's surface.

(a) Outline the energy transformations involved as solar energy converts into wind energy. [2]

(b) List **one** advantage and **one** disadvantage of the use of wind generators. [2]

The expression for the maximum theoretical power, P, available from a wind generator is

$$P = \frac{1}{2}A\rho v^3$$

where A is the area swept out by the blades,
ρ is the density of air and
v is the wind speed.

(c) Calculate the maximum theoretical power, P, for a wind generator whose blades are 30 m long when a 20 m s^{-1} wind blows. The density of air is 1.3 kg m^{-3}. [2]

(d) In practice, under these conditions, the generator only provides 3 MW of electrical power.
(i) Calculate the efficiency of this generator. [2]
(ii) Give **two** reasons explaining why the actual power output is less than the maximum theoretical power output. [2]

2 This question is about energy sources.

(a) Give **one** example of a renewable energy source and **one** example of a non-renewable energy source and explain why they are classified as such. [4]

(b) A wind farm produces 35 000 MWh of energy in a year. If there are ten wind turbines on the farm show that the average power output of **one** turbine is about 400 kW. [3]

(c) State **two** disadvantages of using wind power to generate electrical power. [2]

3 This question is about energy transformations.

Wind power can be used to generate electrical energy.

Construct an energy flow diagram which shows the energy transformations, starting with solar energy and ending with electrical energy, generated by windmills. Your diagram should indicate where energy is degraded. [7]

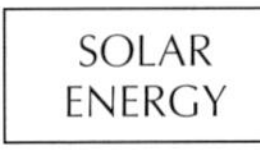

ELECTRICAL ENERGY

4 This question is about a coal-fired power station which is water cooled.

Data:

Electrical power output from the station	= 200 MW
Temperature at which water enters cooling tower	= 288 K
Temperature at which water leaves cooling tower	= 348 K
Rate of water flow through tower	= 4000 kg s^{-1}
Energy content of coal	= 2.8×10^7 J kg^{-1}
Specific heat of water	= 4200 J kg^{-1} K^{-1}

Calculate

(a) the energy per second carried away by the water in the cooling tower; [2]

(b) the energy per second produced by burning the coal; [2]

(c) the overall efficiency of the power station; [2]

(d) the mass of coal burnt each second. [1]

5 This question is about tidal power systems.

(a) Describe the principle of operation of such a system. [2]

(b) Outline **one** advantage and **one** disadvantage of using such a system. [2]

(c) A small tidal power system is proposed. Use the data in the table below to calculate the total energy available and hence estimate the useful output power of this system.

Height between high tide and low tide	4 m
Trapped water would cover an area of	1.0×10^6 m^2
Density of water	1.0×10^3 kg m^{-3}
Number of tides per day	2

[4]

6 This question is about a hydroelectric power scheme using tidal energy.

The diagram shows a hydroelectric scheme constructed in the ocean near the shore. Built into the dam wall is a system of pipes and adjustable valves (not shown), to allow water to flow one way or the other, and a turbine connected to an electric generator.

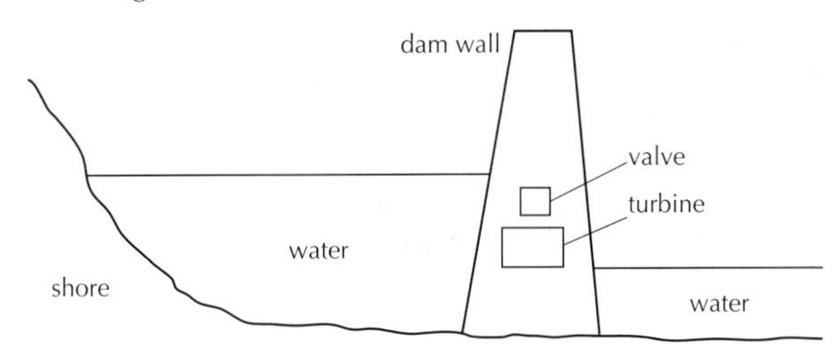

(a) Explain in some detail how such a system might work to provide electrical energy from the tides. [5]

(b) Tidal systems work only on a small scale and in certain places. Suggest **two** factors which make it impractical for such systems to provide electrical energy on a large scale or a widespread scale. [2]

(c) The tides are the immediate source of energy for this hydroelectric system.
(i) Where does the energy of the tides come from? [1]
(ii) Discuss briefly whether tidal energy systems give us something for nothing and whether the source of tidal energy can eventually be used up. [2]

HL Projectile motion

COMPONENTS OF PROJECTILE MOTION

If two children are throwing and catching a tennis ball between them, the path of the ball is always the same shape. This motion is known as **projectile motion** and the shape is called a **parabola**.

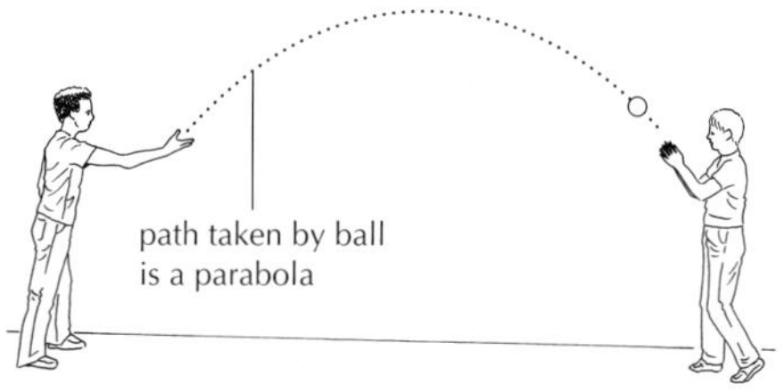

The only forces acting during its flight are gravity and friction. In many situations, air resistance can be ignored.

It is moving horizontally and vertically **at the same time** but the horizontal and vertical components of the motion are **independent** of one another. Assuming the gravitional force is constant, this is always true.

Horizontal component

There are no forces in the horizontal direction, so there is no horizontal acceleration. This means that the horizontal velocity must be constant.

ball travels at a constant horizontal velocity

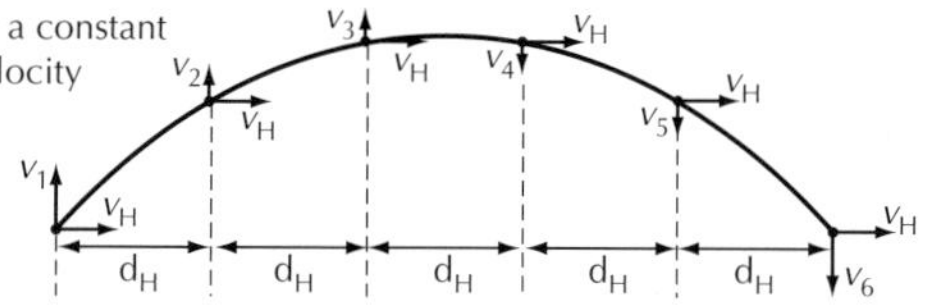

Vertical component

There is a constant vertical force acting down, so there is a constant vertical acceleration. The value of the vertical acceleration is 10 m s^{-2} – the acceleration due to gravity.

vertical velocity changes

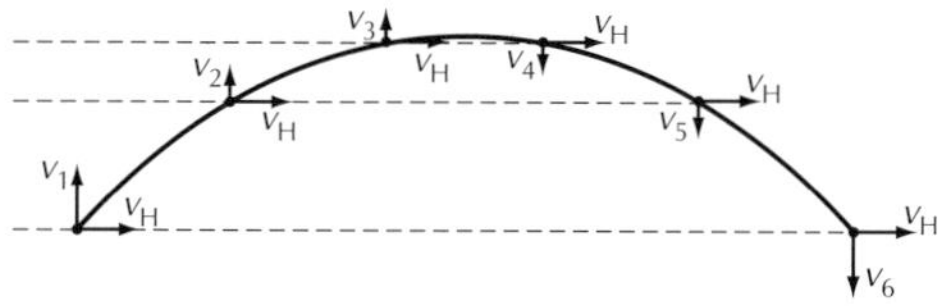

MATHEMATICS OF PARABOLIC MOTION

The graphs of the components of parabolic motion are shown below.

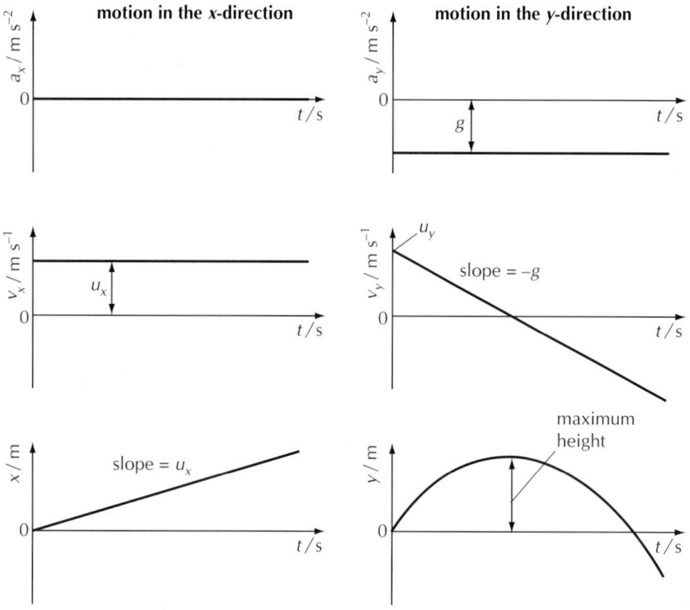

Once the components have been worked out, the actual velocities (or displacements) at any time can be worked out by vector addition.

The solution of any problem involving projectile motion is as follows:
- use the angle of launch to resolve the initial velocity into components.
- the time of flight will be determined by the vertical component of velocity.
- the range will be determined by the horizontal component (and the time of flight).
- the velocity at any point can be found by vector addition.

Useful 'shortcuts' in calculations include the following facts:
- for a given speed, the greatest range is achieved if the launch angle is 45°.
- if two objects are released together, one with a horizontal velocity and one from rest, they will both hit the ground together.

EXAMPLE

A projectile is launched horizontally from the top of a cliff.

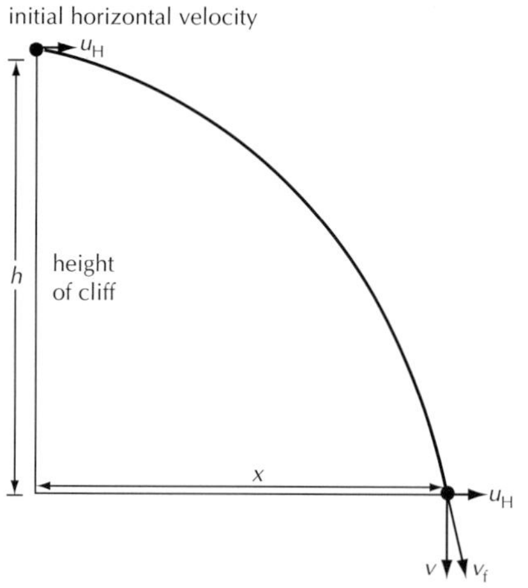

vertical motion	horizontal motion
$u = 0$	$u = u_H$
$v = ?$	$v = u_H$
$a = 10$ m s^{-2}	$a = 0$
$s = h$	$s = x$
$t = ?$	$t = ?$

$$s = ut + \tfrac{1}{2}at^2$$

so $h = 0 + \tfrac{1}{2} \times 10 \times t^2$

$\therefore$ $t^2 = \frac{2h}{10}$

$$t = \sqrt{\frac{2h}{10}} \text{ s}$$

Since $v = u + at$

$v = 0 + 10\sqrt{\frac{2h}{10}}$ m $\qquad x = u_H \times t$

$= \sqrt{20h}$ m s^{-1} $\qquad = u_H \times \sqrt{\frac{2h}{10}}$ m

The final velocity v_f is the vector addition of v and u_H.

Gravitational potential energy and potential

GRAVITATIONAL POTENTIAL ENERGY

It is easy to work out the difference in gravitational energy when a mass moves between two different heights near the Earth's surface.

The difference in energies = $m\,g\,(h_2 - h_1)$

There are two important points to note:

- this derivation has assumed that the gravitational field strength g is constant. However, Newton's theory of universal gravitation states that the field MUST CHANGE with distance. **This equation can only be used if the vertical distance we move is not very large.**
- the equation assumes that the gravitational potential energy gives zero PE at the surface of the Earth. This works for everyday situations but it is not fundamental.

The true zero of gravitational potential energy is taken at infinity.

If the potential energy of the mass, m, was zero at infinity, and it lost potential energy moving in towards mass M, the potential energy must be **negative** at a given point, P.

The value of gravitational potential energy of a mass at any point in space is defined as the work done in moving it from infinity to that point. The mathematics needed to work this out is not trivial since the force changes with distance.

It turns out that

$$\text{Gravitational potential energy of mass } m \text{ (due to } M) = -\frac{G\,M\,m}{r}$$

This is a scalar quantity (measured in joules) and is independent of the path taken from infinity.

potential energy decreases as gravitational force does work

zero of potential energy taken to be at infinity

M — F_4 m — F_3 m — F_2 m — F_1 m

as m moves towards M in the force on m increases

GRAVITATIONAL POTENTIAL

We can define a new quantity, the **gravitational potential V_g** that measures the energy per unit test mass.

$$V_g = \frac{W}{m} \quad \frac{\text{(work done)}}{\text{(test mass)}}$$

The SI units of gravitational potential are J kg^{-1}. It is a scalar quantity.

Using Newton's law of universal gravitation, we can work out the gravitational potential at a distance r from any point mass.

V_g, r, $V_g = -\frac{GM}{r}$

This formula and the graph also work for spherical masses (planets etc.). The gravitational potential as a result of lots of masses is just the addition of the individual potentials. This is an easy sum since potential is a scalar quantity.

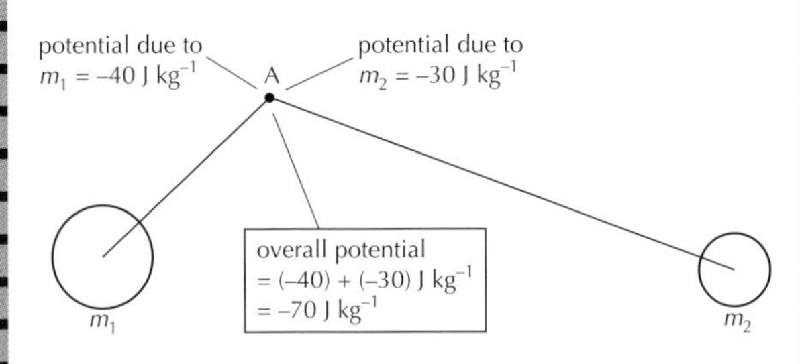

Once you have the potential at one point and the potential at another, the difference between them is the energy you need to move a unit mass between the two points. It is independent of the path taken.

ESCAPE SPEED

The escape speed of a rocket is the speed needed to be able to escape the gravitational attraction of the planet. This means getting to an infinite distance away.

We know that gravitational potential at the surface of a planet = $\frac{-GM}{R_p}$.

(where R_p is the radius of the planet)

This means that for a rocket of mass m, the difference between its energy at the surface and at infinity = $\frac{GMm}{R_p}$

Therefore the minimum kinetic energy needed = $\frac{GMm}{R_p}$

In other words,

$$\frac{1}{2}m\,(v_{escape})^2 = \frac{GMm}{R_p}$$

so

$$v_{escape} = \sqrt{\left(\frac{2\,GM}{R_p}\right)}$$

This derivation assumes the planet is isolated.

EXAMPLE

The escape speed from an isolated planet like Earth (radius of Earth $R_E = 6.37\times10^6$ m) is calculated as follows:

$$v_{escape} = \sqrt{\left(\frac{2\times6.67\times10^{-11}\times5.98\times10^{24}}{6.37\times10^{6}}\right)}\ \text{m s}^{-1}$$

$$= \sqrt{(1.25\times10^{8})}\ \text{m s}^{-1}$$

$$= 1.12\times10^{4}\ \text{m s}^{-1}$$

$$\approx 11\ \text{km s}^{-1}$$

The vast majority of rockets sent into space are destined to orbit the Earth so they leave with a speed that is less than the escape speed.

Electric potential energy and potential

POTENTIAL AND POTENTIAL DIFFERENCE

The concept of electrical potential difference between two points was introduced on page 46. As the name implies, potential difference is just the difference between the potential at one point and the potential at another. Potential is simply a measure of the total electrical energy per unit charge at a given point in space. The definition is very similar to that of gravitational potential.

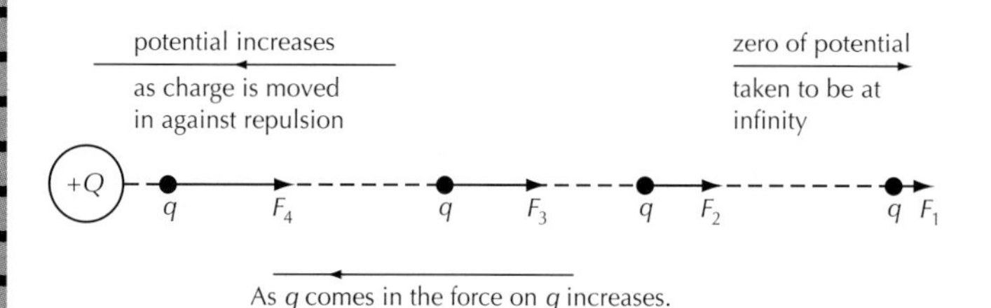

If the total work done in bringing a positive test charge q from infinity to a point in an electric field is W, then the electric potential at that point, V, is defined to be

$$V = \frac{W}{q}$$

The units for potential are the same as the units for potential difference: $J\,C^{-1}$ or volts.

$$V = \frac{Q}{4\pi\varepsilon_o r}$$

This equation only applies to a single point charge.

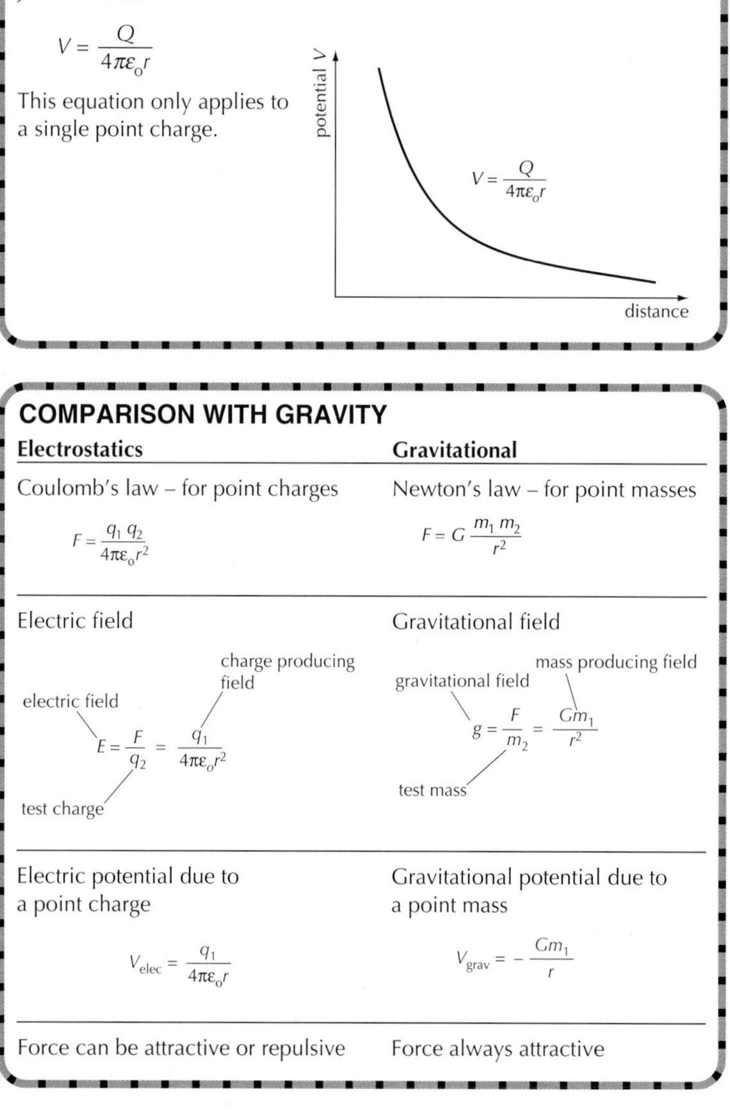

$$V = \frac{Q}{4\pi\varepsilon_o r}$$

COMPARISON WITH GRAVITY

Electrostatics	Gravitational
Coulomb's law – for point charges $F = \frac{q_1 q_2}{4\pi\varepsilon_o r^2}$	Newton's law – for point masses $F = G\frac{m_1 m_2}{r^2}$
Electric field: $E = \frac{F}{q_2} = \frac{q_1}{4\pi\varepsilon_o r^2}$ (electric field; test charge; charge producing field)	Gravitational field: $g = \frac{F}{m_2} = \frac{Gm_1}{r^2}$ (gravitational field; test mass; mass producing field)
Electric potential due to a point charge $V_{elec} = \frac{q_1}{4\pi\varepsilon_o r}$	Gravitational potential due to a point mass $V_{grav} = -\frac{Gm_1}{r}$
Force can be attractive or repulsive	Force always attractive

POTENTIAL DUE TO MORE THAN ONE CHARGE

If several charges all contribute to the total potential at a point, it can be calculated by adding up the individual potentials due to the individual charges.

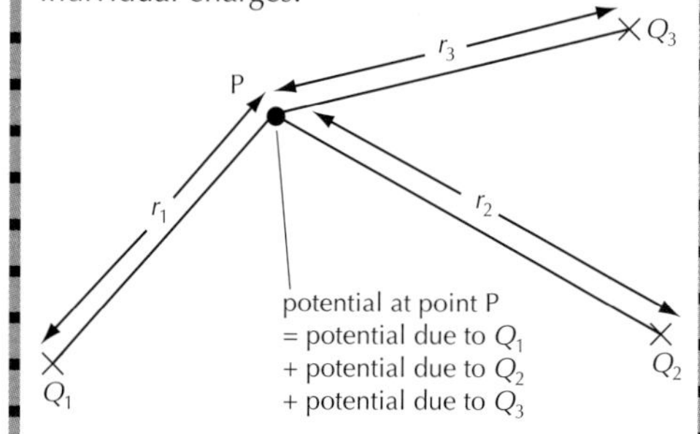

The electric potential at any point outside a charged conducting sphere is exactly the same as if all the charge had been concentrated at its centre.

POTENTIAL AND FIELD STRENGTH

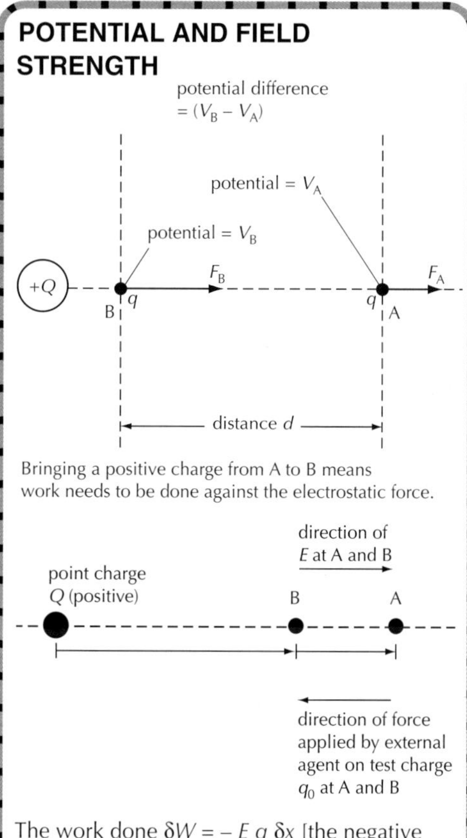

The work done $\delta W = -E\,q\,\delta x$ [the negative sign is because the direction of the force needed to do the work is opposite to the direction of E]

Therefore $E = -\frac{1}{q}\frac{\delta W}{\delta x}$

$$= -\frac{\delta V}{\delta x} \quad [\text{since } \delta V = \frac{\delta W}{q}]$$

In words,

electric field = – potential gradient

Units = $\frac{\text{volt}}{\text{metre}}$ (Vm^{-1})

Equipotentials

EQUIPOTENTIAL SURFACES

The best way of representing how the electric potential varies around a charged object is to identify the regions where the potential is the same. These are called **equipotential** surfaces. In two dimensions they would be represented as lines of equipotential. A good way of visualising these lines is to start with the contour lines on a map.

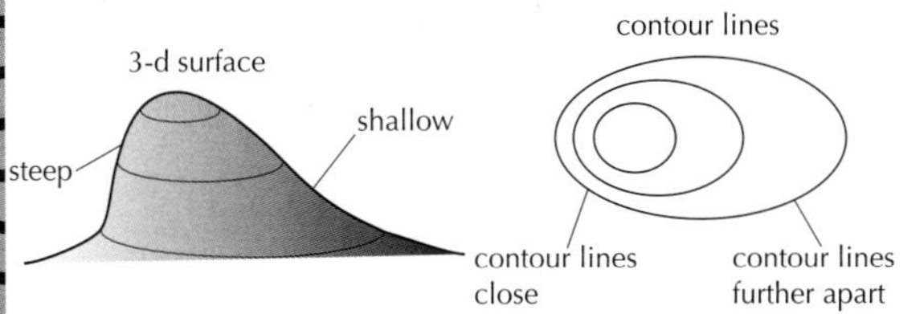

The contour diagram on the right represents the changing heights of the landscape on the left. Each line joins up points that are at the same height. Points that are high up represent a high value of gravitational potential and points that are low down represent a low gravitational potential. Contour lines are lines of equipotential in a gravitational field.

The same can be done with an electric field. Lines are drawn joining up points that have the same electric potential. The situation below shows the equipotentials for an isolated positive point charge.

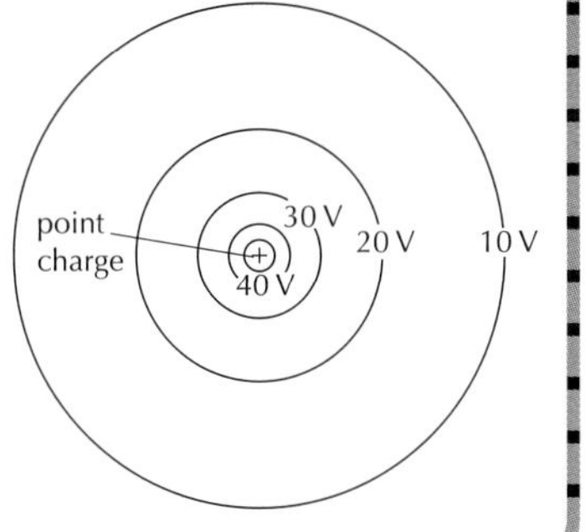

RELATIONSHIP TO FIELD LINES

There is a simple relationship between electric field lines and lines of equipotential – they are always at right angles to one another. Imagine the contour lines. If we move along a contour line, we stay at the same height in the gravitational field. This does not require work because we are moving at right angles to the gravitational force. Whenever we move along an electric equipotential line, we are moving between points that have the same electric potential – in other words, no work is being done. Moving at right angles to the electric field is the only way to avoid doing work in an electric field. Thus equipotential lines must be at right angles to field lines as shown below.

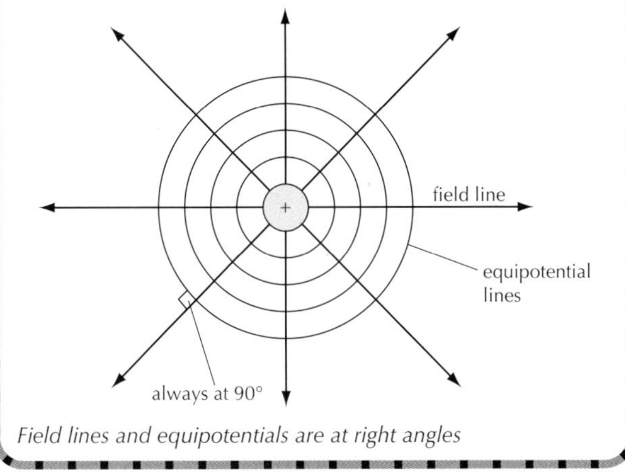

Field lines and equipotentials are at right angles

EXAMPLES OF EQUIPOTENTIALS

The diagrams below show equipotential lines for various situations.

Equipotentials outside a charge-conducting sphere and a point mass.

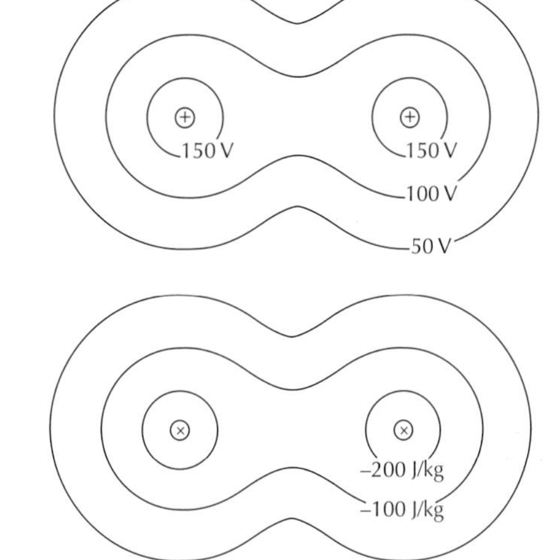

Equipotentials for two point charges (same charge) and two point masses.

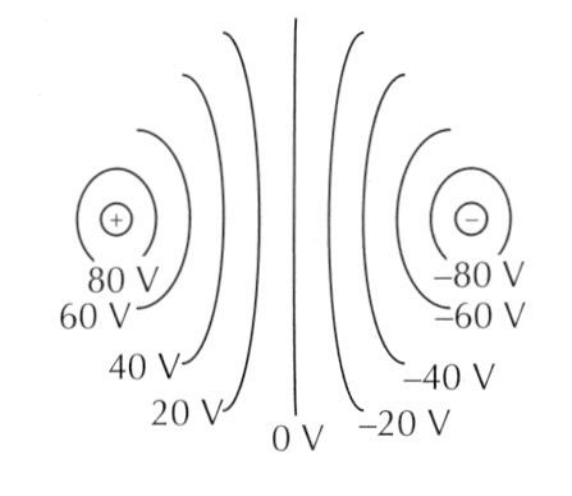

Equipotentials for two point charges (equal and opposite charges).

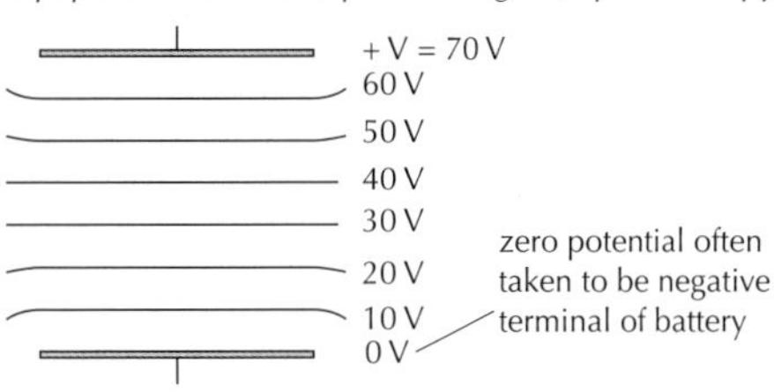

Equipotentials lines between charged parallel plates.

It should be noted that although the correct definition of zero potential is at infinity, most of the time we are not really interested in the actual value of potential, we are only interested in the value of the difference in potential. This means that in some situations (such as the parallel conducting plates) it is easier to imagine the zero at a different point. This is just like setting sea level as the zero for gravitational contour lines rather than correctly using infinity for the zero.

Orbital motion

KEPLER'S THIRD LAW

The are hundreds of artificial satellites in orbit around the Earth. These satellites do not rely on any engines to keep them in orbit – the gravitational force from the Earth provides the centripetal force required.

Gravitational attraction = centripetal force

Equations for both of these quantities have been worked out elsewhere in this book

Gravitational attraction = $\frac{GMm}{r^2}$ [page 51]

Centripetal force = $\frac{m\,v^2}{r}$ [page 25]

Therefore $\frac{GMm}{r^2} = \frac{m\,v^2}{r}$ where r = radius of orbit

$$GM = v^2 r \quad (1)$$

$$v = \sqrt{\left(\frac{GM}{r}\right)} \quad (2)$$

Do not confuse v (velocity of orbit) with v (escape speed).

Since the satellite does one orbit (one circumference) in time T,

$$\text{Speed } v = \frac{\text{circumference}}{T} = \frac{2\pi r}{T}$$

This can be substituted into equation (1) to give

$$GM = \left(\frac{2\pi r}{T}\right)^2 r = \frac{4\pi^2 r^3}{T^2}$$

G, M and $(4\pi^2)$ are all constants so $\frac{r^3}{T^2}$ = constant

This is an important relationship. It is known as Kepler's third law. Although we derived it for artificial satellites in **circular** orbit around the Earth, it actually applies to ANY closed orbit.

ENERGY OF AN ORBITING SATELLITE

We already know that the gravitational energy = $\frac{-GMm}{r}$

The kinetic energy = $\frac{1}{2}\,m\,v^2$ but $v = \sqrt{\left(\frac{GM}{r}\right)}$

[equation (2), above]

$\therefore$ kinetic energy = $\frac{1}{2}\,m\,\frac{GM}{r} = \frac{1}{2}\,\frac{GMm}{r}$

So total energy = KE + PE

$$= \frac{1}{2}\,\frac{GMm}{r} - \frac{GMm}{r} = -\frac{1}{2}\,\frac{GMm}{r}$$

Note that

- In the orbit the magnitude of the KE = $\frac{1}{2}$ magnitude of the PE
- The overall energy of the satellite is negative. (A satellite must have a total energy less than zero otherwise it would have enough energy to escape the Earth's gravitional field.)
- In order to move from a small radius orbit to a large radius orbit, the total energy must increase. To be precise, an increase in orbital radius makes the total energy go from a large negative number to a smaller negative number – this is an increase.

This can be summarised in graphical form.

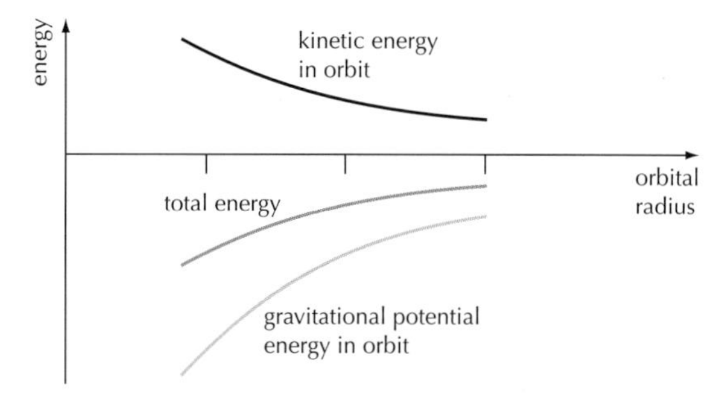

WEIGHTLESSNESS

One way of defining the weight of a person is to say that it is the value of the force recorded on a supporting scale.

If the scales were set up in a lift, they would record different values depending on the **acceleration** of the lift.

An extreme version of these situations occurs if the lift cable breaks and the lift (and passenger) accelerates down at 10 m s^{-2}.

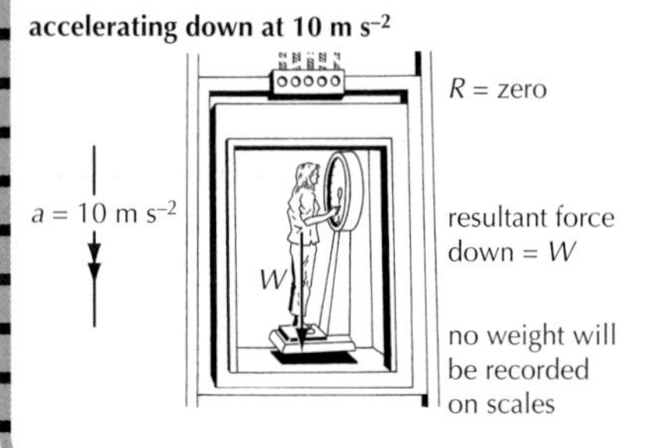

The person would appear to be weightless for the duration of the fall. Given the possible ambiguity of the term 'weight', it is better to call this situation the **apparent weightlessness** of objects in free fall together.

An astronaut in an orbiting space station would also appear weightless. The space station and the astronaut are in free fall together.

In the space station, the gravitational pull on the astronaut provides the centripetal force needed to stay in the orbit. This resultant force causes the centripetal acceleration. The same is true for the gravitational pull on the satellite and the satellite's acceleration. There is no contact force between the satellite and the astronaut so, once again, we have apparent weightlessness.

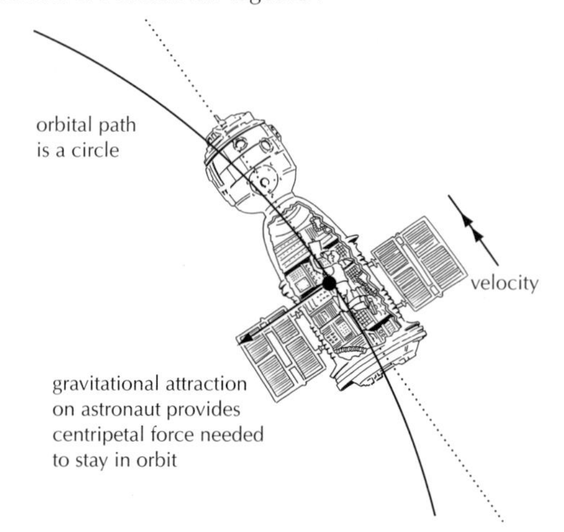

IB QUESTIONS – MOTION IN FIELDS

1 Which **one** of the following graphs best represents the variation of the kinetic energy, KE, and of the gravitational potential energy, GPE, of an orbiting satellite at a distance r from the centre of the Earth?

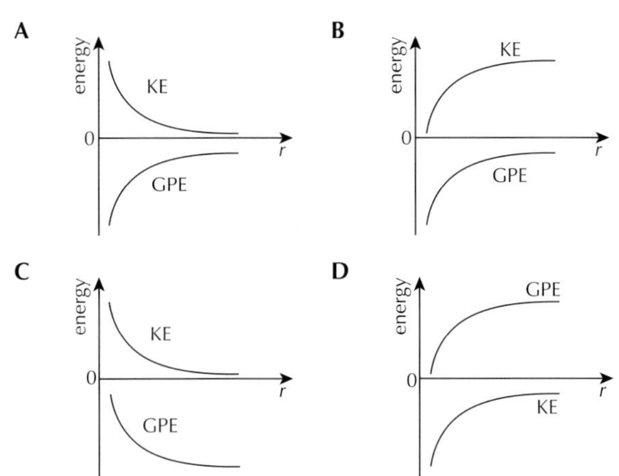

2 The diagram below illustrates some equipotential lines between two charged parallel metal plates.

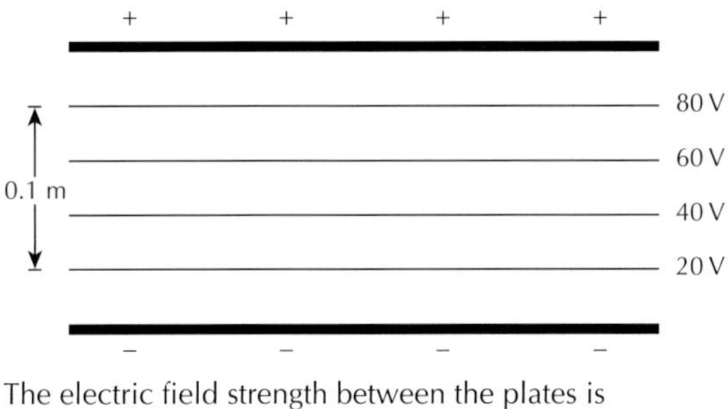

The electric field strength between the plates is

A 6 NC^{-1} **C** 600 NC^{-1}
B 8 NC^{-1} **D** 800 NC^{-1}

3 A particle of mass m moves with constant speed v in a circle of radius r. The work done on the particle by the centripetal force in one complete revolution is

A $2\pi m v^2$ **B** $\frac{2\pi v^2}{m}$ **C** $\frac{2\pi m}{v^2}$ **D** zero

4 A projectile is launched **horizontally** from a high tower. Which **one** of the following graphs best represents the **vertical component** of the projectile's velocity from the time it is launched to the time it hits the ground? Assume negligible air resistance.

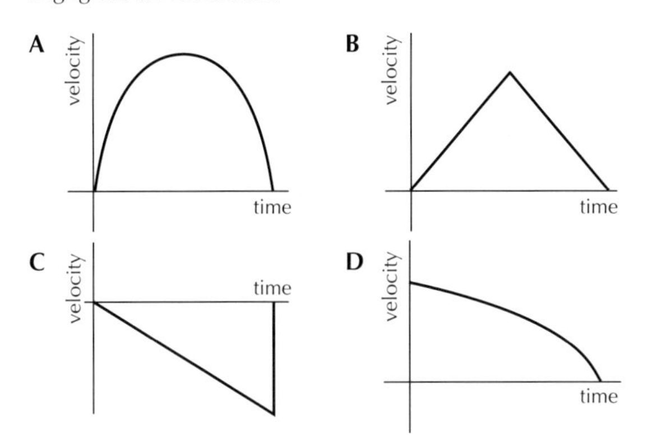

5 The Space Shuttle orbits about 300 km above the surface of the Earth. The shape of the orbit is circular, and the mass of the Space Shuttle is 6.8×10^4 kg. The mass of the Earth is 6.0×10^{24} kg, and radius of the Earth is 6.4×10^6 m.

(a) **(i)** Calculate the change in the Space Shuttle's gravitational potential energy between its launch and its arrival in orbit. [3]

(ii) Calculate the speed of the Space Shuttle whilst in orbit. [2]

(iii) Calculate the energy needed to put the Space Shuttle into orbit. [2]

(b) **(i)** What forces, if any, act on the astronauts inside the Space Shuttle whilst in orbit? [1]

(ii) Explain why astronauts aboard the Space Shuttle feel weightless. [2]

(c) Imagine an astronaut 2 m outside the exterior walls of the Space Shuttle, and 10 m from the centre of mass of the Space Shuttle. By making appropriate assumptions and approximations, calculate how long it would take for this astronaut to be pulled back to the Space Shuttle by the force of gravity alone. [7]

6 **(a)** The diagram below shows a planet of mass M and radius R_p.

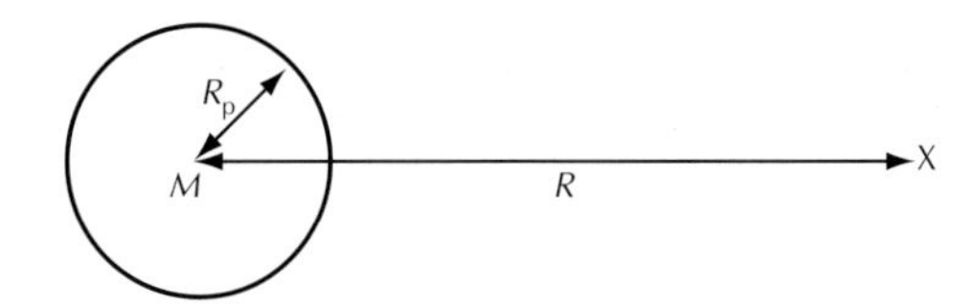

The gravitational potential V due to the planet at point X distance R from the centre of the planet is given by

$$V = -\frac{GM}{R}$$

where G is the universal gravitational constant.

Show that the gravitational potential V can be expressed as

$$V = -\frac{g_0 R_p^2}{R}$$

where g_0 is the acceleration of free fall at the surface of the planet. [3]

(b) The graph below shows how the gravitational potential V due to the planet varies with distance R from the centre of the planet for values of R greater than R_p, where $R_p = 2.5 \times 10^6$ m.

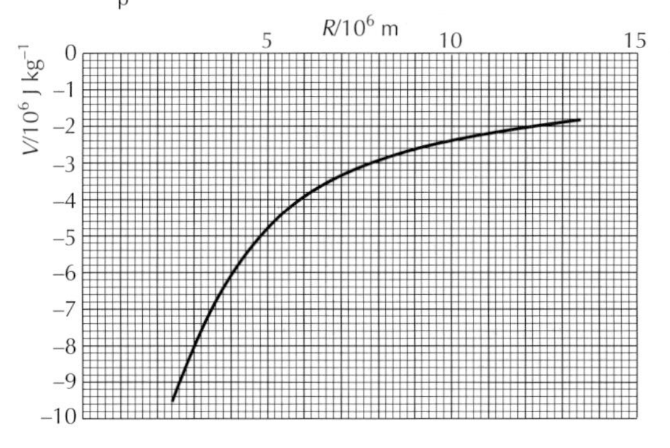

Use the data from the graph to

(i) determine a value of g_0. [2]

(ii) show that the minimum energy required to raise a satellite of mass 3000 kg to a height 3.0×10^6 m above the **surface** of the planet is about 1.7×10^{10} J. [3]

HL Thermodynamic systems and concepts

DEFINITIONS

Historically, the study of the behaviour of ideal gases led to some very fundamental concepts that are applicable to many other situations. These laws, otherwise known as the laws of **thermodynamics,** provide the modern physicist with a set of very powerful intellectual tools.

The terms used need to be explained.

Thermodynamic system — Most of the time when studying the behaviour of an ideal gas in particular situations, we focus on the macroscopic behaviour of the gas as a whole. In terms of work and energy, the gas can gain or lose thermal energy and it can do work or work can be done on it. In this context, the gas can be seen as a **thermodynamic system**.

The surroundings — If we are focusing our study on the behaviour of an ideal gas, then everything else can be called its **surroundings**. For example the expansion of a gas means that work is done by the gas on the surroundings (see below).

Heat ΔQ — In this context heat refers to the transfer of a quantity of thermal energy between the system and its surroundings. This transfer must be as a result of a temperature difference.

HOT
thermal energy flow
HOT
thermal energy flow
COLD
HOT
thermal energy flow

Work ΔW — In this context, work refers to the macroscopic transfer of energy. For example

1. work done = force×distance

F
F
compression

When a gas is compressed, work is done on the gas

When a gas is compressed, the surroundings do work on it. When a gas expands it does work on the surroundings.

2. work done = potential difference×current×time

heater
V
A
power supply

This is just another example of work being done on the gas.

Internal energy U
ΔU = change in internal energy

The internal energy can be thought of as the energy held within a system. It is the sum of the PE due to the intermolecular forces and the kinetic energy due to the random motion of the molecules. See page 28.

This is different to the total energy of the system, which would also include the overall motion of the system and any PE due to external forces.

In thermodynamics, it is the changes in internal energy that are being considered. If the internal energy of a gas is increased, then its temperature must increase. A change of phase (e.g. liquid → gas) also involves a change of internal energy.

system with internal energy U
v velocity (system also has kinetic energy)
h height (system also has gravitational potential energy)

The total energy of a system is not the same as its internal energy

WORK DONE DURING EXPANSION AT CONSTANT PRESSURE

Whenever a gas expands, it is doing work on its surroundings. If the pressure of the gas is changing all the time, then calculating the amount of work done is complex. This is because we cannot assume a constant force in the equation of work done (work done = force × distance). If the pressure changes then the force must also change. If the pressure is constant then the force is constant and we can calculate the work done.

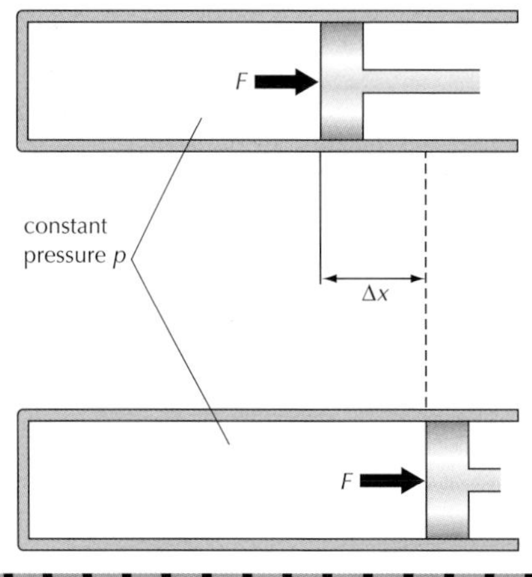

Work done ΔW = force×distance

$= F\,\Delta x$

Since pressure $= \dfrac{\text{force}}{\text{area}}$

$F = pA$

therefore

$\Delta W = pA\Delta x$

but $A\Delta x = \Delta V$

so work done $= p\,\Delta V$

So if a gas increases its volume (ΔV is positive) then the gas does work (ΔW is positive)

The gas laws

GAS LAWS

For a given sample of a gas, the pressure, volume and temperature are all related to one another.

The graphs below outline what might be observed experimentally.

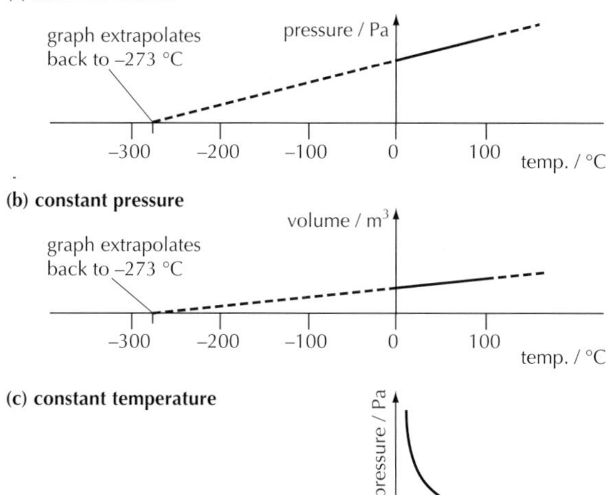

Points to note:

- Although pressure and volume both vary linearly with Celsius temperature, neither pressure nor volume is proportional to Celsius temperature.
- A different sample of gas would produce a different straight-line variation for pressure (or volume) against temperature but both graphs would extrapolate back to the same low temperature, –273 °C. This temperature is known as **absolute zero**.
- As pressure increases, the volume decreases. In fact they are inversely proportional.

The trends can be seen more clearly if this information is presented in a slightly different way.

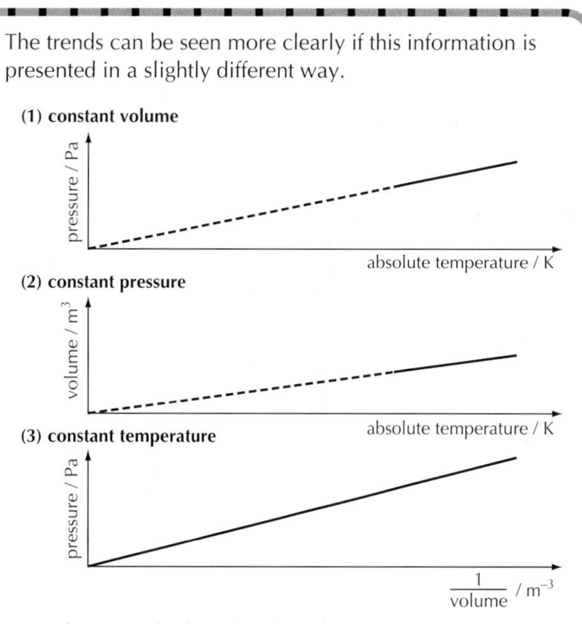

From these graphs for a fixed mass of gas we can say that

1. At constant V, $p \propto T$ or $\frac{p}{T}$ = constant (the pressure law)
2. At constant p, $V \propto T$ or $\frac{V}{T}$ = constant (Charles's law)
3. At constant T, $p \propto \frac{1}{V}$ or $p\,V$ = constant (Boyle's law)

These relationships are known as the **ideal gas laws**. The temperature is always expressed in kelvin. (These laws do not always apply to experiments done with real gases. A real gas is said to 'deviate' from ideal behaviour under certain conditions (e.g. high pressure).)

EQUATION OF STATE

The three ideal gas laws can be combined together to produce one mathematical relationship.

$$\frac{pV}{T} = \text{constant.}$$

This constant will depend on the mass and type of gas.

If we compare the value of this constant for different masses of different gases, it turns out to depend on the number of molecules that are in the gas – not their type. In this case we use the definition of the mole to state that for n moles of ideal gas

$$\frac{pV}{nT} = \text{a universal constant.}$$

The universal constant is called the **molar gas constant** R.

The SI unit for R is J mol^{-1} K^{-1}

R = 8.314 J mol^{-1} K^{-1}

Summary: $\frac{pV}{nT} = R$ Or $p\,V = n\,R\,T$

EXAMPLE

What volume will be occupied by 8 g of helium (mass number 4) at room temperature (20°C) and atmospheric pressure (1.0×10^5 Pa)

$$n = \frac{8}{4} = 2 \text{ moles}$$

$$T = 20 + 273 = 293 \text{ k}$$

$$V = \frac{nRT}{p} = \frac{2 \times 8.314 \times 293}{1.0 \times 10^5} = 0.049 \text{ m}^3$$

IDEAL GASES AND REAL GASES

An ideal gas is a one that follows the gas laws for all values of p, V and T and thus ideal gases cannot be liquefied. The microscopic description of an ideal gas is given on page 31. Real gases, however, can approximate to ideal behaviour providing that the intermolecular forces are small enough to be ignored. For this to apply, the pressure /density of the gas must be low and the temperature must be reasonably high.

The first law of thermodynamics

FIRST LAW OF THERMODYNAMICS

There are three fundamental laws of thermodynamics. The first law is simply a statement of the principle of energy conservation as applied to the system. If an amount of thermal energy ΔQ is given to a system, then one of two things must happen (or a combination of both). The system can increase its internal energy ΔU or it can do work ΔW.

As energy is conserved

$$\Delta Q = \Delta U + \Delta W$$

It is important to remember what the signs of these symbols mean. They are all taken from the system's 'point of view'.

ΔQ If this is **positive**, then thermal energy is going into the system.
If it is **negative**, the thermal energy is going out of the system.

ΔU If this is **positive**, then the internal energy of the system is **increasing**.
(The temperature of the gas is increasing.)
If it is **negative**, the internal energy of the system is **decreasing**.
(The temperature of the gas is decreasing.)

ΔW If this is **positive**, then the **system is doing work** on the surroundings.
(The gas is expanding.)
If it is **negative**, the **surroundings are doing work** on the system.
(The gas is contracting.)

p–V DIAGRAMS AND WORK DONE

It is often useful to represent the changes that happen to a gas during a thermodynamic process on a *p-V* diagram. An important reason for choosing to do this is that the area under the graph represents the work done. The reasons for this are shown below.

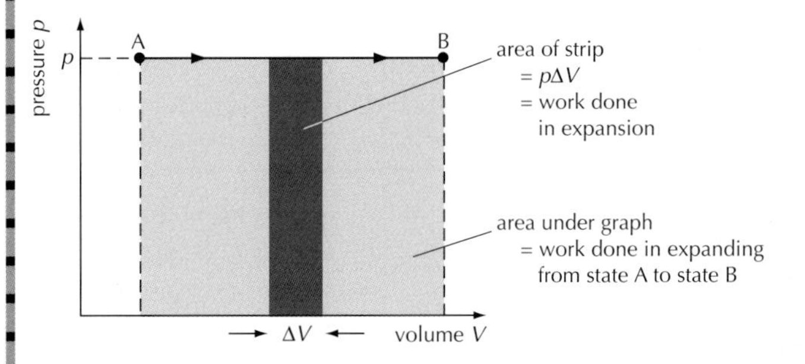

This turns out to be generally true for any thermodynamic process.

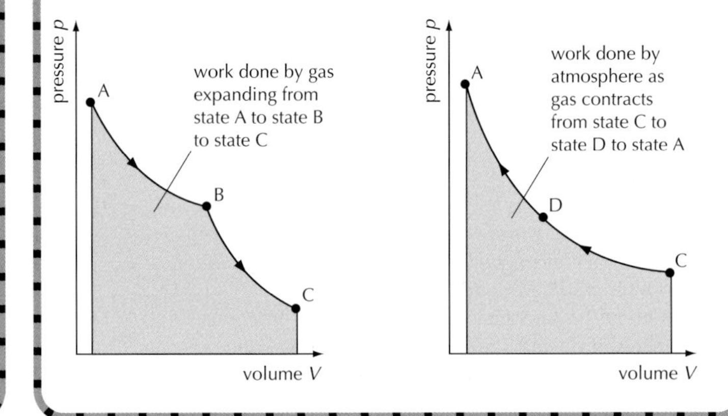

IDEAL GAS PROCESSES

A gas can undergo any number of different types of change or process. Four important process are considered below. In each case the changes can be represented on a pressure–volume diagram and the first law of thermodynamics must apply. To be precise, these diagrams represent a type of process called a reversible process.

1. Isochoric (isovolumetric)

In an isochoric process, also called an isovolumetric process, the gas has a constant volume. The diagram below shows an **isochoric decrease** in pressure.

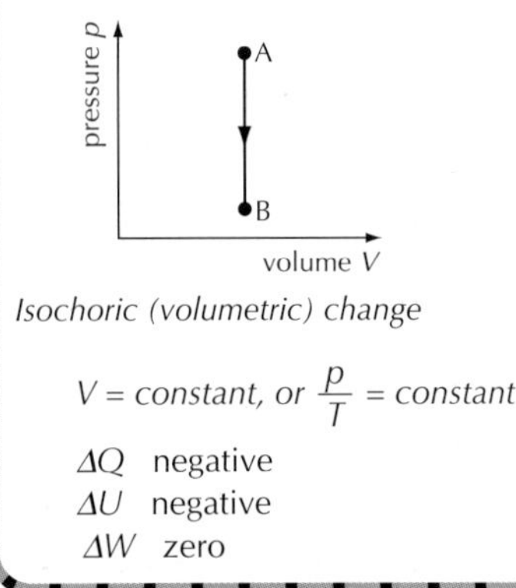

Isochoric (volumetric) change

V = constant, or $\frac{p}{T}$ = constant

ΔQ negative
ΔU negative
ΔW zero

2. Isobaric

In an isobaric process the gas has a constant pressure. The diagram below shows an **isobaric expansion**.

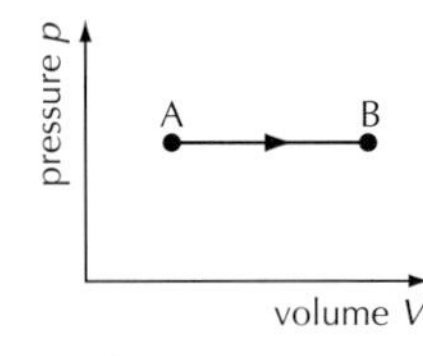

Isobaric change

p = constant, or $\frac{V}{T}$ = constant

ΔQ positive
ΔU positive
ΔW positive

3. Isothermal

In an isothermal process the gas has a constant temperature. The diagram below shows an **isothermal expansion**.

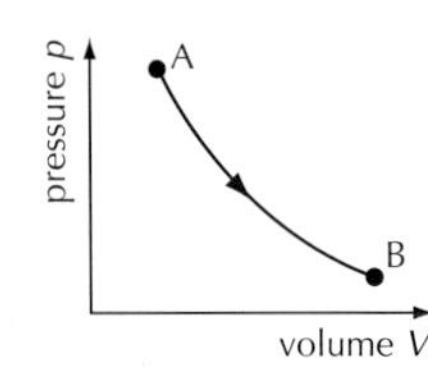

Isothermal change

T = constant, or *pV* = constant

ΔQ positive
ΔU zero
ΔW positive

4. Adiabatic

In an adiabatic process there is no thermal energy transfer between the gas and the surrounds. This means that if the gas does work it must result in a decrease in internal energy. A rapid compression or expansion is approximately adiabatic. This is because done quickly there is not sufficient time for thermal energy to be exchanged with the surroundings. The diagram below shows an **adiabatic expansion**.

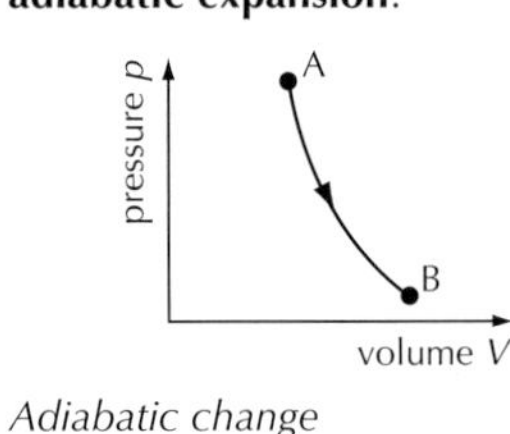

Adiabatic change

ΔQ zero
ΔU negative
ΔW positive

Second law of thermodynamics and entropy

SECOND LAW OF THERMODYNAMICS

Historically the **second law of thermodynamics** has been stated in many different ways. All of these versions can be shown to be equivalent to one another.

In principle there is nothing to stop the complete conversion of thermal energy into useful work. In practice, a gas can not continue to expand forever – the apparatus sets a physical limit. Thus **the continuous conversion of thermal energy into work requires a cyclical process – a heat engine.**

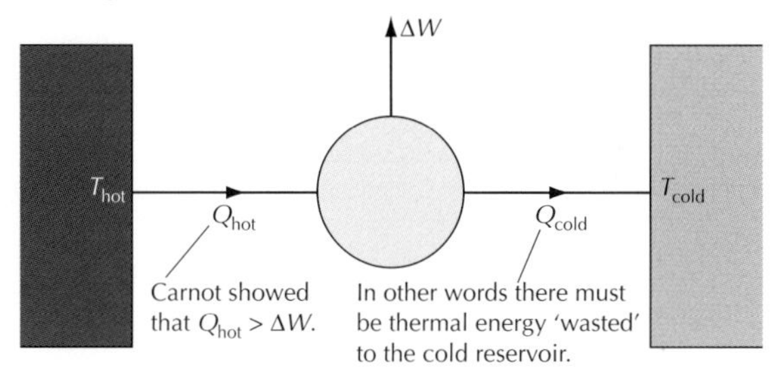

ΔW
T_{hot}
Q_{hot}
Q_{cold}
T_{cold}
Carnot showed that $Q_{hot} > \Delta W$.
In other words there must be thermal energy 'wasted' to the cold reservoir.

This realisation leads to possibly the simplest formulation of the second law of thermodynamics (the **Kelvin-Planck** formulation).

No heat engine, operating in a cycle, can take in heat from its surroundings and totally convert it into work

Other possible formulations include the following:

No heat pump can transfer thermal energy from a low-temperature reservoir to a high-temperature reservoir without work being done on it (Clausius)

Heat flows from hot objects to cold objects

The concept of **entropy** leads to one final version of the second law.

The entropy of the universe can never decrease

EXAMPLES

The first and second laws of thermodynamics both must apply to all situations. Local decreases of entropy are possible so long as elsewhere there is a corresponding increase.

1. A refrigerator is an example of a heat pump.

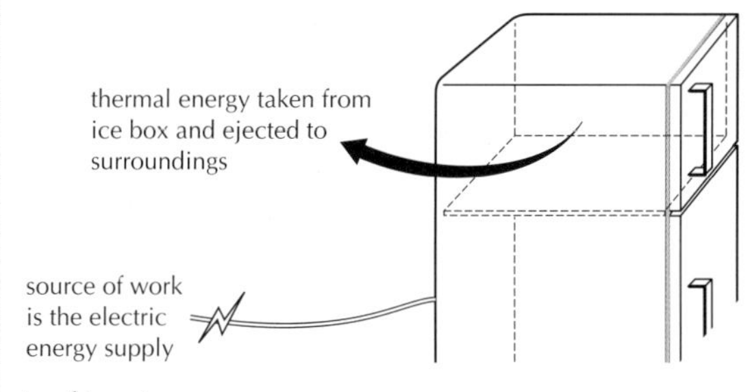

thermal energy taken from ice box and ejected to surroundings
source of work is the electric energy supply

A refrigerator

2. It should be possible to design a theoretical system for propelling a boat based around a heat engine. The atmosphere could be used as the hot reservoir and cold water from the sea could be used as the cold reservoir. The movement of the boat through the water would be the work done. This is possible BUT it cannot continue to work for ever. The sea would be warmed and the atmosphere would be cooled and eventually there would be no temperature difference.

ENTROPY AND ENERGY DEGRADATION

Entropy is a property that expresses the disorder in the system.

The details are not important but the entropy S of a system is linked to the number of possible arrangements W of the system.

Because molecules are in random motion, one would expect roughly equal numbers of gas molecules in each side of a container.

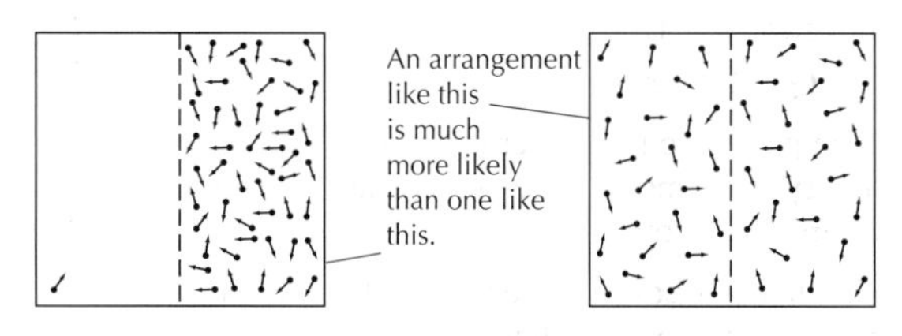

An arrangement like this is much more likely than one like this.

The number of ways of arranging the molecules to get the set-up on the right is greater than the number of ways of arranging the molecules to get the set-up on the left. This means that the entropy of the system on the right is greater than the entropy of the system on the left.

In any random process the amount of disorder will tend to increase. In other words, the total entropy will always increase. The entropy change ΔS is linked to the thermal energy change ΔQ and the temperature T. ($\Delta S = \Delta Q/T$.)

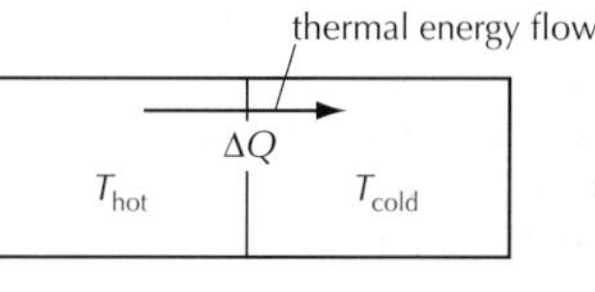

thermal energy flow
ΔQ
T_{hot}
T_{cold}

decrease of entropy $= \dfrac{\Delta Q}{T_{hot}}$ increase of entropy $= \dfrac{\Delta Q}{T_{cold}}$

When thermal energy flows from a hot object to a colder object, overall the total entropy has increased.

In many situations the idea of energy **degradation** is a useful concept. The more energy is shared out, the more degraded it becomes – it is harder to put it to use. For example, the internal energy that is 'locked' up in oil can be released when the oil is burned. In the end, all the energy released will be in the form of thermal energy – shared among many molecules. It is not feasible to get it back.

3. Water freezes at 0 °C because this is the temperature at which the entropy increase of the surroundings (when receiving the latent heat) equals the entropy decrease of the water molecules becoming more ordered. It would not freeze at a higher temperature because this would mean that the overall entropy of the system would decrease.

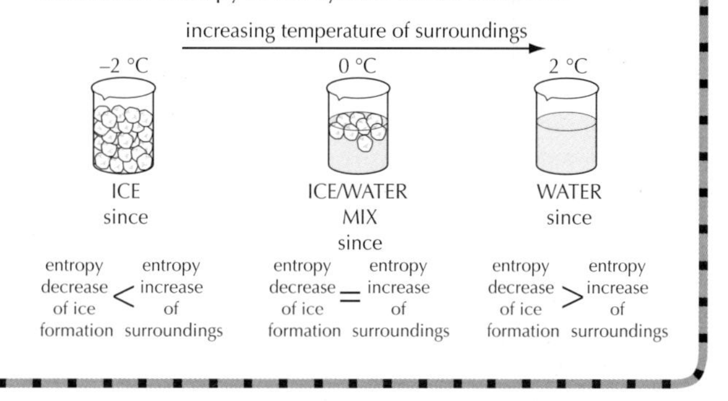

increasing temperature of surroundings
−2 °C
0 °C
2 °C
ICE since
ICE/WATER MIX since
WATER since
entropy decrease of ice formation < entropy increase of surroundings
entropy decrease of ice formation = entropy increase of surroundings
entropy decrease of ice formation > entropy increase of surroundings

Heat engines and heat pumps

HEAT ENGINES

A central concept in the study of thermodynamics is the **heat engine**. A heat engine is any device that uses a source of thermal energy in order to do work. It converts heat into work. The internal combustion engine in a car and the turbines that are used to generate electrical energy in a power station are both examples of heat engines. A block diagram representing a generalised heat engine is shown below.

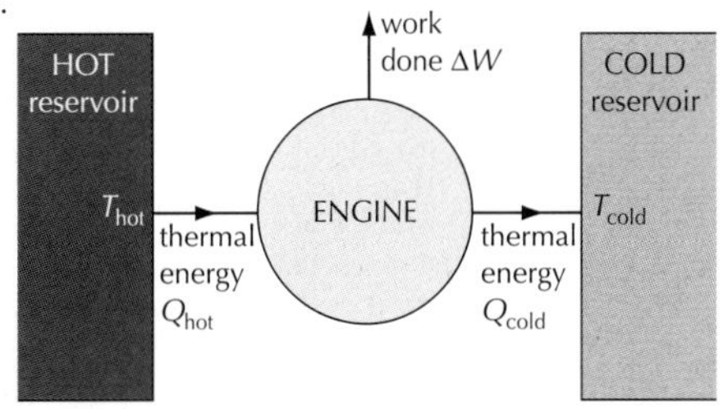

Heat engine

In this context, the word **reservoir** is used to imply a constant temperature source (or sink) of thermal energy. Thermal energy can be taken from the hot reservoir without causing the temperature of the hot reservoir to change. Similarly thermal energy can be given to the cold reservoir without increasing its temperature.

An ideal gas can be used as a heat engine. The p-V diagram right represents a simple example. The four-stage cycle returns the gas to its starting conditions, but the gas has done work. The area enclosed by the cycle represents the amount of work done.

In order to do this, some thermal energy must have been taken from a hot reservoir (during the isovolumetric increase in pressure and the isobaric expansion). A different amount of thermal energy must have been ejected to a cold reservoir (during the isovolumetric decrease in pressure and the isobaric compression).

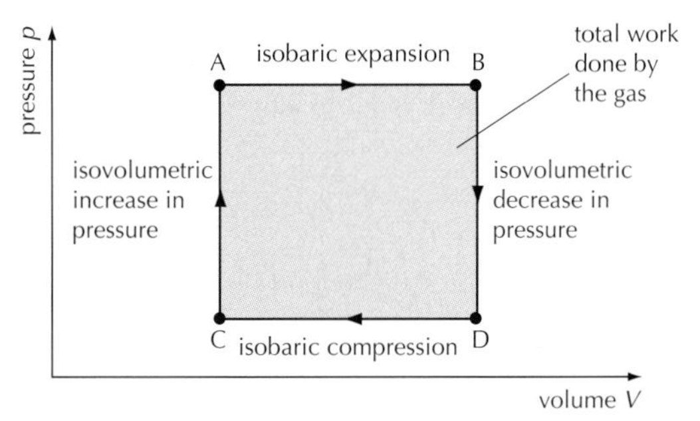

The thermal efficiency of a heat engine is defined as

$$\text{Efficiency} = \frac{\text{work done}}{\text{(thermal energy taken from hot reservoir)}}$$

This is equivalent to

$$\text{Efficiency} = \frac{\text{rate of doing work}}{\text{(thermal power taken from hot reservoir)}}$$

In symbols, efficiency $= \frac{\Delta W}{Q_{hot}}$

The cycle of changes that results in a heat engine with the maximum possible efficiency is called the **Carnot cycle**.

HEAT PUMPS

A **heat pump** is a heat engine being run in reverse. A heat pump causes thermal energy to be moved from a cold reservoir to a hot reservoir. In order for this to be achieved, mechanical work must be done.

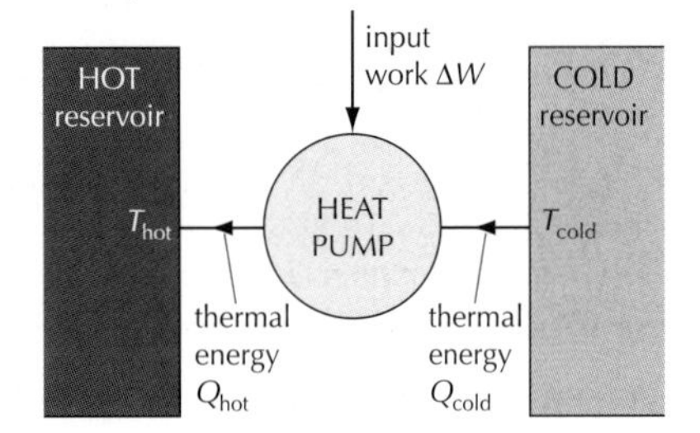

Heat pump

Once again an ideal gas can be used as a heat pump. The thermodynamic processes can be exactly the same ones as were used in the heat engine, but the processes are all opposite. This time an anticlockwise circuit will represent the cycle of processes.

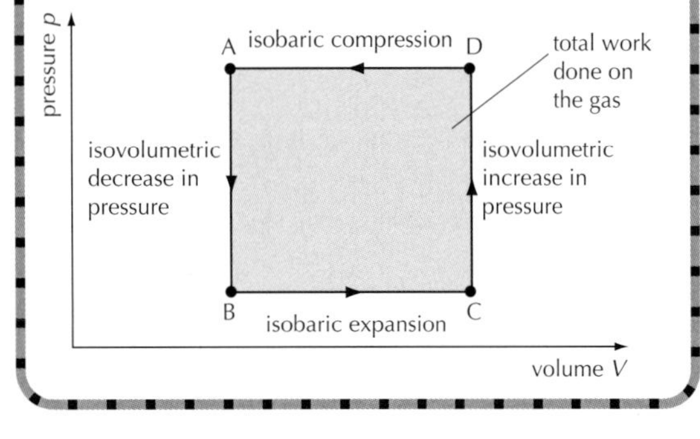

CARNOT CYCLES AND CARNOT THEOREM

The Carnot cycle represents the cycle of processes for a theoretical heat engine with the maximum possible efficiency. Such an idealised engine is called a **Carnot engine**.

It consists of an ideal gas undergoing the following processes.

- Isothermal expansion ($A \to B$)
- Adiabatic expansion ($B \to C$)
- Isothermal compression ($C \to D$)
- Adiabatic compression ($D \to A$)

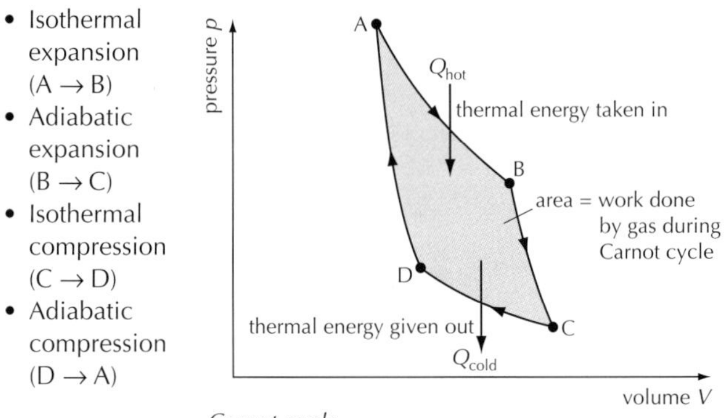

Carnot cycle

The temperatures of the hot and cold reservoirs fix the maximum possible efficiency that can be achieved.

The efficiency of a Carnot engine can be shown to be

$$e_c = 1 - \left(\frac{T_{cold}}{T_{hot}}\right) \text{ (where } T \text{ is in kelvin)}$$

An engine operates at 300 °C and ejects heat to the surroundings at 20 °C. The maximum possible theoretical efficiency

$$= 1 - \frac{293}{573}$$

$$= 0.49 = 49\%$$

IB QUESTIONS – THERMAL PHYSICS 

1 An ideal gas expands **isothermally**, absorbing a certain amount of energy, Q, in the process. It then returns to its original volume **adiabatically**. During the adiabatic process, the internal energy change of gas will be

A zero. **C** equal to Q.
B smaller than Q. **D** greater than Q.

2 When the volume of a gas increases it does work. The work done would be greatest for which one of the following processes?

A Isobaric.
B Adiabatic.
C Isothermal.
D The work done would be the same for all of the above.

3 It is proposed to build a heat engine that would operate between a hot reservoir at a temperature of 400 K and a cold reservoir at 300 K. See the diagram below. In each cycle it would take 100 J from the hot reservoir, lose 25 J to the cold reservoir and do 75 J of work.

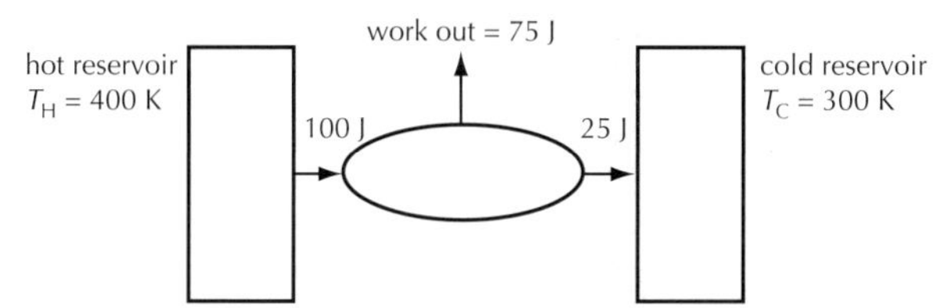

This proposed heat engine would violate

A both the first and the second laws of theormodynamics.
B the first but not the second law of thermodynamics.
C the second but not the first law of thermodynamics.
D neither the first nor the second law of thermodynamics.

4 The energy absorbed by an ideal gas during an isothermal expansion is equal to

A the work done by the gas.
B the work done on the gas.
C the change in the internal energy of the gas.
D zero.

5 A fixed mass of a gas undergoes various changes of temperature, pressure and volume such that it is taken round the p–V cycle shown in the diagram below.

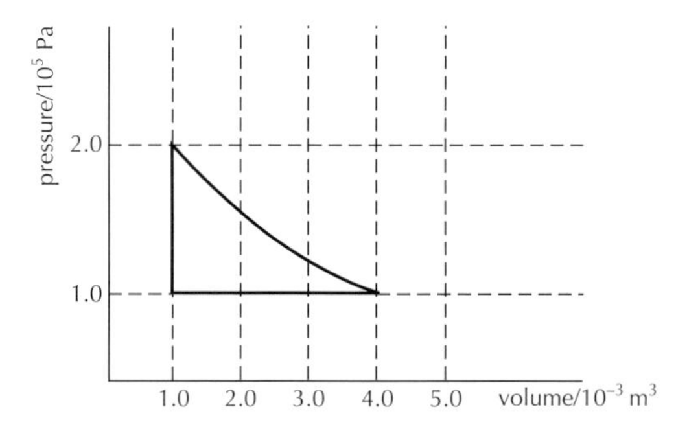

The following sequence of processes takes place during the cycle.

X → Y the gas expands at constant temperature and the gas absorbs energy from a reservoir and does 450 J of work.

Y → Z the gas is compressed and 800 J of thermal energy is transferred from the gas to a reservoir.

Z → X the gas returns to its initial stage by absorbing energy from a reservoir.

(a) Is there a change in internal energy of the gas during the processes X → Y? Explain. [2]

(b) Is the energy absorbed by the gas during the process X → Y less than, equal to or more than 450 J? Explain. [2]

(c) Use the graph to determine the work done on the gas during the process Y → Z. [3]

(d) What is the change in internal energy of the gas during the process Y → Z? [2]

(e) How much thermal energy is absorbed by the gas during the process Z → X? Explain your answer. [2]

(f) What quantity is represented by the area enclosed by the graph? Estimate its value. [2]

(g) The overall efficiency of a heat engine is defined as

$$\text{Efficiency} = \frac{\text{net work done by the gas during a cycle}}{\text{total energy absorbed during a cycle}}$$

If this p–V cycle represents the cycle for a particular heat engine determine the efficiency of the heat engine. [2]

6 In a **diesel** engine, air is initially at a pressure of 1×10^5 Pa and a temperature of 27 °C. The air undergoes the cycle of changes listed below. At the end of the cycle, the air is back at its starting conditions.

1 An **adiabatic compression** to 1/20th of its original volume.
2 A brief **isobaric expansion** to 1/10th of its original volume.
3 An **adiabatic expansion** back to its original volume.
4 A cooling down at constant volume.

(a) Sketch, with labels, the cycle of changes that the gas undergoes. Accurate values are not required. [3]

(b) If the pressure after the **adiabatic compression** has risen to 6.6×10^6 Pa, calculate the temperature of the gas. [2]

(c) In which of the four processes:
(i) is work done **on** the gas? [1]
(ii) is work done **by** the gas? [1]
(iii) does ignition of the air-fuel mixture take place? [1]

(d) Explain how the 2nd law of thermodynamics applies to this cycle of changes. [2]

7 **(a)** State what is meant by *an ideal* gas. [2]

(b) The internal volume of a gas cylinder is 2.0×10^{-2} m^3. An ideal gas is pumped into the cylinder until the pressure becomes 20 MPa at a temperature of 17°C.

Determine
(i) the number of moles of gas in the cylinder [2]
(ii) the number of gas atoms in the cylinder. [2]

(c) **(i)** Using your answers in **(b)**, determine the average volume occupied by one gas atom. [1]
(ii) Estimate a value for the average separation of the gas atoms. [2]

Nature and production of standing (stationary) waves

STANDING WAVES

A special case of interference occurs when two waves meet that are:

- of the same amplitude
- of the same frequency
- travelling in opposite directions.

In these conditions a **standing wave** will be formed.

The conditions needed to form standing waves seem quite specialised, but standing waves are in fact quite common. They often occur when a wave reflects back from a boundary along the route that it came. Since the reflected wave and the incident wave are of (nearly) equal amplitude, these two waves can interfere and produce a standing wave.

Perhaps the simplest way of picturing a standing wave would be to consider two transverse waves travelling in opposite directions along a stretched rope. The series of diagrams below shows what happens.

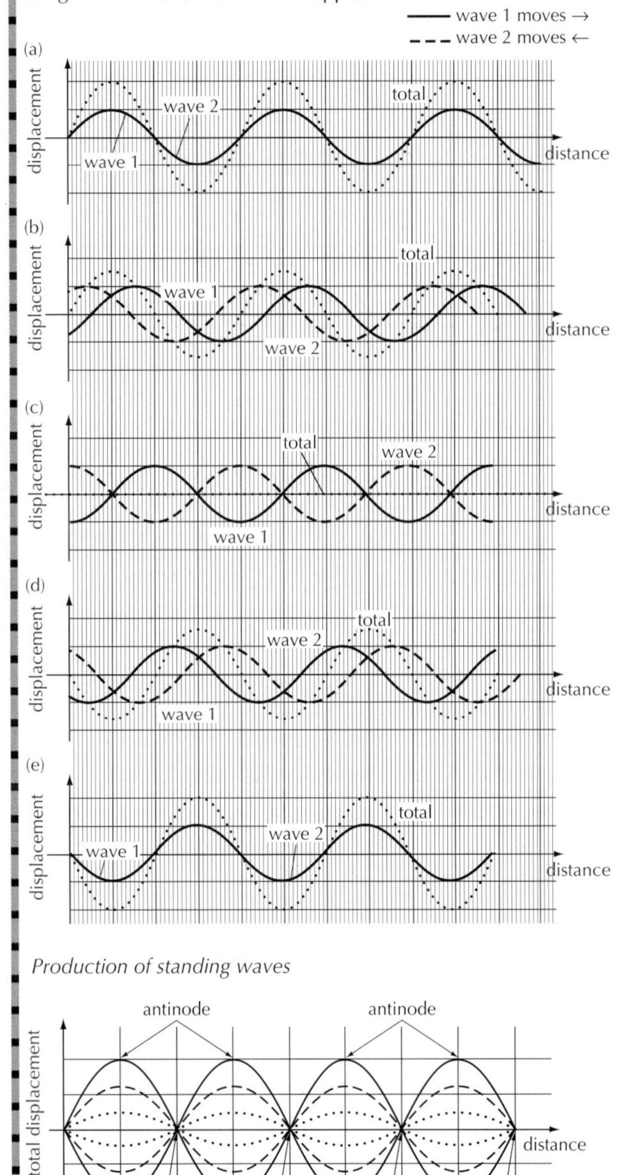

Production of standing waves

A standing wave – the pattern remains fixed.

There are some points on the rope that are always at rest. These are called the **nodes**. The points where the maximum movement takes place are called **antinodes**. The resulting standing wave is so called because the wave pattern remains fixed in space – it is its amplitude that changes over time. A comparison with a normal (travelling) wave is given below.

	Stationary wave	**Normal (travelling) wave**
Amplitude	All points on the wave have different amplitudes. The maximum amplitude is $2A$ at the antinodes. It is zero at the nodes.	All points on the wave have the same amplitude.
Frequency	All points oscillate with the same frequency.	All points oscillate with the same frequency.
Wavelength	This is **twice** the distance from one node (or antinode) to the next node (or antinode).	This is the shortest distance (in metres) along the wave between two points that are in phase with one another.
Phase	All points between one node and the next node are moving in phase.	All points along a wavelength have different phases.
Energy	Energy is not transmitted by the wave, but it does have an energy associated with it.	Energy is transmitted by the wave.

Although the example above involved transverse waves on a rope, a standing wave can also be created using sound or light waves. All musical instruments involve the creation of a standing sound wave inside the instrument. The production of laser light involves a standing light wave. Even electrons in hydrogen atoms can be explained in terms of standing waves.

A standing longitudinal wave can be particularly hard to imagine. The diagram below attempts to represent one example – a standing sound wave.

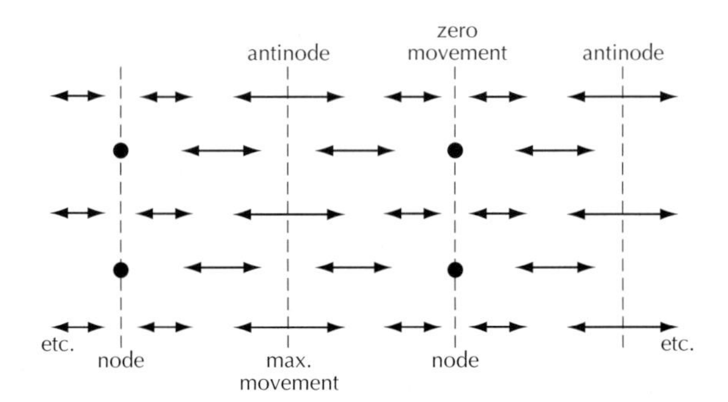

A longitudinal standing wave

Boundary conditions and resonance

RESONANCE

The concept of **resonance** is a very general concept – examples can be found in many areas of physics. These include the motion of a child on a swing (when the swing is being pushed by an adult) and the tuning of a radio circuit.

In outline, any system that can oscillate can be made to resonate. A system will have its own natural frequency (or frequencies) of oscillation, but it can be forced to oscillate at any frequency if it is driven by another oscillator. Resonance occurs when the driving frequency is equal to the system's own natural frequency. Under these conditions, the amplitude of the oscillations grows and the energy of the system reaches a maximum.

BOUNDARY CONDITIONS

The boundary conditions of the system specify the conditions that must be met at the edges (the boundaries) of the system when standing waves are taking place. Any standing wave that meets these boundary conditions will be a possible resonant mode of the system.

1. Transverse waves on a string.
If the string is fixed at each end, the ends of the string cannot oscillate. Both ends of the string would reflect a travelling wave and thus a standing wave is possible. The only standing waves that fit these boundary conditions are ones that have nodes at each end. The diagrams below show the possible resonant modes.

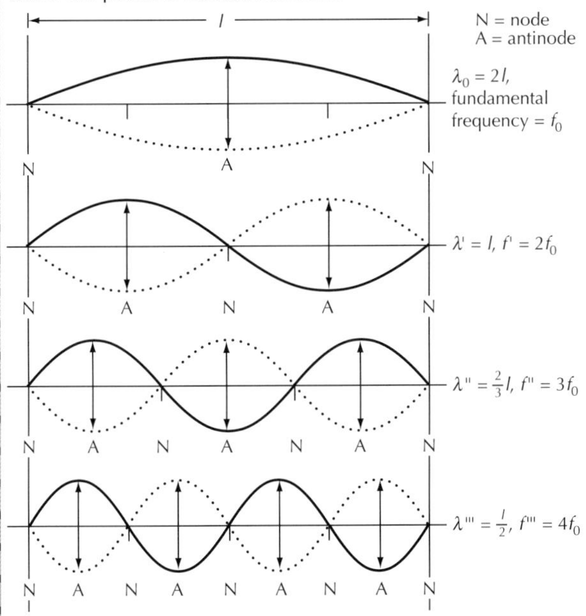

Fundamental and higher resonant modes for a string.

The resonant mode that has the lowest frequency is called the **fundamental** or the **first harmonic**. Higher resonant modes are called **harmonics**. Many musical instruments (e.g. piano, violin, guitar etc) involve similar oscillations of metal 'strings'.

2. Longitudinal sound waves in a pipe.
A longitudinal standing wave can be set up in the column of air enclosed in a pipe. As in the example above, this results from the reflections that take place at both ends.

As before, the boundary conditions determine the standing waves that can exist in the tube. A closed end must be a displacement node. An open end must be an antinode. Possible standing waves are shown for a pipe open at both ends and a pipe closed at one end.

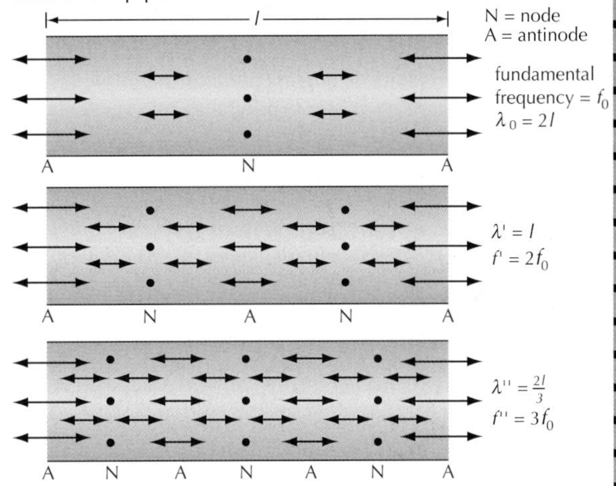

Fundamental and higher resonant modes for a pipe open at both ends.

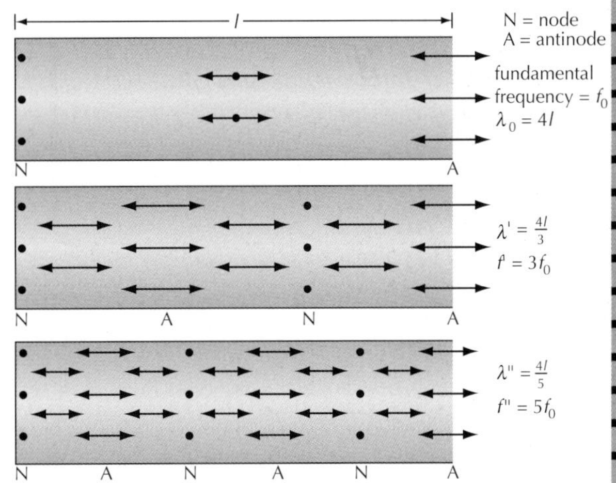

Fundamental and higher resonant modes for a pipe closed at one end.

Musical instruments that involve a standing wave in a column of air include the flute, the trumpet, the recorder, organ pipes etc.

EXAMPLE

An organ pipe (open at one end) is 1.2 m long.
Calculate its fundamental frequency.
The speed of sound is 330 m s^{-1}.

$l = 1.2$ m $\quad \therefore \frac{\lambda}{4} = 1.2$ m (fundamental)

$\therefore \lambda = 4.8$ m

$v = f\lambda$

$f = \frac{330}{4.8} \simeq 69$ Hz

The Doppler effect

DOPPLER EFFECT

The Doppler effect is the name given to the change of frequency of a wave as a result of the movement of the source or the movement of the observer.

When a source is moving:

- Sound waves are emitted at a particular frequency from the source.
- The speed of the sound wave in air does not change, but the motion of the source means that the wave fronts are all 'bunched up' ahead of the source.
- This means that the stationary observer receives sound waves of reduced wavelength.
- Reduced wavelength corresponds to an increased frequency of sound.

The overall effect is that the observer will hear sound at a higher frequency than it was emitted by the source. This applies when the source is moving towards the observer. A similar analysis quickly shows that if the source is moving away from the observer, sound of a lower frequency will be received. A change of frequency can also be detected if the source is stationary, but the observer is moving.

- When a police car or ambulance passes you on the road, you can hear the pitch of the sound change from high to low frequency. It is high when it is approaching and low when it is going away.
- Radar detectors can be used to measure the speed of a moving object. They do this by measuring the change in the frequency of the reflected wave.
- For the Doppler effect to be noticeable with light waves, the source (or the observer) needs to be moving at high speed. If a source of light of a particular frequency is moving away from an observer, the observer will receive light of a lower frequency. Since the red part of the spectrum has lower frequency than all the other colours, this is called a **red shift**.
- If the source of light is moving towards the observer, there will be a **blue shift**.

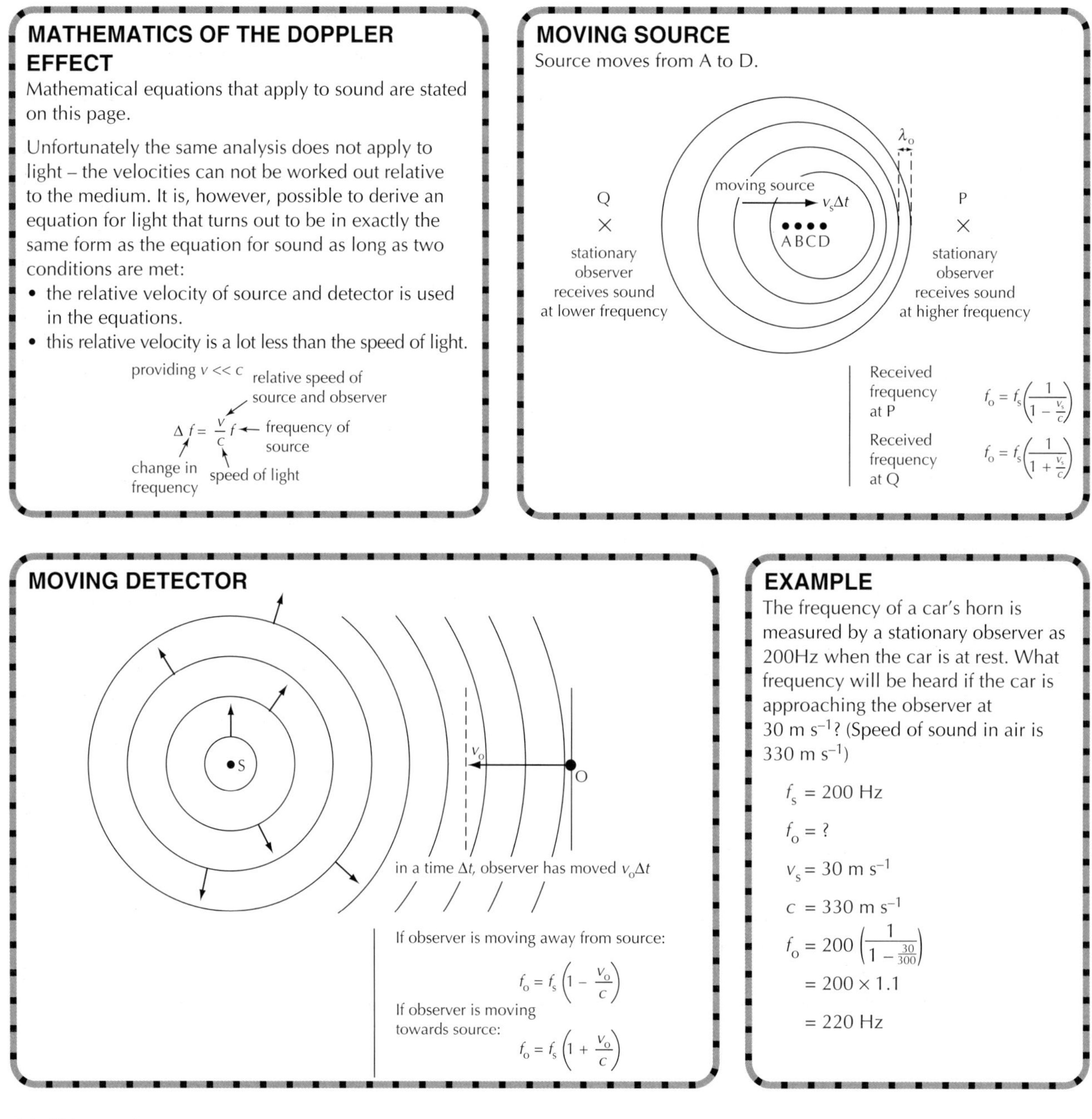

MATHEMATICS OF THE DOPPLER EFFECT

Mathematical equations that apply to sound are stated on this page.

Unfortunately the same analysis does not apply to light – the velocities can not be worked out relative to the medium. It is, however, possible to derive an equation for light that turns out to be in exactly the same form as the equation for sound as long as two conditions are met:

- the relative velocity of source and detector is used in the equations.
- this relative velocity is a lot less than the speed of light.

providing $v << c$

$$\Delta f = \frac{v}{c} f$$

MOVING SOURCE

Source moves from A to D.

Received frequency at P $\quad f_o = f_s\left(\frac{1}{1 - \frac{v_s}{c}}\right)$

Received frequency at Q $\quad f_o = f_s\left(\frac{1}{1 + \frac{v_s}{c}}\right)$

MOVING DETECTOR

If observer is moving away from source:

$$f_o = f_s\left(1 - \frac{v_o}{c}\right)$$

If observer is moving towards source:

$$f_o = f_s\left(1 + \frac{v_o}{c}\right)$$

EXAMPLE

The frequency of a car's horn is measured by a stationary observer as 200Hz when the car is at rest. What frequency will be heard if the car is approaching the observer at 30 m s^{-1}? (Speed of sound in air is 330 m s^{-1})

$f_s = 200$ Hz

$f_o = ?$

$v_s = 30$ m s^{-1}

$c = 330$ m s^{-1}

$f_o = 200\left(\frac{1}{1 - \frac{30}{300}}\right)$

$= 200 \times 1.1$

$= 220$ Hz

HL

Diffraction

BASIC OBSERVATIONS

Diffraction is a wave effect. The objects involved (slits, apertures etc.) have a size that is of the same order of magnitude as the wavelength of visible light.

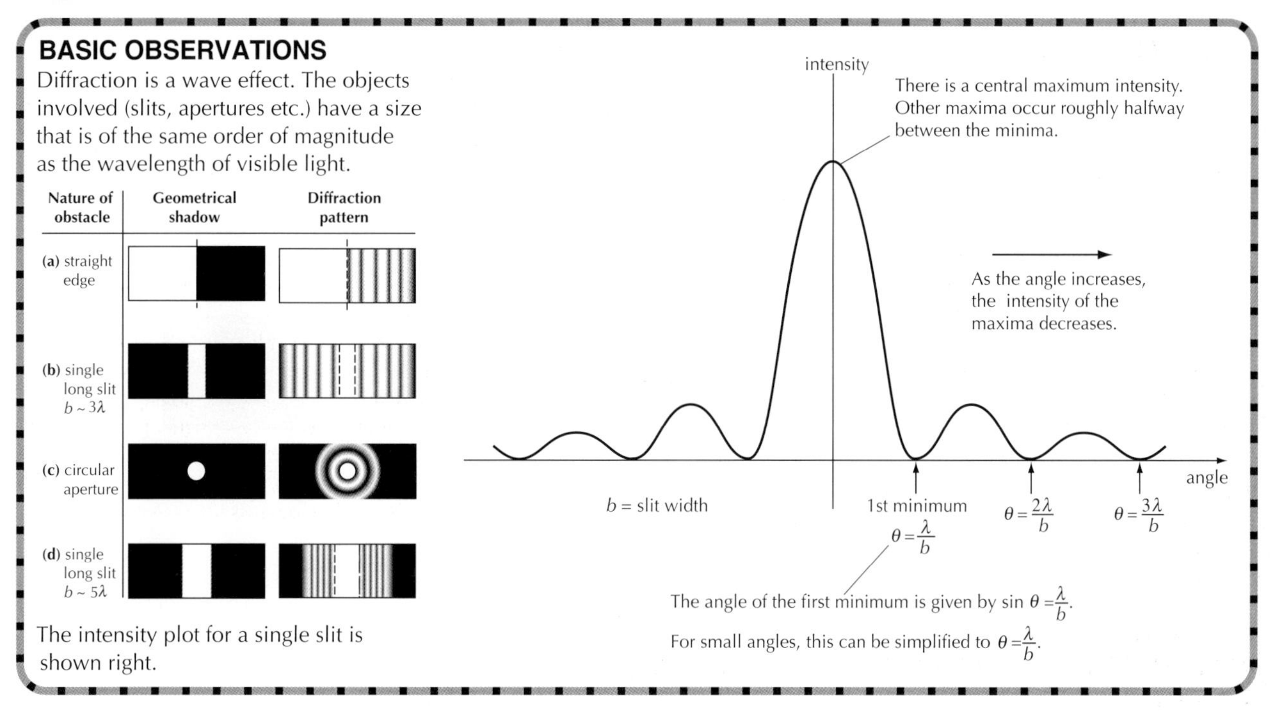

Nature of obstacle
Geometrical shadow
Diffraction pattern
(a) straight edge
(b) single long slit $b \sim 3\lambda$
(c) circular aperture
(d) single long slit $b \sim 5\lambda$
intensity
There is a central maximum intensity. Other maxima occur roughly halfway between the minima.
As the angle increases, the intensity of the maxima decreases.
angle
b = slit width
1st minimum $\theta = \frac{\lambda}{b}$
$\theta = \frac{2\lambda}{b}$
$\theta = \frac{3\lambda}{b}$
The angle of the first minimum is given by $\sin\theta = \frac{\lambda}{b}$.
For small angles, this can be simplified to $\theta = \frac{\lambda}{b}$.

The intensity plot for a single slit is shown right.

EXPLANATION

The shape of the relative intensity versus angle plot can be derived by applying an idea called **Huygens' principle**. We can treat the slit as a series of secondary wave sources. In the forward direction (θ = zero) these are all in phase so they add up to give a maximum intensity. At any other angle, there is a path difference between the rays that depends on the angle.

The overall result is the addition of all the sources. The condition for the first minimum is that the angle must make all of the sources across the slit cancel out.

b, θ

path difference across slit = $b\sin\theta$

For 1st minimum: $b\sin\theta = \lambda$

$\therefore \sin\theta = \frac{\lambda}{b}$

Since angle is small, $\sin\theta \approx \theta$

$\therefore \theta = \frac{\lambda}{b}$

for 1st minimum

PRACTICAL SIGNIFICANCE OF DIFFRACTION

Whenever an observer receives information from a source of electromagnetic waves, diffraction causes the energy to spread out. This spreading takes place as a result of any obstacle in the way and the width of the device receiving the electromagnetic radiation. Two sources of electromagnetic waves that are angularly close to one another will both spread out and interfere with one another. This can affect whether or not they can be resolved (see page 96).

Diffraction effects mean that it is impossible ever to see atoms because they are smaller than the wavelength of visible light, meaning that light will diffract around the atoms. It is, however, possible to image atoms using smaller wavelengths. Practical devices where diffraction needs to be considered include:

- CDs and DVDs – the maximum amount of information that can be stored depends on the size and the method used for recording information, see page 115.
- The electron microscope – resolves items that cannot be resolved using a light microscope. The electrons have an effective wavelength that is much smaller than the wavelength of visible light (see page 105).
- Radio telescopes – the size of the dish limits the maximum resolution possible. Several radio telescopes can be linked together in an array to create a virtual radio telescope with a greater diameter and with a greater ability to resolve astronomical objects.

Resolution

DIFFRACTION AND RESOLUTION

If two sources of light are very close in angle to one another, then they are seen as one single source of light. If the eye can tell the two sources apart, then the sources are said to be **resolved**. The diffraction pattern that takes place at apertures affects the eye's ability to resolve sources. The examples to the right show how the appearance of two line sources will depend on the diffraction that takes place at a slit. The resulting appearance is the addition of the two overlapping diffraction patterns. The graph of the resultant relative intensity of light at different angles is also shown.

These examples look at the situation of a line source of light and the diffraction that takes place at a slit. A more common situation would be a point source of light, and the diffraction that takes place at a circular aperture. The situation is exactly the same, but diffraction takes place all the way around the aperture. As seen on the previous page, the diffraction pattern of the point source is thus concentric circles around the central position. The geometry of the situation results in a slightly different value for the first minimum of the diffraction pattern.

For a **slit**, the first minimum was at the angle

$$\theta = \frac{\lambda}{b}$$

For a **circular aperture**, the first minimum is at the angle

$$\theta = \frac{1.22\,\lambda}{b}$$

If two sources are just resolved, then the first minimum of one diffraction pattern is located on top of the maximum of the other diffraction pattern. This is known as the **Rayleigh criterion**.

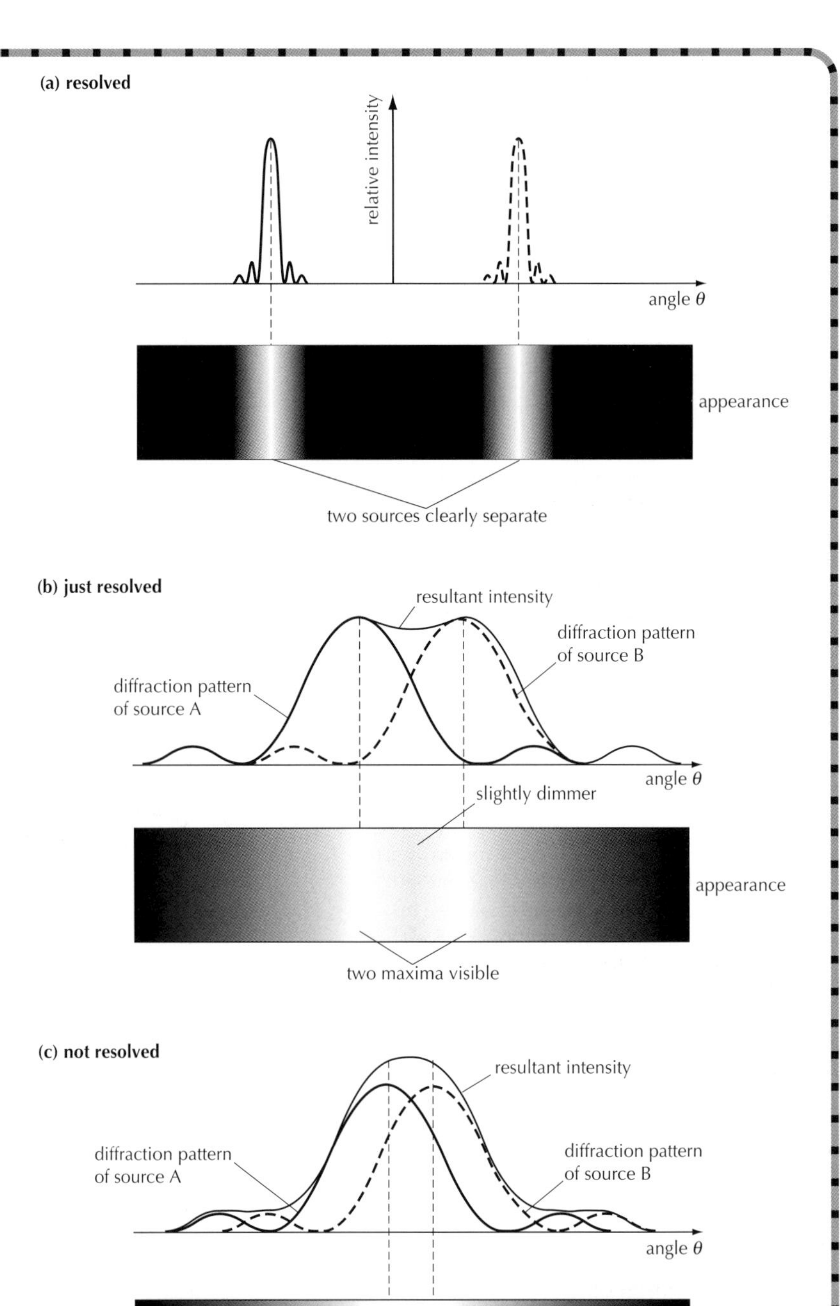

EXAMPLE

Late one night, a student was observing a car approaching from a long distance away. She noticed that when she first observed the headlights of the car, they appeared to be one point of light. Later, when the car was closer, she became able to see two separate points of light. If the wavelength of the light can be taken as 500 nm and the diameter of her pupil is approximately 4 mm, calculate how far away the car was when she could first distinguish two points of light. Take the distance between the headlights to be 1.8 m.

When just resolved

$$\theta = \frac{1.22 \times \lambda}{b}$$

$$= \frac{1.22 \times 5 \times 10^{-7}}{0.004}$$

$$= 1.525 \times 10^{-4}$$

Since θ small

$$\theta = \frac{1.8}{x} \quad [x \text{ is distance to car}]$$

$$\Rightarrow x = \frac{1.8}{1.525 \times 10^{-4}}$$

$$= 11.803$$

$$\simeq 12 \text{ km}$$

Polarisation

POLARISED LIGHT

Light is part of the electromagnetic spectrum. It is made up of oscillating electric and magnetic fields that are at right angles to one another (for more details see page 155). They are transverse waves; both fields are at right angles to the direction of propagation. The **plane of vibration** of electromagnetic waves is defined to be the plane that contains the electric field and the direction of propagation.

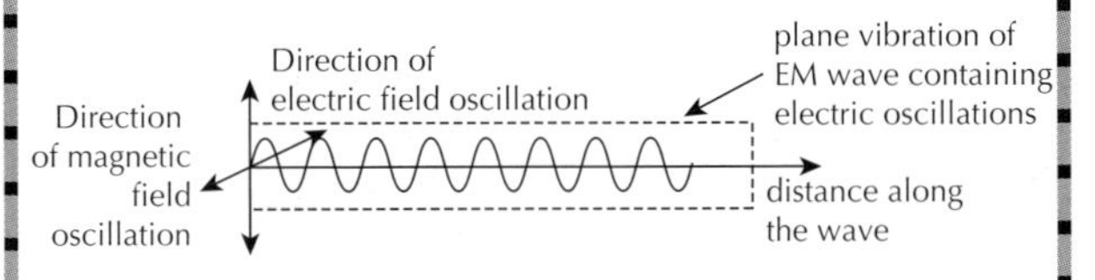

There are an infinite number of ways for the fields to be oriented. Light (or any EM wave) is said to be **unpoloarised** if the plane of vibration varies randomly whereas **plane-polarised** light has a fixed plane of vibration. The diagrams below represent the electric fields of light when being viewed "head on".

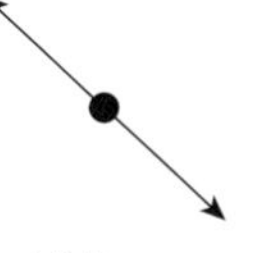

polarised light: over a period of time, the electric field only oscillates in one direction

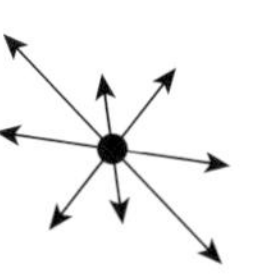

unpolarised light: over a period of time, the electric field oscillates in random directions

A mixture of polarised light and unpolarised light is **partially plane-polarised**. If the plane of polarisation rotates uniformly the light is said to be **circularly polarised**.

Most light sources emit unpolarised light whereas radio waves, radar and laboratory microwaves are often plane-polarised as a result of the processes that produce the waves. Light can polarised as a result of reflection or selective absorption. In addition, some crystals exhibit **double refraction** or **birefringence** where an unpolarised ray that enters a crystal is split into two plane-polarised beams that have mutually perpendicular planes of polarisation.

A **polariser** is any device that produces plane-polarised light from an unpolarised beam. An **analyser** is a polariser used to detect polarised light.

BREWSTER'S LAW

A ray of light incident on the boundary between two media will, in general, be reflected and refracted. The reflected ray is always partially plane-polarised. If the reflected ray and the refracted ray are at right angles to one another, then the reflected ray is totally plane-polarised. The angle of incidence for this condition is known as the **polarising angle**.

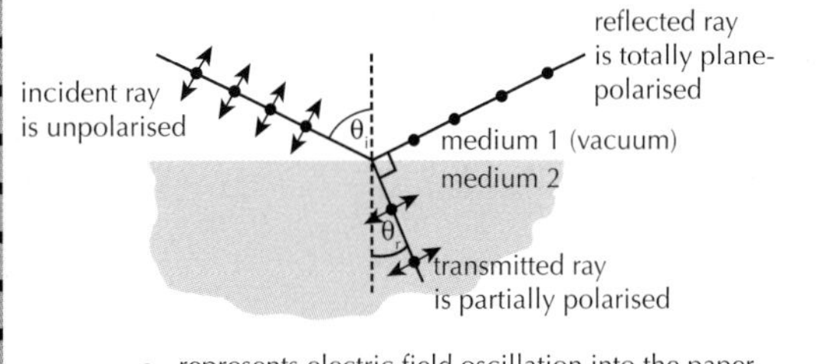

represents electric field oscillation into the paper

represents electric field oscillation in the plane of the paper

$\theta_i + \theta_r = 90°$

Brewster's law relates the refractive index of medium 2, n, to the incident angle θ_i:

$$n = \frac{\sin \theta_i}{\sin \theta_r} = \frac{\sin \theta_i}{\cos \theta_i} = \tan \theta_i$$

MALUS' LAW

When plane-polarised light is incident on an analyser, its preferred direction will allow a component of the light to be transmitted:

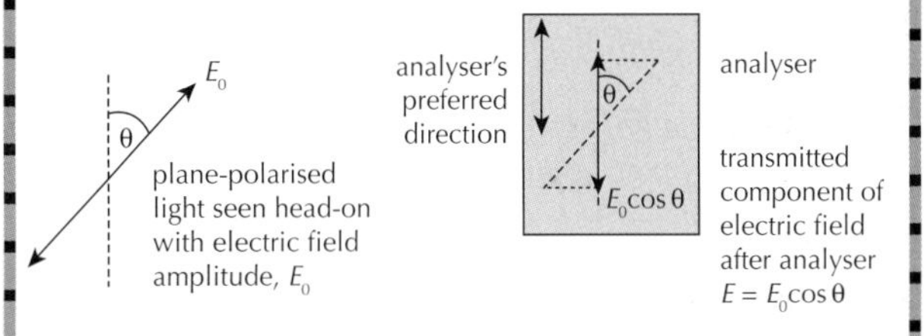

The intensity of light is proportional to the (amplitude)^2.

Transmitted intensity $I \propto E^2$

$\therefore\ I \propto E_0{}^2 \cos^2\theta$ as expressed by **Malus's law:**

$$I = I_0 \cos^2\theta$$

I is transmitted intensity of light in W m^{-2}

I_0 is incident intensity of light in W m^{-2}

θ is the angle between the plane of vibration and the analyser's preferred direction

OPTICALLY ACTIVE SUBSTANCES

An **optically active** substance is one that rotates the plane of polarisation of light that passes through it. Many solutions (e.g. sugar solutions of different concentrations) are optically active.

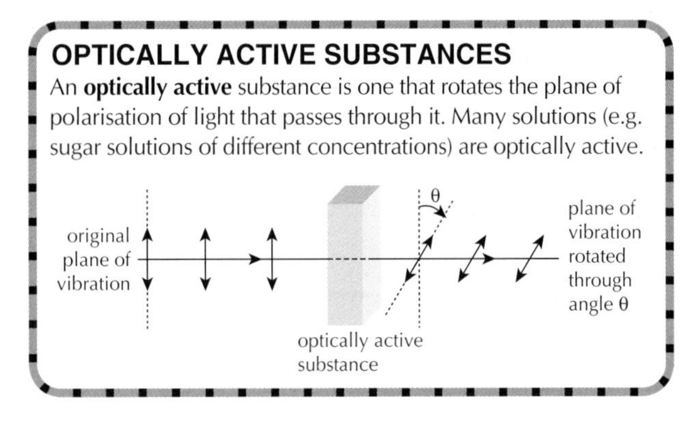

Uses of polarisation

POLAROID SUNGLASSES

Polaroid is a material containing long chain molecules. The molecules selectively absorb light that have electric fields aligned with the molecules in the same way that a grid of wires will selectively absorb microwaves.

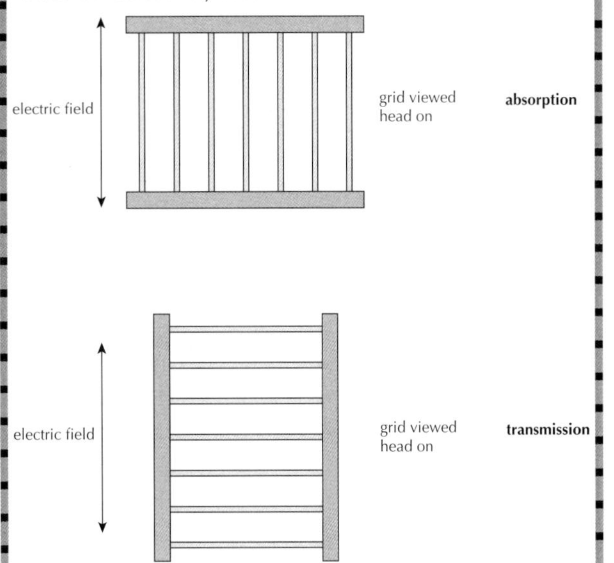

When worn normally by a person standing up, Polaroid dark glasses allow light with vertically oscillating electric fields to be transmitted and absorb light with horizontally oscillating electric fields.

- The absorption will mean that the overall light intensity is reduced.
- Light that has reflected from horizontal surfaces will be horizontally plane polarised to some extent.
- Polaroid sunglasses will preferentially absorb reflected light, reducing 'glare' from horizontal surfaces.

CONCENTRATION OF SOLUTIONS

For a given optically active solution, the angle θ through which the plane of polarisation is rotated is proportional to:

- The length of the solution through which the plane-polarised light passes.
- The concentration of the solution.

A polarimeter is a device that measures θ for a given solution. It consists of two polarisers (a polariser and an analyser) that are initially aligned. The optically active solution is introduced between the two and the analyser is rotated to find the maximum transmitted light.

STRESS ANALYSIS

Glass and some plastics become birefringent (see page 97) when placed under stress. When polarised white light is passed through stressed plastics and then analysed, bright coloured lines are observed in the regions of maximum stress.

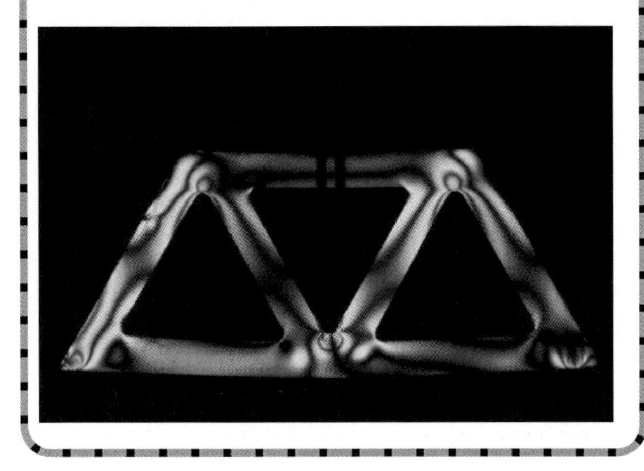

LIQUID-CRYSTAL DISPLAYS (LCDS)

LCDs are used in a wide variety of different applications that include calculator displays and computer monitors. The liquid crystal is sandwiched between two glass electrodes and is birefringent. One possible arrangement with crossed polarisers surrounding the liquid crystal is shown below:

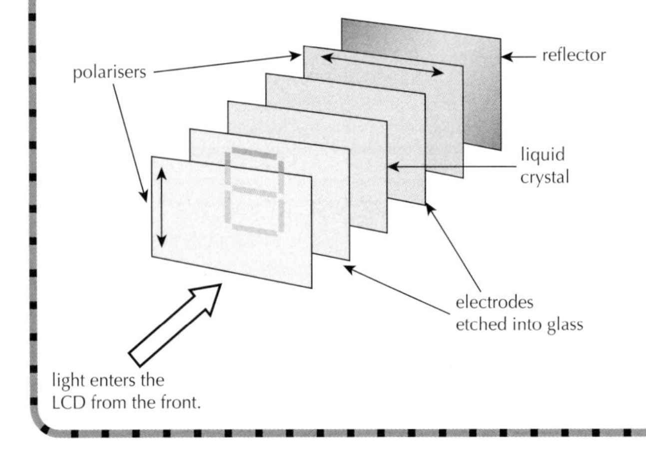

- With no liquid crystal between the electrodes, the second polariser would absorb all the light that passed through the first polariser. The screen would appear black.
- The liquid crystal has a twisted structure and, in the absence of a potential difference, causes the plane of polarisation to rotate through 90°.
- This means that light can pass through the second polariser, reach the reflecting surface and be transmitted back along its original direction.
- With no p.d. between the electrodes, the LCD appears light.
- A p.d. across the liquid crystal causes the molecules to align with the electric field. This means less light will be transmitted and this section of the LCD will appear darker.
- The extent to which the screen appears grey or black can be controlled by the p.d.
- Coloured filters can be used to create a colour image.
- A picture can be built up from individual picture elements.

IB QUESTIONS – WAVE PHENOMENA

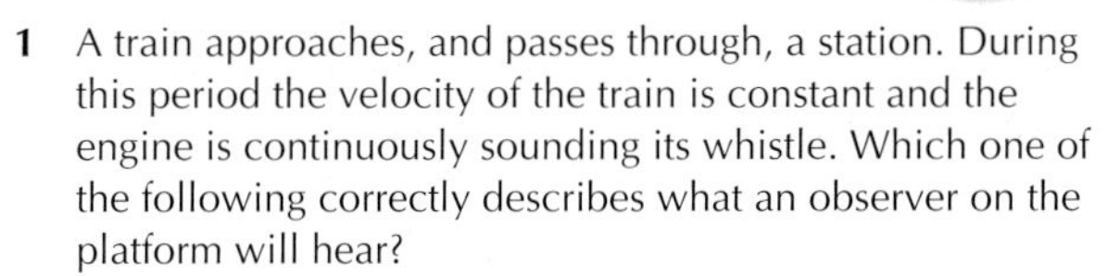

1 A train approaches, and passes through, a station. During this period the velocity of the train is constant and the engine is continuously sounding its whistle. Which one of the following correctly describes what an observer on the platform will hear?

	Sound heard as the train is **approaching** the station	Sound heard as the train is **passing through** the station
A	Constant frequency	Increasing frequency
B	Increasing frequency	Decreasing frequency
C	Decreasing frequency	Increasing frequency
D	Constant frequency	Decreasing frequency

2 When a train travels towards you sounding its whistle, the pitch of the sound you hear is different from when the train is at rest. This is because

A the sound waves are travelling faster toward you.

B the wavefronts of the sound reaching you are spaced closer together.

C the wavefronts of the sound reaching you are spaced further apart.

D the sound frequency emitted by the whistle changes with the speed of the train.

3 A sound source emits a note of constant frequency. An observer is travelling in a straight line towards the source at a constant speed. As she approaches the source she will hear a sound that

A gets higher and higher in frequency.

B gets lower and lower in frequency.

C is of constant frequency but of a frequency higher than that of the sound from the source.

D is of constant frequency but of a frequency lower than that of the sound from the source.

4 A car is travelling at constant speed towards a stationary observer whilst its horn is sounded. The frequency of the note emitted by the horn is 660 Hz. The observer, however, hears a note of frequency 720 Hz.

(a) With the aid of a diagram, explain why a higher frequency is heard. [2]

(b) If the speed of sound is 330 m s^{-1}, calculate the speed of the car. [2]

5 A bright source of light is viewed through two polarisers whose preferred directions are initially parallel. Calculate the angle through which one sheet should be turned to reduce the transmitted intensity to half its original value.

6 An unpolarised beam of monochromatic light is incident on a plane air–glass interface. The refractive index of the glass is 1.50. Calculate the angle of incidence for the reflected ray to be completely plane polarised.

7 A tube that is open at both ends is placed in a deep tank of water, as shown below.

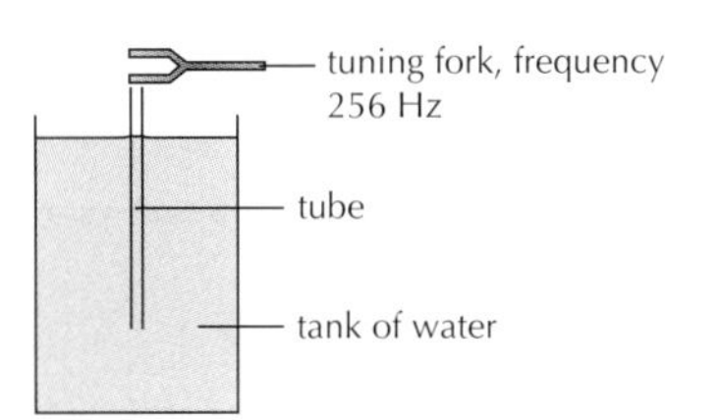

(a) A tuning fork of frequency 256 Hz is sounded continuously above the tube. The tube is slowly raised out of the water and, at one position of the tube, a maximum loudness of sound is heard. Explain the formation of a standing wave in the tube. [2]

(b) The tube is raised a further small distance. Explain, by reference to resonance, why the loudness of the sound changes. [4]

(c) The tube is gradually raised from a position of maximum loudness until the next position of maximum loudness is reached. The length of the tube above the water surface is increased by 65.0 cm. Calculate the speed of sound in the tube. [4]

8 This question is about optical resolution.

The two point sources shown in the diagram below (not to scale) emit light of the same frequency. The light is incident on a rectangular, narrow slit and after passing through the slit, is brought to a focus on the screen.

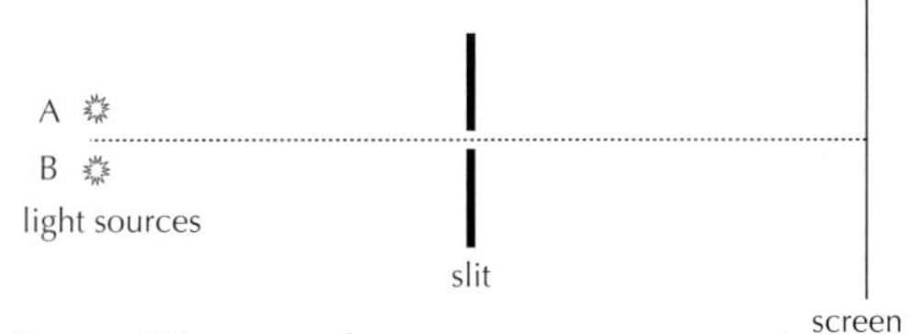

Source B is covered.

(a) Draw a sketch graph to show how the intensity I of the light from A varies with distance along the screen. Label the curve you have drawn A. [2]

Source B is now uncovered. The images of A and B on the screen are just resolved.

(b) Using the same axes as in **(a)**, draw a sketch graph to show how the intensity I of the light from B varies with distance along the screen. Label this curve B. [1]

The bright star Sirius A is accompanied by a much fainter star, Sirius B. The mean distance of the stars from Earth is 8.1×10^{16} m. Under ideal atmospheric conditions, a telescope with an objective lens of diameter 25 cm can just resolve the stars as two separate images.

(c) Assuming that the average wavelength emitted by the stars is 500 nm, estimate the apparent, linear separation of the two stars. [3]

9 A student look at two distant point sources of light. The wavelength of each source is 590 nm. The angular separation between these two sources is 3.6×10^{-4} radians subtended at the eye. At the eye, images of the two sources are formed by the eye on the retina.

(a) State the Rayleigh criterion for the two images on the retina to be just resolved. [2]

(b) Estimate the diameter of the circular apertue of the eye. [1]

(c) Use your estimate in **(b)** to determine whether the student can resolve these two sources. Explain your answer. [2]

Induced electromotive force (e.m.f.)

INDUCED E.M.F.

When a conductor moves through a magnetic field, an e.m.f. is induced. The e.m.f. induced depends on:

- The speed of the wire
- The strength of the magnetic field
- The length of the wire in the magnetic field

We can calculate the magnitude of the induced e.m.f. by considering an electron at equilibrium in the middle of the wire.

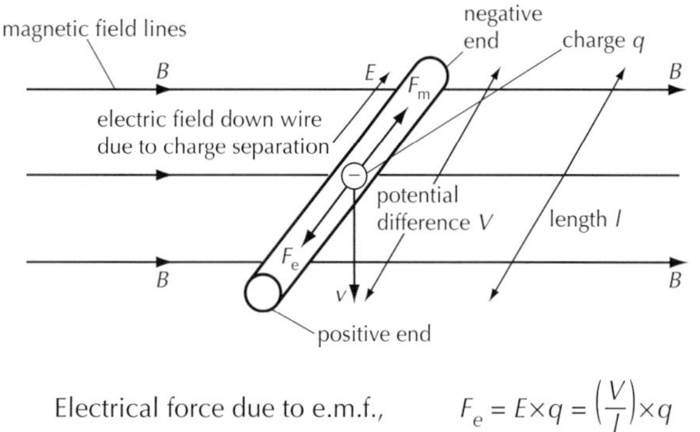

Electrical force due to e.m.f., $F_e = E \times q = \left(\frac{V}{l}\right) \times q$

Magnetic force due to movement, $F_m = B\,q\,v$

So $B\,q\,v = \left(\frac{V}{l}\right) q$

$V = B\,l\,v$

As no current is flowing, the e.m.f. = potential difference

e.m.f. = $B\,l\,v$

If the wire was part of a complete circuit (outside the magnetic field), the e.m.f. induced would cause a current to flow.

LENZ'S LAW

Lenz's law states that

'The direction of the induced e.m.f. is such that if an induced current were able to flow, it would oppose the change which caused it.'

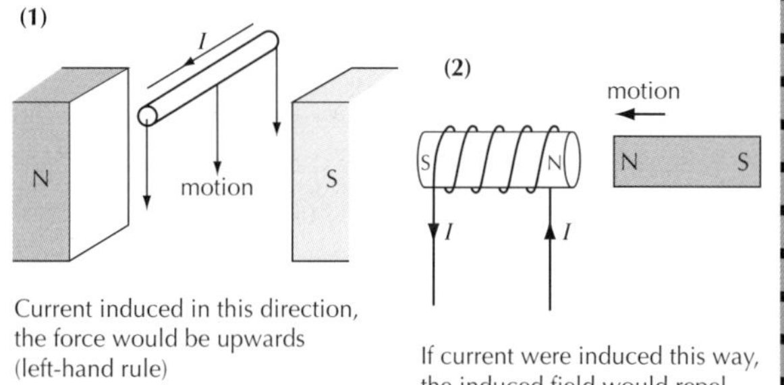

Current induced in this direction, the force would be upwards (left-hand rule) ∴ original motion would be opposed.

If current were induced this way, the induced field would repel the magnet – opposing motion.

EXAMPLE

An aeroplane flies at 200 m s^{-1}. Estimate the maximum p.d. that can be generated across its wings.

Vertical component of Earth's magnetic field = 10^{-5} T (approximately)

Length across wings = 30 m (estimated)

e.m.f. = $10^{-5} \times 30 \times 200$

$= 6 \times 10^{-2}$ V

$= 0.06$ V

PRODUCTION OF INDUCED E.M.F. BY RELATIVE MOTION

An e.m.f. is induced in a conductor whenever flux is cut. But flux is more than just a way of picturing the situation.

If the magnetic field is perpendicular to the surface, the magnetic flux $\Delta\phi$ passing through the area ΔA is defined in terms of the magnetic field strength B as follows.

$$\Delta\phi = B\,\Delta A, \text{ so } B = \frac{\Delta\phi}{\Delta A}$$

In a uniform field, $B = \frac{\phi}{A}$

An alternative name for 'magnetic field strength' is '**flux density**'.

If the area is not perpendicular, but at an angle θ to the field lines, the equation becomes

$\phi = B\,A\cos\theta$ (units: T m^{-2})

θ is the angle between **B** and the normal to the surface.

Flux can also be measured in webers (Wb), defined as follows.

1 Wb = 1T m^{-2}

These relationships allow us to calculate the induced e.m.f. ε in terms of flux.

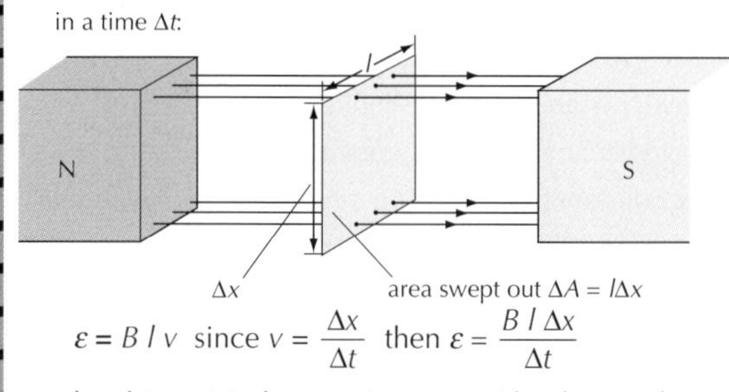

$\varepsilon = B\,l\,v$ since $v = \frac{\Delta x}{\Delta t}$ then $\varepsilon = \frac{B\,l\,\Delta x}{\Delta t}$

but $l\,\Delta x = \Delta A$, the area 'swept out' by the conductor in a time Δt so $\varepsilon = \frac{B\,\Delta A}{\Delta t}$

but $B\,\Delta A = \Delta\phi$ so $\varepsilon = \frac{\Delta\phi}{\Delta t}$

In words, 'the e.m.f. induced is equal to the rate of cutting of flux'. If the conductor is kept stationary and the magnets are moved, the same effect is produced.

TRANSFORMER–INDUCED E.M.F.

An e.m.f. is also produced in a wire if the magnetic field changes with time.

If the amount of flux passing through one turn of a coil is ϕ, then the total **flux linkage** with all N turns of the coil is given by

Flux linkage = $N\,\phi$

The universal rule that applies to all situations involving induced e.m.f. can now be stated as

'The magnitude of an induced e.m.f. is proportional to the rate of change of flux linkage.'

This is known as **Faraday's law**

e.m.f. = $N\,\frac{\Delta\phi}{\Delta t}$

Alternating current (1)

COIL ROTATING IN A MAGNETIC FIELD – A.C. GENERATOR

The structure of a typical a.c. generator is shown below.

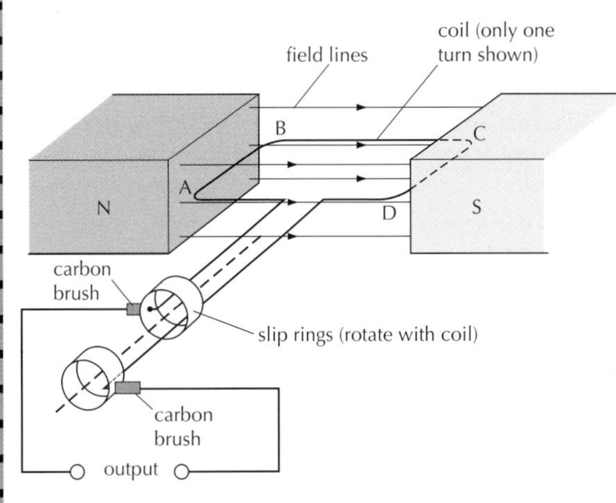

a.c. generator

The coil of wire rotates in the magnetic field due to an external force. As it rotates the flux linkage of the coil changes with time and induces an e.m.f. (Faraday's law) causing a current to flow. The sides AB and CD of the coil experience a force opposing the motion (Lenz's law). The work done rotating the coil generates electrical energy.

A coil rotating at constant speed will produce a sinusoidal induced e.m.f. Increasing the speed of rotation will reduce time period of the oscillation *and* increase the amplitude of induced e.m.f. (as rate of change of flux linkage is increased.)

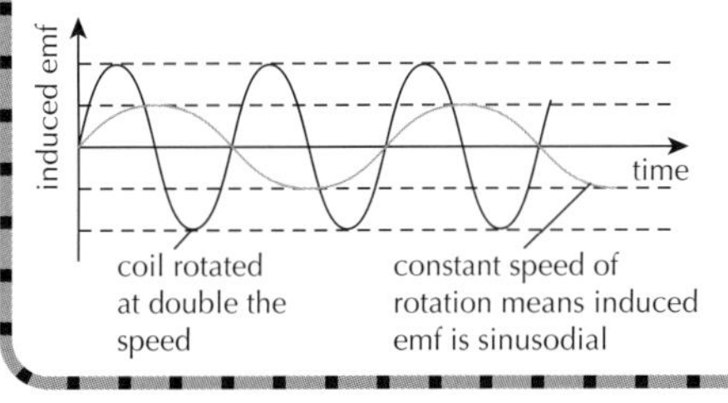

R.M.S. VALUES

If the output of an a.c. generator is connected to a resistor an alternating current will flow. A sinusoidal potential difference means a sinusoidal current.

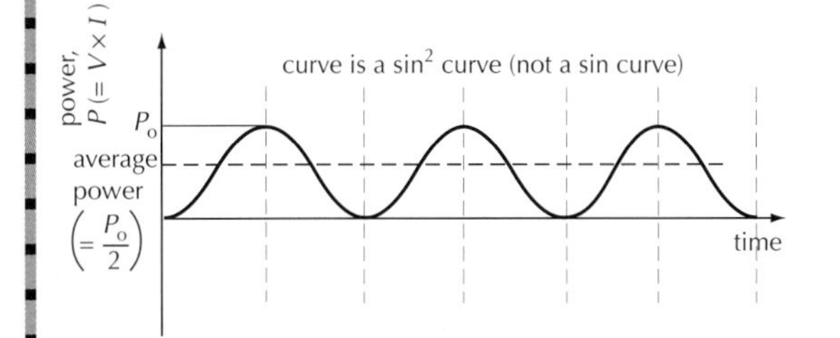

The graph shows that the average power dissipation is half the peak power dissipation for a sinusoidal current.

$$\text{Average power} = \frac{I_0^2 R}{2} = \left(\frac{I_0}{\sqrt{2}}\right)^2 R$$

Thus the effective current through the resistor is √(mean value of I^2) and it is called the **root mean square** current or **r.m.s.** current, $I_{r.m.s.}$.

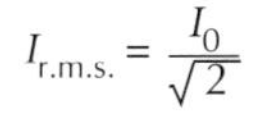

$$I_{r.m.s.} = \frac{I_0}{\sqrt{2}}$$

When a.c. values for voltage or current are quoted, it is the root mean square value that is being used. In Europe this value is 230 V, whereas in the USA it is 110 V.

TRANSFORMER OPERATION

An alternating potential difference is put into the transformer, and an alternating potential difference is given out. The value of the output potential difference can be changed (increased or decreased) by changing the **turns ratio**. A **step-up** transformer increases the voltage, whereas a **step-down** transformer decreases the voltage.

The following sequence of calculations provides the correct method for calculating all the relevant values.

- The output voltage is fixed by the input voltage and the turns ratio.
- The value of the load that you connect fixes the output current (using $V = I R$).
- The value of the output power is fixed by the values above ($P = V I$).
- The value of the input power is equal to the output power for an ideal transformer.
- The value of the input current can now be calculated (using $P = V I$).

So how does the transformer manage to alter the voltages in this way?

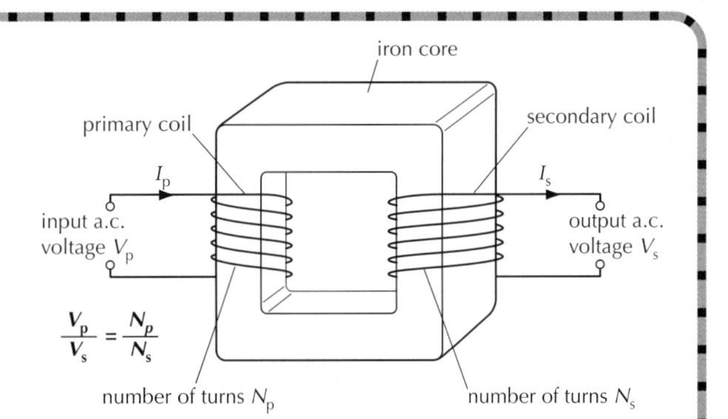

Transformer structure

- The alternating p.d. across the primary creates an a.c. within the coil and hence an alternating magnetic field in the iron core.
- This alternating magnetic field links with the secondary and induces an e.m.f. The value of the induced e.m.f. depends on the rate of change of flux linkage, which increases with increased number of turns on the secondary. The input and output voltages are related by the turns ratio.

Alternating current (2)

LOSSES IN THE TRANSMISSION OF POWER

In addition to power losses associated with the resistance of the power supply lines, which cause the power lines to warm up, there are also losses associated with non-ideal transformers:

- **Resistance of the windings** of a transformer result in the transformer warming up.
- **Eddy currents** are unwanted currents induced in the iron core. The currents are reduced by **laminating** the core into individually electrically insulated thin strips.
- **Hysteresis** losses cause the iron core to warm up as a result of the continued cycle of changes to its magnetism.
- **Flux** losses are caused by magnetic 'leakage'. A transformer is only 100% efficient if all of the magnetic flux that is produced by the primary links with the secondary.

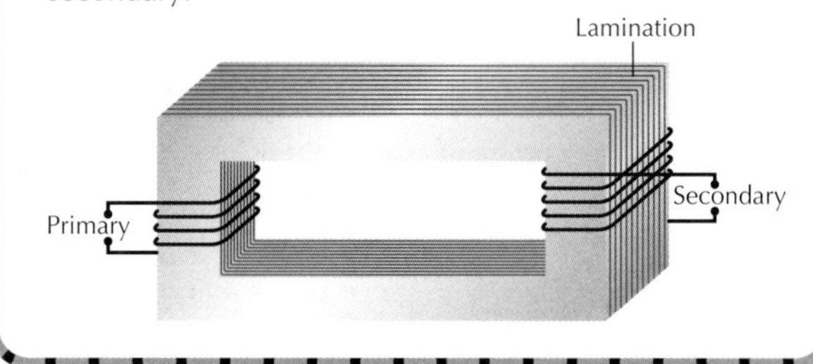

POSSIBLE RISKS

Electrical power lines on pylons are not insulated along their length and are thus extremely dangerous if they become unattached from the pylon. In addition, some statistical evidence exists which suggests that there are regions (near some power lines) where more children are diagnosed with leukaemia, a cancer of the blood, than would be expected if the causes were random. The reasons for these **leukaemia clusters**, if they exist, are not understood.

Electrical power lines carry alternating currents, which means they produce changing extra-low-frequency electromagnetic fields. These changing fields are theoretically able to induce currents within any conductor, including human bodies, nearby. There is, however, no known mechanism by which such fields could cause leukaemia and current experimental evidence suggests that low-frequency fields do not preferentially harm genetic material. Any risks are likely to be dependent on three factors:

- Current (density).
- A.C. frequency.
- Length of exposure.

TRANSMISSION OF ELECTRICAL POWER

Transformers play a very important role in the safe and efficient transmission of electrical power over large distances.

- If large amounts of power are being distributed, then the currents used will be high. (Power = $V\,I$)
- The wires cannot have zero resistance. This means they must dissipate some power
- Power dissipated is $P = I^2 R$. If the current is large then the $(\text{current})^2$ will be very large.
- Over large distance, the power wasted would be very significant.
- The solution is to choose to transmit the power at a very high potential difference.
- Only a small current needs to flow.
- A very high potential difference is much more efficient, but very dangerous to the user.
- Use step-up transformers to increase the voltage for the transmission stage and then use step-down transformers for the end user.

IB QUESTIONS – ELECTROMAGNETIC INDUCTION 

1 The **primary** of an ideal transformer has 1000 turns and the **secondary** 100 turns. The current in the primary is 2 A and the input power is 12 W.

Which **one** of the following about the **secondary current** and the **secondary power output** is true?

	secondary current	secondary power output
A	20 A	1.2 W
B	0.2 A	12 W
C	0.2 A	120 W
D	20 A	12 W

2 This question is about electromagnetic induction.

A small coil is placed with its plane parallel to a long straight current-carrying wire, as shown below.

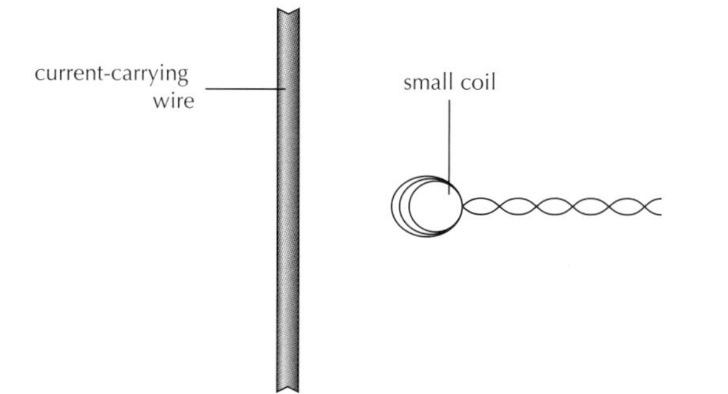

(a) **(i)** State Faraday's law of electromagnetic induction. [2]

(ii) Use the law to explain why, when the current in the wire changes, an e.m.f. is induced in the coil. [1]

3 The diagram shows a simple generator with the coil rotating between magnetic poles. Electrical contact is maintained through two brushes, each touching a slip ring.

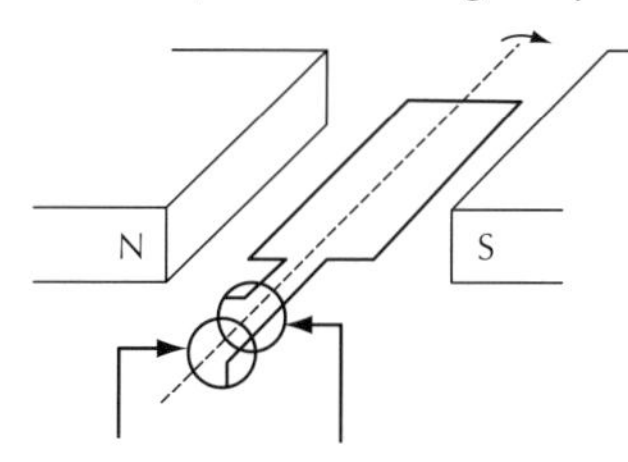

At the instant when the rotating coil is oriented as shown, the voltage across the brushes

A is zero.

B has its maximum value.

C has the same constant value as in all other orientations.

D reverses direction.

4 An alternating voltage of peak value 300 V is applied across a 50 Ω resistor. The average power dissipated in watts will be

A zero **B** $\frac{(300)^2}{50}$ **C** $\frac{(300)^2}{50\sqrt{2}}$ **D** $\frac{(300)^2}{50 \times 2}$

5 Two loops of wire are next to each other as shown here. There is a source of alternating e.m.f. connected to loop 1 and an ammeter in loop 2.

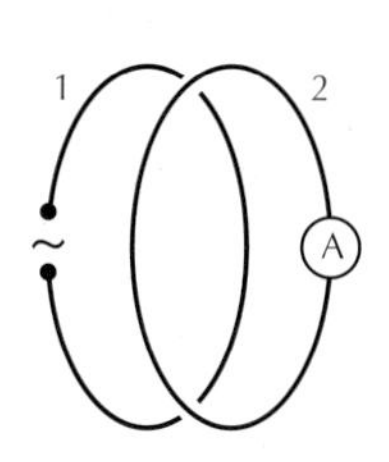

The variation with time of the current in loop 1 is shown as line 1 in each of the graphs below. In which graph does line 2 best represent the current in loop 2?

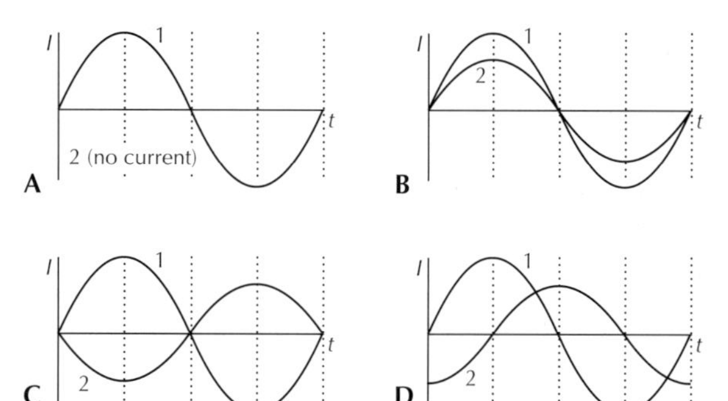

6 A loop of wire of negligible resistance is rotated in a magnetic field. A 4 Ω resistor is connected across its ends. A cathode ray oscilloscope measures the varying induced potential difference across the resistor as shown below.

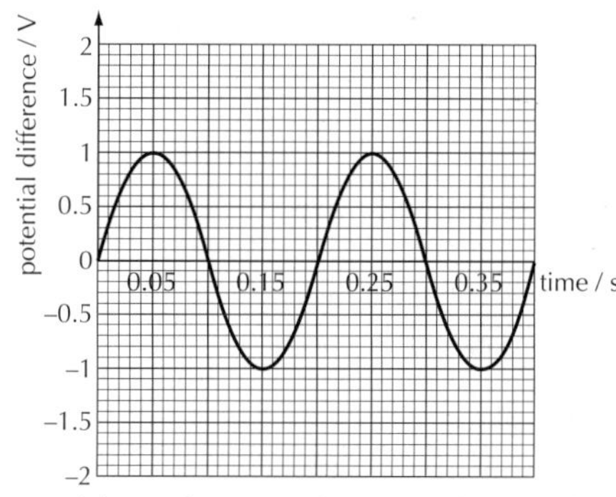

(a) If the coil is rotated at twice the speed, show on the axes above how potential difference would vary with time. [2]

(b) What is the r.m.s. value of the induced potential difference, $V_{r.m.s.}$, at the **original** speed of rotation? [1]

(c) On the axes below, draw a graph showing how the power dissipated in the resistor varies with time, at the **original** speed of rotation. [3]

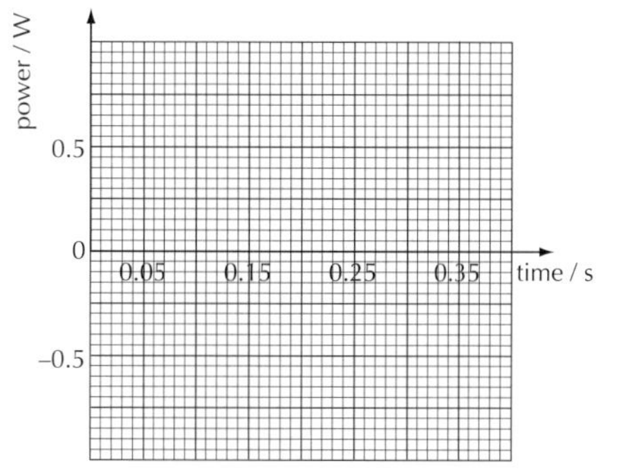

The quantum nature of radiation

PHOTOELECTRIC EFFECT

Under certain conditions, when light (ultra-violet) is shone onto a metal surface (such as zinc), electrons are emitted from the surface.

More detailed experiments (see below) showed that:

- Below a certain **threshold frequency**, f_0, no photoelectrons are emitted, no matter how long one waits.
- Above the threshold frequency, the maximum kinetic energy of these electrons depends on the frequency of the incident light.
- The number of electrons emitted depends on the intensity of the light and does not depend on the frequency.
- There is no noticeable delay between the arrival of the light and the emission of electrons.

These observations cannot be reconciled with the view that light is a wave. A wave of any frequency should eventually bring enough energy to the metal plate.

STOPPING POTENTIAL EXPERIMENT

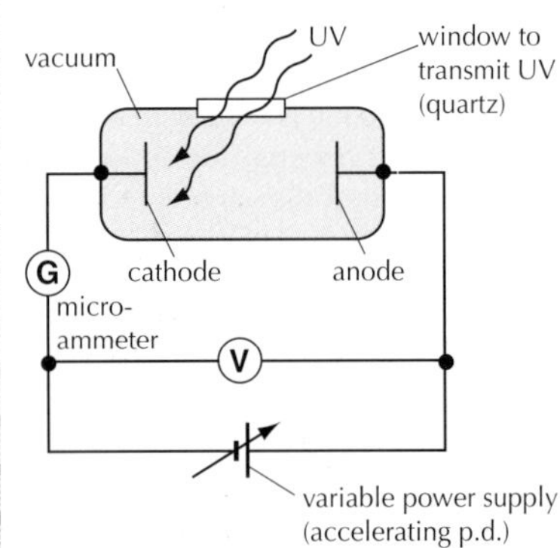

In the apparatus above, photoelectrons are emitted by the cathode. They are then accelerated across to the anode by the potential difference.

The potential between cathode and anode can also be reversed.

In this situation, the electrons are decelerated. At a certain value of potential, the stopping potential, V_s, no more photocurrent is observed. The photoelectrons have been brought to rest before arriving at the anode.

photocurrent
high-intensity UV
low-intensity UV of same frequency
V_s
potential

The stopping potential depends on the frequency of UV light in the linear way shown in the graph below.

stopping potential V_s
threshold frequency, f_0
frequency

The stopping potential is a measure of the maximum kinetic energy of the electrons.

Max KE of electrons = $V_s e$

[since p.d. = $\frac{\text{energy}}{\text{charge}}$

and e = charge on an electron]

$$\therefore \frac{1}{2}mv^2 = V_s e \quad \therefore v = \frac{2V_s e}{m}$$

EINSTEIN MODEL

Einstein introduced the idea of thinking of light as being made up of particles.

His explanation was that

- Electrons at the surface need a certain minimum energy in order to escape from the surface. This minimum energy is called the **work function** of the metal and given the symbol ϕ.
- The UV light energy arrives in lots of little packets of energy – the packets are called photons.
- The energy in each packet is fixed by the frequency of UV light that is being used, whereas the number of packets arriving per second is fixed by the intensity of the source.
- The energy carried by a photon is given by

$$E = hf$$

Planck's constant 6.63×10^{-34} J s; energy in joules; frequency of light in Hz

- Different electrons absorb different photons. If the energy of the photon is large enough, it gives the electron enough energy to leave the surface of the metal.
- Any "extra" energy would be retained by the electron as kinetic energy.
- If the energy of the photon is too small, the electron will still gain this amount of energy but it will soon share it with other electrons.

Above the threshold frequency, incoming energy of photons = energy needed to leave the surface + kinetic energy

in symbols,

$$hf = \phi + KE_{max} \quad \text{or} \quad hf = \phi + V_s e$$

This means that a graph of frequency against stopping potential should be a straight line of gradient $\frac{e}{h}$.

EXAMPLE

What is the maximum velocity of electrons emitted from a zinc surface (ϕ = 4.2 eV) when illuminated by EM radiation of wavelength 200 nm?

$\phi = 4.2 \text{ eV} = 4.2 \times 1.6 \times 10^{-19} \text{ J} = 6.72 \times 10^{-19} \text{ J}$

$$\text{Energy of photon} = h\frac{c}{\lambda} = \frac{6.63 \times 10^{-34} \times 3 \times 10^{8}}{2 \times 10^{-7}}$$

$$= 9.945 \times 10^{-19} \text{ J}$$

$$\therefore \text{KE of electron} = (9.945 - 6.72) \times 10^{-19} \text{ J}$$

$$= 3.225 \times 10^{-19} \text{ J}$$

$$\therefore v = \sqrt{\frac{2\,KE}{m}}$$

$$= \sqrt{\frac{2 \times 3.225 \times 10^{-19}}{9.1 \times 10^{-31}}}$$

$$= 8.4 \times 10^{5} \text{ m s}^{-1}$$

The wave nature of matter

WAVE–PARTICLE DUALITY

The photoelectric effect of light waves clearly demonstrates that light can behave like particles, but its wave nature can also be demonstrated – it reflects, refracts, diffracts and interferes just like all waves. So what exactly is it? It seems reasonable to ask two questions.

1. *Is light a wave or is it a particle?*
The correct answer to this question is "yes"! At the most fundamental and even philosophical level, light is just light. Physics tries to understand and explain what it is. We do this by imagining models of its behaviour. Sometimes it helps to think of it as a wave and sometimes it helps to think of it as a particle, but neither model is complete. Light is just light. This dual nature of light is called **wave-particle duality**.

2. *If light waves can show particle properties, can particles such as electrons show wave properties?*
Again the correct answer is "yes". Most people imagine moving electrons as little particles having a definite size, shape, position and speed. This model does not explain why electrons can be diffracted through small gaps. In order to diffract they must have a wave nature. Once again they have a dual nature. See the experiment below.

DE BROGLIE HYPOTHESIS

If matter can have wave properties and waves can have matter properties, there should be a link between the two models. The de Broglie hypothesis is that all moving particles have a "matter wave" associated with them. The wavelength of this matter wave is given by the de Broglie equation:

$$\lambda = \frac{h}{p}$$

λ is the wavelength in m

h is Planck's constant = 6.63×10^{-34} J s

p is the momentum in kg m s^{-1}

This matter wave can be thought of as a probability function associated with the moving particle. The (amplitude)^2 of the wave at any given point is a measure of the probability of finding the particle at that point.

DAVISSON AND GERMER EXPERIMENT

The diagram below shows the principle behind the Davisson and Germer electron diffraction experiment.

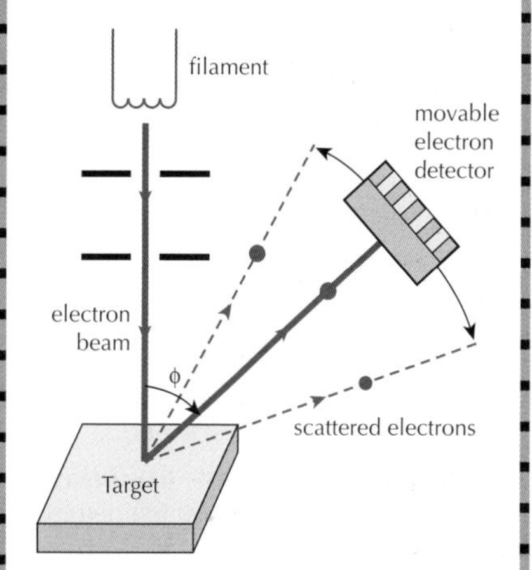

A beam of electrons strikes a target nickel crystal. The electrons are scattered from the surface. The intensity of these scattered electrons depends on the speed of the electrons (as determined by their accelerating potential difference) and the angle.

A maximum scattered intensity was recorded at an angle that quantitatively agrees with the constructive interference condition from adjacent atoms on the surface.

ELECTRON DIFFRACTION EXPERIMENT

In order to show diffraction, an electron 'wave' must travel through a gap of the same order as its wavelength. The atomic spacing in crystal atoms provides such gaps. If a beam of electrons impinges upon powdered carbon then the electrons will be diffracted according to the wavelength.

If accelerating p.d. = 1000 V

KE of electrons $= eV$
$= 1.6 \times 10^{-19} \times 1000$ J
$= 1.6 \times 10^{-16}$ J

For non-relativistic speeds, the link between momentum p and KE is given by

$$KE = \frac{1}{2}mv^2 = \frac{p^2}{2m}$$

$$p = \sqrt{2m \times \text{KE}}$$

$$= \sqrt{2 \times 9.1 \times 10^{-31} \times 1.6 \times 10^{-16}} \text{ kg m s}^{-1}$$

$$= 1.7 \times 10^{-23} \text{ kg m s}^{-1}$$

$$\lambda = \frac{h}{p}$$

$$= \frac{6.6 \times 10^{-34}}{1.7 \times 10^{-23}} \text{ m}$$

$$= 3.9 \times 10^{-11} \text{ m}$$

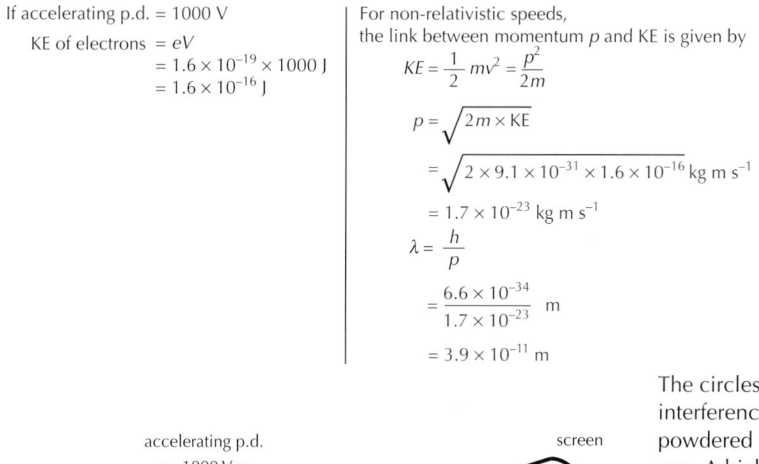

The circles correspond to the angles where constructive interference takes place. They are circles because the powdered carbon provides every possible orientation of gap. A higher accelerating potential for the electrons would result in a higher momentum for each electron. According to the de Broglie relationship, the wavelength of the electrons would thus decrease. This would mean that the size of the gaps is now proportionally bigger than the wavelength so there would be less diffraction. The circles would move in to smaller angles. The predicted angles of constructive interference are accurately verified experimentally.

Atomic spectra and atomic energy states

INTRODUCTION

As we have already seen, atomic spectra (emission and absorption) provide evidence for the quantization of the electron energy levels. See page 59 for the laboratory setup.

Different atomic models have attempted to explain these energy levels. Schrödinger's model describes the electrons by using wave functions. A simpler model, the 'electron in a box' explains the origin of atomic energy levels.

'ELECTRON IN A BOX' MODEL

If electrons can be described by waves, then the discrete energy levels that exist should be related to different possible wave descriptions. In Schrödinger's model the electron wave functions are essentially standing waves that fit the boundary conditions in the atom. The situation is very hard to imagine as we are talking about a wave in all three dimensions. It is probably better to simplify the situation by considering only one dimension and start by remembering the physical standing waves that can be fitted onto a stretched string (see Topic 11).

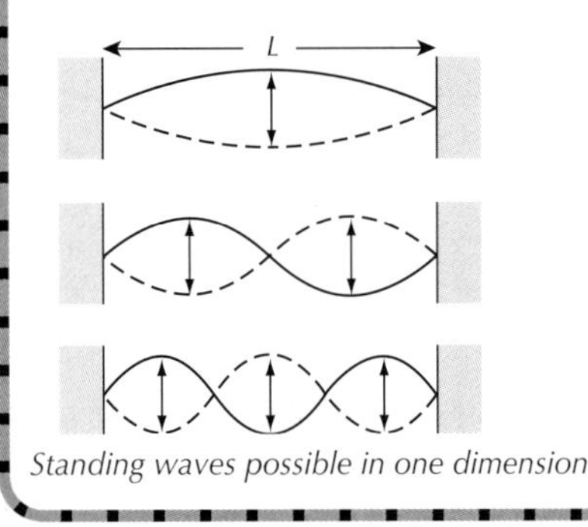

Standing waves possible in one dimension

The possible wavelength λ of standing waves that fit into this box

$$= 2L,\ L,\ \frac{2L}{3},\ \frac{L}{2},\ \text{etc}$$

$$= \frac{2L}{n} \text{ where } n \text{ is an integer (1, 2, 3, 4 etc).}$$

The de Bröglie relationship, $\lambda = \frac{h}{p}$, can be used to calculate the momentum of the 'electron in a box':

$$p = \frac{h}{\lambda} = \frac{nh}{2L}$$

The kinetic energy E_K of the electron of mass, m_e, is thus:

$$E_K = \frac{1}{2}m_e v^2 = \frac{1}{2}\frac{p^2}{m_e} = \frac{1}{2m_e}\left(\frac{nh}{L2}\right)^2 = \frac{n^2h^2}{8m_eL^2}$$

Thus different energy levels are predicted for the electron which correspond to the different possible standing waves.

EXAMPLE

The diagram below represents some of the electron energy levels in the hydrogen atom. Calculate the wavelength of the photon emitted when an electron falls from $n = 3$ to $n = 2$.

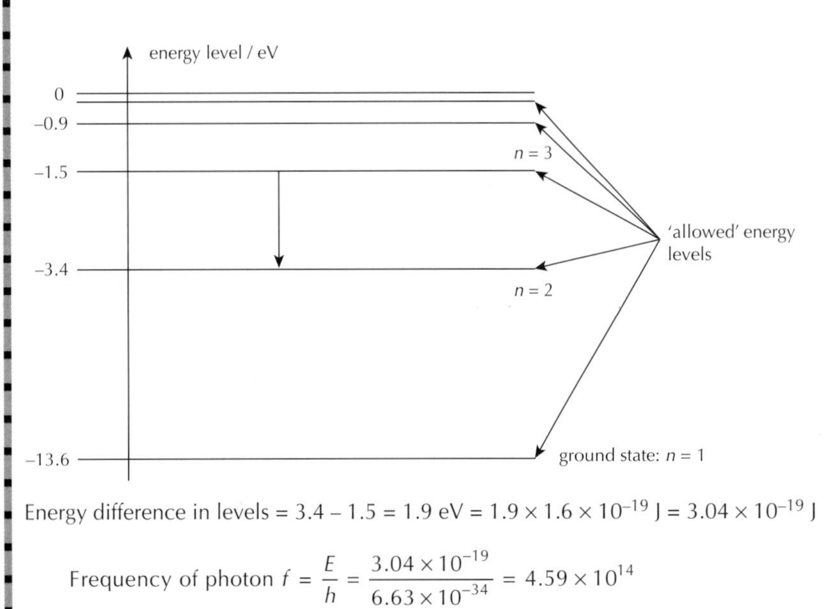

Energy difference in levels = 3.4 – 1.5 = 1.9 eV = $1.9 \times 1.6 \times 10^{-19}$ J = 3.04×10^{-19} J

$$\text{Frequency of photon } f = \frac{E}{h} = \frac{3.04 \times 10^{-19}}{6.63 \times 10^{-34}} = 4.59 \times 10^{14}$$

$$\text{Wavelength of photon } \lambda = \frac{c}{f} = \frac{3.00 \times 10^{8}}{4.59 \times 10^{14}} = 6.54 \times 10^{-7} \text{ m} = 654 \text{ nm}$$

This is in the visible part of the electromagnetic spectrum.

The Schrödinger model of the atom

SCHRÖDINGER MODEL

Erwin Schrödinger (1887–1961) built on the concept of matter waves and proposed an alternative model of the hydrogen atom using wave mechanics. The **Copenhagen interpretation** is a way to give a physical meaning to the mathematics of wave mechanics.

- The description of particles (matter and /or radiation) in quantum mechanics is in terms of a **wave function** ψ. This wave function has no physical meaning but the square of the wave function does.
- At any instant of time, the wave function has different values at different points in space.
- The mathematics of how this wave function develops with time and interacts with other wave functions is like the mathematics of a travelling wave.
- The probability of finding the particle (electron or photon etc.) at any point in space within the atom is given by the square of the amplitude of the wave function at that point.
- When an observation is made the wave function is said to **collapse**, and the complete physical particle (electron or photon etc.) will be observed to be at one location.

The standing waves on a string have a fixed wavelength but for energy reasons the same is not true for the electron wave functions. As an electron moves away from the nucleus it must lose kinetic energy because they have opposite charges. Lower kinetic energy means that it would be travelling with a lower momentum and the de Broglie relationship predicts a longer wavelength. This means that the possible wave functions that fit the boundary conditions have particular shapes.

The wave function provides a way of working out the probability of finding an electron at that particular radius. The $(\text{amplitude})^2$ of the wave at any given point is a measure of the probability of finding the electron at that distance away from the nucleus – in any direction.

The wave function exists in all three dimensions, which makes it hard to visualise. Often the electron orbital is pictured as a cloud. The exact position of the electron is not known but we know where it is more likely to be.

In Schrödinger's model there are different wave functions depending on the total energy of the electron. Only a few particular energies result in wave functions that fit the boundary conditions – electrons can only have these particular energies within an atom. An electron in the ground state has a total energy of –13.6 eV, but its position at any given time is undefined in this model. The wave function for an electron of this energy can be used to calculate the probability of finding it at a given distance away from the nucleus.

- The resulting **orbital** for the electron can be described in terms of the probability of finding the electron at a certain distance away. The probability of finding the electron at a given distance away is shown in the graph below.

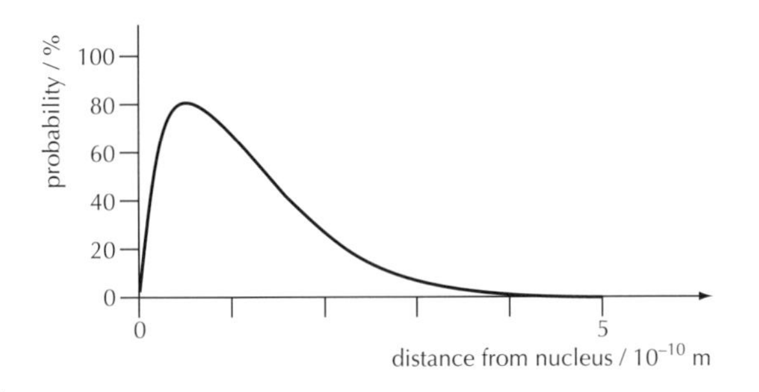

- The electron in this orbital can be visualised as a 'cloud' of varying electron density. It is more likely to be in some places than other places, but its actual position in space is undefined.

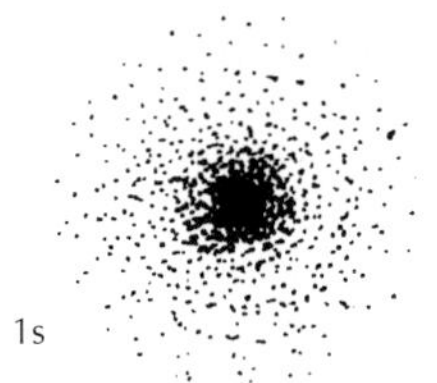

Electron cloud for the 1s orbital in hydrogen

There are other fixed total energies for the electron that result in different possible orbitals. In general as the energy of the electron is increased it is more likely to be found at a further distance away from the nucleus.

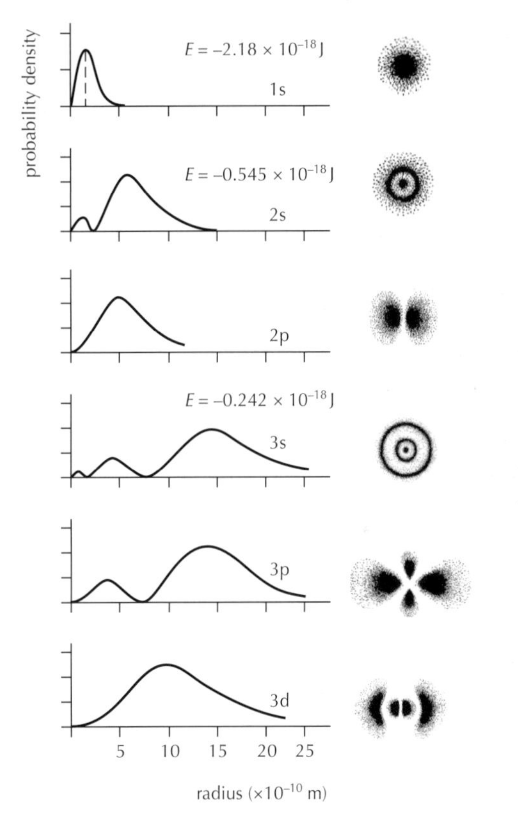

Probability density functions for some orbitals in the hydrogen atom. The scale on the vertical axis is different from graph to graph.

The Heisenberg uncertainty principle and the loss of determinism

HEISENBERG UNCERTAINTY PRINCIPLE

The **Heisenberg uncertainty principle** identifies a fundamental limit to the possible accuracy of any physical measurement. This limit arises because of the nature of quantum mechanics and not as a result of the ability (or otherwise) of any given experimenter. He showed that it was impossible to measure exactly the position **and** the momentum of a particle simultaneously. The more precisely the position is determined, the less precisely the momentum is known in this instant, and vice versa. They are linked variables and are called **conjugate quantities**.

There is a mathematical relationship linking these uncertainties.

$$\Delta x \Delta p \geq \frac{h}{4\pi}$$

Δx The uncertainty in the measurement of position

Δp The uncertainty in the measurement of momentum

Measurements of energy and time are also linked variables.

$$\Delta E \Delta t \geq \frac{h}{4\pi}$$

ΔE The uncertainty in the measurement of energy

Δt The uncertainty in the measurement of time

The implications of this lack of precision are profound. Before quantum theory was introduced, the physical world was best described by deterministic theories – e.g. Newton's laws. A deterministic theory allows us (in principle) to make absolute predictions about the future.

Quantum mechanics is not deterministic. It cannot ever predict exactly the results of a single experiment. It only gives us the probabilities of the various possible outcomes. The uncertainty principle takes this even further. Since we cannot know the precise position and momentum of a particle at any given time, its future can never be determined precisely. The best we can do is to work out a range of possibilities for its future.

It has been suggested that science would allow us to calculate the future so long as we know the present exactly. As Heisenberg himself said, it is not the conclusion that is wrong but the premise.

The nucleus

THE NUCLEUS – SIZE

In the example below, alpha particles are allowed to bombard gold atoms.

As they approach the gold nucleus, they feel a force of repulsion. If an alpha particle is heading directly for the nucleus, it will be reflected straight back along the same path. It will have got as close as it can. Note that none of the alpha particles actually collides with the nucleus – they do not have enough energy.

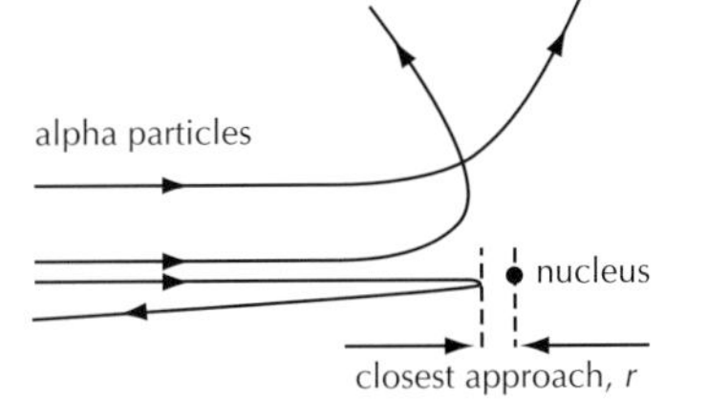

Alpha particles are emitted from their source with a known energy. As they come in they gain electrostatic potential energy and lose kinetic energy (they slow down). At the closest approach, the alpha particle is temporarily stationary and all its energy is potential.

Since electrostatic energy = $\frac{q_1 q_2}{4\pi\varepsilon_0 r}$, and we know q_1, the charge on an alpha particle and q_2, the charge on the gold nucleus we can calculate r.

ENERGY LEVELS

The energy levels in a nucleus are higher than the energy levels of the electrons but the principle is the same. When an alpha particle or a gamma photon is emitted from the nucleus only discrete energies are observed. These energies correspond to the difference between two **nuclear energy levels** in the same way that the photon energies correspond to the difference between two **atomic energy levels**.

The decay of ^{226}Ra into ^{222}Rn

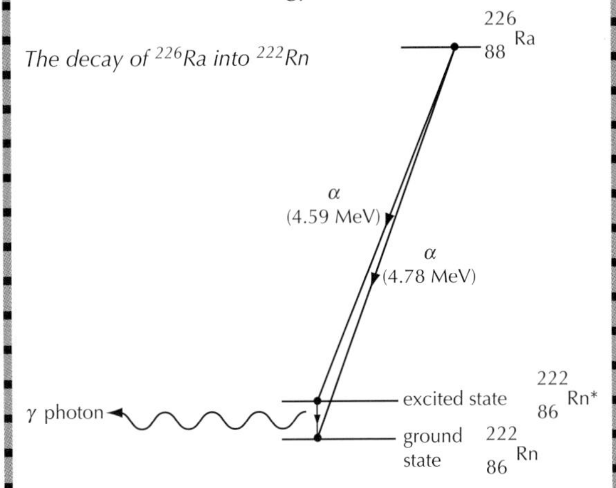

Beta particles are observed to have a continuous spectrum of energies. In this case there is another particle (the antineutrino in the case of beta minus decay) that shares the energy. Once again the amount of energy released in the decay is fixed by the difference between the *nuclear* energy levels involved. The beta particle and the antineutrino can take varying proportions of the energy available. The antineutrino, however, is very difficult to observe (see next page).

EXAMPLE

If the α particles have an energy of 4.2 MeV, the closest approach to the gold nucleus (Z = 79) is given by

$$\frac{(2 \times 1.6 \times 10^{-19})\,(79 \times 1.6 \times 10^{-19})}{4 \times \pi \times 8.85 \times 10^{-12} \times r} = 4.2 \times 10^6 \times 1.6 \times 10^{-19}$$

$$\therefore\ r = \frac{2 \times 1.6 \times 10^{-19} \times 79}{4 \times \pi \times 8.85 \times 10^{-12} \times 4.2 \times 10^6}$$

$$= 5.4 \times 10^{-14}\ \text{m}$$

MASS

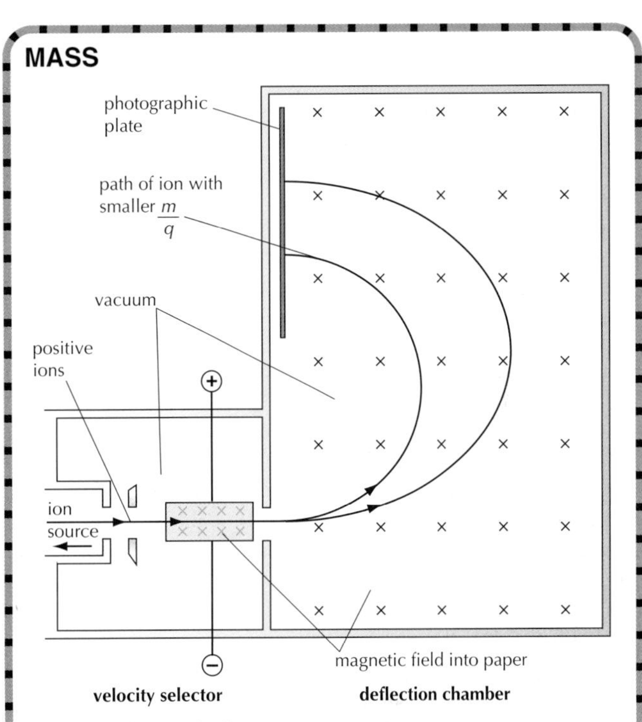

Essentials of a Bainbridge mass spectrometer

Isotopes of the same element have the same chemical properties so they cannot be separated using chemical reactions. The mass spectrometer provides a way of determining the masses of individual nuclei. The principle is to use a magnetic field to deflect moving ions of a substance. If a moving ion enters a constant magnetic field, B, it will follow a circular path (see page 55). The magnetic force provides the centripetal force required.

$$Bqv = \frac{mv^2}{r}$$

The radius of the circle is thus $r = \frac{mv}{Bq}$

If the ions have the same charge, q, and they are all selected to be travelling at the same speed, v, then the radius of the circle will be depend on the mass of the ion. A larger-mass ion will travel in a larger circle. The mass spectrometer can thus easily demonstrate that there are different values of nuclear mass for a given element. In other words, isotopes exist.

Radioactive decay

NEUTRINOS AND ANTINEUTRINOS

Understanding beta decay properly requires accepting the existence of a virtually undetectable particle, the neutrino. It is needed to account for the "missing" energy and (angular) momentum when analysing the decay mathematically. Calculations involving mass difference mean that we know how much energy is available in beta decay. For example, an isotope of hydrogen, tritium, decays as follows:

$${}^{3}_{1}H \rightarrow {}^{3}_{2}He + {}^{0}_{-1}\beta$$

The mass difference for the decay is 19.5 k eV c^{-2}. This means that the beta particles should have 19.5 k eV of kinetic energy. In fact, a few beta particles are emitted with this energy, but all the others have less than this. The average energy is about half this value and there is no accompanying gamma photon. All beta decays seemed to follow a similar pattern.

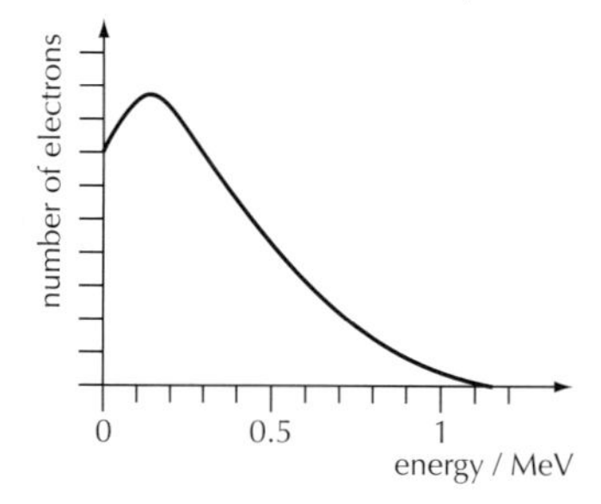

The energy distribution of the electrons emitted in the beta decay of bismuth-210. The kinetic energy of these electrons is between zero and 1.17 MeV.

The neutrino (and antineutrino) must be electrically neutral. Its mass would have to be very small, or even zero. It carries away the excess energy but it is very hard to detect. One of the triumphs of the particle physics of the last century was to be able to design experiments that confirmed its existence. The full equation for the decay of tritium is:

$${}^{3}_{1}H \rightarrow {}^{3}_{2}He + {}^{0}_{-1}\beta + \bar{\nu}$$

where $\bar{\nu}$ is an antineutrino

As has been mentioned before, another form of radioactive decay can also take place, namely positron decay. In this decay, a proton within the nucleus decays into a neutron and the antimatter version of an electron, a positron, is emitted.

$${}^{1}_{1}p \rightarrow {}^{1}_{0}n + {}^{0}_{+1}\beta^{+} + \nu$$

In this case, the positron, β^{+}, is accompanied by a neutrino.

The antineutrino is the antimatter form of the neutrino.

e.g. $${}^{19}_{10}Ne \rightarrow {}^{19}_{9}F + {}^{0}_{+1}\beta^{+} + \nu$$

$${}^{14}_{6}C \rightarrow {}^{14}_{7}N + {}^{0}_{-1}\beta^{-} + \bar{\nu}$$

MATHEMATICS OF EXPONENTIAL DECAY

The basic relationship that defines exponential decay as a random process is expressed as follows:

$$\frac{dN}{dt} \propto -N$$

The constant of proportionality between the rate of decay and the number of nuclei available to decay is called the decay constant and given the symbol λ. Its units are time^{-1} i.e. s^{-1} or yr^{-1} etc.

$$\mathbf{\frac{dN}{dt} = -\lambda N}$$

The solution of this equation is:

$$\mathbf{N = N_0 e^{-\lambda t}}$$

The activity of a source, A, is:

$$\mathbf{A = A_0 e^{-\lambda t}}$$

It is useful to take natural logarithms:

$$\ln(N) = \ln(N_0 e^{-\lambda t})$$
$$= \ln(N_0) + \ln(e^{-\lambda t})$$
$$= \ln(N_0) - \lambda t \ln(e)$$
$$\therefore \ln(N) = \ln(N_0) - \lambda t \quad \text{(since } \ln(e) = 1\text{)}$$

This is of the form $y = c + mx$ so a graph of ln N vs. t will give a straight-line graph.

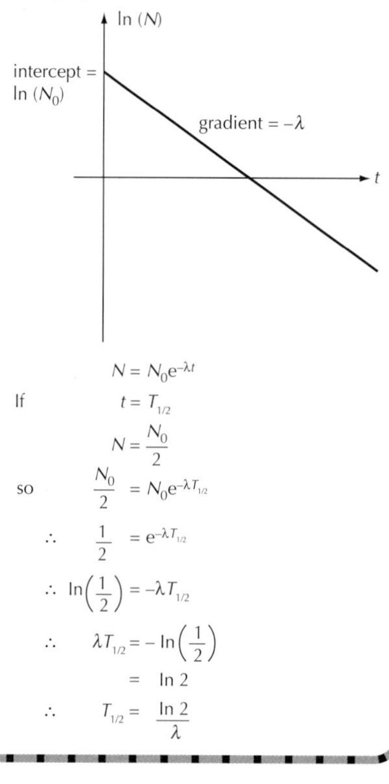

$$N = N_0 e^{-\lambda t}$$

If $$t = T_{1/2}$$

$$N = \frac{N_0}{2}$$

so $$\frac{N_0}{2} = N_0 e^{-\lambda T_{1/2}}$$

$$\therefore \frac{1}{2} = e^{-\lambda T_{1/2}}$$

$$\therefore \ln\left(\frac{1}{2}\right) = -\lambda T_{1/2}$$

$$\therefore \lambda T_{1/2} = -\ln\left(\frac{1}{2}\right)$$
$$= \ln 2$$

$$\therefore T_{1/2} = \frac{\ln 2}{\lambda}$$

METHODS FOR MEASURING HALF-LIFE

When measuring the activity of a source, the background rate should be subtracted.

- If the half-life is short, then readings can be taken of activity against time.
  → A simple graph of activity against time would produce the normal exponential shape. Several values of half-life could be read from the graph and then averaged. This method is simple and quick but not the most accurate.
  → A graph of ln (activity) against time could be produced. This should give a straight line and the decay constant can be calculated from the gradient.
- If the half-life is long, then the activity will effectively be constant over a period of time. In this case one needs to find a way to calculate the number of nuclei present and then use

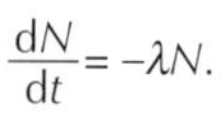

$$\frac{dN}{dt} = -\lambda N.$$

EXAMPLE

The half-life of a radioactive isotope is 10 days. Calculate the fraction of a sample that remains after 25 days.

$$T_{1/2} = 10 \text{ days}$$

$$\lambda = \frac{\ln 2}{T_{1/2}}$$
$$= 6.93 \times 10^{-2} \text{ day}^{-1}$$

$$N = N_0 e^{-\lambda t}$$

$$\text{Fraction remaining} = \frac{N}{N_0}$$
$$= e^{-(6.92 \times 10^{-2} \times 25)}$$
$$= 0.187$$
$$= 18.7\%$$

IB QUESTIONS – QUANTUM AND NUCLEAR PHYSICS 

1 The diameter of a proton is of the order

A 10^{-9} m **B** 10^{-11} m **C** 10^{-13} m **D** 10^{-15} m

2 When bullets leave the barrel of a rifle they are not observed to diffract because

A the de Broglie hypothesis only applies to electrons.
B the speed of the bullet is not great enough.
C the de Broglie wavelength of the bullet is too small.
D the de Broglie wavelength of the bullet is too large.

3 The diagram represents the available energy levels of an atom. How many emission lines could result from electron transitions between these energy levels?

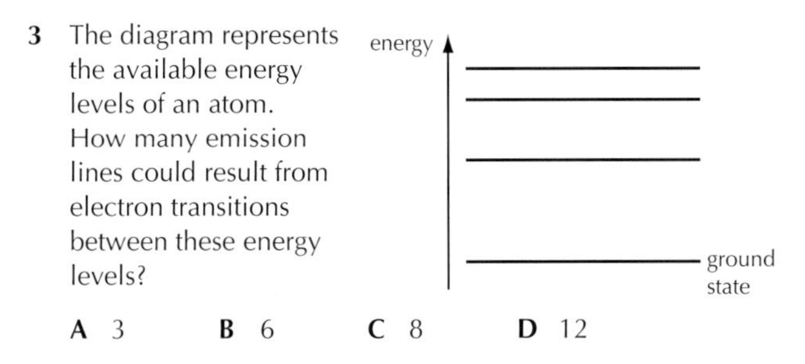

A 3 **B** 6 **C** 8 **D** 12

4 A medical physicist wishes to investigate the decay of a radioactive isotope and determine its decay constant and half-life. A Geiger–Müller counter is used to detect radiation from a sample of the isotope, as shown.

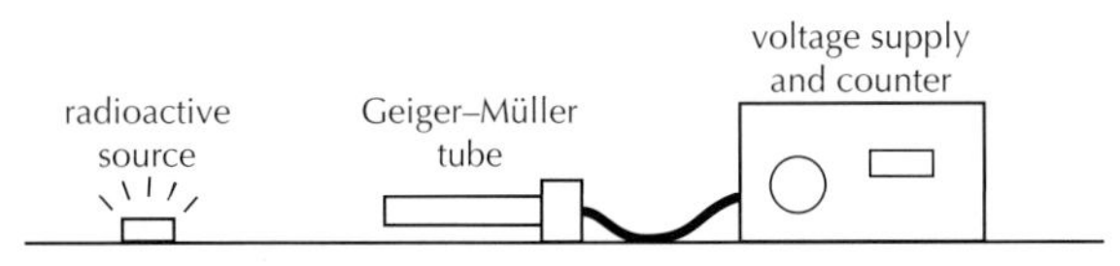

(a) Define the activity of a radioactive sample. [1]

Theory predicts that the activity A of the isotope in the sample should decrease exponentially with time t according to the equation $A = A_0 e^{-\lambda t}$, where A_0 is the activity at $t = 0$ and λ is the decay constant for the isotope.

(b) Manipulate this equation into a form which will give a straight line if a semi-log graph is plotted with appropriate variables on the axes. State what variables should be plotted. [2]

The Geiger-counter detects a proportion of the particles emitted by the source. The physicist records the count-rate R of particles detected as a function of time t and plots the data as a graph of ln R versus t, as shown below.

(c) Does the plot show that the experimental data are consistent with an *exponential* law? Explain. [1]

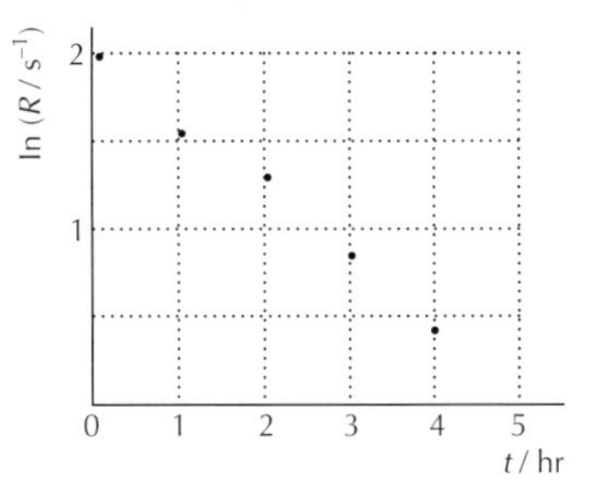

(d) The Geiger-counter does not measure the total activity A of the sample, but rather the count-rate R of those particles that enter the Geiger tube. Explain why this will not matter in determining the decay constant of the sample. [1]

(e) From the graph, determine a value for the decay constant λ. [2]

The physicist now wishes to calculate the half-life.

(f) Define the half-life of a radioactive substance. [1]

(g) Derive a relationship between the decay constant λ and the half-life τ. [2]

(h) Hence calculate the half-life of this radioactive isotope. [1]

5 This question is about the quantum concept.

A biography of Schrödinger contains the following sentence: "Shortly after de Broglie introduced the concept of *matter waves* in 1924, Schrödinger began to develop a new atomic theory."

(a) Explain the term *'matter waves'*. State what quantity determines the wavelength of such waves. [2]

(b) Electron diffraction provides evidence to support the existence of matter waves. What is electron diffraction? [2]

(c) Calculate the de Broglie wavelength of electrons with a kinetic energy 30 eV. [3]

(d) How does the concept of *matter waves* apply to the electrons within an atom? [2]

6 Light is incident on a clean metal surface in a vacuum. The maximum kinetic energy KE_{max} of the electrons ejected from the surface is measured for different values of the frequency f of the incident light.

The measurements are shown plotted below.

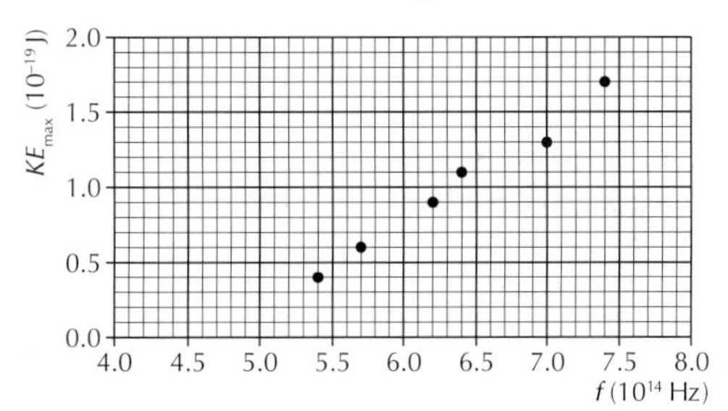

(a) Draw a line of best fit for the plotted data points. [1]

(b) Use the graph to determine

(i) the Planck constant [2]

(ii) the minimum energy required to eject an electron from the surface of the metal (the *work function*). [3]

(c) Explain briefly how Einstein's photoelectric theory accounts for the fact that no electrons are emitted from the surface of this metal if the frequency of the incident light is less than a certain value. [3]

7 Thorium-227 (Th-227) undergoes α-decay with a half-life of 18 days to form radium-223 (Ra-223). A sample of Th-227 has an initial activity of 3.2×10^5 Bq.

Determine the activity of the remaining thorium-227 after 50 days.

HL Analogue and digital techniques

ANALOGUE AND DIGITAL

Information comes in many forms, for example:

- text
- pictures
- speech/sound/music
- video
- a mixture of all of the above.

The transfer of information between two people requires the use of agreed codes. In a language, the sounds of different words are an agreed code for different concepts. Different alphabets are agreed codes for the sound of different words in written form.

The use of technology for the storage, processing and transfer of information involves converting the original information into a different form (often an electric or electronic **signal**), which can be converted back as required.

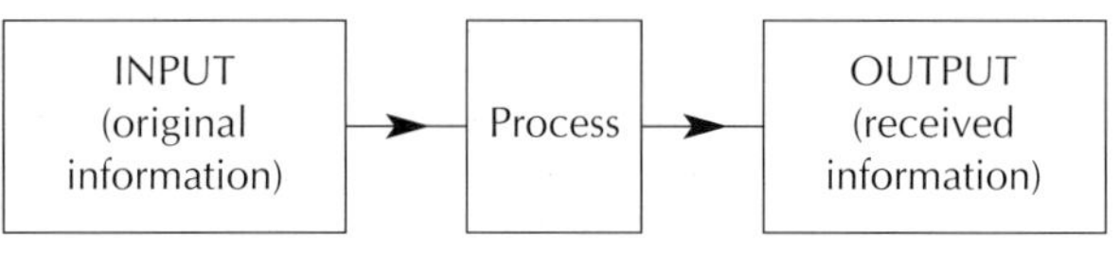

This can be achieved using **analogue** or **digital** techniques. Analogue techniques involve codes and signals that can take a large number of different values between given limits. The analogue signal continuously varies with time.

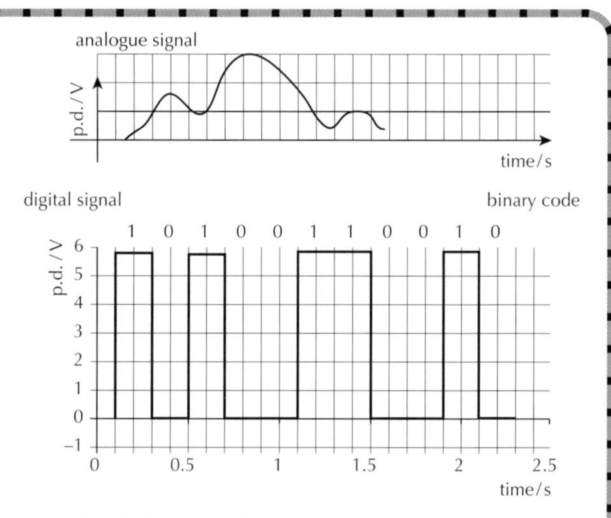

Digital techniques involve codes and signals involve a large number of **b**inary dig**its**, or bits. Each bit can only take one of two possible values. (1 or 0; High or Low; ON or OFF; True or False). Eight separate bits of information are called a **byte**. Larger quantities of information are measured in kilobytes, megabytes, gigabytes etc. A digital signal repeatedly changes between the two available levels.

Method of storage	Typical source information	Overview of process	Comments	Analogue or digital
Photocopying	Text or pictures	Optical and electrostatic processes are used to create a hard copy of the original document.	The process is not 100% accurate. A photocopy of a photocopy will be further reduced in quality.	Analogue
Microfiche	Text or pictures	Optical process used to photograph a large number of original documents to be held miniaturized form.	An optical microfiche reader is used to access the information.	Analogue
LPs ('vinyl')	Music or speech	Sound variations are stored as variations in a track on the LP.	Dust and/or scratches alter the output quality.	Analogue
Cassette tapes	Music or speech	Sound variations are stored as variations in the magnetic field orientations in the tape.	Each time the tape is played, the quality is slightly reduced. The tape can be stretched or damaged.	Analogue
Floppy discs	Text	Different characters are stored as series of magnetic variations in the disc. Only two possible variations are utilized.	Magnetic fields can corrupt the data stored on a floppy disc.	Digital
Computer memory (microchips)	All forms	Variations are stored using large numbers of transistors and capacitors within the chip. Only two possible variations are utilized.	Some designs have fixed information stored (ROM) whereas other have the flexibility to change the information (RAM).	Digital
Hard discs	All forms	Variations are stored as series of magnetic variations in the disc. Only two possible variations are utilized.	Hard discs operate in the same way as floppy disc. Large amounts of microchip memory can be equivalent to a hard disc.	Digital
DVDs	Video	Variations of light and sound are stored as a series of optical "bumps" or "pits" on the DVD track.	Minor damage to the disc can be accommodated while the data is being accessed. Major damage can prevent any data from being accessible.	Digital

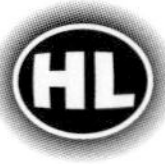

Numbers in different bases

BINARY

The conversion between analogue and digital involves the conversion from normal numbers (decimal, or base 10) to binary (base 2). Normal numbers are represented by counting in powers of 10 so each digit represents 1, 10, 100, 1000 etc. Binary numbers are represented by counting in powers of 2 so each bit represents 1, 2, 4, 8, 16 etc.

Numbers are read from left to right. The left hand digit represents the largest power and the right hand digit represents the smallest power. In binary notation, the largest power is called the **M**ost **S**ignificant **B**it (**MSB**) and the smallest power is called the **L**east **S**ignificant **B**it (**LSB**).

e.g. $19 = 16 + 2 + 1 = 2^4 + 2^1 + 2^0$

19 in decimal = 10011 in binary

Most Significant Bit — Least Significant Bit

Base 10	2^4	2^3	2^2	2^1	2^0	Base 2
0	0	0	0	0	0	00000
1	0	0	0	0	1	00001
2	0	0	0	1	0	00010
3	0	0	0	1	1	00011
4	0	0	1	0	0	00100
5	0	0	1	0	1	00101
6	0	0	1	1	0	00110
7	0	0	1	1	1	00111
8	0	1	0	0	0	01000
9	0	1	0	0	1	01001
10	0	1	0	1	0	01010
11	0	1	0	1	1	01011
12	0	1	1	0	0	01100
13	0	1	1	0	1	01101
14	0	1	1	1	0	01110
15	0	1	1	1	1	01111
16	1	0	0	0	0	10000
17	1	0	0	0	1	10001
18	1	0	0	1	0	10010
19	1	0	0	1	1	10011
20	1	0	1	0	0	10100
etc						

ASCII

A given number of bits can represent a fixed number of different values. Each additional bit doubles the number of different possible values that can be represented.

Number of bits used	Number of different possible values
1	2
2	4
3	8
4	16
5	32
6	64
7	128
8	256
9	512
10	1024

Text written in English can contain a mixture of letters, numbers, symbols and punctuation marks. Additional characters are also needed to represent the formatting that takes place (new line, new paragraph etc). Computers often use a code known as ASCII (American Standard Code for Information Interchange). It is an 8-bit code representing 256 different possible characters and formatting codes.

Advantages and disadvantages of digital techniques (1)

CONVERSION BETWEEN ANALOGUE AND DIGITAL FORMS

The conversion of everyday analogue information (e.g. music) into digital form (e.g. mp3 format on a computer) involves:

- Sampling the input information at regular intervals.
- Converting each sampled signal into one value from a fixed range of possible values (**quantum levels**).
- Converting each sampled quantum level into digital form (a binary number).

The accuracy of digital information can be improved by:

- Increasing the sampling frequency.
- Increasing the number of available quantum levels.

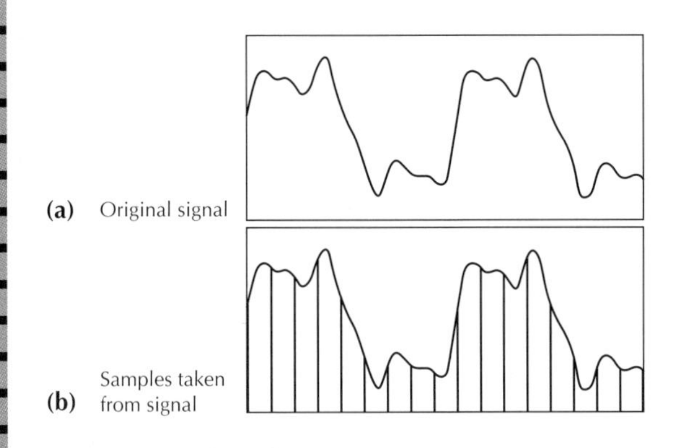

Problem with sampling

Sampling too slowly misses high frequency detail in the original signal

The output process involves converting the series of bits back into an analogue signal. Limitations in human sensing ability (ears and eyes) mean that some improvements of accuracy would not be distinguishable.

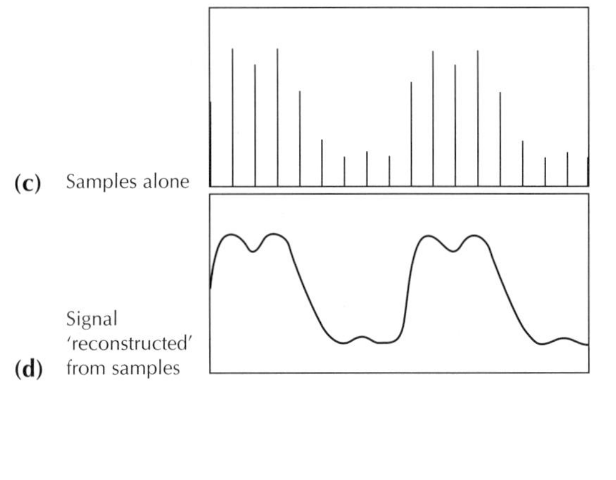

Advantages and disadvantages of digital techniques

COMPARISON BETWEEN DIGITAL AND ANALOGUE TECHNIQUES

All processes introduce some amount of unwanted variations. For example unwanted additional variations (**noise**) can be added and the original signal can also be altered.

In analogue processes this affects the quality of the final signal. In digital processes only two possible signals are ever used so minor variations can be irrelevant to the final signal.

Electronic techniques can correct for missing or corrupted signals.

The process used for the storage and retrieval of information can also affect quality. Some retrieval processes (e.g. mechanical ones such as vinyl and tapes) damage the data that is being stored, making each subsequent retrieval lower in quality. In comparison other process (e.g. the optical techniques in a DVD) can be used to ensure that the original is not altered when data is accessed.

	Digital	**Analogue**
Complexity of code	There must be a complex set of rules for the conversion of input into digital signal and from digital to output.	Can be simple e.g. direct parallel between pressure variation of a sound and electrical p.d.
Quality	If the sample frequency and number of quantum levels are sufficiently high, then the output can be indistinguishable from the input.	Quality can be virtually indistinguishable from input but very liable to damage or corruption.
Reproducibility	Optical techniques can ensure that each subsequent retrieval is virtually identical.	Process of retrieval often affects quality of future retrievals.
Retrieval speed	Text and simple data can be retrieved at great speed. More complex data (video etc.) takes longer but selecting different sections of information often does not add significant time.	Often the retrieval process requires a significant time. Movement between different sections of the data can take significant time.
Portability of stored data	Modern miniaturization techniques have ensured that a large quantities of data can be stored in a very small device.	Although stored data can be compact, many analogue storage systems occupy significantly larger volume compared with digital alternative.
Manipulation of data	Manipulation of data can be easily achieved without significant corruption.	All manipulation will increase the possibility of data corruption.

IMPLICATIONS FOR SOCIETY OF EVER-INCREASING DATA STORAGE

The increasing ease with which data can be electronically stored means that more information is being saved and this information is being accessed and shared by more people. This has several implications:

Moral / Ethical	• Digital technologies mean that information that is potentially problematic for society can be easily recorded and shared. • Issues concerning the access and ownership of electronic data. • Should there be any limits on what information an adult chooses to access or share with other adults? • Does a society have the right to control individuals' access to digital information? • If you are the subject of a stored piece of information, do you have any rights of access and/or control over its use? • To what extent does the state have the right to acquire identity data from the citizen? • The role of the historian is important, as is the need to create an archive policy. If only particular information is stored, then the information record will be biased.
Social	• Digital technology can be used to record and highlight abuses of human rights. • As more data is created, the access and storage of this information can control societies' opinions and views. • Individuals who do not have access to the Internet will be disadvantaged. • Since records are not on paper, what needs to be done to ensure the record is still available to subsequent generations?
Economic	• The cost of electronic data storage and transfer is very different from the cost of storage and transfer of more traditional techniques. • There will be an impact on the process of economic decision making which is often based on imperfect information. The distribution of data implies more perfect information and thus more competitive markets. • The above implies very quick price comparisons being possible and thus more stable markets, better control and a tendency for inflation to reduce. • It should result in better predictions for the future.
Environmental	• Electronic data storage could replace traditional techniques, thus resulting in a saving in resources. • In reality much paper is still wasted as information is printed off several times, creating a waste issue. • The resources needed for the manufacture of electronic data storage will be more sought after.

CDs and DVDs

HOW DATA IS STORED ON A CD

The track on a CD starts in the centre and works outwards in a spiral. It is made up of small 'bumps' or 'lands' and 'pits':

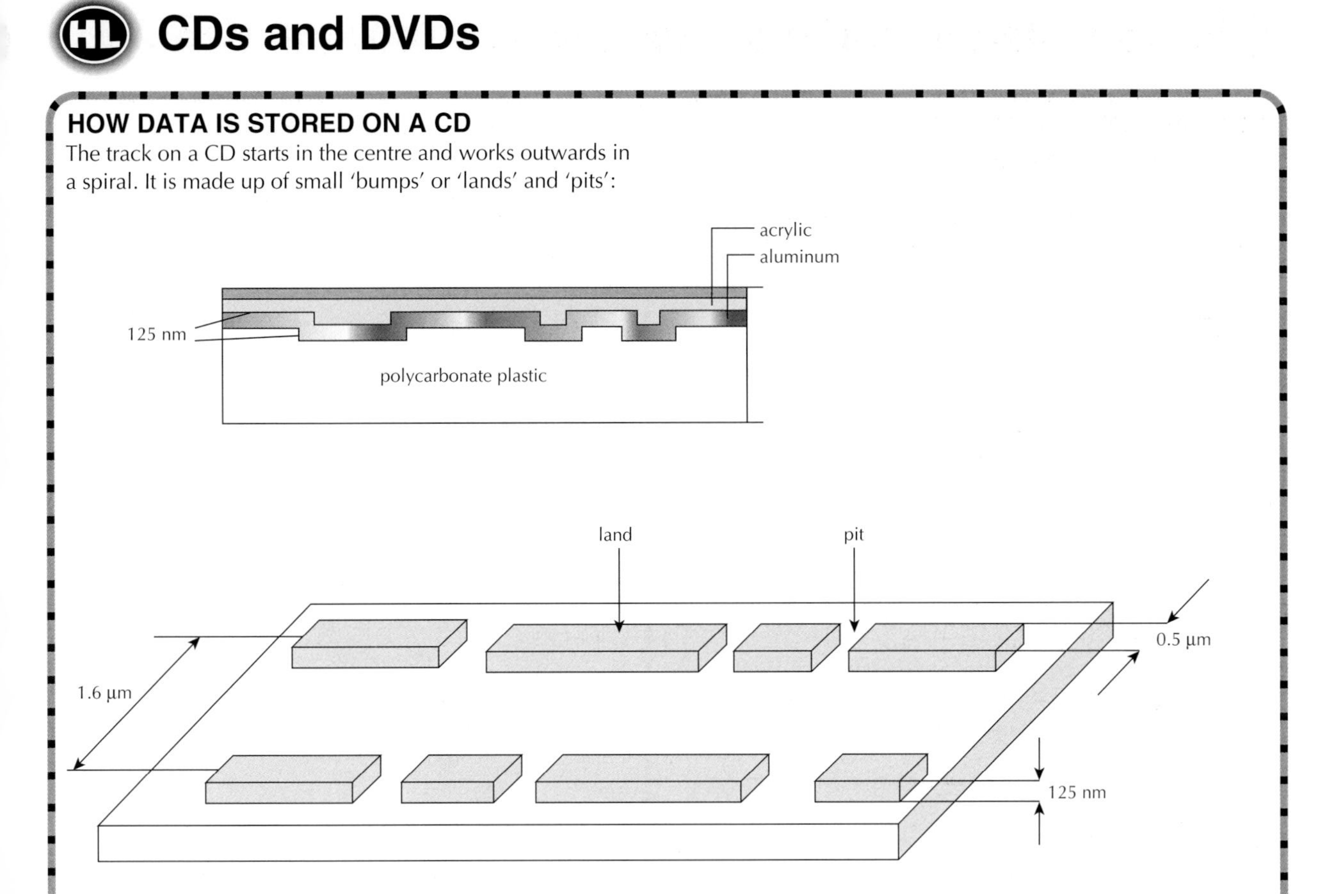

The information is read by sensing the amplitude of the reflection of a laser beam reflecting off the bumps and pits:

- The speed of rotation of the disc is controlled so that a constant length of track is scanned in a given time.
- The CD has a higher speed of revolution when the laser is reading near the centre compared with the outer edge.
- The laser beam is focused onto the track.
- When the beam reflects from a land or a pit, a strong signal is received.
- When the beam reflects from the edge between a land and a pit, destructive interference takes place and a weak signal is received.
- A strong signal represents 0 and a weak signal represents 1.

laser beam

constructive interference

destructive interference

$\frac{\lambda}{4}$ = bump height

EXAMPLES

1. Laser light of frequency 6×10^{14} Hz is used in a laser. Calculate an appropriate depth of a pit on a CD.

$$\lambda = \frac{c}{f} = \frac{3 \times 10^8}{6 \times 10^{14}} = 0.5 \times 10^{-6} = 500 \times 10^{-9} \text{ m} = 500 \text{ nm}$$

$$\text{depth of pit} \approx \frac{\lambda}{4} = \frac{500}{4} = 125 \text{ nm} \approx 1 \times 10^{-7} \text{ m}$$

2. A CD track moved from a radius of 25 mm to 58 mm with an average radius of 40 mm. The distance between sprials on the track is 1.6 μm. (a) Estimate the length of the track. (b) The scanning velocity is 1.2 m s^{-1}, estimate how long the CD will last. (c) It can store 700 Mbytes of information. What is the average length of track per bit of information?

a) Number of turns = 33 mm/1.6 μm = 20 625

Track length = 20 625 × 2 × π × 0.04 = 5184 m ≈ 5.2 km

b) CD playing time = 5184 m/1.2 = 4320 s = 72 minutes

c) Average length per bit = 5184 m/($7 \times 10^8 \times 8$) = 0.9 μm

3. Estimate the playing time of a 700 Mbyte CD storing stereo music using 16 bit sampling.

Max audio frequency = 20 kHz

∴ sampling frequency ≈ 40 kHz

Number of bits every second for each channel = 40 000 × 16 = 6.4×10^5 bits

Total number of bits per second for stereo = 1.28×10^6 bits

Total storage capacity of CD = 7×10^8 bytes = 56×10^8 bits

Maximum time for CD = (56×10^8)/(1.28×10^6) = 4375 s ≈ 73 minutes

HL Capacitance and charge-coupled devices (CCDs)

CAPACITANCE

Capacitors are devices that can store charge. The charge stored q is proportional to the p.d. across the capacitor V and the constant of proportionality is called the capacitance C.

Symbol: (capacitor symbol) C

$$C = \frac{q}{V}$$

capacitance in farads; charge in coulombs; p.d. in volts

The farad (F) is a very large unit and practical capacitances are measured in μF, nF or pF.

$1 \text{ F} = 1 \text{ C V}^{-1}$

A measurement of the p.d. across a capacitance allows the charge stored to be calculated.

IMAGE PARAMETERS

Quantum efficiency

The pixels in a perfect CCD would each emit one photoelectron for each photon incident on its surface. Practical CCDs do not achieve this level of efficiency. Quantum efficiency QE is the ratio of the number of photoelectrons emitted to the number of photons incident on the pixel.

$$QE = \frac{N_e}{N_p}$$

Quantum efficiencies depend on the design on the CCD and the wavelengths involved and typically vary between 20% and 90%.

Magnification and resolution

Lenses are used to focus an image of the object onto the CCD. Magnification is the ratio of the length of the image on the CCD to the length of the object.

Two points on an object may be just resolved on a CCD if the images of the points are two pixels apart.

EXAMPLE

A digital camera is used to photograph an object. Two points on the object are separated by 0.0020 cm. The CCD in the camera has a collecting area of 16 cm^2 and contains 4.0 megapixels. The magnification of the camera is 1.5. Can the images of the points be resolved?

Area corresponding to each pixel $16 \times 10^{-4}/4 \times 10^{6} = 4.0 \times 10^{-10}$ m^2

Separation of pixels = $\sqrt{(4.0 \times 10^{-10})} = 2.0 \times 10^{-5}$ m

Equivalent separation on the object = $2.0 \times 10^{-5}/1.5 =$ 0.0013 cm

Distance between two pixels < 0.0020 cm so image can be resolved

CHARGE-COUPLED DEVICE (CCD)

A CCD is a silicon (semiconductor) microchip that can be used to electronically record an image focused onto its surface. The surface is divided into a large number of small areas called **pixels**.

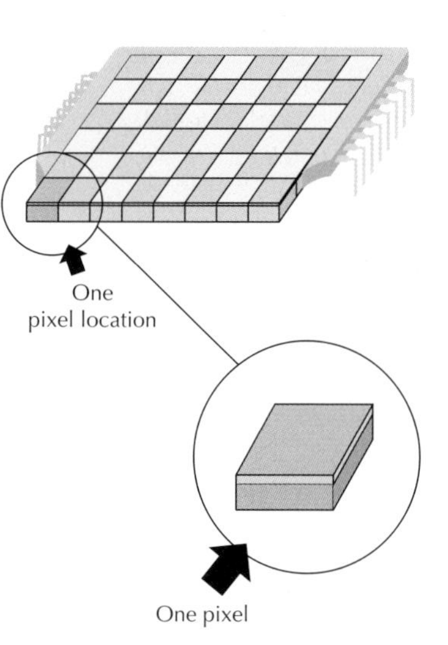

There are four stages to recording a digital image using a CCD:

- During a photo exposure, each element within the CCD generates a charge proportional the incident light as a result of the photoelectric effect.
  - Each pixel converts light energy into electrical energy using the photoelectric effect.
  - As more photons are incident on a pixel, more electrons are emitted.
- The charge is collected in different pixels.
  - The pixel behaves as a capacitor and a charge builds up on each pixel.
  - The amount of charge is proportional to the number of photons (which is also related to the intensity of the light) incident on the pixel.
  - The p.d. across each pixel is proportional to the number of photons incident on the pixel.
- The charge collected from each pixel is transferred in turn by 'coupling' charges from one pixel to the next in turn.
  - Storage changes can be transferred along a line of pixels in sequence.
  - The signal processing takes place line by line to ensure the charge on each pixel is recorded.
- Individual charge packets are converted to an output voltage and then digitally encoded.
  - The value of the p.d is measured and converted into a digital signal in binary code.
  - The light intensity information from each pixel can be stored along with another digital signal representing the position of the pixel on the surface.
  - The signals from each pixel can be stored.
  - The information can be used to reconstruct the image as the information from each pixel has recorded the different light intensities in different parts of the image.

Uses of CDs

PRACTICAL USES OF CCDS

CCDs are used for image capturing in a large range of the electromagnetic spectrum. The following list provides some examples.

Device	Comment
Digital cameras	Very convenient to take and share photographs, but image quality can be lower than traditional film unless camera is of high quality (which is more expensive). In order to create a colour photograph, the image is analysed three times – once each for red, green and blue.
Video cameras	Digitized images are usually better quality than analogue images stored on magnetic videotape and are easier to store and transport. It is possible to continuously record video without interruption during playback. Searches are faster and easier to perform. Digital storage is fast and utilizes re-usable media – useful for security recordings.
Telescopes	Sensitivity of CCDs is better than traditional film and allows for detailed analysis over a range of frequencies. Exposure times are reduced. CCDs also allow for remote operation of telescopes. The Hubble space telescope is in orbit around the Earth.
Medical X-ray imaging	Traditional X-rays involve the use of film. Digital X-rays can have better contrast and have the advantage of being able to be processed, allowing the images to be easily compared using imaging techniques. Information can be quickly and easily shared between distant hospitals and storage and access can be improved.
Scanners	Everyday images can be scanned into digital form using a scanner. Fax machines scan text and transmit the image along telephone lines.

	Quality of the processed image
Quantum efficiency	The greater the QE, the greater the sensitivity of the device.
Magnification	A greater magnification means that more pixels are used for a given section of the image. The image will be more detailed.
Resolution	The greater the resolution, the greater the amount of detail recorded. An improvement in resolution will mean a given image will occupy more memory.

Advantages of CCDs compared with use of film:

- Each photo does not require the use of film and is thus cheaper.
- Traditional films have a QE of less than 10% whereas values of over 90% are achievable for some frequencies for CCDs. This means that very faint objects can be photographed.
- The image is digital so it can be enhanced and edited using electronic processing techniques.
- Image can be viewed virtually immediately – there is no processing time.
- Storage and archiving of a large number of photographs is easy.

IB QUESTIONS – DIGITAL TECHNOLOGY

1 Estimate the minimum number of bits that are needed

(a) to represent each character in a simple sentence written in English

(b) to write this question. [4]

2 Two friends talk on the telephone for 15 minutes. The sound is transmitted by using 8 bit sampling at a sampling frequency of 8 kHz. Calculate the total number of bits of information that have been transmitted. [4]

3 Outline how the interference of light is used to recover information stored on a CD. [6]

4 The wavelength of light used in a CD player is 500 nm. Calculate the bump heights on the CD. [2]

5 State and explain five advantages of storing information in digital rather than analogue form. [5]

6 **(a)** A 3μF capacitor is charged to 240 V. Calculate the charge stored. [1]

(b) Estimate the amount of time it would take for the charge you have calculated in **(a)** to flow through a 60 W light bulb connected to the 240 V mains electricity. [2]

(c) The charged capacitor in **(a)** is discharged through a 60 W 240 V light bulb.

(i) Explain why the current during its discharge will not be constant. [2]

(ii) Will the bulb light during discharge? Explain your answer. [2]

7 A satellite system takes an image of the surface of the Earth. Each image covers 100 km^2 and is recorded by a CCD of area 25 cm^2.

(a) Calculate the magnification of system. [2]

(b) The resolution of the system is 10 m.

(i) Calculate the separation of the pixels on the CCD. [2]

(ii) Find the number of pixels on the CCD. [2]

8 Outline how

(i) the image on a CCD is digitized

(ii) the image stored in a CCD is retrieved. [6]

The eye and sight

HUMAN EYE

The structure of the human eye is shown below:

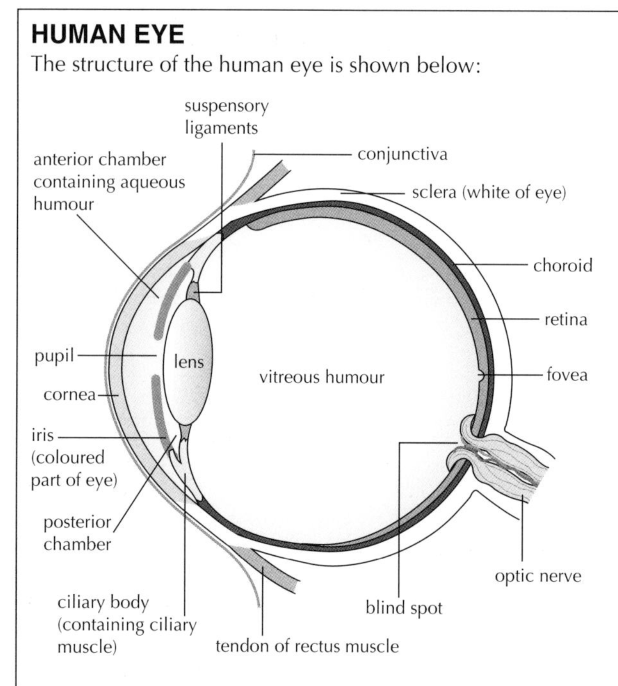

Refraction takes place as light enters the eye. Light passes through the cornea, aqueous humour, lens and vitreous humour in turn before striking the retina. The retina contains two different types of light-sensitive cells – the rods and cones. The light-response curves for the three different types of cones (labelled S, M and L) and the rods (labelled R) are shown top right.

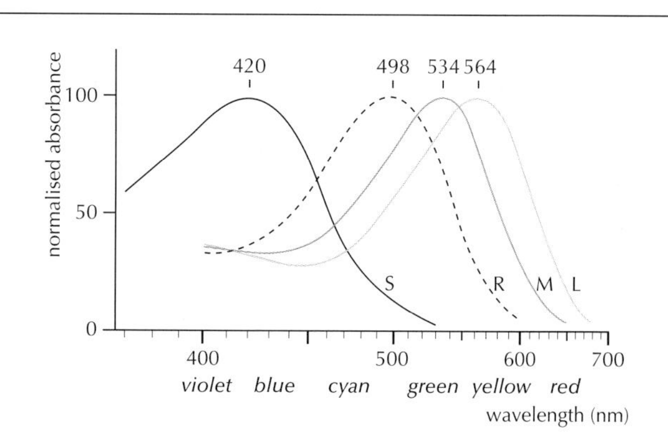

Photopic vision is the colour vision that takes place at normal light levels. It is provided by three different **cone** cells which have peak sensitivity in the short (S), medium (M) and long (L) visible wavelengths, respectively.

Scotopic vision is the black and white vision that takes place in dim light. It is provided by the **rod** cells as shown in the above response curve (R). The chemical necessary for this "night vision" can take several minutes to be synthesized once light levels have been reduced.

The rods and cones are not evenly distributed across the retina. The density of the cones is at a maximum in the centre whereas rods peak in density at an angle of approximately 20° away from the centre. There are no rods or cones in the region where the optic nerve leaves the back of the eye. This is known as the **blind spot**. The graph below shows the vertical variation of the density of the cells.

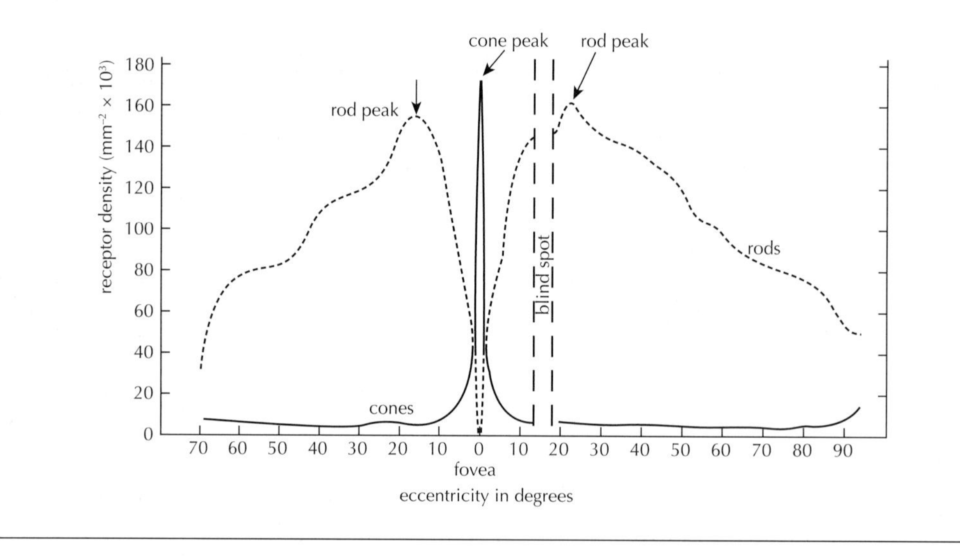

POSSIBLE CHANGES TO THE EYE

Accommodation is name given to the process by which the eye can focus on different objects. The eye lens is naturally "short and fat" but can be pulled to be "long and thin" by taut **suspensory ligaments** which attach the lens to the circular **ciliary muscle**. The ciliary muscle controls the tension in the ligaments. A relaxed circular ciliary muscle means the lens is thin so the eye is focused on infinity.

The pupil contracting or expanding controls the amount of light that enters the eye.

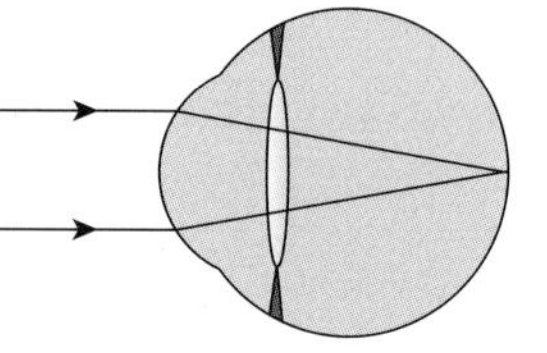

distant object is focused with thin lens – the circular ciliary muscles are relaxed, making the suspensory ligaments taut

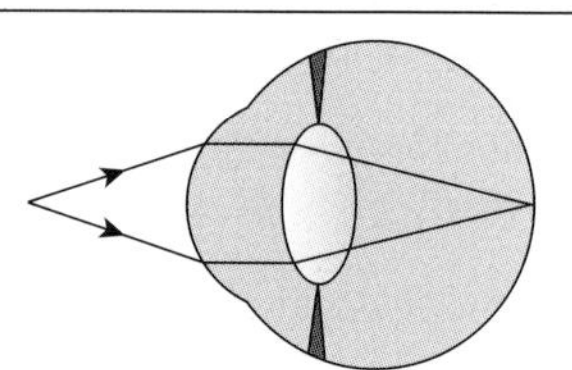

close object is focused with fatter lens – the circular ciliary muscles are contracted, making the suspensory ligaments slack

Perception

PERCEPTION OF COLOUR

Every monochromatic frequency in the visible portion of the electromagnetic spectrum is perceived by the eye as a different colour of the spectrum (red, orange, yellow, green, blue, indigo, violet). When two or more frequencies of light enter the eye at the same time this can also be perceived as one colour as represented below.

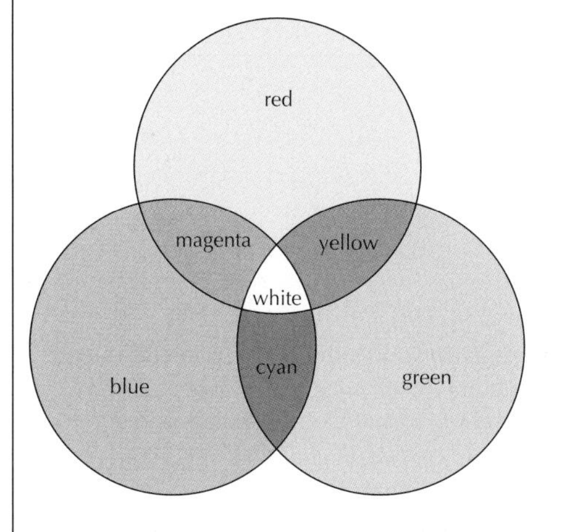

In this context of mixing different frequencies of light, red, green and blue are known as primary colours. In combination they produce the secondary colours: magenta (purple), cyan and yellow.

A filter placed in front of a source of light absorbs most frequencies and only lets through a particular combination of frequencies. A coloured surface absorbs most frequencies and only scatters the colour seen.

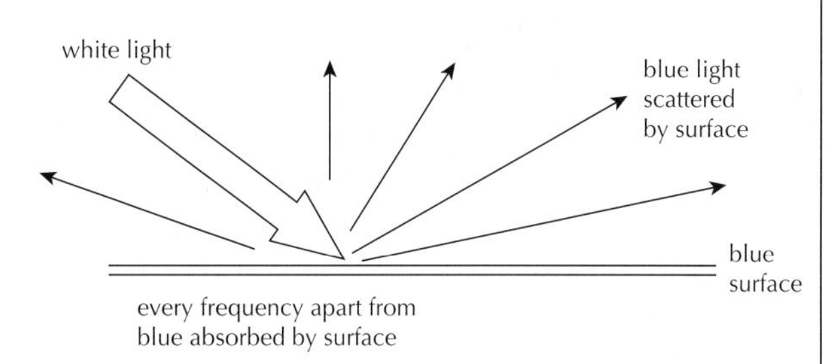

A blue surface will appear black when illuminated by red light.

The **trichromatic** theory of colour vision explains our perception of colour by the brain interpreting the different information received from the three different types of cones. The cones that are most responsive to short, medium and long wavelength are sometimes known as the blue, green and red sensitive cones, although these colours do not exactly correspond the wavelengths at which the cones are most sensitive.

Colour blindness is often caused by the failure of one or more types of cones to respond. Red–green colour blindness is the most common hereditary problem and is much more common in males than in females. The genes responsible for the red and green proteins are on the X-chromosome so males only have one copy. A defect on either gene causes red–green colour blindness.

PERCEPTION OF VISUAL DEPTH

There are many visual clues that are used by our eye/brain system that result in our sense of three dimensions. One very important process in the perception of depth is the brain's interpretation of the different images that are seen by each of our two eyes. Each one views any given scene from a slightly different perspective. When an object is far away, the relative difference between the locations of these two images will be small whereas a close object will have a greater difference between its two image locations. This process of stereoscopic vision requires both eyes to be functioning properly.

Other clues of the location for an unknown object come from the extent to which eyes are 'crossed' when focused on it, a comparison between objects of known size and an analysis of the order in which object must be located gained from knowing that closer objects can 'get in the way' of seeing more distant ones.

A normal human eye can focus on objects located between the **near point** (the closest point that can be focused upon without straining or optical aids – taken to be at a distance of 25 cm) and the **far point** (the furthest point that can be focused upon - taken to be at infinity).

PERCEPTION OF LIGHT AND SHADOW

As well as stereoscopic vision, the brain also interprets a wide variety of other visual signals to complete the perception of an object. For example:

- Architectural effects can be created by the use of light and shadow; deep shadow gives the impression of massiveness.
- The brain is very good at fitting blocks of colour into an outline picture. When a line picture is 'coloured in' using crayons, colour that spreads over a line tends to be ignored.
- We perceive the colour of an object to remain essentially constant even when the illumination used changes from sunlight to artificial light.
- Colour can be used to:
  - give an impression of 'warmth' (e.g. blue tints are often perceived as 'cold')
  - change the perceived size of a room (e.g. light-coloured ceilings seem to heighten the room).

The rest of Standard level Option A is identical to Chapter 11.

Options B, C and D

- Standard level option B is identical to Chapter 13.
- Standard level option C is identical to Chapter 14 in addition to pages 149–153 in Chapter 17.
- Standard level option D is identical to pages 171–5 in Chapter 19 in addition to pages 198–201 and 205–6.

The Solar System and beyond (1)

SOLAR SYSTEM

We live on the Earth. This is one of eight planets that orbit the Sun – collectively this system is known as the Solar System. Each planet is kept in its elliptical orbit by the gravitational attraction between the Sun and the planet. Other smaller masses such as **dwarf planets** like Pluto or planetoids also exist.

	Mercury	Venus	Earth	Mars	Jupiter	Saturn	Uranus	Neptune
diameter / km	4880	12 104	12 756	6787	142 800	120 000	51 800	49 500
distance to Sun /$\times 10^8$ m	58	107.5	149.6	228	778	1427	2870	4497

Relative positions of the planets

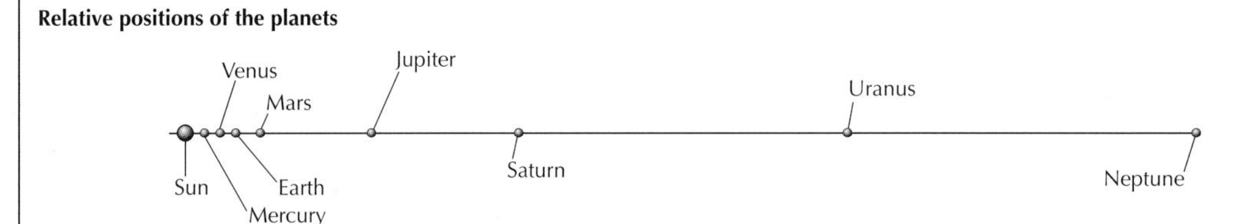

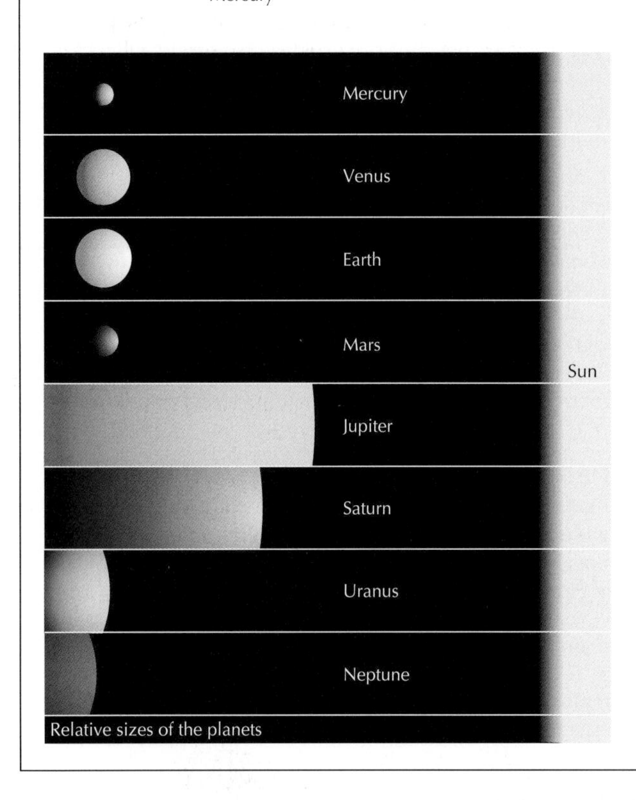

Relative sizes of the planets

Some of these planets (including the Earth) have other small planets orbiting around them called moons. Our Moon's diameter is about 1/4 of the Earth's and it is 3.8×10^8 m away.

An **asteroid** is a small rocky body that drifts around the Solar System. There are many orbiting the Sun between Mars and Jupiter – the asteroid belt. An asteroid on a collision course with another planet is known as a meteoroid.

Small meteors can be vaporised due to the friction with the atmosphere ('shooting stars') whereas larger ones can land on Earth. The bits that arrive are called **meteorites**.

Comets are mixtures of rock and ice (a 'dirty snowball') in very elliptical orbits around the Sun. Their 'tails' always point away from the Sun.

VIEW FROM EARTH

If we look up at the night sky we see the stars – many of these 'stars' are, in fact, other galaxies but they are very far away. The stars in our own galaxy appear as a band across the sky – the Milky Way.

Patterns of stars have been identified and 88 different regions of the sky have been labelled as the different **constellations**. Stars in a constellation are not necessarily close to one anothers.

Over the period of a night, the constellations seem to rotate around one star. This apparent rotation is a result of the rotation of the Earth about its own axis.

On top of this nightly rotation, there is a slow change in the stars and constellations that are visible from one night to the next. This variation over the period of one year is due to the rotation of the Earth about the Sun.

FROM PLACE TO PLACE

If you move from place to place around the Earth, the section of the night sky that is visible over a year changes with latitude. The total pattern of the constellations is always the same, but you will see different sections of the pattern.

The Solar System and beyond (2)

DURING ONE DAY

The most important observation is that the pattern of the stars remains the same from one night to the next. Patterns of stars have been identified and 88 different regions of the sky have been labelled as the different **constellations**. A particular pattern is not always in the same place, however. The constellations appear to move over the period of one night. They appear to rotate around one direction. In the Northern Hemisphere everything seems to rotate about the pole star.

It is common to refer measurements to the 'fixed stars' the patterns of the constellations. The fixed background of stars always appears to rotate around the pole star. During the night, some stars rise above the horizon and some stars set beneath it.

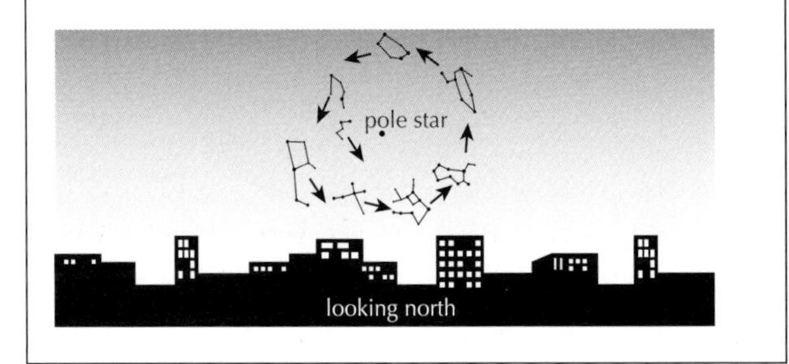

The same movement is continued during the day. The Sun rises in the East and sets in the West, reaching its maximum height at midday. At this time in the Northern Hemisphere the Sun is in a southerly direction.

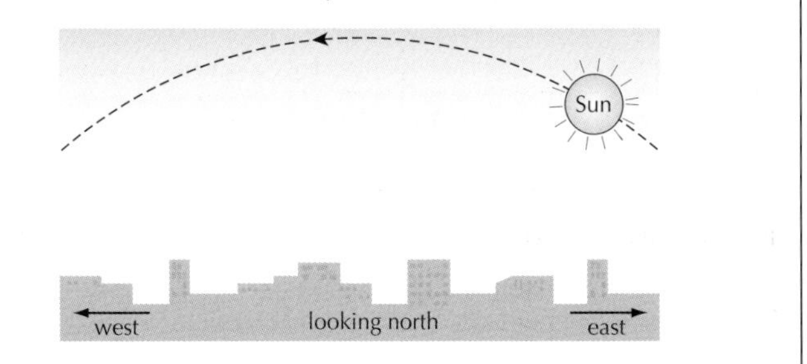

DURING THE YEAR

Every night, the constellations have the same relative positions to each other, but the location of the pole star (and thus the portion of the night sky that is visible above the horizon) changes slightly from night to night. Over the period of a year this slow change returns back to the exactly the same position

The Sun continues to rise in the East and set in the West, but as the year goes from winter into summer, the arc gets bigger and the Sun climbs higher in the sky.

UNITS

When comparing distances on the astronomical scale, it can be quite unhelpful to remain in SI units. Possible other units include the **astromonical unit (AU)**, the **parsec (pc)** or the **light year (ly)**. See page 126 for the definition of the first two of these.

The light year is the distance travelled by light in one year (9.5×10^{15} m). The nearest star to our Sun is about 4 light years away. Our galaxy is about 100 000 light years across. The nearest galaxy is about a million light years away and the observable Universe is about 1.5×10^{9} light years across.

THE MILKY WAY GALAXY

When observing the night sky a faint band of light can be seen crossing the constellations. This 'path' (or 'way') across the night sky became known as the Milky Way. What you are actually seeing is some of the millions of stars that make up our own galaxy but they are too far away to be seen as individual stars. The reason that they appear to be in a band is that our galaxy has a spiral shape.

The centre of our galaxy lies in the direction of the constellation Sagittarius. The galaxy is rotating – all the stars are orbiting the centre of the galaxy as a result of their mutual gravitational attraction. The period of orbit is about 250 million years.

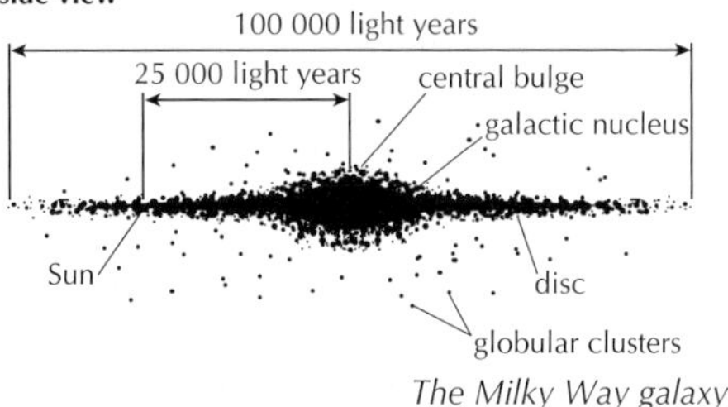

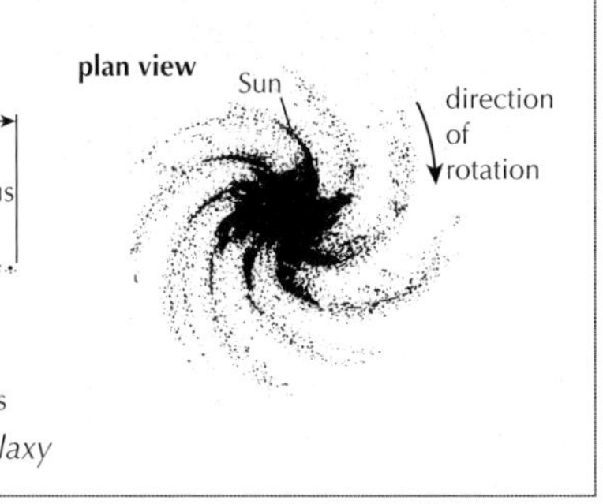

The Milky Way galaxy

THE UNIVERSE

Stars are grouped together in **stellar clusters**. Our Sun is just one of the billions of stars in our **galaxy** (the Milky Way galaxy). The galaxy rotates with a period of about 2.5×10^{8} years.

Beyond our galaxy, there are billions of other galaxies. Some of them are grouped together into **clusters** or **super clusters** of galaxies, but the vast majority of space (like the gaps between the planets or between stars) appears to be empty – essentially a vacuum. Everything together is known as the **Universe**.

	1.5×10^{26} m (= 15 billion light years)	the visible Universe
	5×10^{22} m (= 5 million light years)	local group of galaxies
	10^{21} m (= 100 000 light years)	our galaxy
	10^{13} m (= 0.001 light years)	our Solar System

Energy source

ENERGY FLOW FOR STARS

The stars are emitting a great deal of energy. The source for all this energy is the fusion of hydrogen into helium. See page 64. Sometimes this is referred to as 'hydrogen burning' but it this is not a precise term. The reaction is a nuclear reaction, not a chemical one (such as combustion). Overall the reaction is

$$4\,{}^{1}_{1}\text{p} \rightarrow {}^{4}_{2}\text{He} + 2\,{}^{0}_{1}\text{e}^{+} + 2\nu$$

The mass of the products is less than the mass of the reactants. Using $E = m\,c^2$ we can work out that the Sun is losing mass at a rate of 4×10^{9} kg s^{-1}. This takes place in the core of a star. Eventually all this energy is radiated from the surface – approximately 10^{26} J every second. The structure inside a star does not need to be known in detail.

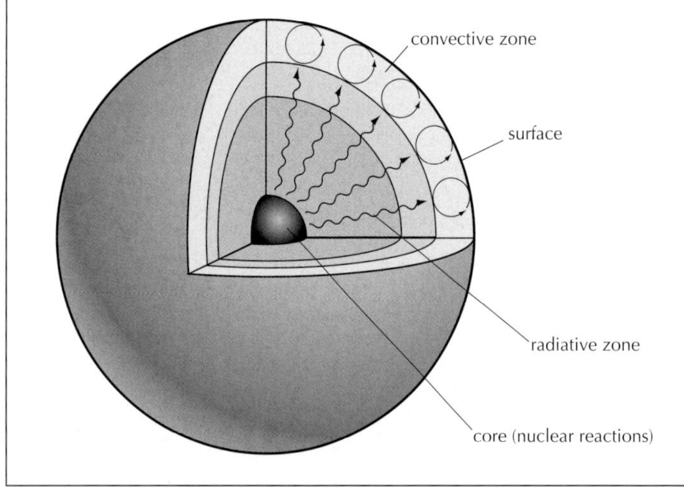

EQUILIBRIUM

The Sun has been radiating energy for the past 4½ billion years. It might be imagined that the powerful reactions in the core should have forced away the outer layers of the Sun a long time ago. Like other stars, the Sun is stable because there is an equilibrium between this outward pressure and the inward gravitational force.

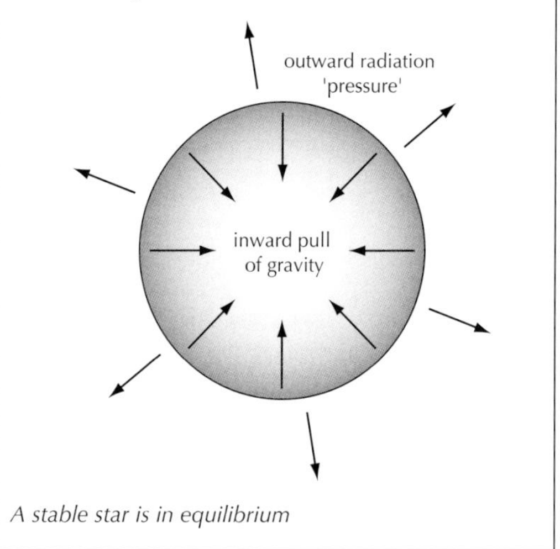

A stable star is in equilibrium

Luminosity

LUMINOSITY AND APPARENT BRIGHTNESS

The total power **radiated** by a star is called its **luminosity** (L). The SI units are watts. This is very different to the power **received** by an observer on the Earth. The power received per unit area is called the **apparent brightness** of the star. The SI units are W m^{-2}.

If two stars were at the **same distance** away from the Earth then the one with the greatest luminosity would be brighter. Stars are, however, at different distances from the Earth. The brightness is inversely proportional to the (distance)^{2}.

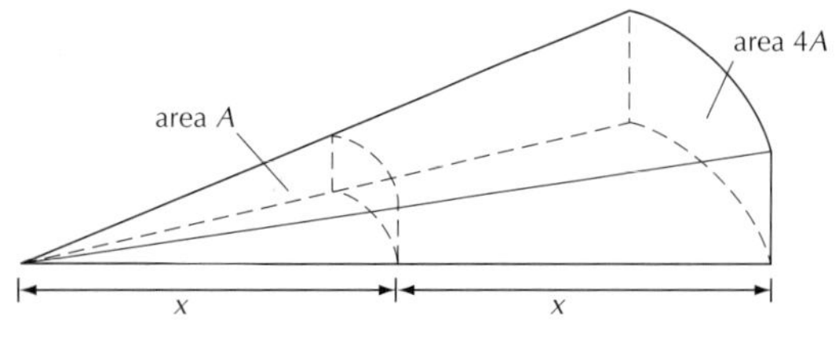

As distance increases, the brightness decreases since the light is spread over a bigger area.

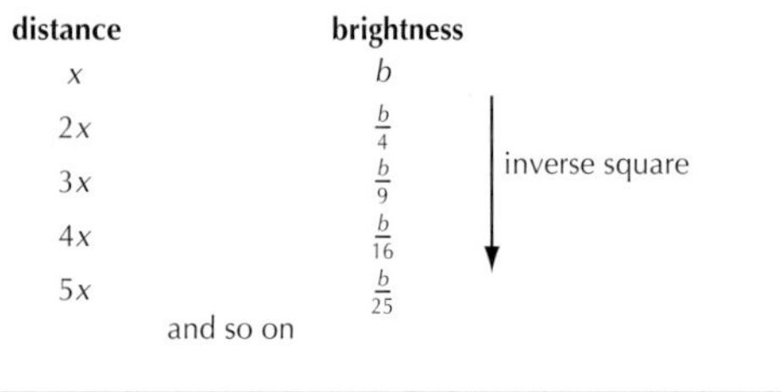

distance	brightness
x	b
$2x$	$\frac{b}{4}$
$3x$	$\frac{b}{9}$
$4x$	$\frac{b}{16}$
$5x$	$\frac{b}{25}$
and so on	

inverse square

$$\text{Brightness } b = \frac{L}{4\pi r^2}$$

It is thus possible for two very different stars to have the same apparent brightness. It all depends on how far away the stars are.

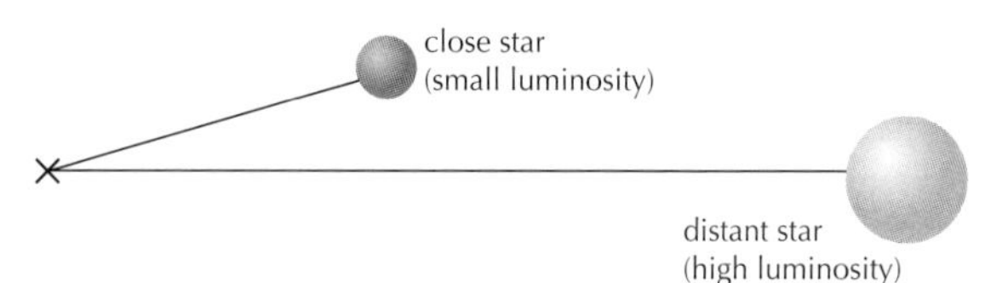

Two stars can have the same apparent brightness even if they have different luminosities

UNITS

The SI units for luminosity and brightness have already been introduced. In practice astronomers compare the brightness of stars using the **apparent magnitude** scale. This scale is defined on page 127. A magnitude 1 star is brighter than a magnitude 3 star. This brightness is often shown on star maps.

The magnitude scale can also be used to compare the luminosity of different stars, provided the distance to the star is taken into account. Astronomers quote values of **absolute magnitude** in order to compare luminosities on an easy and familiar scale.

Stellar spectra

ABSORPTION LINES

The radiation from stars is not a perfect continuous spectrum – there are particular wavelengths that are 'missing'.

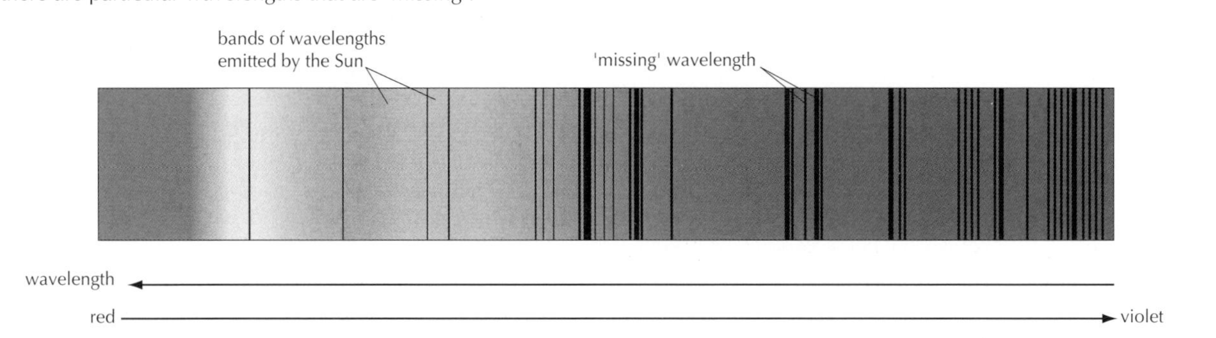

The missing wavelengths correspond to the absorption spectrum of a number of elements. Although it seems sensible to assume that the elements concerned are in the Earth's atmosphere, this assumption is incorrect. The wavelengths would still be absent if light from the star was analysed in space.

The absorption is taking place in the outer layers of the star. This means that we have a way of telling what elements exist in the star – at least in its outer layers.

A star that is moving relative to the Earth will show a Doppler shift in its absorption spectrum. Light from stars that are receeding will be **red-shifted** whereas light from approaching stars will be **blue-shifted**.

CLASSIFICATION OF STARS

Different stars give out different spectra of light. This allows us to classify stars by their **spectral class**. Stars that emit the same type of spectrum are allocated to the same spectral class. Historically these were just given a different letter, but we now know that these different letters also correspond to different surface temperatures.

The seven main spectral classes (in order of **decreasing** surface temperature) are O, B, A, F, G, K and M. This is sometimes remembered as 'Oh Be A Fine Girl/Guy, Kiss Me'. The main spectral classes can be subdivided.

Class	Effective surface temperature/K	Colour
O	28 000–50 000	blue
B	9900–28 000	blue-white
A	7400–9900	white
F	6000–7400	yellow-white
G	4900–6000	yellow
K	3500–4900	orange
M	2000–3500	orange-red

BLACK BODY RADIATION

Stars can be analysed as perfect emitters, or black bodies. The luminosity of a star is related to its brightness, surface area and temperature according to the Stefan-Boltzmann law, Wien's law can be used to relate the wavelength at which the intensity is a maximum to its temperature. See page 74 for more details.

SUMMARY

If we know the distance to a star (see page 127 for the techniques) we can analyse the light from the star and work out:

- the chemical composition (by analysing the absorption spectrum).
- the surface temperature (using a measurement of λ_{max} and Wien's law).
- the luminosity (using measurements of the brightness and the distance away).
- the surface area of the star (using the luminosity, the surface temperature and the Stefan–Boltzmann law).

Types of star

SINGLE STARS

The source of energy for our Sun is the fusion of hydrogen into helium. This is also true for many other stars. There are however, other types of stars that are known to exist in the Universe.

Type of star	Description
Red giant stars	As the name suggests, these stars are large in size and red in colour. Since they are red, they are comparatively cool. They turn out to be one of the later possible stages for a star. The source of energy is the fusion of some elements other than hydrogen. Red **supergiants** are even larger.
White dwarf stars	As the name suggests, these stars are small in size and white in colour. Since they are white, they are comparatively hot. They turn out to be one of the final stages for some stars. Fusion is no longer taking place, and a white dwarf is just a hot remnant that is cooling down. Eventually it will cease to give out light when it becomes sufficiently cold. It is then known as a brown dwarf.
Cepheid variables	These are stars that are a little unstable. They are observed to have a regular variation in brightness and hence luminosity. This is thought to be due to an oscillation in the size of the star. They are quite rare but are very useful as there is a link between the period of brightness variation and their average luminosity. This means that astronomers can use them to help calculate the distance to some galaxies.

BINARY STARS

Our Sun is a single star. Many 'stars' actually turn out to be two (or more) stars in orbit around each other. (To be precise they orbit around their common centre of mass). These are called **binary stars**.

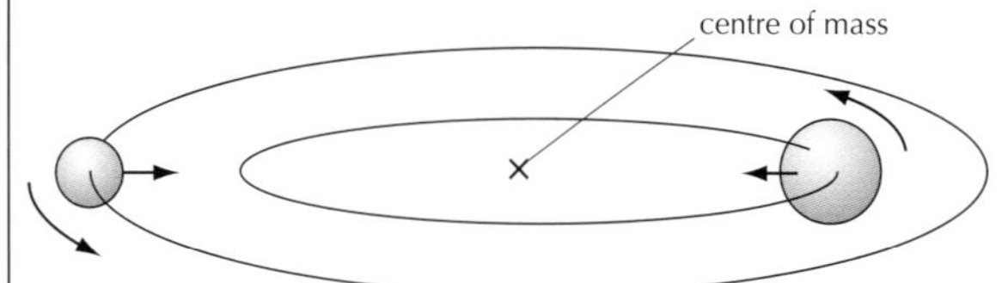

binary stars – two stars in orbit around their common centre of mass

There are different categories of binary star – **visual**, **spectroscopic** and **eclipsing**.

1. A visual binary is one that can be distinguished as two separate stars using a telescope.

2. A spectroscopic binary star is identified from the analysis of the spectrum of light from the 'star'. Over time the wavelengths show a periodic shift or splitting in frequency. An example of this is shown below.

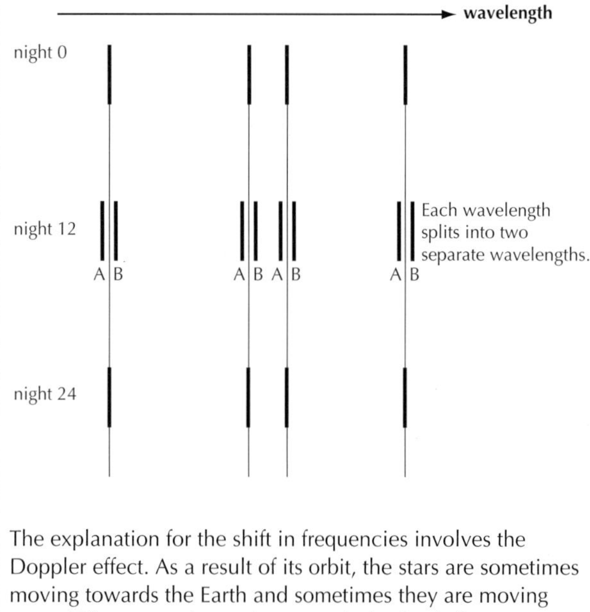

The explanation for the shift in frequencies involves the Doppler effect. As a result of its orbit, the stars are sometimes moving towards the Earth and sometimes they are moving away. When a star is moving towards the Earth, its spectrum will be blueshifted. When it is moving away, it will be redshifted.

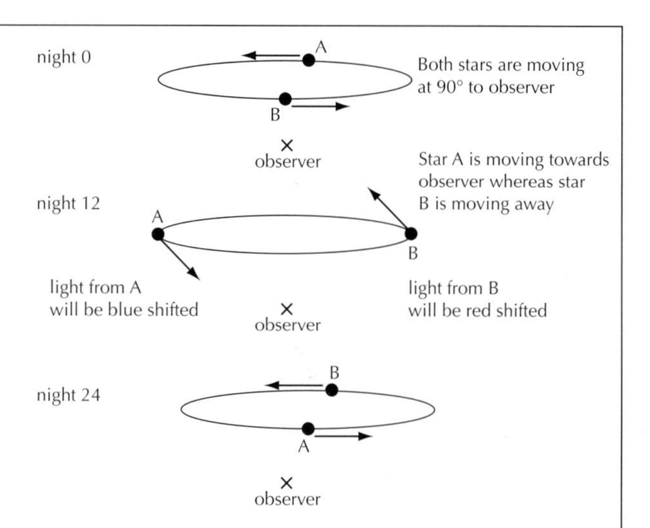

3. An eclipsing binary star is identified from the analysis of the brightness of the light from the 'star'. Over time the brightness shows a periodic variation. An example of this is shown below.

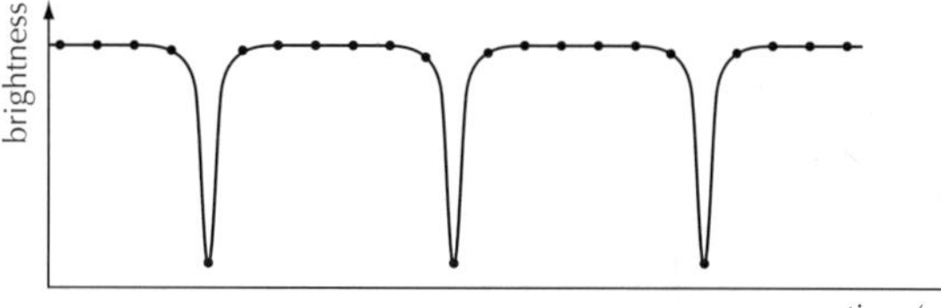

The explanation for the 'dip' in brightness is that as a result of its orbit, one star gets in front of the other. If the stars are of equal brightness, then this would cause the total brightness to drop to 50%.

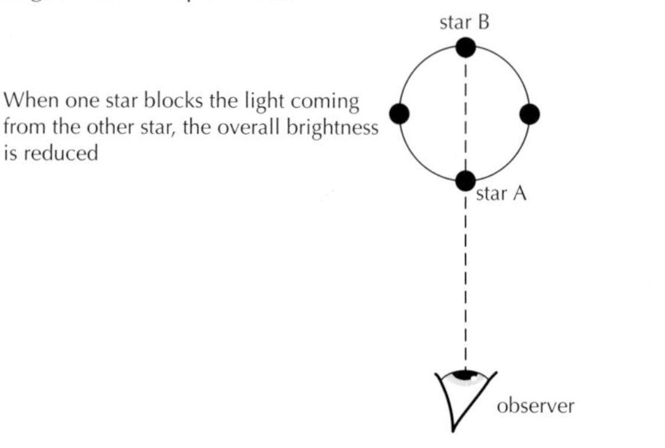

The Hertzsprung–Russell diagram

H–R DIAGRAM

The point of classifying the various types of stars is to see if any patterns exist. A useful way of making this comparison is the **Hertzsprung–Russell diagram**. Each dot on the diagram represents a different star. The following axes are used to position the dot.

- The vertical axis is the luminosity (or absolute magnitude) of the star. It should be noted that the scale is logarithmic.
- The horizontal axis is the spectral class of the star in the order OBAFGKM. This is the same as a scale of **decreasing** temperature. Once again, the scale is not a linear one.

The result of such a plot is shown below.

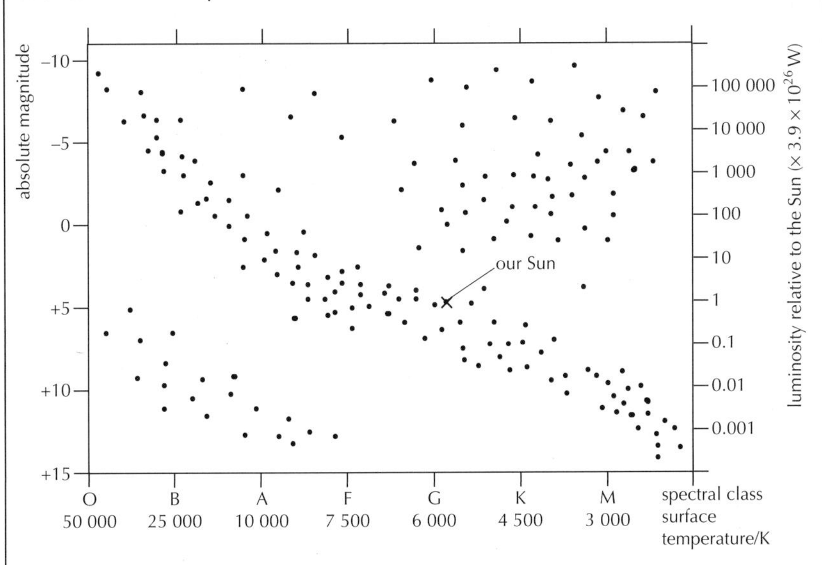

A large number of stars that fall on a line that (roughly) goes from top left to bottom right. This line is known as the **main sequence** and stars that are on it are known as main sequence stars. Our Sun is a main sequence star. These stars are 'normal' stable stars – the only difference between them is their mass. They are fusing hydrogen to helium. The stars that are not on the main sequence can also be broadly put into categories.

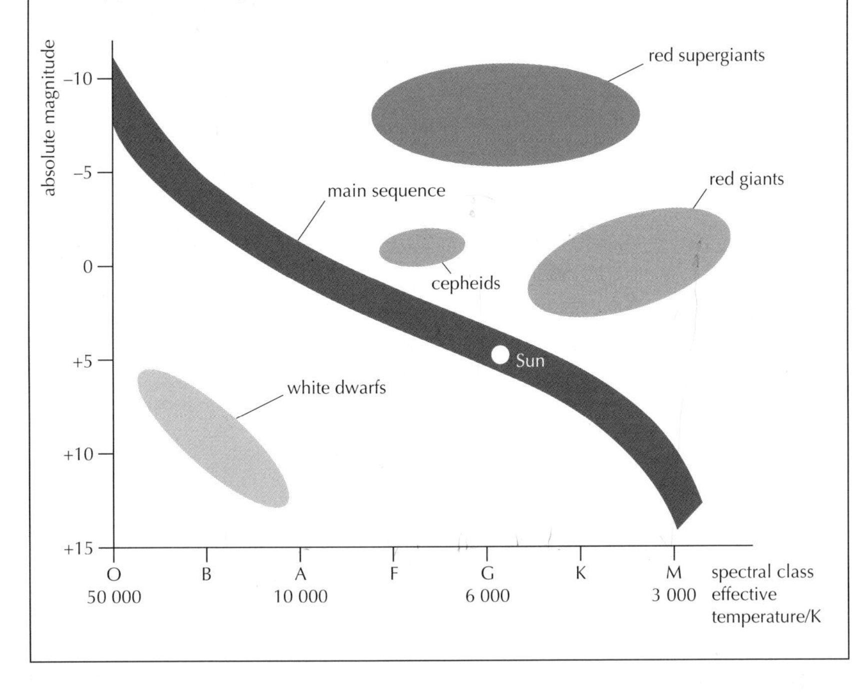

Parallax method

PRINCIPLES OF MEASUREMENT

As you move from one position to another objects change their relative positions. As far as you are concerned, near objects appear to move when compared with far objects. Objects that are very far away do not appear to move at all. You can demonstrate this effect by closing one eye and moving your head from side to side. An object that is near to you (for example the tip of your finger) will appear to move when compared with objects that are far away (for example a distant building).

This apparent movement is known as **parallax** and the effect can used to measure the distance to some of the stars in our galaxy. All stars appear to move over the period of a night, but some stars appear to move in relation to other stars over the period of a year.

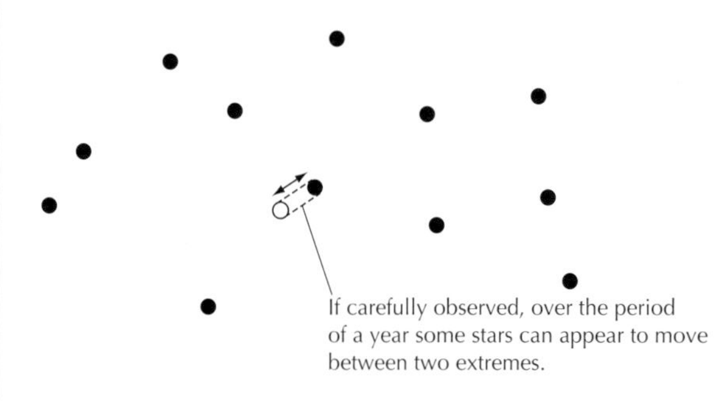

If carefully observed, over the period of a year some stars can appear to move between two extremes.

The reason for this apparent movement is that the Earth has moved over the period of a year. This change in observing position has meant that a close star will have an apparent movement when compared with a more distant set of stars. The closer a star is to the Earth, the greater will be the parallax shift.

Since all stars are very distant, this effect is a very small one and the parallax angle will be very small. It is usual to quote parallax angles not in degrees, but in seconds. An angle of 1 second of arc ('') is equal to one sixtieth of 1 minute of arc (') and 1 minute of arc is equal to one sixtieth of a degree.

In terms of angles, $3600'' = 1°$

$360° = 1$ full circle.

EXAMPLE

The star alpha Eridani (Achemar) is 1.32×10^{18} m away. Calculate its parallax angle.

$$d = 1.32 \times 10^{18} \text{ m}$$

$$= \frac{1.32 \times 10^{18}}{3.08 \times 10^{16}} \text{ pc}$$

$$= 42.9 \text{ pc}$$

$$\text{parallax angle} = \frac{1}{42.9}$$

$$= 0.023''$$

MATHEMATICS – UNITS

The situation that gives rise to a change in apparent position for close stars is shown below.

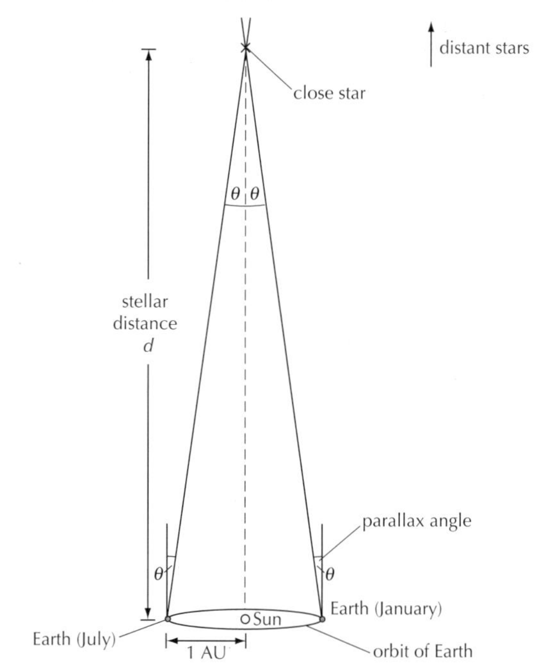

The parallax angle, θ, can be measured by observing the changes in a star's position over the period of a year. From trigonometry, if we know the distance from the Earth to the Sun, we can work out the distance from the Earth to the star, since

$$\tan \theta = \frac{\text{(distance from Earth to Sun)}}{\text{(distance from Sun to Star)}}$$

Since θ is a very small angle, $\tan \theta \approx \sin \theta \approx \theta$ (in radians)

This means that $\theta \propto \dfrac{1}{\text{(distance from Earth to star)}}$

In other words, parallax angle and distance away are inversely proportional. If we use the right units we can end up with a very simple relationship. The units are defined as follows.

The distance from the Sun to the Earth is defined to be one **astronomical unit (AU)**. It is 1.5×10^{11} m. Calculations show that a star with a parallax angle of exactly one second of arc must be 3.08×10^{16} m away (3.26 light years). This distance is defined to be one **parsec (pc)**. The name 'parsec' represents '**par**allel angle of one **sec**ond'.

If distance = 1 pc, $\theta = 1$ second

If distance = 2 pc, $\theta = 0.5$ second etc.

Or, distance in pc = $\dfrac{1}{\text{(parallax angle in seconds)}}$

$$d = \frac{1}{p}$$

The parallax method can be used to measure stellar distances that are less than **about 100 parsecs**. The parallax angle for stars that are at greater distances becomes too small to measure accurately. It is common, however, to continue to use the unit. The standard SI prefixes can also be used even though it is not strictly an SI unit.

1000 parsecs = 1 kpc

10^6 parsecs = 1 Mpc etc.

Absolute and apparent magnitudes

THE MAGNITUDE SCALE

The everyday scale used by astronomers to compare the brightnesses of stars is the magnitude scale. The scale was introduced over 2000 years ago as a way of classifying stars. Stars were all assigned to one of six classifications according to their brightness as seen by the naked eye. Very bright stars were called magnitude 1 stars, whereas the faintest stars were called magnitude 6.

With the aid of telescopes, we can now see stars that are fainter than the magnitude 6 stars. We can also accurately measure the difference in the power being received from the stars. A magnitude 1 star is 100 times brighter than a magnitude 6 star and the scale is logarithmic. This is now used to define the magnitude scale.

The scale can seem strange at first. As the magnitude numbers get bigger and bigger, the stars are getting dimmer and dimmer. Magnitudes are negative for very bright stars.

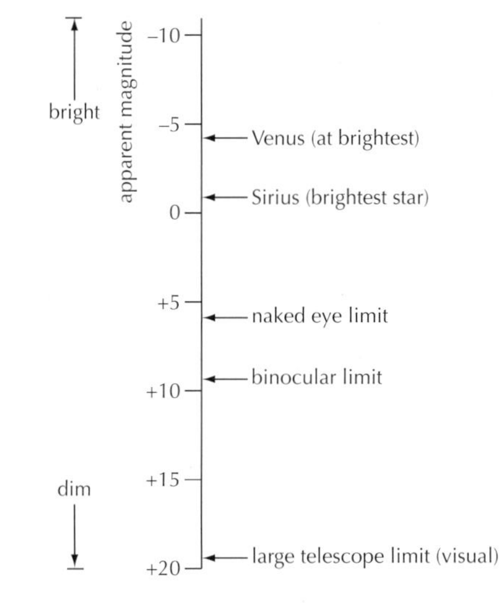

The difference between a magnitude 1 star and a magnitude 6 star is 5 'steps' on the magnitude scale and the scale is logarithmic. This means that each 'step' equates to a brightness decrease of 2.512 since

$$2.512 \times 2.512 \times 2.512 \times 2.512 \times 2.512 = 100$$

or

$$(2.512)^5 = 100$$

EXAMPLE

Use the information in the table to compare the power received from the two stars Sirius and Betelgeuse.

Difference in apparent magnitudes

$= 0.50 - (-1.46)$

$= 1.96$

$$\therefore \frac{\text{power received from Sirius}}{\text{power received from Betelgeuse}} = (2.512)^{1.96}$$

$$= 6.08$$

ABSOLUTE MAGNITUDES

As has been mentioned before, the observed brightness of a star depends on its luminosity **and its distance** from Earth. If two different stars have the same magnitude this does not mean they are the same size. In order to be able to compare stars in this way, the concept of **absolute magnitude** has been introduced.

The absolute magnitude of a star is the apparent magnitude that it would have IF it were observed from a distance of 10 parsecs. Since most stars are much further than 10 parsecs away, they would be brighter if observed from a distance of 10 parsecs. This means their absolute magnitudes are more negative than their apparent magnitudes.

The relationship between absolute magnitude M, apparent magnitude m and distance away d (as measured in parsecs) is given by the following formula.

$$m - M = 5 \log\left(\frac{d}{10}\right)$$

Star	Apparent magnitude m	Distance/pc	Absolute magnitude M
Sirius	−1.46	2.65	+1.4
Canopus	−0.72	70	−4.9
Alpha Centauri	−0.10	1.32	+4.3
Procyon	+0.38	3.4	+2.7
Betelgeuse	+0.50	320	−7.0

SPECTROSCOPIC PARALLAX

The term '**spectroscopic parallax**' can be confusing because it does not involve parallax at all! It is the procedure whereby the luminosity of a star can be estimated from its spectrum.

The assumption made is that the spectra from distant stars are the same as the spectra from nearby stars. If this is the case, we can use the H–R diagram for nearby stars to estimate the luminosity of the star that is further away.

Once the luminosity has been estimated, the distance to the star can be calculated from a measurement of apparent brightness. This involves using the inverse square law.

$$b = \frac{L}{4\pi d^2}$$

This procedure involves quite a lot of uncertainty. Matter between the star and the observer (for example, dust) can affect the light that is received. It would absorb some of the light and make the star's apparent brightness less than it should be. In addition, dust can scatter the different frequencies in different ways, making the identification of spectral class harder.

As the stellar distance increases, the uncertainty in the luminosity becomes greater and so the uncertainty in the distance calculation becomes greater. For this reason spectroscopic parallax is limited to measuring stellar distances up to about 10 Mpc.

Cepheid variables

PRINCIPLES

At distances greater than 10 Mpc, neither parallax nor spectroscopic parallax can be relied upon to measure the distance to a star. The essential difficulty is that when we observe the light from a star at these distances, we do not know the difference between a bright source that is far away and a dimmer source that is closer. This is the principal problem in the experimental determination of astronomical distances to other galaxies.

When we observe another galaxy, all of the stars in that galaxy are approximately the same distance away from the Earth. What we really need is a light source of known luminosity in the galaxy. If we had this then we could make comparisons with the other stars and judge their luminosities. In other words we need a 'standard candle' – that is a star of known luminosity. Cepheid variable stars provide such a 'standard candle'.

A Cepheid variable star is quite a rare type of star. Its outer layers undergo a periodic compression and contraction and this produces a periodic variation in its luminosity.

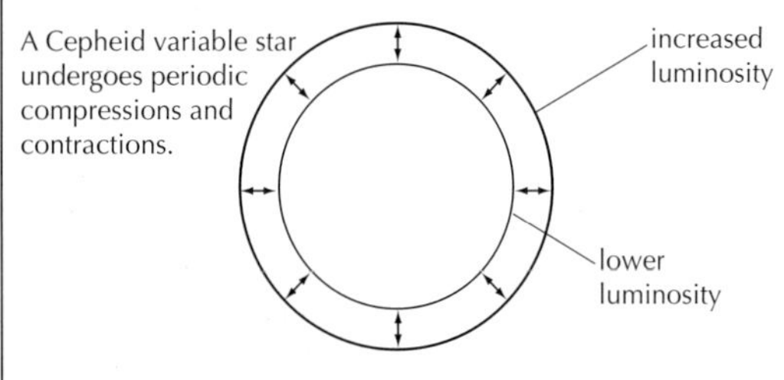

These stars are useful to astronomers because the period of this variation in luminosity turns out to be related to the average absolute magnitude of the Cepheid. Thus the luminosity of a Cepheid can be calculated by observing the variations in brightness.

MATHEMATICS

The process of estimating the distance to a galaxy (in which the individual stars can be imaged) might be as follows:

- locate a Cepheid variable in the galaxy.
- measure the variation in brightness over a given period of time.
- use the luminosity–period relationship for Cepheids to estimate the average luminosity.
- use the average luminosity, the average brightness and the inverse square law to estimate the distance to the star.

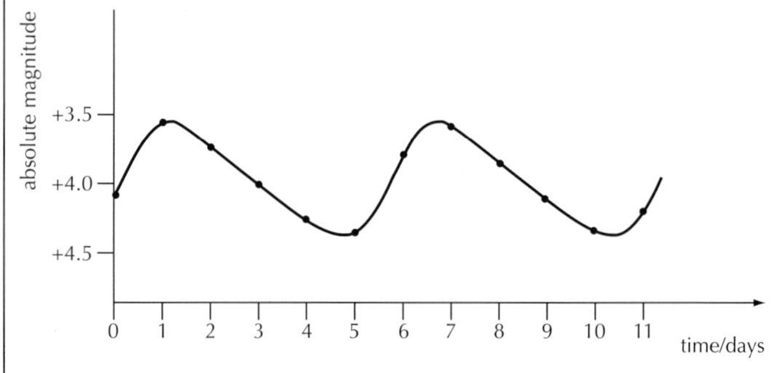

Variation of absolute magnitude for a particular Cepheid variable

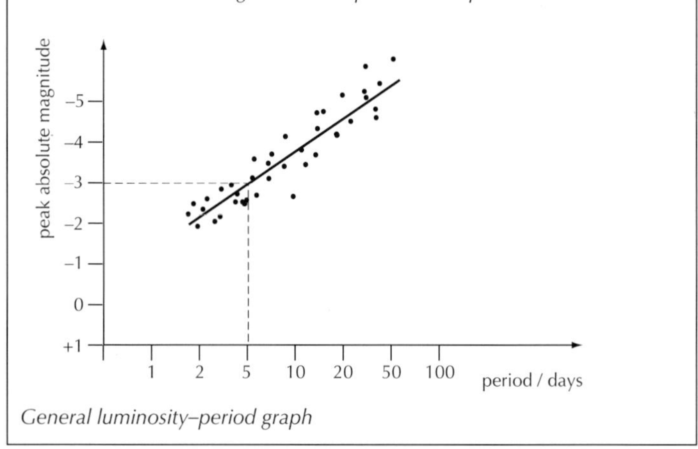

General luminosity–period graph

EXAMPLE

A Cepheid variable star has a period of 5.0 days and a maximum apparent magnitude of +8.0.

Calculate its distance away.

Using the luminosity–period graph (above)

$\Rightarrow$ absolute magnitude = –3.0

1) Using $m - M = 5 \log\left(\frac{d}{10}\right)$

$$\log \frac{d}{10} = \frac{m - M}{5}$$

$$= \frac{8.0 - (-3.0)}{5}$$

$$= 2.2$$

$$\therefore \frac{d}{10} = 10^{2.2} = 158.5 \text{ pc}$$

$$d = 1585 \text{ pc}$$

or

2) Difference between apparent and absolute magnitude = 11.0

$$\therefore \frac{\text{power received 10 pc away}}{\text{power received at Earth}} = (2.512)^{11}$$

$$= 25\,130$$

$$\therefore \left(\frac{d}{10}\right)^2 = 25\,130 \text{ [inverse square law]}$$

$$\frac{d}{10} = \sqrt{25\,130} = 158.5 \text{ pc}$$

$$d = 1585 \text{ pc}$$

Olbers' paradox

WHY IS IT A PARADOX?

The night sky is dark – this is an obvious experimental observation. Stars are sources of light but if we look up in a random direction we do not necessarily see a star. There seems to be nothing wrong with this but that depends on the model of the Universe that you are using.

If we look out into space, we do not see an 'edge' – it seems to carry on much the same in all directions. The constellations seem always to remain the same. From these observations Newton thought it was reasonable to assume three things about the Universe. He thought it was:

- infinite
- uniform
- static

Each of these assumptions seem reasonable enough. It is hard to imagine that the Universe is infinite, but it is worse trying to imagine that it isn't! Remember the Universe means everything – so you can't have something 'outside' the Universe.

The problem is that if the Universe really is like this then the night sky should be bright! Whatever direction you look, you should eventually come across a star. Unfortunately it is not good enough to think that the star might be so far away that it is too dim to see. If you analyze the problem in a little detail you soon find out that this explanation does not work. It really is a paradox.

MATHEMATICS

If all the stars were identical, then those at a given distance away would be the same brightness. The stars that are further away would be dimmer but **there would be more of them**.

If we start with Newton's assumptions we can imagine there to be an average number of stars in a given volume of space density ρ. Consider the stars that are at a distance between r and $r + \delta r$ – in other words the stars that are in the 'shell' of thickness δr as shown below.

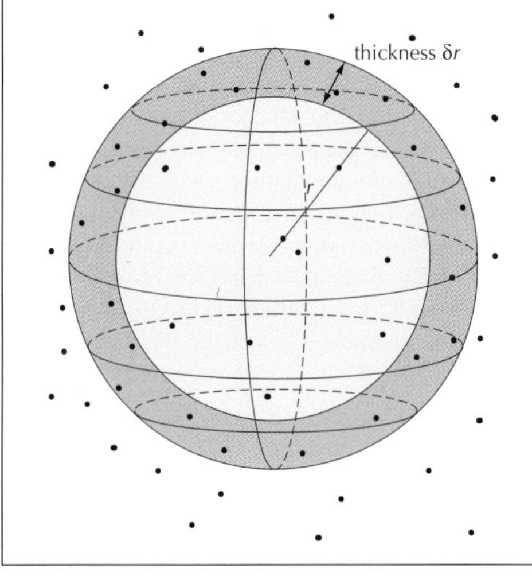

The volume of this shell is $4 \pi r^2 \delta r$.

So the number of stars that are in this shell is $4 \pi r^2 \delta r \rho$

The important thing to notice is that the number of stars in the shell $\propto r^2$.

If we move out to a shell at a greater distance, the stars in that shell will all be dimmer than the stars in the closer shell. This is due to the inverse square nature of radiation.

Brightness of one star in the shell $\propto \frac{1}{r^2}$, but the number of stars in the shell $\propto r^2$.

Overall this means that the contribution made to the overall brightness by the closer shell of stars is exactly the same as the contribution made by the more distant shell of stars.

To work out the overall brightness of the night sky, we need to add up the contributions from all the shells out from the Earth. Since the Universe is assumed to be infinite the number of shells will be infinite so the night sky should be infinitely bright!

POSSIBLE SOLUTIONS TO THE PARADOX

There have been many possible attempts at explaining the paradox.

1. Perhaps the Universe is not infinite (or is non-uniform). Surprisingly a lot of people are happy to imagine that there is a 'cosmic edge' – i.e. the stars do not carry on for ever but there is a limit to the Universe. Most people who imagine this ignore the question of what is beyond the edge. This does not seem to agree with observation (essentially the Universe is the same everywhere and in all directions). The current model of the Universe is that it is infinite.

2. Perhaps the light is absorbed before it gets to us. This sounds very possible, but unfortunately if something was in the way and it absorbed the light, it would warm up. Eventually it would get hot enough to reradiate the energy.

3. Perhaps the Universe is not static (in terms of speed). The Universe is now known to be expanding – see next page. This movement means that the frequency of the light that we receive is shifted due to the Doppler effect. This shift does not resolve the paradox, but it does mean that our model of the Universe is not correct. If it is currently expanding then what does this imply about the Universe in the past?

4. Perhaps the Universe is not static (i.e. it has changed). When we added up the contributions made by different 'shells' of stars, we assumed that the light from the far shells would add to the light from the near shells. But light has a finite velocity, so it would take time for this light to travel from the further shells. Stars are now known to have a finite lifetime – we cannot assume that they have been around forever.

On top of this, our current model of the Universe imagines that it has a limited history – it was created approximately 15 billion years ago. In this case we would only be able to see the light from stars that were less than 15 billion light years away. If we are receiving light from a finite number of stars, then the night sky will be dark.

The Big Bang model

EXPANSION OF THE UNIVERSE

If a galaxy is moving away from the Earth, the light from it will be redshifted. The surprising fact is that light from almost all galaxies shows red shifts – almost all of them are moving away from us. The Universe is expanding.

At first sight, this expansion seems to suggest that we are in the middle of the Universe, but this is a mistake. We only seem to be in the middle because it was we who worked out the velocities of the other galaxies. If we imagine being in a different galaxy, we would get exactly the same picture of the Universe.

A good way to picture this expansion is to think of the Universe as a sheet of rubber stretching off into the distance. The galaxies are placed on this huge sheet. If the tension in the rubber is increased, everything on the sheet moves away from everything else.

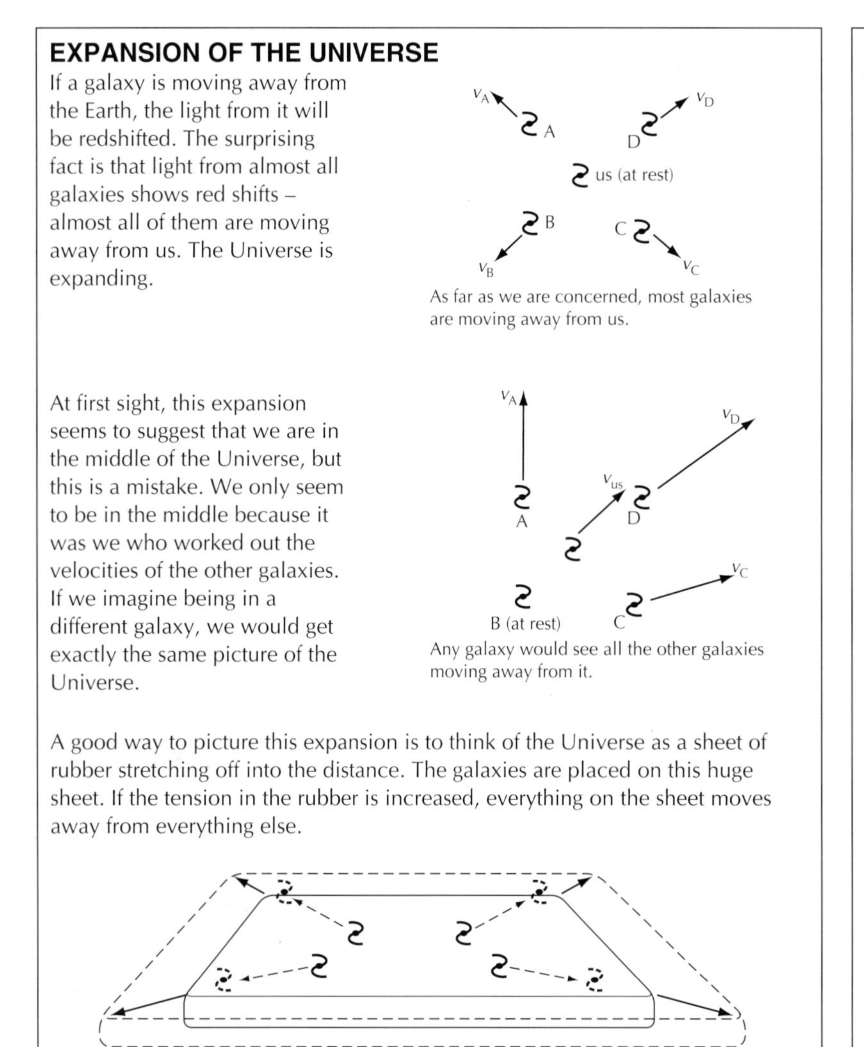

As far as we are concerned, most galaxies are moving away from us.

Any galaxy would see all the other galaxies moving away from it.

As the section of rubber sheet expands, everything moves away from everything else.

THE UNIVERSE IN THE PAST – THE BIG BANG

If the Universe is currently expanding, at some time in the past all the galaxies would have been closer together. If we examine the current expansion in detail we find that all the matter in the observable universe would have been together at the SAME point approximately 15 billion years ago.

This point, the creation of the Universe, is known as the Big Bang. It pictures all the matter in the Universe being created crushed together (very high density) and being very hot indeed. Since the Big Bang, the Universe has been expanding – which means that, on average, the temperature and density of the Universe have been decreasing. The rate of expansion has been decreasing as a result of the gravitational attraction between all the masses in the Universe.

Note that this model does not attempt to explain how the Universe was created, or by Whom. All it does is analyse what happened after this creation took place. The best way to imagine the expansion is to think of the expansion of space itself rather than the galaxies expanding into a void. The Big Bang was the creation of space and time. Einstein's theory of relativity links the measurements of space and time so properly we need to imagine the Big Bang as the creation of space **and time**. It does not make sense to ask about what happened before the Big Bang, because the notion of before and after (i.e. time itself) was created in the Big Bang.

BACKGROUND MICROWAVE RADIATION

A further piece of evidence for the Big Bang model came with the discovery of the **background microwave radiation** by Penzias and Wilson.

They discovered that microwave radiation was coming towards us from all directions in space. The strange thing was that the radiation was the same in all directions and did not seem to be linked to a source. Further analysis showed that this radiation was a very good match to theoretical black-body radiation produced by an extremely cold object – a temperature of just 3 K.

This is in perfect agreement with the predictions of Big Bang. There are two ways of understanding this.

1. All objects give out electromagnetic radiation. The frequencies can be predicted using the theoretical model of black-body radiation. The background radiation is the radiation from the Universe itself and is now cooled down to an average temperature of 3 K.

2. Some time after the Big Bang, radiation became able to travel through the Universe (see page 135 for details). It has been travelling towards us all this time. During this time the Universe has expanded – this means that the wavelength of this radiation will have increased (space has stretched).

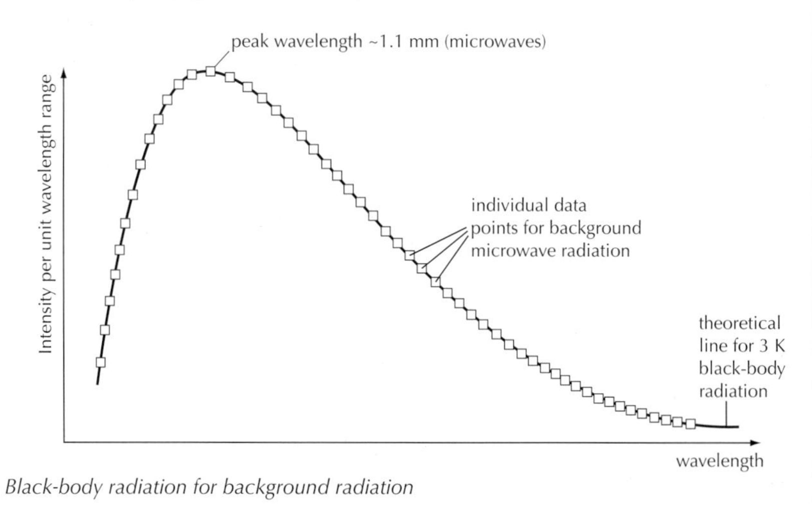

Black-body radiation for background radiation

The development of the Universe

FUTURE OF THE UNIVERSE: THEORY

If the Universe is expanding at the moment, what is it going to do in the future? Remember that our current model of the Universe is that it is infinite. We cannot talk about the size of the Universe, but we can talk about the size of the *observable* Universe. At the moment the furthest object that we can see is about 12 billion light years away. What is going to happen to the size of the observable Universe?

As a result of the Big Bang, other galaxies are moving away from us. If there were no forces between the galaxies, then this expansion could be thought of as being constant.

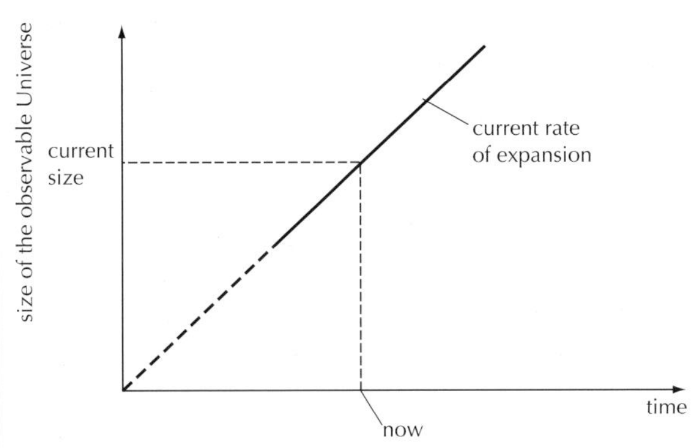

The expansion of the Universe cannot, however, have been uniform. The force of gravity acts between all masses. This means that if two masses are moving apart from one another there is a force of attraction pulling them back together. This force must have slowed the expansion down in the past. What it is going to do in the future depends on the current rate of expansion and the density of matter in the Universe.

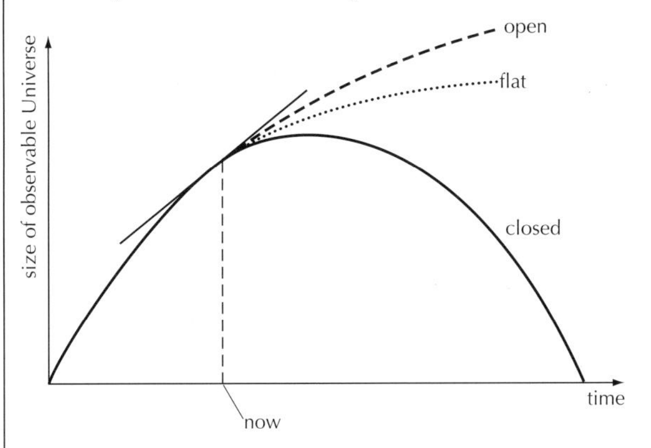

An **open Universe** is one that continues to expand forever. The force of gravity slows the rate of recession of the galaxies down a little bit but it is not strong enough to bring the expansion to a halt. This would happen if the density in the Universe were low.

A **closed Universe** is one that is brought to a stop and then collapses back on itself. The force of gravity is enough to bring the expansion to an end. This would happen if the density in the Universe were high.

A **flat Universe** is the mathematical possibility between open and closed. The force of gravity keeps on slowing the expansion down but it takes an infinite time to get to rest. This would only happen if universe were exactly the right density. One electron more, and the gravitational force would be a little bit bigger. Just enough to start the contraction and make the Universe closed.

CRITICAL DENSITY

The theoretical value of density that would create a flat Universe is called the **critical density**. Its value is not certain because the current rate of expansion is not easy to measure. It is round about 5×10^{-26} kg m^{-3} or 30 proton masses every cubic metre. If this sounds very small remember that enormous amounts of space exist that contain little or no mass at all.

The density of the Universe is not an easy quantity to measure. It is reasonably easy to estimate the mass in a galaxy by estimating the number of stars and their average mass. This calculation results in a galaxy mass which is too small. We know this because we can use the mathematics of orbital motion to work out how much mass there must be keeping the outer stars in orbit around the galactic centre.

We think we can see a maximum of 10% of the matter that must exist in the galaxy. This means that much of the mass of a galaxy and indeed the Universe itself must be **dark matter** – in other words we cannot observe it because it is not radiating sufficiently for us to detect it.

MACHOS, WIMPS AND OTHER THEORIES

Astrophysicists are attempting to come up with theories to explain why there is so much dark matter and what it consists of. There are a number of possible theories:

- the matter could be found in **M**assive **A**stronomical **C**ompact **H**alo **O**bjects or **MACHOs** for short. There is some evidence that lots of ordinary matter does exist in these groupings. These can be thought of as low-mass 'failed' stars or high-mass planets. They could even be black holes. These would produce little or no light.
- some fundamental particles (neutrinos) are known to exist in huge numbers. It is not known if their masses are zero or just very very small. If they turn out to be the latter then this could account for a lot of the missing mass.
- there could be new particles that we do not know about. These are the **W**eakly **I**nteracting **M**assive **P**articles. Many experimenters around the world are searching for these so-called **WIMPs**.
- perhaps our current theories of gravity are not completely correct. Some theories try to explain the missing matter as simply a failure of our current theories to take everything into account.

FUTURE OF THE UNIVERSE: CURRENT OBSERVATIONS

Current scientific evidence suggests that the universe is open. There is also evidence that the rate of expansion may have increased. If this observation is verified then new theories are needed to explain the process.

Nucleosynthesis

MAIN SEQUENCE STARS

The general name for the creation of nuclei of different elements as a result of fission reactions is **nucleosynthesis**. Details of how this overall reaction takes place in the Sun do not need to be recalled.

The whole process is known as the **proton–proton cycle** or **p–p cycle**.

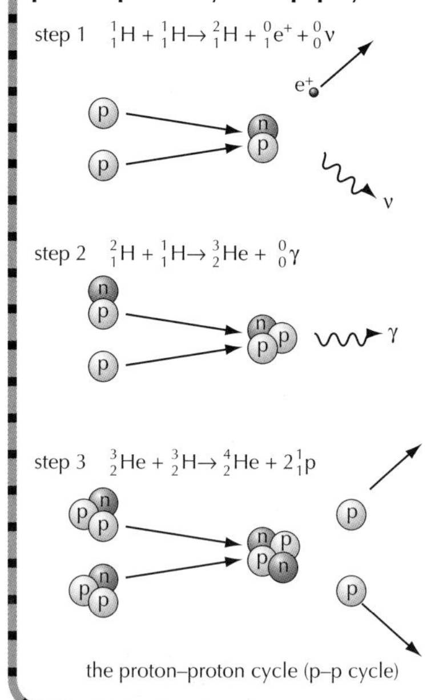

the proton–proton cycle (p–p cycle)

In order for any of these reactions to take place, two positively charged particles (hydrogen or helium nuclei) need to come close enough for interactions to take place. Obviously they will repel one another.

This means that they must be at a high temperature.

If a large cloud of hydrogen is hot enough, then these nuclear reactions can take place spontaneously. The power radiated by the star is balanced by the power released in these reactions – the temperature is effectively constant. The star remains a stable size because the outward pressure of the radiation is balanced by the inward gravitational pull.

But how did the cloud of gas get to be at a high temperature in the first place? As the cloud comes together, the loss of gravitational potential energy must mean an increase in kinetic energy and hence temperature. In simple terms the gas molecules speed up as they fall in towards the centre to form a proto-star.

Once ignition has taken place, the star can remain stable for billions of years.

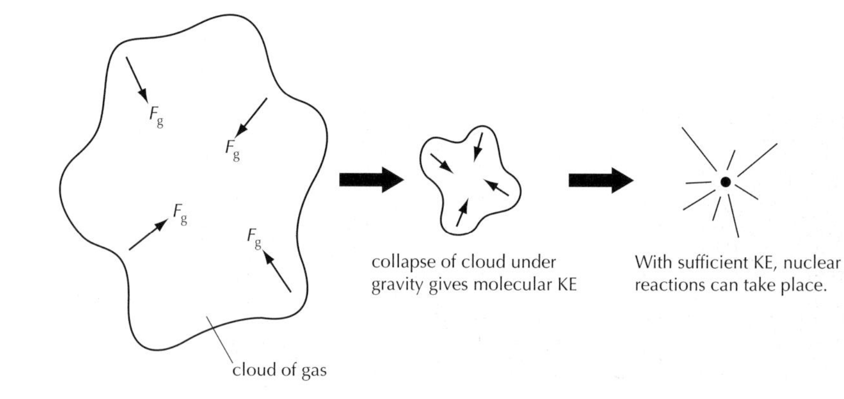

AFTER THE MAIN SEQUENCE

The star cannot continue in its main sequence state forever. It is fusing hydrogen into helium and at some point hydrogen in the core will become rare. The fusion reactions will happen less often. This means that the star is no longer in equilibrium and the gravitational force will, once again, cause the core to collapse.

This collapse increases the temperature of the core still further and helium fusion is now possible. The net result is for the star to increase massively in size – this expansion means that the outer layers are cooler. It becomes a red giant star.

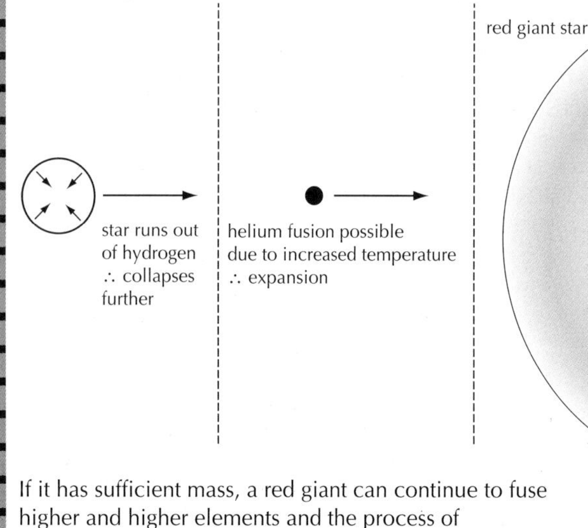

If it has sufficient mass, a red giant can continue to fuse higher and higher elements and the process of nucleosynthesis can continue.

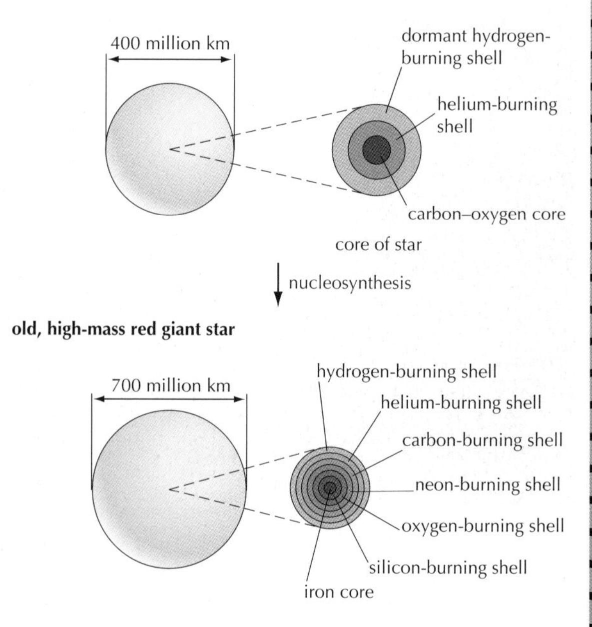

This process of fusion as a source of energy must come to an end with the nucleosythesis of iron. The iron nucleus has the greatest binding energy per nucleon of all nuclei. In other words the fusion of iron to form a higher mass nucleus would need to take in energy rather than release energy. The star cannot continue to shine. What happens next is outlined on the following page.

Evolutionary paths of stars and stellar processes

POSSIBLE FATES FOR A STAR (AFTER RED GIANT PHASES)

The previous page showed that the red giant phase for a star must eventually come to an end. There are essentially two possible routes with different final states. The route that is followed depends on the initial mass of the star. An important 'critical' mass is called the **Chandrasekhar limit** and it is equal to approximately 1.4 times the mass of our Sun. Below this limit a process called electron degeneracy pressure prevent the further collapse of the remnant.

If a star has a mass less than 4 Solar masses, its remnant will be less than 1.4 Solar masses and so it is below the Chandrasekhar limit. It this case the red giant forms a **planetary nebula** and becomes a **white dwarf** which ultimately becomes invisible. The name 'planetary nebula' is another term that could cause confusion. The ejected material would not be planets in the same sense as the planets in our Solar System.

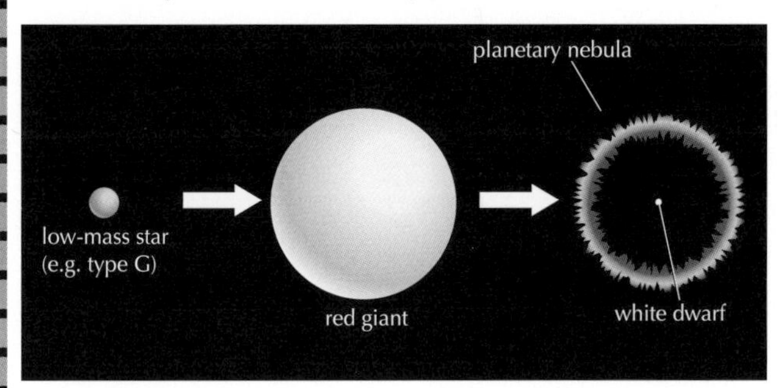

If a star is greater than 4 Solar masses, its remnant will have a mass greater than 1.4 Solar masses. It is above the Chandrasekhar limit. In this case the red supergiant experiences a **supernova**. It then becomes a **neutron star** or collapses to a **black hole**. See page 183. The final state again depends on mass.

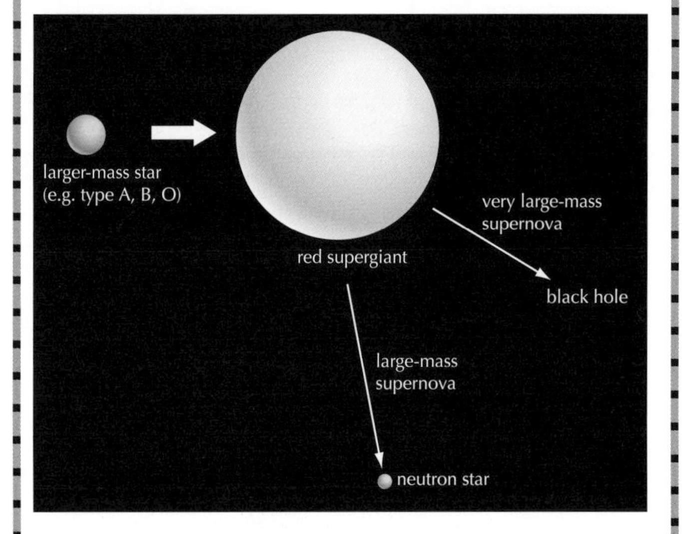

A neutron star is stable due to neutron degeneracy pressure. It should be emphasised that white dwarfs and neutron stars do not have a source of energy to fuel their radiation. They must be losing temperature all the time. The fact that these stars can still exist for many millions of years shows that the temperatures and masses involved are enormous.

H – R DIAGRAM INTERPRETATION

All of the possible evolutionary paths for stars that have been described here can be represented on a H – R diagram. A common mistake in examinations is for candidates to imply that a star somehow moves along the line that represents the main sequence. It does not. Once formed it stays at a stable luminosity and spectral class – i.e. it is represented by one fixed point in the H – R diagram.

evolution of a low-mass star

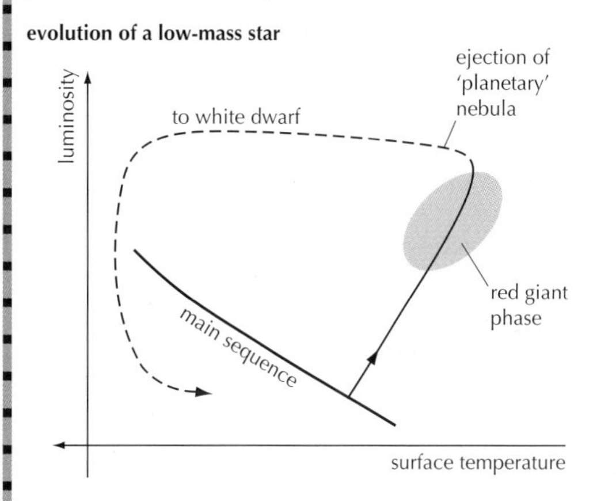

evolution of a high-mass star

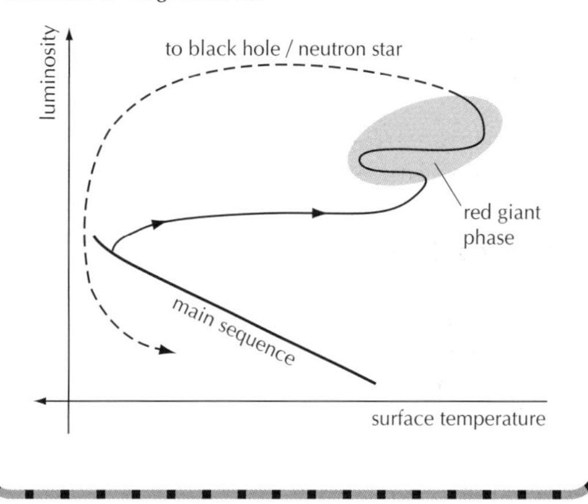

PULSARS AND QUASARS

Pulsars are cosmic sources of very weak radio wave energy that pulsate at a very rapid and precise frequency. These have now been theoretically linked to rotating neutron stars. A rotating neutron star would be expected to emit an intense beam of radio waves in a specific direction. As a result of the star's rotation, this beam moves around and causes the pulsation that we receive on Earth.

Quasi-stellar objects or quasars appear to be point-like sources of light and radio waves that are very far away. Their redshifts are very large indeed, which places them at the limits of our observations of the Universe. If they are indeed at this distance they must be emitting a great deal of power for their size (approximately 10^{40} W!). The process by which this energy is released is not well understood, but some theoretical models have been developed that rely on the existence of super-massive black holes. The energy radiated is as a result of whole stars 'falling' into the black hole.

Galactic motion

DISTRIBUTIONS OF GALAXIES

Galaxies are not distributed randomly throughout space. They tend to be found clustered together. For example, in the region of the Milky Way there are twenty or so galaxies in less than 2½ million light years.

The Virgo galactic cluster (50 million light years away from us) has over 1000 galaxies in a region 7 million light years across. On an even larger scale, the galactic clusters are grouped into huge **superclusters** of galaxies. In general, these superclusters often involve galaxies arranged together in joined 'filaments' (or bands) that are arranged as though randomly throughout empty space.

MOTION OF GALAXIES

As has been seen on page 130 it is a surprising observational fact that the vast majority of galaxies are moving away from us. The general trend is that the more distant galaxies are moving away at a greater speed as the Universe expands. This does not, however, mean that we are at the centre of the Universe – this would be observed wherever we are located in the Universe.

As explained on page 130, a good way to imagine this expansion is to think of space itself expanding. It is the expansion of space (as opposed to the motion of the galaxies through space) that results in the galaxies' relative velocities. In this model, the redshift of light can be thought of as the expansion of the wavelength due to the 'stretching' of space.

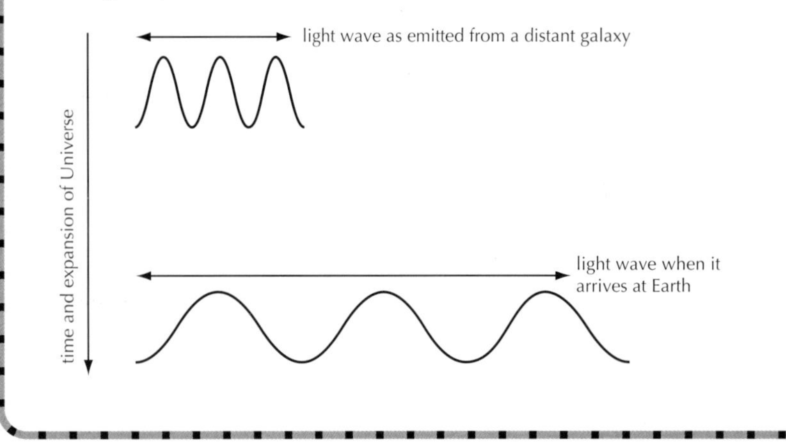

MATHEMATICS

If a star or a galaxy moves away from us, then the wavelength of the light will be altered as predicted by the Doppler effect (see page 94). If a galaxy is going away from the Earth, the speed of the galaxy with respect to an observer on the Earth can be calculated from the redshift of the light from the galaxy. As long as the velocity is small when compared to the velocity of light, a simplified redshift equation can be used.

$$\frac{\Delta\lambda}{\lambda} = \frac{v}{c}$$

Where

$\Delta\lambda$ = change in wavelength of observed light (positive if wavelength is increased)

λ = wavelength of light emitted

v = relative velocity of source of light

c = speed of light

Example

A characteristic absorption line often seen in stars is due to ionized helium. It occurs at 468.6 nm. If the spectrum of a star has this line at a measured wavelength of 499.3 nm, what is the recession speed of the star?

$$\frac{\Delta\lambda}{\lambda} = \frac{(499.3 - 468.6)}{468.6}$$

$$= 6.55 \times 10^{-2}$$

$$\therefore\ v = 6.55 \times 10^{-2} \times 3 \times 10^{-8}\ \text{m s}^{-1}$$

$$= 1.97 \times 10^{7}\ \text{m s}^{-1}$$

Hubble's law

EXPERIMENTAL OBSERVATIONS

Although the uncertainties are large, the general trend for galaxies is that the recessional velocity is proportional to the distance away from Earth. This is Hubble's law.

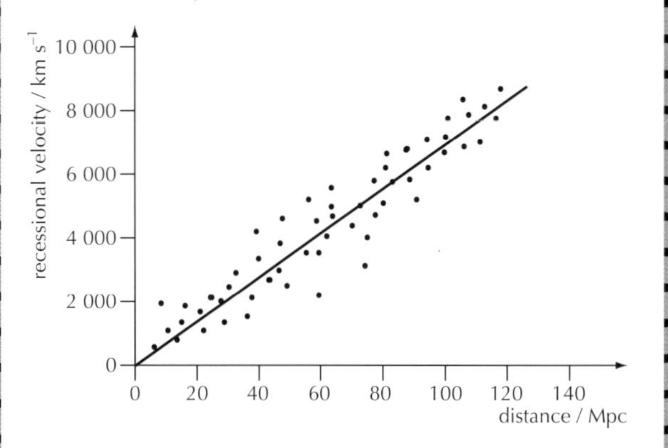

Mathematically this is expressed as

$$v \propto d$$

or

$$v = H_0 d$$

where H_0 is a constant known as the **Hubble constant**. The uncertainties in the data mean that the value of H_0 is not known to any degree of precision. The SI units of the Hubble constant are s^{-1}, but the unit of km s^{-1} Mpc^{-1} is often used.

HISTORY OF THE UNIVERSE

If a galaxy is at a distance x, then Hubble's law predicts its velocity to be H_0x. If it has been travelling at this constant speed since the beginning of the Universe, then the time that has elapsed can be calculated from

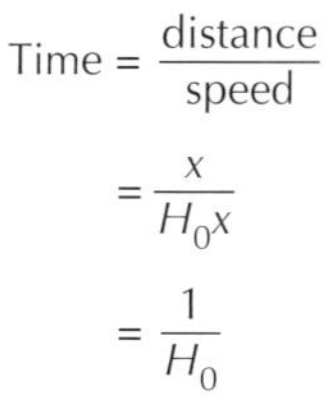

$$\text{Time} = \frac{\text{distance}}{\text{speed}} = \frac{x}{H_0 x} = \frac{1}{H_0}$$

This is an upper limit for the age of the Universe. The gravitational attraction between galaxies means that the speed of recession decreases all the time.

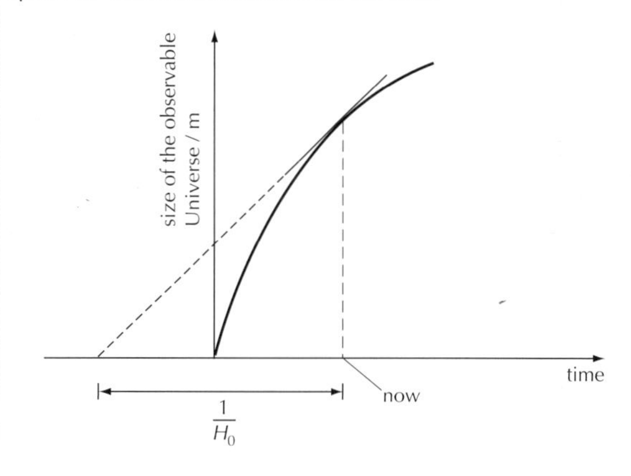

We can 'work backwards' and imagine the process that took place soon after the Big Bang.

- Very soon after the Big Bang, the Universe must have been very hot.
- As the Universe expanded it cooled. It had to cool to a certain temperature before atoms and molecules could be formed.
- The Universe underwent a short period of huge expansion that would have taken place from about 10^{-35} s after the Big Bang to 10^{-32} s.

Time	What is happening	Comments
10^{-45} s → 10^{-35} s	Unification of forces	This is the starting point
10^{-35} s → 10^{-32} s	Inflation	A rapid period of expansion – the so-called **inflationary epoch**. The reasons for this rapid expansion are not fully understood.
10^{-32} s → 10^{-5} s	Quark-lepton era	Matter and antimatter (quarks and leptons) are interacting all the time. There is slightly more matter than antimatter.
10^{-5} s → 10^{-2} s	Hadron era	At the beginning of this short period it is just cool enough for hadrons (e.g. protons and neutrons) to be stable.
10^{-2} s → 10^{2} s	Helium synthesis	During this period some of the protons and neutrons have combined to form helium nuclei. The matter that now exists is the 'small amount' that is left over when matter and antimatter have interacted.
10^{2} s → 3×10^{5} years	Plasma era (radiation era)	The formation of light nuclei has now finished and the Universe is in the form of a plasma with electrons, protons, neutrons, helium nuclei and photons all interacting.
3×10^{5} years → 10^{9} years	Formation of atoms	At the beginning of this period, the Universe has become cool enough for the first atoms to exist. Under these conditions, the photons that exist stop having to interact with the matter. It is these photons that are now being received as part of the background microwave radiation. The Universe is essentially 75% hydrogen and 25% helium.
10^{9} years → now	Formation of stars, galaxies and galactic clusters	Some of the matter can be brought together by gravitational interactions. If this matter is dense enough and hot enough, nuclear reactions can take place and stars are formed.

Astrophysics research

ASTROPHYSICS RESEARCH

Much of the current fundamental research that is being undertaken in astrophysics involves close international collaboration and the sharing of resources. Scientist can be proud of their record of international collaboration. For example, at the time this book was being published, the Cassini spacecraft had been in orbit around Saturn for several years sending information about the planet back to Earth and is designed to continue doing so for many more years.

The Cassini-Huygens spacecraft was funded jointly by ESA (the European Space Agency), NASA (the National Aeronautics and Space Administration of the United States of American) and ASI (Agenzia Spaziale Italiana – the Italian space agency). As well as general information about Saturn, an important focus of the mission was a moon of Saturn called Titan. The Huygens probe was released and sent back information as it descended towards the surface. The information discovered is shared among the entire scientific community.

All countries have a limited budget available for the scientific research that they can undertake. There are arguments both for, and against, investing significant resources into researching the nature of the universe.

Arguments for:

- Understanding the nature of the universe sheds light on fundamental philosophical questions like:
  - Why are we here?
  - Is there (intelligent) life elsewhere in the universe?
- It is one of the most fundamental, interesting and important areas for mankind as a whole and it therefore deserves to be properly researched.
- All fundamental research will give rise to technology that may eventually improve the quality of life for many people.
- Life on Earth will, at some time in the distant future, become an impossibility. If mankind's descedents are to exist in this future, we must be able to travel to distant stars and colonise new planets.

Arguments against:

- The money could be more usefully spent providing food, shelter and medical care to the many millions of people who are suffering from hunger, homelessness and disease around the world.
- If money is to be allocated on research, it is much more worthwhile to invest limited resources into medical research. This offers the immediate possibility of saving lives and improving the quality of life for some sufferers.
- It is better to fund a great deal of small diverse research rather than concentrating all funding into one expensive area. Sending a rocket into space is expensive, thus funding space research should not be a priority.
- Is the information gained really worth the cost?

IB QUESTIONS – OPTION E – ASTROPHYSICS

1 The table below gives information about two nearby stars.

Star	Apparent magnitude	distance away / ly
Fomalhaut (α-Piscis Austrini)	1.2	22
Aldebaran (α-Tauri)	0.9	68

(a) To an observer on Earth which star would appear brighter? Justify your answer. [2]

(b) Explain the difference between **apparent** and **absolute** magnitudes. [2]

(c) Which star would have the *lowest numerical value* for **absolute** magnitude? Explain your answer. [2]

(d) The parallax angle for Fomalhaut is 0.148 arcseconds. Confirm that its distance away is 22 ly. [2]

(e) Would you expect Aldebaran to have a greater or smaller parallax angle than Fomalhaut? Explain your answer. [2]

Another method of determining stellar distances involves a class of variable stars called *Cepheid variables*. One of the first Cepheid variable stars to be studied is δ-Cephei. Its apparent magnitude varies over time as shown below:

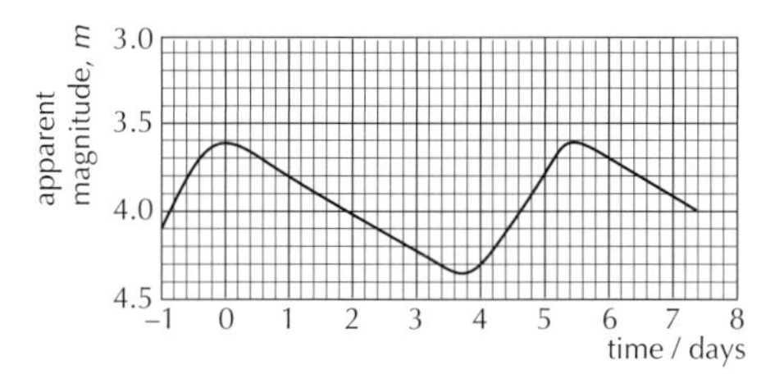

There is a relationship between peak absolute magnitude, *M*, and the time period of the variation in magnitude as shown below:

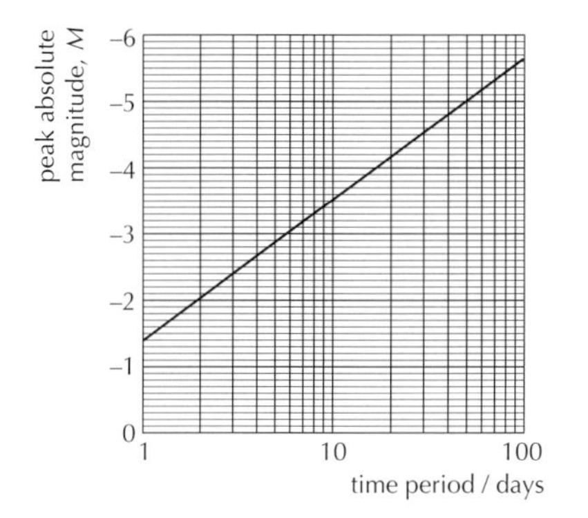

(f) Use the data above to estimate the peak **absolute** magnitude *M* for δ-Cephei. [2]

(g) The relationship between peak absolute magnitude *M*, apparent magnitude *m* and distance *d* is given by the following equation:

$$m - M = 5\log_{10}\left(\frac{d}{10\text{pc}}\right)$$

Use this equation to calculate the distance to δ-Cephei. [3]

2 (a) The spectrum of light from the Sun is shown below.

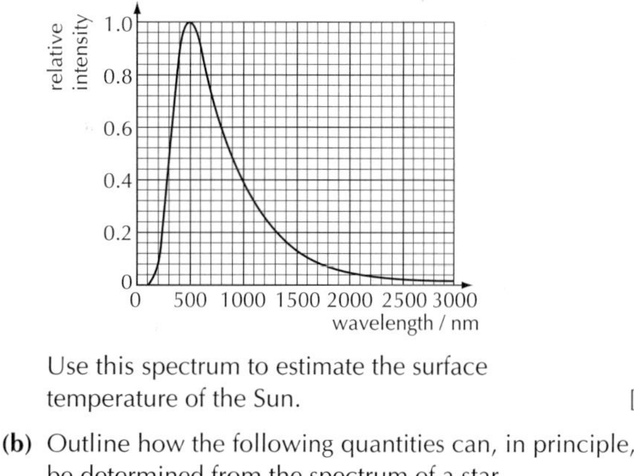

Use this spectrum to estimate the surface temperature of the Sun. [2]

(b) Outline how the following quantities can, in principle, be determined from the spectrum of a star.

(i) The elements present in its outer layers. [2]

(ii) Its speed relative to the Earth. [2]

HL

3 (a) Explain how Hubble's law supports the Big Bang model of the Universe. [2]

(b) Outline **one** other piece of evidence for the model, saying how it supports the Big Bang. [3]

(c) The Andromeda galaxy is a relatively close galaxy, about 700 kpc from the Milky Way, whereas the Virgo nebula is 2.3 Mpc away. If Virgo is moving away at 1200 km s^{-1}, show that Hubble's law predicts that Andromeda should be moving away at roughly 400 km s^{-1}. [1]

(d) Andromeda is in fact moving **towards** the Milky Way, with a speed of about 100 km s^{-1}. How can this discrepancy from the prediction, in both magnitude and direction, be explained? [3]

(e) If light of wavelength 500 nm is emitted from Andromeda, what would be the wavelength observed from Earth? [3]

4 A star viewed from the Earth is not always a single, constant object. Many stars in the *main sequence* are, in fact, *binary stars*. For example, β-Persei is an *eclipsing binary*. Over time, stars are known to change. Some will end up as *neutron stars* or even *black holes*.

(a) What is meant by:

(i) main sequence; [3]
(ii) binary stars; [1]
(iii) eclipsing binary; [1]
(iv) neutron star; [1]
(v) black hole. [1]

(b) Identify the physical processes by which a main sequence star develops into a neutron star. [4]

(c) What evidence is there for the existence of the neutron stars? [2]

(d) What property determines whether a star might develop into a neutron star or a black hole? Outline how this property can be used to predict the outcome. [2]

Radio communication and amplitude modulation (AM) (1)

WAVE MODULATION PRINCIPLES

Any communication system takes in information (the **signal wave**), uses a communication link to send a different type of wave between source and receiver (the **carrier wave**) and then attempts to recreate the signal wave. The carrier wave needs to be continually altered (using a particular chosen technique) in order for the signal information to be transmitted from source to receiver. See page 146 for examples of different channels of communication.

BANDWIDTH

When different communication processes are analysed in terms of their frequency spectra, each process will require a different range of frequencies to be used for a given signal to be transmitted successfully. The range of frequencies in the transmitted signal is known as the **bandwidth.** It is an advantage for a given process to have a small bandwidth as this would allow a given frequency range to contain a large number of independent signals.

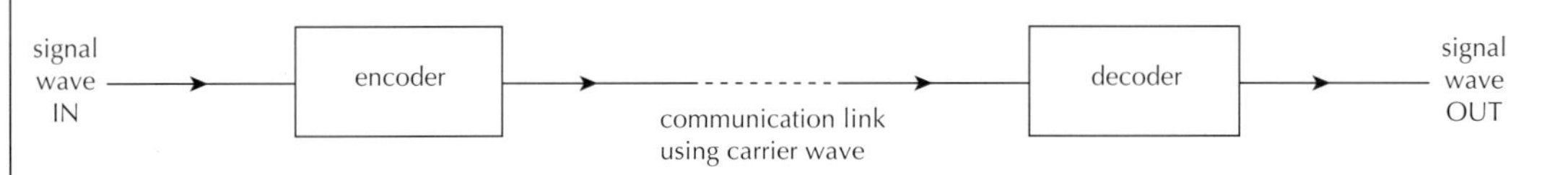

In a perfect communication link, the signal wave IN and the signal wave OUT are identical.

For radio broadcasts that involve speech and/or music, the signal wave into the system is a longitudinal sound wave (in the audible frequency range of 20 Hz to 20 kHz).

- A microphone could be used to convert the longitudinal sound wave into an electrical signal. The electrical oscillations match the pressure variations of the sound wave.
- An **encoder** changes the carrier wave so that this electrical signal can be recreated at the receiver.
- The modified carrier wave is broadcast from source to receiver.
- The electrical system block that recreates the signal wave at is called a **decoder**.
- The recreated electrical signal can then converted back into sound waves using a loudspeaker.

The carrier wave for radio broadcasts is at the lower frequency end of the electromagnetic spectrum. The order of magnitude for the frequencies used is from about 10^5 Hz (long wave broadcasting) up to 10^9 Hz (TV and satellite broadcasting).

Two common ways in which the carrier wave is altered (or **modulated**) in order for the signal information to be broadcast are **amplitude modulation** (AM) or **frequency modulation** (FM). The recreation of the signal wave is called **demodulation**.

FREQUENCY ANALYSIS

Any signal (no matter how complicated) can be mathematically analysed and shown be to equivalent to a mixture of a large number of different sine and cosine waves of different frequencies and different amplitudes. A plot of the comparative amplitudes of the different component frequencies for any signal is called its **frequency spectrum** or **power spectrum**. The analysis necessary to calculate these values is extremely complex.

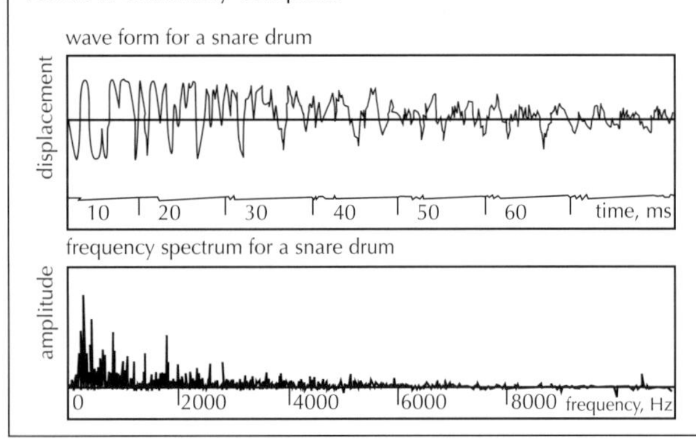

OVERVIEW OF THE AM PROCESS

In the AM process, it is the amplitude of the carrier wave that is modified in order to encode the signal wave:

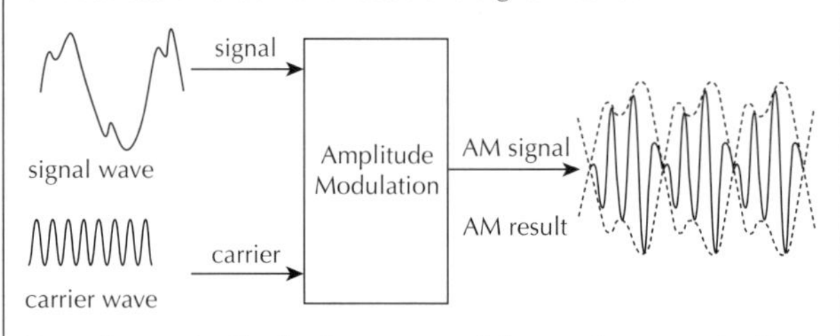

Note that typically the frequency of the carrier wave (radio frequencies) is very much larger than the frequency of the signal wave (audio frequencies). This difference is not accurately represented in the diagram above. The resulting AM signal has a fixed frequency and varying amplitude.

AM RADIO RECEIVER

The diagram below represents the main system blocks in a simple AM radio receiver. The AM radio broadcast contains speech or music at audio frequencies.

System blocks are explained on the next page.

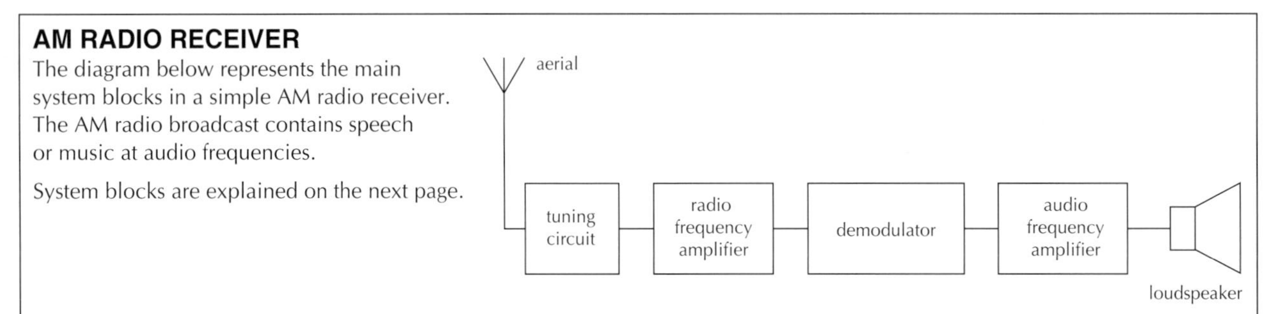

Radio communication and amplitude modulation (AM) (2)

System block (page 138)	Function
Aerial	This receives the electromagnetic radio waves. Each AM radio signal received causes tiny alternating currents to be induced in the aerial.
Tuning circuit	This circuit element preferentially selects a chosen radio frequency. When the natural frequency of the tuning circuit equals the driving frequency of the radio signal, resonance occurs and the amplitude of these oscillations is increased.
Radio frequency amplifier	The amplitude of the selected AM radio frequency signal can then be increased further with an amplifier designed to operate at radio frequencies. A perfect amplifier increases the amplitude of the wave without introducing any distortions. It also ensures that sufficient power is available for subsequent electronic systems to operate properly.
Demodulator	This removes the radio frequency component and reconstructs the original audio frequency input signal.
Audio frequency amplifier	The amplitude of the audio frequency signal is increased with an amplifier designed to operate at audio frequencies. The amplifier ensures that enough power can be provided to the loudspeaker.
Loudspeaker	The audio signal is fed into a loudspeaker that converts the electrical oscillations into sound waves. In a perfect system, these reconstructed sound waves would be identical to the sound waves originally detected by the microphone at the broadcasting station.

POWER SPECTRUM OF AM SIGNAL

The power spectrum of an AM signal depends on the range of frequencies contained in the original signal wave. When a single frequency signal, f_m, is used to modulate the carrier wave, the resulting power spectrum contains just three spikes. The central spike is at the carrier frequency, f_c , with two other bands (the **sidebands**) as shown below:

signal wave:

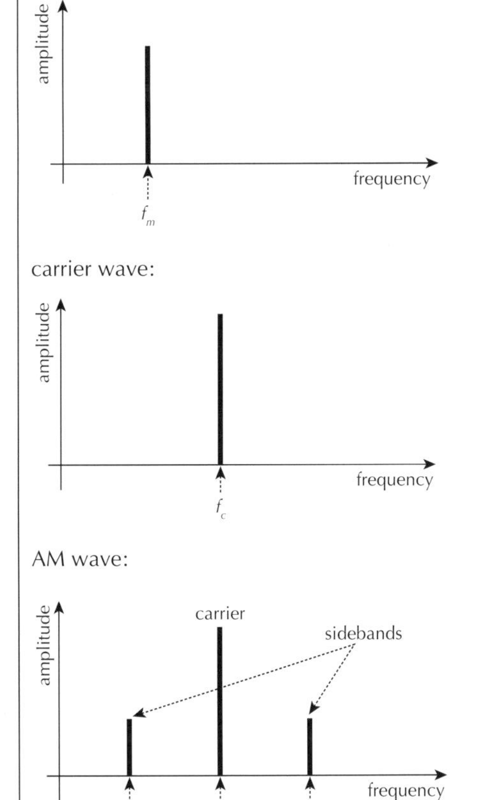

The bandwidth for this channel is twice the signal frequency ($2 f_m$).

In general, if the original signal wave contains a range of frequencies then the power spectrum of the AM signal will contain upper and lower sidebands. The size of the sidebands will be dictated by the range of frequencies in the original signal.

For example, some medium wave radio signals are designed to broadcast frequencies in the range from 50 Hz to 4.5 kHz (a narrower range than the 'complete' audio range of 20 Hz to 20 kHz). The result is minimal distortion for spoken broadcasts but the loss of higher frequencies is noticeable when music is broadcast. The power spectrum for these AM radio signals is shown below:

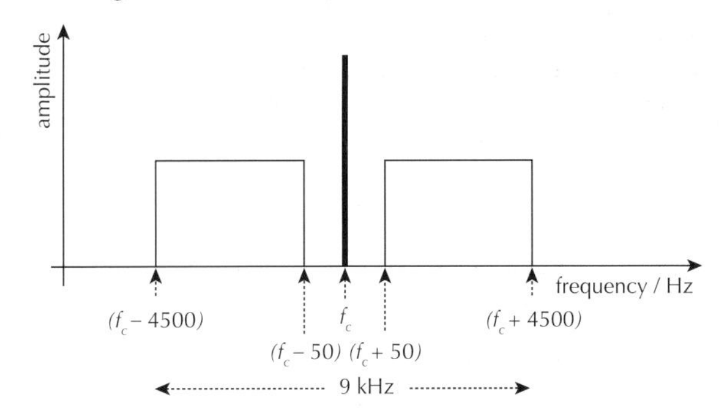

The bandwidth for this type of channel is 9 kHz.

Frequency modulation (FM)

OVERVIEW OF THE FM PROCESS

In the FM process, it is the frequency of the carrier wave that is modified in order to encode the signal wave:

audio signal
0
time

FM carrier
0
time

Note that the amplitude of the carrier wave is not affected by the signal wave. As far as the receiver is concerned, the amplitude is irrelevant. This makes the FM system less susceptible to the effects of noise (see page 141).

POWER SPECTRUM AND BANDWIDTH OF FM SIGNAL

The power spectrum of an FM signal is complex. Side frequencies are produced that are multiples of the modulating frequency. The example below shows the frequency spectrum of a carrier frequency optimally modulated by a square signal wave of frequency f_s = 5 kHz:

Input signal:

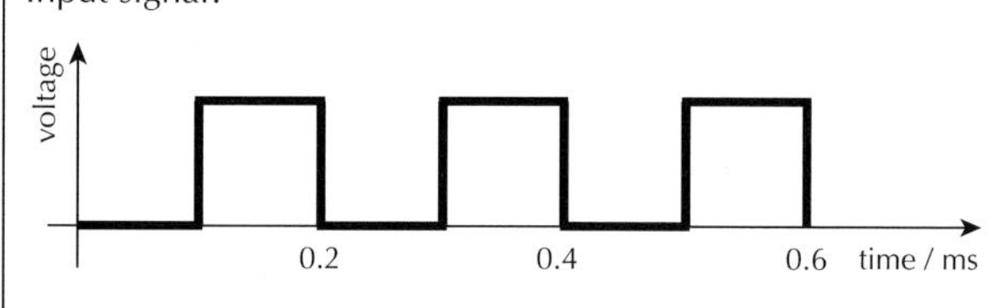

Frequency spectrum:

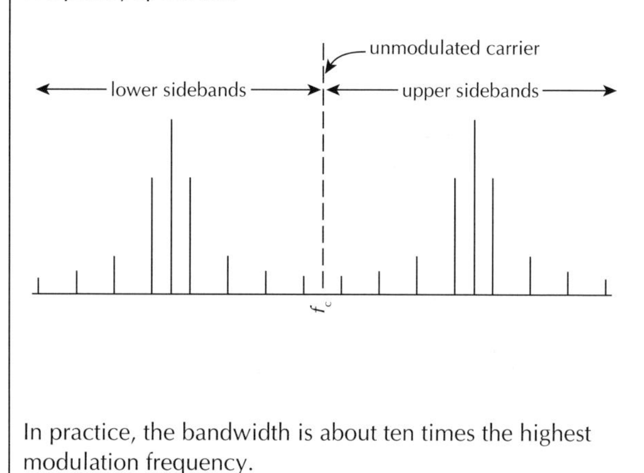

In practice, the bandwidth is about ten times the highest modulation frequency.

COMPARISON BETWEEN AM AND FM

	Advantage	Disadvantage
Amplitude modulated (AM)	• Cheap electronic circuitry • Fewer components to go wrong • Broadcast range large so fewer transmitters needed	• Channels with a bandwidth of 9 kHz are satisfactory for spoken broadcast but quality of music is affected • Electrical noise (see page 141) affects the output signal
Frequency modulated (FM)	• Not as susceptible to effects of noise • Quality of broadcast superior to AM	• Complex circuitry required so increased cost • Limited range of broadcast • Large bandwidth needed (typically 180 kHz or more and approximately five times the bandwidth needed for AM)

Analogue and digital signals

BINARY AND DECIMAL CONVERSION

Further information is shown on page 113 (Topic 14).

Decimal numbers

These are "normal" numbers based on the number ten. In this system, any number can be thought as being made up of a series of symbols in different columns. For example, in this system, the number "two hundred and thirty four" is represented as "234". As you work through the columns from the right to the left, each digit represents a higher power or ten:

Power	10^4	10^3	10^2	10^1	10^0
Number:	10000 ten thousands	1000 thousands	100 hundreds	10 tens	1 units
Symbols:			2	3	4

In the example above, the number "two hundred and thirty four" means two hundreds *plus* three tens *plus* 4 units:

2 × (hundred)

plus 3 × (ten)

plus 4 × (unit)

There are ten different symbols for each value that is possible in each column. The symbols are: 0, 1, 2, 3, 4, 5, 6, 7, 8 and 9. We do not bother to including "leading" zeros so the number is written as "234" and not "00234".

Binary numbers

These are numbers based on the number two. Again any number can be thought as being made up of a series of symbols in different columns. As you work through the columns from the right to the left, each digit represents a higher power of two.

Power	2^4	2^3	2^2	2^1	2^0
Number:	16 sixteens	8 eights	4 four	2 twos	1 units
Symbols:	1	0	1	0	1

The binary number 10101 represents 16 + 4 + 1 = 21. There are only two different symbols possible in each column: 1 or 0. The digit on the far right represents the smallest of all the numbers represented and is called the **least significant bit** or **LSB**. The digit on the far left represents the largest number and is called the **most significant bit** or **MSB**.

ANALOGUE AND DIGITAL

An **analogue** signal is one that varies continuously with time between a minimum and a maximum value. At any given moment of time, the signal can have any value between these two limits. A **digital** signal, however, can only have one of two fixed values represented by ON or OFF, TRUE or FALSE, 1 or 0. These 0s and 1s are called **b**inary dig**its**, or **bits**.

example of an analogue signal:

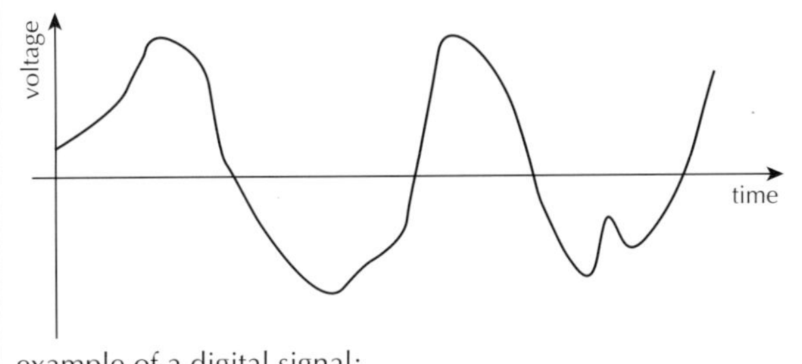

example of a digital signal:

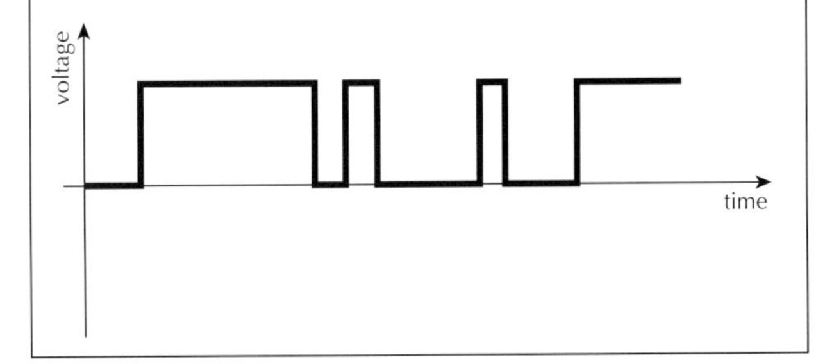

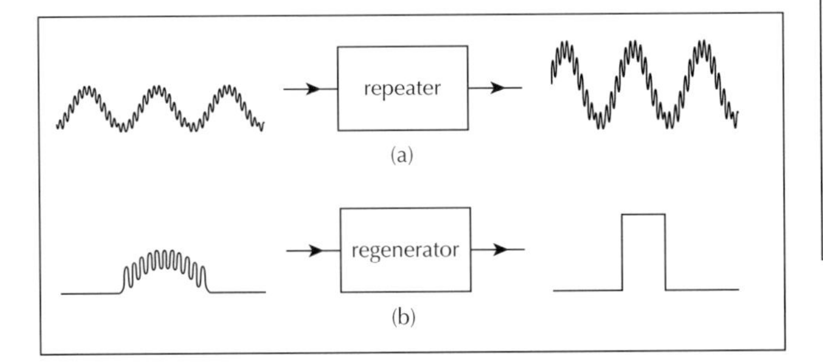

ADVANTAGES OF DIGITAL SIGNALS

Information (e.g. speech, musical sounds, pictures, videos etc.) can be stored and/or transmitted in either digital or analogue format. Many everyday situations involve both formats at some stage in the overall process.

An analogue process is often perceived as a more straightforward process. For example, the pressure variations of a particular sound can be directly represented by a varying analogue electrical signal. The conversion between the sound and the electrical signal that represents the sound involves only a single process.

Representing pressure variations with a digital signal requires a complex code. A series of 1s and 0s is all that is used to recreate the pressure variations after transmission. This means that a large number of **bits** are involved.

All transmitted signals can suffer from two major problems:

- **Attenuation.** This is the loss of signal power as a result of energy being lost to the surroundings. The more the signal is attenuated, the weaker it becomes. If a signal is to be transmitted over a significant distance, **repeater** circuits can be used increase the signal strength.
- **Noise.** This is the addition of any random additional electrical signal to the original electrical signal. This can result in the transmitted signal becoming distorted.

Despite its complexity, digital transmission has advantages over analogue transmission:

- When analogue signals suffer from noise, this directly affects the final quality. Digital signals that suffer from noise can be electronically processed using a **regenerator** or **shaper** circuit that recreates the original digital signal. The overall quality of digital broadcasting can be much better than analogue broadcasting.
- The detection systems only need to distinguish between two possible states – 1 or 0. A reasonably large variation in amplitude of the signal during transmission will not affect the final quality.

Digital system blocks

SYSTEM BLOCKS

Complex circuits can be designed and analysed in terms of different systems blocks. The details of how these circuits are designed and operate are not required and thus details of the power requirements of each system have been ignored. The following circuits are important in the transmission and reception of digital signals.

System block	Inputs and outputs	Function
Clock	One output (the clock signal)	All digital signals are processed in a sequence of steps. A clock pulse is used to control and coordinate the different steps. A regular square wave pulse is generated by an oscillator and a rising or falling edge is used to trigger each step.
Sample-and-hold	One analogue input One digital input (control signal) One analogue output	An analogue signal will typically be varying all the time. The analogue output of this system block follows its analogue input. A control signal into the system stops the variation of the output. The input has been **sampled** and the last value of the input is constantly outputted from the system block. This system forms an important section in the conversion of an analogue signal into digital form (see page 112).
Analogue-to-digital converter (ADC)	One analogue input One digital input (control signal) Several digital output lines depending on the number of bits involved: 4-bit, 8-bit or higher	This system converts an analogue input into a digital output using predetermined conversion rules. There must be a range of possible analogue inputs that converts to each of the possible digital outputs. The digital outputs should be read at the same time (they are **parallel**) ranging from the LSB to the MSB.
Parallel-to-serial converter	Several digital inputs clock input control input One digital output	This system converts a set of digital inputs that are being read at the same time (in parallel) into a series of pulses on a single output that can be read one after another (in **series**). The output signal is coordinated by the clock pulse.
Serial-to-parallel converter	One digital input clock input control input Several digital outputs	This system is the reverse of the parallel-to-series converter. A single digital stream at the input is converted into a parallel set of digital outputs.
Digital-to-analogue converter (DAC)	Several digital inputs clock input control input One analogue output	This system is the reverse of the analogue-to-digital converter. A set of parallel digital inputs is converted into a single analogue output.

EXAMPLE OF A DIGITAL SYSTEM

The blocks below represent the system blocks required to convert a single sample of an input analogue signal into a serial digital output signal ready for transmission. The control and clock lines have been omitted in order to make the diagram easier to follow.

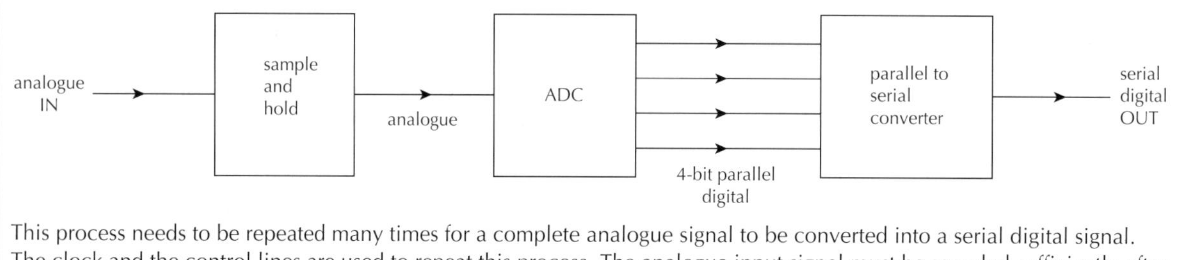

This process needs to be repeated many times for a complete analogue signal to be converted into a serial digital signal. The clock and the control lines are used to repeat this process. The analogue input signal must be sampled sufficiently often in order for it to be recreated at the receiver.

Digital multiplexing and the Internet

FACTORS AFFECTING QUALITY OF TRANSMITTED SIGNAL

When an analogue signal is converted into a digital signal, the signal is sampled at a certain rate and each sample is converted into digital bits. The quality of the final signal will be improved by increasing the rate of sampling, or in other words increasing amount of information that is sent per cycle of original signal. The amount of information that can be sent depends on several factors:

- The greater the sampling frequency, the greater the amount of information that is being transmitted and hence the greater the quality of the final signal. As a rule of thumb, the sampling frequency needs to be at least twice the highest frequency in the original signal.
- The greater the number of bits used in each analogue-to-digital conversion, the greater the precision with which the original signal can be reproduced.

  bit rate = number of bits per sample × sampling frequency
- There will, however, be a limit on the rate at which digital information can be processed and transmitted. The greater the maximum bit-rate transfer, the more possibility there is of improving the final signal by using either of the above techniques or introducing checking and correction processes.

TIME-DIVISION MULTIPLEXING

Multiplexing is the name given to any process which increases the number of messages that can be sent along any single communication line. The result is that multiple messages are sent on one communication line at the same time without interfering with each other. **Time-division multiplexing** uses the time between samples of one message to send other messages.

In a normal (low quality) telephone conversation, the maximum frequency allowed is approximately 4 kHz and so the sampling frequency is 8 kHz. This means that a sample will be sent every 125 μs. The number of bits per sample is, however, only 8 and each bit could take only 1 μs. The whole sample can be sent in 8 μs leaving 117 μs of wasted time until the next sample is due to be sent.

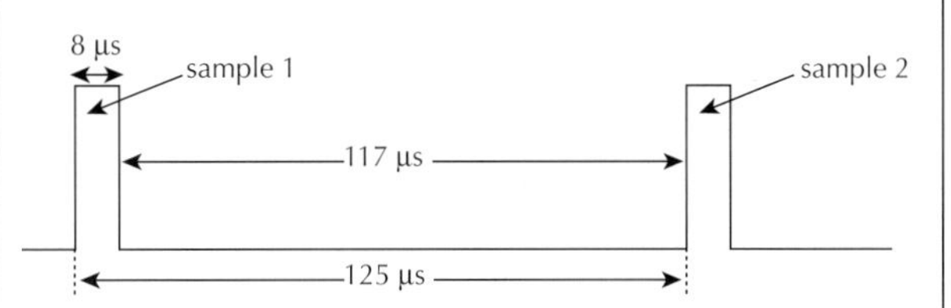

Time-division multiplexing sends several different messages down the same line at once. Complex electronics is needed to separate out the different samples and this requires some additional transmission of data in order for each separate channel to be properly decoded at the receiving end. In practice at least 10 separate channels could use this single line of communication.

CONSEQUENCES OF DIGITAL COMMUNICATIONS

More and more communication channels are switching over to digital processes. For example, many countries are in the process of replacing analogue television broadcasts with digital systems. In addition to this there are many free communication packages available on the internet. This offers the possibility of very cheap voice and video digital communication for anybody who has access to the internet.

Possible consequences

- Digital communication techniques offer the possibility of better quality communication. Sound communication has already been improved and high quality music and video links are likely to become more accessible.
- As digital techniques become more common, mass-production of the equipment necessary means that the cost of communication will tend to fall.
- Free, worldwide communication promises to be a possibility for all those who have access to the internet. This will have a huge impact on how citizens of different countries interact with one another.
- Although many would welcome the increased ability to communicate, this ability will only be in the hands of those who have access to the Internet. The poorest members of society may find it harder to effectively voice their opinions.
- The ability to communicate worldwide using the Internet will impact on telecommunication companies' profits and may cause redundancies or unemployment in these sectors.

Possible issues

- The ability to share information immediately will have an impact on how news is transmitted around the world and on the extent to which information flow can be controlled by governments.
- Access to the Internet can entertain, advise, inform and educate the individual.
- Given the diverse peoples and societies that all have access to the Internet, there has to be a cultural judgment involved in deciding the content that is allowable. Information and images that are acceptable to some individuals will be offensive and even illegal to others.
- Ideally, individuals can be enlightened by the Internet and this knowledge can help to destroy superstitions and break down barriers between countries and religions. In reality it can be a source of friction.
- The greater mobility of information accelerates a general trend towards the existence of a global marketplace. Increased sharing of information means that the reasons for any economic decisions could be more robust and this could lead to more stable economies.
- As markets become more global, it may become easier for international companies to source their labour worldwide. This may cause job losses in some countries.
- Internet conferencing has the potential to change the way in which business is done. Video conferencing could cut costs associated with business flights around the world.

Critical angle and optical fibres

TOTAL INTERNAL REFLECTION AND CRITICAL ANGLE

In general, both reflection and refraction can happen at the boundary between two media.

It is, under certain circumstances, possible to guarantee complete (total) reflection with no transmission at all. This can happen when a ray meets the boundary and it is travelling in the denser medium.

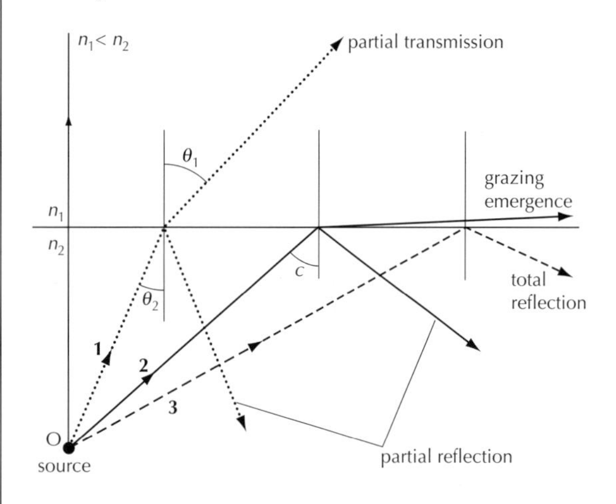

Ray1 This ray is partially reflected and partially refracted.

Ray2 This ray has a refracted angle of nearly 90°. The **critical ray** is the name given to the ray that has a refracted angle of 90°. The **critical angle** is the angle of incidence c for the critical ray.

Ray3 This ray has an angle of incidence **greater** than the critical angle. Refraction cannot occur so the ray must be totally reflected at the boundary and stay inside medium 2. The ray is said to be **totally internally reflected**.

The critical angle can be worked out as follows. For the critical ray,

$$n_1 \sin \theta_1 = n_2 \sin \theta_2$$

$$\theta_1 = 90°$$

$$\theta_2 = \theta_c$$

$$\therefore \sin \theta_c = \frac{1}{n}$$

EXAMPLES

1. What a fish sees under water

entire world outside water is visible in an angle of twice the critical angle

critical angle c

At greater than the critical angle, the surface of the water acts like a mirror. Objects inside the water are seen by reflection.

2. Prismatic reflectors

A prism can be used in place of a mirror. If the light strikes the surface of the prism at greater than the critical angle, it must be totally internally reflected.

Prisms are used in many optical devices. Examples include:

- periscopes – the double reflection allows the user to see over a crowd.
- binoculars – the double reflection means that the binoculars do not have to be too long
- SLR cameras – the view through the lens is reflected up to the eyepiece.

periscope

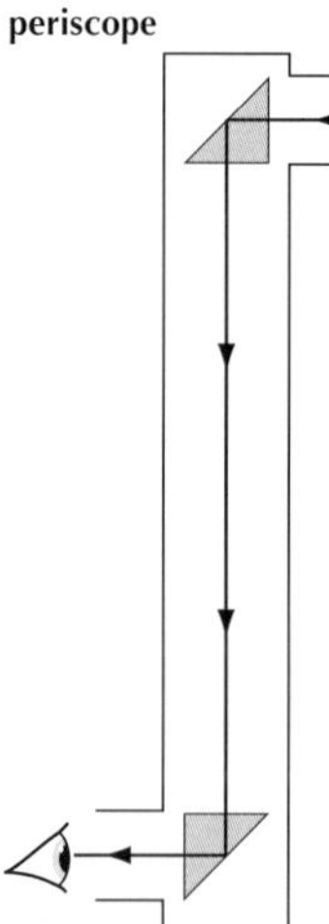

binoculars

The prism arrangement delivers the image to the eyepiece the right way up. By sending the light along the instrument three times, it also allows the binoculars to be shorter.

eyepiece lens

objective lens

3. Optical fibre

Optical fibres can be used to guide light along a certain path. The idea is to make a ray of light travel along a transparent fibre by bouncing between the walls of the fibre. So long as the incident angle of the ray on the wall of the fibre is always greater than the critical angle, the ray will always remain within the fibre even if the fibre is bent (see left).

Two important uses of optical fibres are:

- in the communication industry. Digital data can be encoded into pulses of light that can then travel along the fibres. This is used for telephone communication, cable TV etc.
- in the medical world. Bundles of optical fibres can be used to carry images back from inside the body. This instrument is called an endoscope.

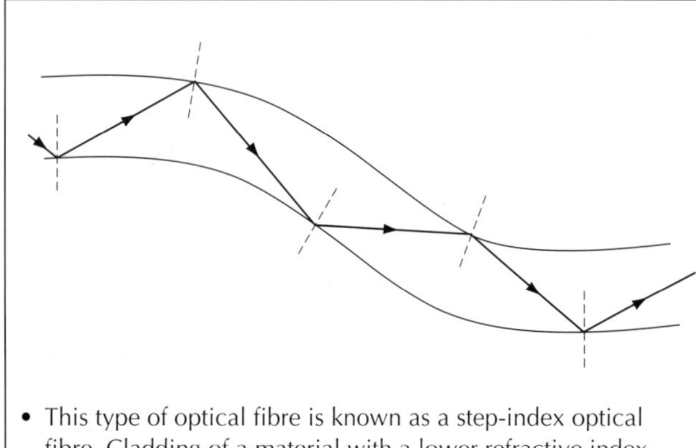

- This type of optical fibre is known as a step-index optical fibre. Cladding of a material with a lower refractive index surrounds the fibre. This cladding protects and strengthens the fibre.

Dispersion, attenuation and noise in optical fibres

MATERIAL DISPERSION

The refractive index of any substance depends on the frequency of electromagnetic radiation considered. This is the reason that white light is dispersed into different colours when it passes through a triangular prism.

As light travels along an optical fibre, different frequencies will travel at slightly different speeds. This means that if the source of light involves a range of frequencies, then a pulse that starts out as a square wave will tend to spread out as it travels along the fibre. This process is known as **material dispersion**.

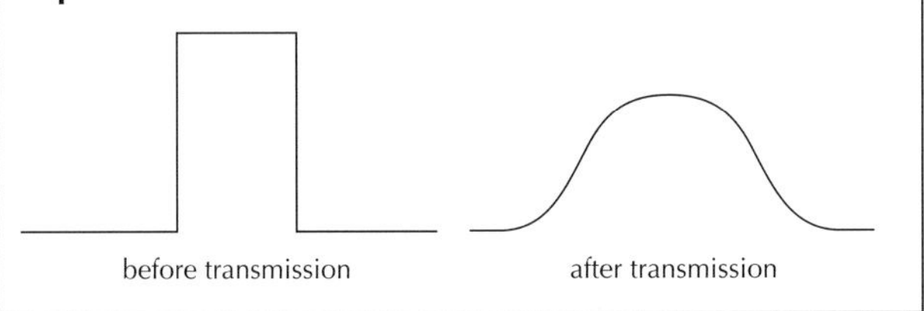

MODAL DISPERSION

If the optical fibre has a significant diameter, another process that can cause the stretching of a pulse is **multipath** or **modal dispersion**. The path length along the centre of a fibre is shorter than a path that involves multiple reflections. This means that rays from a particular pulse will not all arrive at the same time because of the different distances they have travelled.

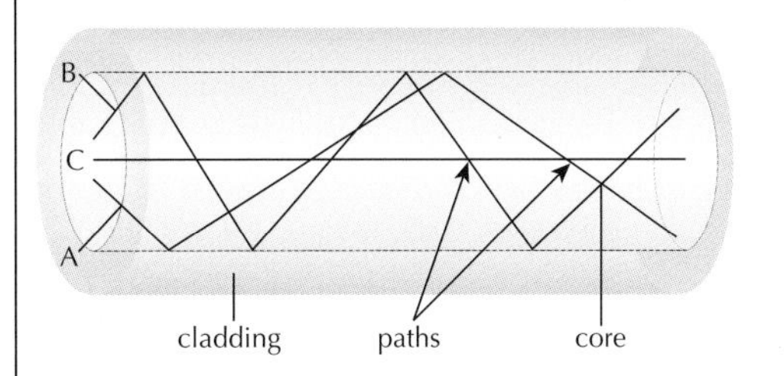

The problems caused by modal dispersion have led to the development of monomode step-index fibres. These optical fibres have very narrow cores (of the same order of magnitude as the wavelength of the light being used (approximately 5 μm) so that there is only one effective transmission path – directly along the fibre.

ATTENUATION

As light travels along an optical fibre, some of the energy can be scattered or absorbed by the glass. The intensity of the light energy that arrives at the end of the fibre is less than the intensity that entered the fibre. The signal is said to be **attenuated**.

The amount of attenuation is measured on a logarithmic scale and is measured in decibels (dB). The attenuation is given by

$$\text{attenuation} = 10\log\frac{I_1}{I_2}$$

I_1 is the intensity of the output power measured in W
I_2 is the intensity of the input power measured in W

A negative attenuation means that the signal has been reduced in power. A positive attenuation would imply that the signal has been amplified.

See page 188 for another example of the use of the decibel scale.

It is common to quote the attenuation per unit length as measured in dB km^{-1}. For example, 5 km of fibre optic cable causes an input power of 100 mW to decrease to 1 mW. The attenuation per unit length is calculated as follows:

$$\text{attenuation} = 10 \log (10^{-3}/10^{-1}) = 10 \log (10^{-2}) = -20 \text{ dB}$$

$$\text{attenuation per unit length} = -20 \text{ dB}/5 \text{ km} = -4 \text{ dB km}^{-1}$$

The attenuation of a 10km length of this fibre optic cable would therefore be –40 dB. The overall attenuation resulting from a series of factors is the algebraic sum of the individual attenuations.

The attenuation in an optical fibre is a result of several processes: those caused by impurities in the glass, the general scattering that takes place in the glass and the extent to which the glass absorbs the light. These last two factors are affected by the wavelength of light used. The overall attenuation is shown below:

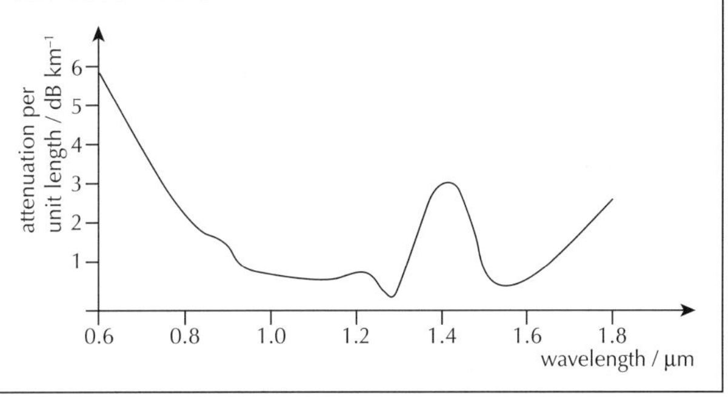

CAPACITY

Attenuation causes an upper limit to the amount of digital information that can be sent along a particular type of optical fibre. This is often stated in terms of its capacity.

capacity of an optical fibre = bit rate × distance

A fibre with a capacity of 80 Mbit s^{-1} km can transmit 80 Mbit s^{-1} along a 1 km length of fibre but only 20 Mbit s^{-1} along a 4 km length.

NOISE, AMPLIFIERS AND RESHAPERS

Noise is inevitable in any electronic circuit. Any dispersions or scatterings that take place within an optical fibre will also add to the noise.

An amplifier increases the signal strength and thus will tend to correct the effect of attenuation – these are also sometimes called regenerators. An amplifier will also increase any noise that has been added to the electrical signal.

A reshaper can reduce the effects of noise on a digital signal by returning the signal to a series of 1s and 0s with sharp transitions between the allowed levels.

Channels of communication (1)

OPTIONS FOR COMMUNICATION

The table below shows some common communication links.

Wire pairs	Two wires can connect the sender and receiver of information. For example a simple link between a microphone, an amplifier and a loudspeaker.
Coaxial cables Copper wire Insulation Copper mesh Outside insulation	This arrangement of two wires reduces electrical interference. A central wire is surrounded by the second wire in the form of an outer cylindrical copper tube or mesh. An insulator separates the two wires. Wire links can carry frequencies up to about 1 GHz but the higher frequencies will be attenuated more for a given length of wire. A typical 100 MHz signal sent down low-loss cable would need repeaters at intervals of approximately 0.5 km. The upper limit for a single coaxial cable is approximately 140 Mbit s^{-1}.
Optical fibres	Laser light can be used to send signals down optical fibres with approximately the same frequency limit as cables – 1 GHz. The attenuation in an optical fibre is less than in a coaxial cable. The distance between repeaters can easily be tens (or even hundreds) of kilometres.
Radio waves ariel radio	A wide range of the electromagnetic spectrum is used to transmit signals and the different sections have different transmission properties. They all travel at the speed of light in a vacuum and the frequency and wavelength are inversely related by the equation $c = f \times \lambda$
Satellite communication Satellite	A artificial satellite can be placed in orbit around the Earth. This can be used to relay information between a sender and a receiver. In general, a satellite would have to be tracked as its position changes in the sky. A particular types of orbit (geostationary orbit – see page 148) means that some satellites remain above the same point on the Earth's surface.

	frequency, f	wavelength, λ
very low frequency, VLF	3 – 30 kHz	100 – 10 km
low frequency, LF	30 – 300 kHz	10 – 1 km
medium frequency, MF	300 – 3000 kHz	1000 – 100 m
high frequency, HF	3 – 30 MHz	100 – 10 m
very high frequency, VHF	30 – 300 MHz	10 – 1 m
ultra high frequency, UHF	300 – 3000 MHz	100 – 10 cm
super high frequency, SHF	3 – 30 GHz	10 – 1 cm
extra high frequency, EHF	30 – 300 GHz	10 – 0.1 cm

Additional possible communication links include:

- **Microwaves** – radio waves with a frequency above 1 GHz (i.e. top end of UHF plus SHF and EHF waves) are referred to as microwaves.
- **Infrared waves** – useful for very short range communications e.g. between a computer and a printer in the same room or between a remote control and a television. IR signals can also be sent down optical fibres.
- **Ultrasonic waves** – also useful for very short range communications but using a high frequency sound wave rather than an electromagnetic wave.

PROPAGATION OF RADIO WAVES

The different sections of the EM spectrum have different transmission properties. The transmission of some waves involves a section of the Earth's atmosphere called the ionosphere. The properties of the layers that make up the ionosphere change depending on the time of day and the weather. It reflects, absorbs or transmits radio waves depending on their frequency.

The three main methods of the propagation of radio waves are:

- **Ground waves** – most important for waves in the VLF, LF and MF ranges.
- **Sky waves** – most important for waves in the HF range. Reflections from the ionosphere (and the Earth's surface) mean that long-range communications are possible.
- **Space waves** – waves in the VHF range (and above) pass through the ionosphere and are not reflected. Thus these frequencies involve "line of sight" transmission. They are used in satellite communication.

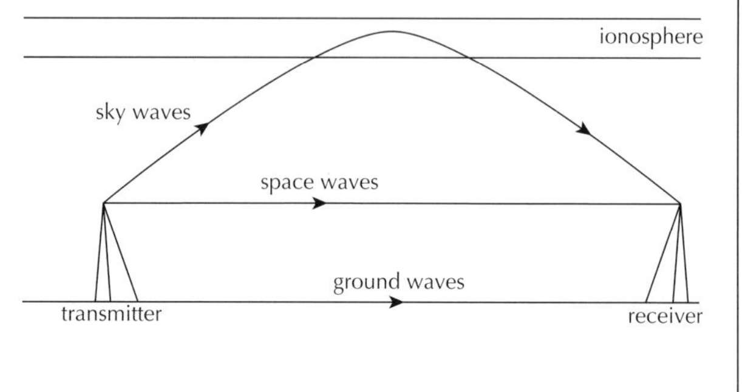

Channels of communication (2)

USES

Wire pairs	Very simple communication systems e.g. intercom
Coaxial cables	Coaxial cables are used to transfer signals from TV aerials to TV receivers. Historically they are standard for underground telephone links.
Optical fibres	Long-distance telecommunication and high volume transfer of digital data including video data.
Radio waves	LF long-distance communication / radio broadcasting MF AM radio broadcasting HF international broadcasting, radio controlled devices, local radio communication VHF FM radio broadcasting, TV, aircraft communication, emergency services communication, radio astronomy UHF satellite communication, TV, mobile telephones, microwave links SHF satellite communication, microwave links.
Satellite	Communication, relaying information for TV, radio, communications etc. Satellites can also be used to for monitoring the weather, remote sensing, navigation and military surveillance.

ADVANTAGES AND DISADVANTAGES

Wire pairs	Very simple and cheap. Susceptible to noise and interference. Unable to transfer information at the highest rates.
Coaxial cables	Simple and straightforward. Less susceptible to noise compared to simple wire pair but noise still a problem.
Optical fibres	Compared to coaxial cables with equivalent capacity, optical fibres: • have a higher transmission capacity • are much smaller in size and weight • cost less • allow for a wider possible spacing of regenerators • offer immunity to electromagnetic interference • suffer from negligible cross talk (signals in one channel affecting another channel) • are very suitable for digital data transmission • provide good security • are quiet – they do not hum even when carrying large volume of data. There are some disadvantages: • the repair of fibres is not a simple task • regenerators are more complex and thus potentially less reliable.
Radio waves	The high frequency end is able to transfer large amounts of data in either digital or analogue form. Terrestrial radio broadcasting equipment can be costly to install but is significantly cheaper than using satellite communication technology. The susceptibility to noise depends on the modulation process used. The attenuation depends on section of the spectrum chosen. Some communications are possible around the world but this can depend on external factors (the ionosphere).
Satellite communication	The attenuation of high frequency radio waves involved is sufficiently small to allow the communication between satellite and Earth receiver. By operating at high radio frequencies, the amount of information transfer possible is large. Many satellite systems are capable of broadcasting hundreds of TV channels all at the same time. The great distance between source and receiver means that there is a significant amount of noise and attenuation. With appropriate modulation techniques, however, digital signals can be transmitted virtually unaffected by noise. The cost of launching a satellite system is huge.

Satellite communications

GEOSTATIONARY SATELLITES

The gravitational attraction between the Earth and a satellite provides the centripetal force necessary for the satellite to stay in orbit. The radius of its orbits fixes the orbital period and speed that is required to stay in orbit. Orbits that are close to the surface of the Earth take less time than those that are further away. Satellites in close orbits are moving faster than those in distant ones.

A **geostationary satellite** maintains the same position relative to a point on the Earth's surface. This is achieved by placing the satellite in orbit above the equator at a certain distance away so that its orbital period is equal to the time take for the Earth to revolve on its axis – 24 hours. The satellite stays directly above a fixed point on the equator.

This orbital distance for geostationary satellites can be calculated using Newton's law of gravitation and is 4.23×10^7 m from the centre of the Earth or roughly 3.6×10^4 km above the surface of the Earth (≈ 5.6 Earth radii). Because the satellite appears to remain stationary, communication between the satellite and Earth is relatively easy.

ISSUES

Potential issues include:

- The potential for a huge rate of transfer of information means that satellites can be used to entertain, advise, inform, educate and potentially break down barriers between countries and religions.
- Cultural judgment is involved in deciding what is and what is not acceptable to broadcast. How does one decide what is ethical and what is not?
- The potential exists for international communication to be more accessible and cheaper.
- If the cost of international calls goes down, how will his price change affect the way we communicate and the wider society?
- If the ways in which we communicate change, will the changes affect all sectors of a given community in the same way?
- Spy satellite raise issues concerned with uncontrolled monitoring. Who has the right to access any information gather in this way and who decides to withdraw freedom of access for 'undesirables'?

POLAR SATELLITES

There are many satellites that are not placed in a geostationary orbit. A polar orbit is a low altitude orbit that passes over the poles. During the orbital time, typically an hour or two, the Earth will have rotated on its axis. During one twenty four hour period, the satellite will complete several orbits.

Extended communication with the satellite will require the satellite to be tracked (i.e. the sending and receiving aerials to point towards the moving satellite). Communication will only be possible for a limited amount of time from any one station located on the surface of the Earth.

Polar satellites can be used for monitoring the weather, remote sensing or military surveillance.

COMMUNICATIONS WITH GEOSTATIONARY SATELLITES

The frequencies used for communication with geostationary satellites are all in the SHF range and above (i.e. in the GHz range).

The uplink frequency (the frequency used to send information to the satellite) and downlink frequency (the frequency used to receive information from the satellite) are different. When the signal is broadcast from the transmitting aerial, it is sent at high power. If the same aerial (or even one nearby) were receiving at the same frequency then the outgoing signal could swamp the incoming signal. An additional advantage of using separate frequencies is that it removes the possibility of positive feedback or resonance occurring.

Possible frequency bands used by satellites include 6/4 GHz, 14/11 GHz and 30/20 GHz. In each case the first number is the approximate uplink frequency and the second number is the approximate downlink frequency.

ADVANTAGES AND DISADVANTAGES FOR COMMUNICATION

	Geostationary	**Polar**
Tracking	A geostationary satellite maintains its relative position so sending or receiving information from the satellite does not involve tracking its position. A permanent communication link with the satellite is possible.	A polar satellite moves in its relative position, so requires the satellite to be tracked. A permanent communication link with the satellite is not possible.
Orbital height	Geostationary orbits are distant which means the signals received will be weak and highly attenuated.	Polar orbits are closer which means that the signals received can be stronger for the same power of transmitter.
Coverage	The large distance to the satellite means that the area 'seen' by the satellite (the **footprint**) is relatively large. Adjustments to the broadcasting aerial can restrict the effective footprint. A large coverage is beneficial for communications. Communication will not be possible with regions on the other side of the Earth.	The close orbit means that the area covered by the satellite is restricted. Communication will be possible with most regions of the Earth at some time. Being closer in means that any monitoring done by the satellite can be done in greater detail.

Electronics – operational amplifier (1)

OP-AMP PROPERTIES

A very important basic analogue circuit component is the **operational amplifier** or op-amp. Op-amps are available singly or in multiple combinations in extremely cheap, miniaturised integrated circuit (IC) format. The op-amp can be adapted for use many different practical circuits.

In order for this component to work, it needs to be connected to the power supply. Typically two supply rails are required which are often fixed to be at +15 and –15 V.

The op-amp has two inputs and one output. For ease of drawing, the power supply rails are often omitted from circuit diagrams involving op-amps but they have been included in the diagram below.

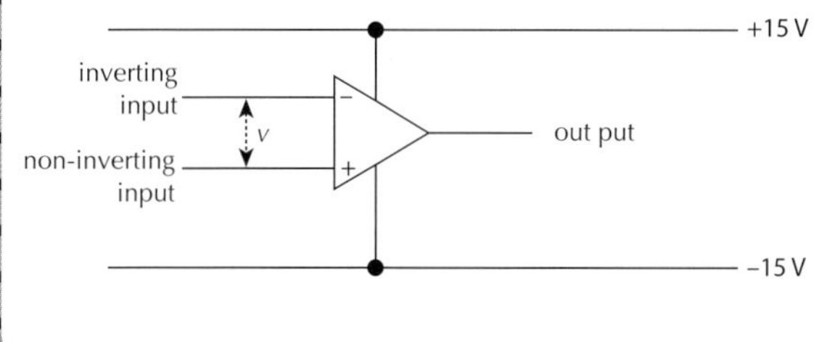

The inverting input is labelled with a '–' and the non-inverting input with a '+' and the output depends on the potential difference, v, between the two inputs. If the non-inverting input is higher than the inverting input, then the output will be positive. If the inverting input is higher than the non-inverting input, then the output will be negative.

The basic property of the op-amp is that, within the voltage limits fixed by its power supply, it is a **high gain differential amplifier**. This means that the output of the device, V_{out}, is equal to the difference between the two inputs multiplied by the **gain** (a very large number). Typically the gain may be 100 000 in which case the circuit above would have:

$$V_{out} = -100\,000 \times v$$

A perfect op-amp has the following properties:

- The gain is infinite.
- The current drawn on the inputs is zero.

In reality a gain of 10^5 and an input a.c. resistance of at least $10^6\ \Omega$ is typical. The maximum output of the op-amp is fixed by the supply rails. With supply rails of ±15 V, the output is typically fixed between ±13 V. When the output is either +13 V or –13 V, the output is said to be **saturated**.

INVERTING AMPLIFIER CIRCUIT

The circuit below shows an op-amp connected to two resistors, R_{IN} and R_F. Note that:

- The power supply rails have been omitted for clarity.
- The bottom rail is not a power rail but at 0 V (the earth potential). This means every potential difference is measured with respect to this potential.
- The circuit involves feedback as the output is connected back to the input (via the resistors).

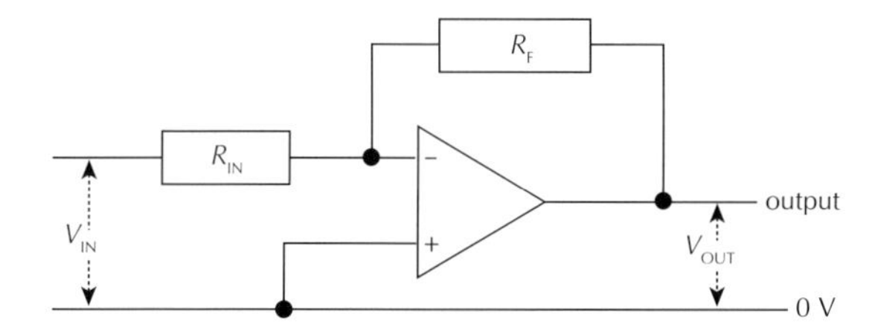

The analysis of the circuit is as follows:

- The non-inverting input of the op-amp is connected to 0 V i.e. it is at earth.
- Providing the op-amp is not saturated, the large gain means that there will be negligible difference between the non-inverting and the inverting input of the op-amp.
- Thus the inverting input can be assumed to be at 0 V as well. It is known as the **virtual earth** approximation.
- The potential difference across the input resistor R_{IN} is from V_{IN} down to the virtual earth.
- The potential difference across the feedback resistor R_F is from the virtual earth down to V_{OUT}.

We can assume that no current flows into the op-amp at either of its inputs.

- The current flowing in the input resistor must equal the current flowing in the feedback resistor.
- The output voltage must be negative as current is flowing from the virtual earth to the right.

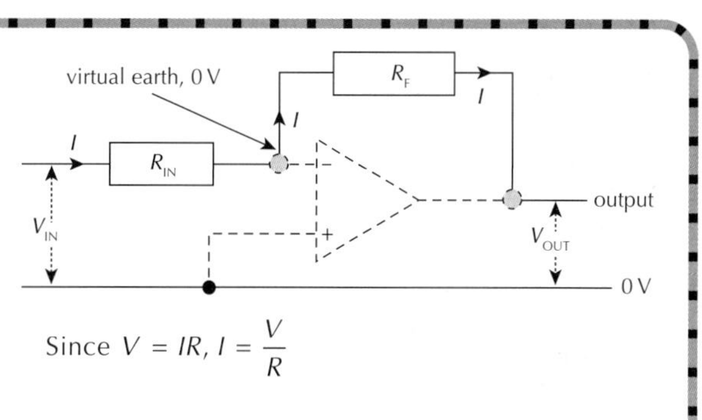

Since $V = IR$, $I = \dfrac{V}{R}$

$$\therefore I = \frac{V_{IN}}{R_{IN}} = \frac{-V_{OUT}}{R_F}$$

$$\therefore V_{OUT} = -\frac{R_F}{R_{IN}} \times V_{IN}$$

The voltage gain, G, of the system overall is given by

$$G = \frac{V_{OUT}}{V_{IN}} = -\frac{R_F}{R_{IN}}$$

Note that:

- This circuit is an inverting amplifier.
- The value of the gain can be chosen by fixing the values of the resistors.
- The derived equation only applies so long as the op-amp is not saturated.

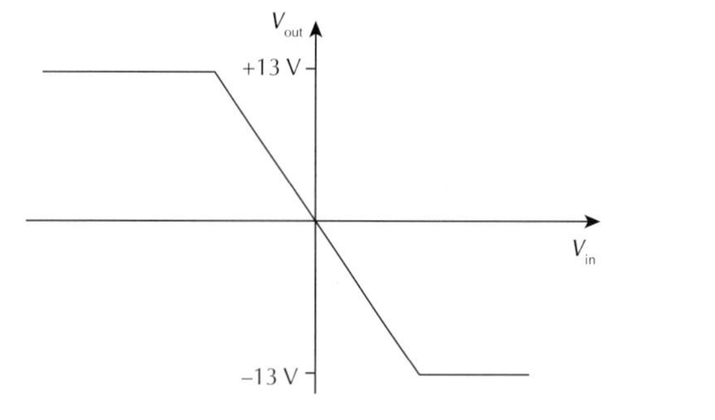

Operational amplifier (2)

NON-INVERTING AMPLIFIER CIRCUIT

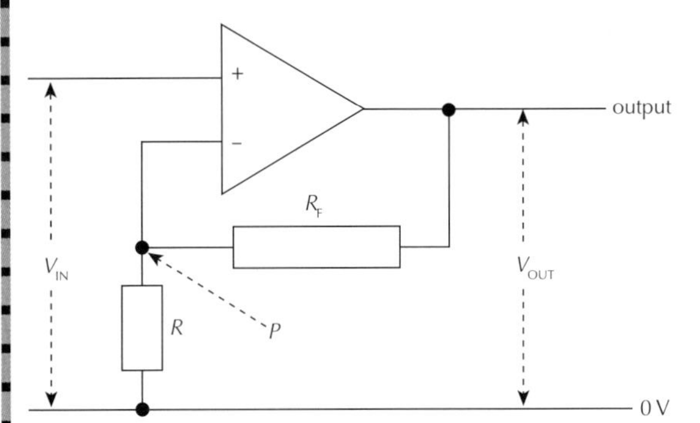

Note that:

- In the diagram above, the inputs to the op-amp are the opposite orientation compared with the diagram for the inverting amplifier circuit.
- The feedback resistor is still connected to the non-inverting input – this is still negative feedback.

Circuit analysis:

- The high gain means that so long as the op-amp is not saturated, the p.d. between the op-amp's inputs will be negligible.
- The potential at point P will thus be V_{IN}.
- No current flows into the non-inverting input so the current through R will be the same as through R_F.

$$\therefore I = \frac{V_{IN}}{R} = \frac{V_{OUT} - V_{IN}}{R_F}$$

$$\therefore V_{OUT} = \frac{R_F}{R} \cdot V_{IN} + V_{IN} = V_{IN}\left(1 + \frac{R_F}{R}\right)$$

- The voltage gain, G, of the system overall is given by

$$G = \frac{V_{OUT}}{V_{IN}} = 1 + \frac{R_F}{R}$$

Note that:

- This circuit is a non-inverting amplifier.
- The value of the gain can be chosen by fixing the values of the resistors.
- The derived equation only applies so long as the op-amp is not saturated.

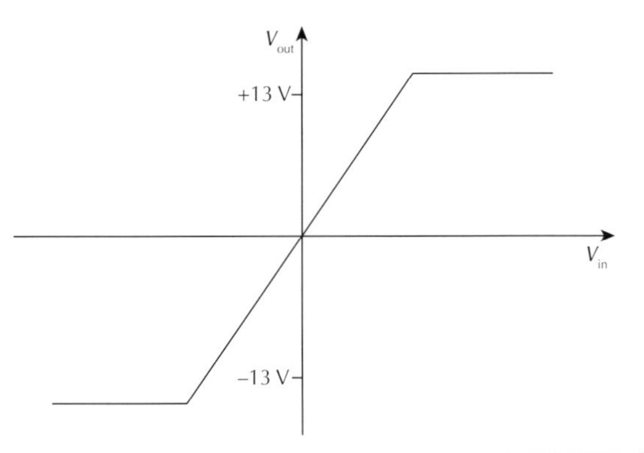

Practical op-amp circuits (1)

COMPARATOR CIRCUIT

A comparator circuit often has:

- One input of an op-amp at a fixed potential as a result of a potential divider circuit.
- The other input of an op-amp at a varying potential as a result of a potential divider that includes a sensor e.g. LDR, thermistor etc.
- The output of the op-amp will be positive or negative saturation and can then control either a light (LED) or a buzzer.

Example 1

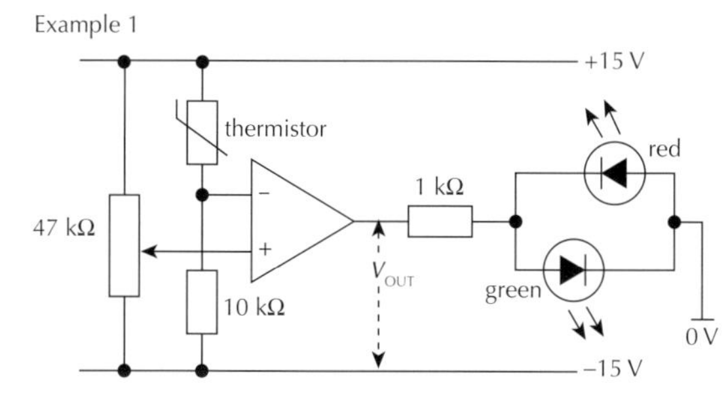

- The circuit detects hot & cold by either lighting a red LED or a green LED.
- The red LED lights when the thermistor is hot, the green when it is cold.
- The 47 kΩ potentiometer fixes the input voltage on the non-inverting input to the op-amp to a value between +15 V and −15 V.
- The position of the slider on the potentiometer fixes the temperature at which the green LED goes off and the red LED goes on.
- The potential divider of the thermistor and 10 kΩ resistor fixes the input voltage to the inverting input of the op-amp.
- The 1 kΩ resistor acts as a **protective resistor** to the LED preventing too large a current from flowing.

When the thermistor is cold:

- its resistance is large
- the inverting input is close to −15 V and lower than the non-inverting input
- the output is positive saturation
- current flows from the output through the green LED to the 0V; the green LED is on.

When the thermistor is hot:

- its resistance is low
- the inverting input is close to +15 V and larger than the non-inverting input
- the output is negative saturation
- current flows from 0V through the red LED to the output; the red LED is on.

Example 2

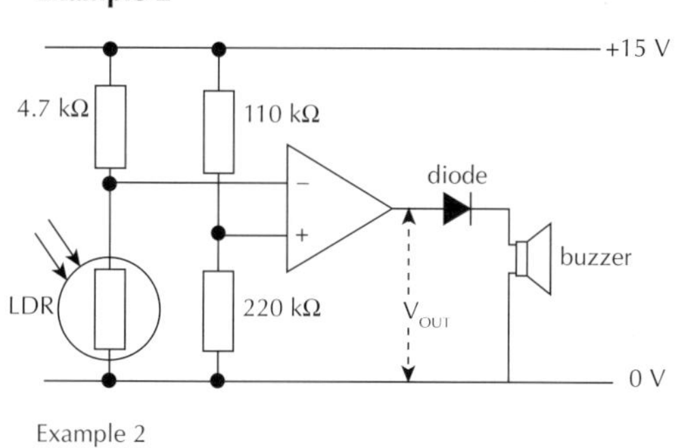

Example 2

This circuit makes the buzzer sound when the light level rises above a certain value.

- The potential divider of the 110 kΩ and 220 kΩ resistors fixes the input voltage to the inverting input of the op-amp

  $\text{input} = \left(\frac{220}{330}\right) \times 15 = 10.3\ \text{V}.$

- The diode is a device that only allows the flow of current from left to right i.e. the buzzer will sound when the op-amp is at positive saturation but not when it is at negative saturation.

When the light level is low:

- the resistance of the LDR is large
- the potential at the inverting input is higher than at the non-inverting input
- the op-amp is at negative saturation
- the buzzer does not sound as the diode prevents a current following from 0 V.

When the light level is high:

- the resistance of the LDR is low
- the potential at the inverting input is lower than at the non-inverting input
- the op-amp is at positive saturation
- the buzzer sounds as a current flows from the op-amp, through the diode and buzzer to 0 V.

Practical op-amp circuits (2)

SCHMITT TRIGGER

A Schmitt trigger has one input and one output:

The output of a Schmitt trigger has two possible values: HIGH and LOW. The potential on the input triggers the change between HIGH and LOW but a different value needs to be reached for a rising signal when compared with a falling signal.

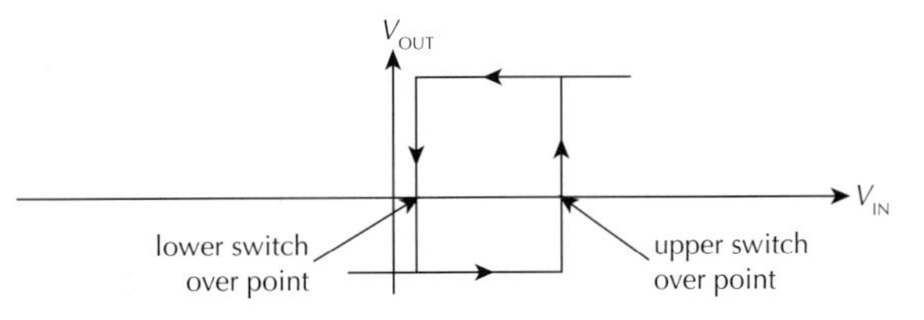

This can be used to reshape digital pulses that have been subjected to noise or dispersion:

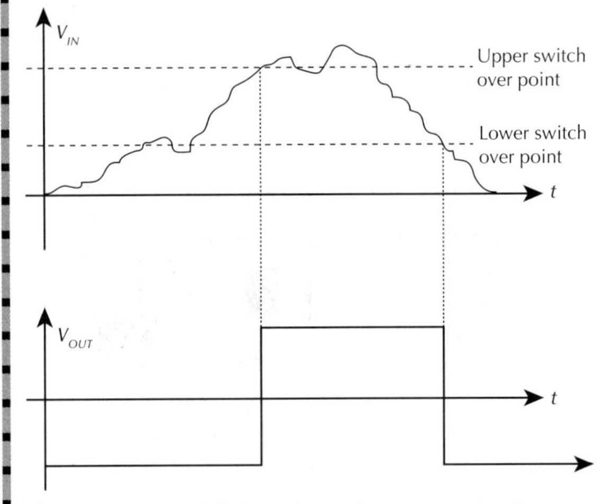

The op-amp circuit below shown is an example of a non-inverting Schmitt trigger.

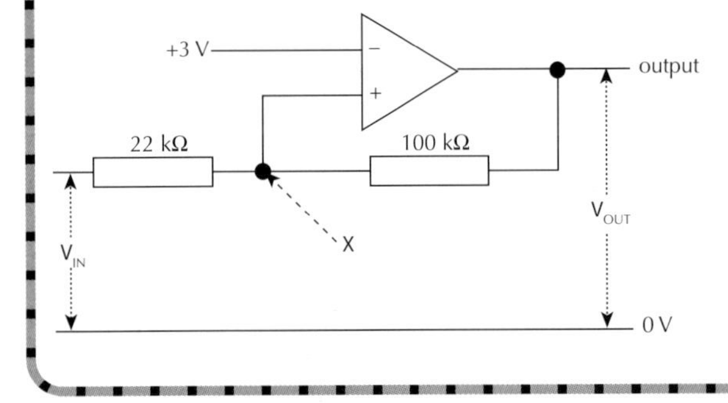

In this example the inverting input is fixed at +3 V. This can be done using a potential divider circuit which has not been included.

When V_{IN} is zero:

- The op-amp is at negative saturation (= –13V)
- The potential at point *X* (the non-inverting input) is $-\left(\frac{22}{122}\right) \times 13 = -2.3$ V.
- The non-inverting input is lower than the inverting input so negative saturation will remain until the potential point *X* rises above 3V.

The upper switchover point for *X* to rise to 3V is calculated as follows:

- The potential difference across 100 kΩ must be from +3 to –13 = 16 V.
- The current through the 100 kΩ = $\frac{16}{100}$ = 0.16 mA.
- The potential difference across 22 kΩ = 0.16 × 22 = 3.5 V.
- Switchover happens when V_{IN} is 3.5 V above X i.e. V_{IN} = +6.5V.
- When V_{IN} rises to +6.5 V, the op-amp switches to positive saturation (= +13V).
- The potential difference across the two resistors is now from +3 to +13.
- Potential at $X = 3 + \left(\frac{22}{122}\right) \times 10 = +4.8$ V, so positive saturation is maintained.

The lower switchover point takes place when V_{IN} is decreased low enough to make the potential at *X* go down to +3 V while the op-amp is at positive saturation. This is calculated as follows:

- The potential difference across 100 kΩ must be from +13 to +3 = 10 V.
- The current through the 100 kΩ = $\frac{10}{100}$ = 0.1 mA.
- The potential difference across 22 kΩ = 0.1 × 22 = 2.2 V.
- Switchover happens when V_{IN} is 2.2 V below *X* i.e. when V_{IN} = +0.8V.

Summary

Upper switchover point = 6.5 V
Lower switchover point = 0.8 V

Mobile phone system

PUBLIC SWITCHED TELEPHONE NETWORK (PSTN)

The simplest way of making a telephone call between two subscribers involves all subscribers being connected to a telephone exchange. The exchange connects the caller with the receiver. When telephones were first invented, an operator manually put these connections in place!

The exchanges were automated with the use of electric relays and digital switching techniques but the main telephone network still uses wires to connect telephones and a significant amount of the telephonic traffic still travels on coaxial cable. The principle remains the same: in order for a telephone conversation to take place, a large number of relays and switches connect different phone lines together.

ISSUES ASSOCIATED WITH MULTIMEDIA COMMUNICATIONS

As technologies improve, the bandwidth of mobile phones and their base stations will also improve, allowing for portable cheap multimedia communications to be widely available. Communications will not be limited to just standard telephonic audio connections. High quality sound and pictures along with TV and video links in real time are likely to become much more accessible.

- The effect on society as a whole as a result of communication becoming more global is unknown. The physical and geographic boundaries between different countries potentially have a reduced effect on the people that an individual is able to contact.
- The new communications techniques have the potential to radically alter the way in which business is done.
- Although significant numbers of users will be able to afford the new methods of communication, this method of communication will still be unavailable to many possibly creating a 'technological underclass'.
- New technologies will not be available, or used in the same way, by those of different ages in a society. This could, in time, affect the way in which the society operates.

CELLULAR EXCHANGE

A mobile phone network involves a large number of low powered transmitters and receivers in the UHF band known as a **cellular network**. A country is divided up into 'cells' and each of these cells has its own transmitter/receiver (the **base station**) and is joined (via an electronic exchange) into the PSTN.

Each base station covers a small area of the country. Often large numbers of small stations are required to cover all areas of a city successfully. Each station operates at different frequencies. The mobile phones are capable of sending and receiving at all of the frequencies being utilized.

Many base stations are connected together via cellular exchanges. When a user makes a mobile telephone call, a connection channel is allocated between the mobile phone and the base station. Using multiplexing techniques, it is possible for a single frequency to provide many different available channels. The different frequencies for the different stations are chosen so that neighbouring cells always utilize different frequencies. If a caller moves between cells, then the frequency is automatically switched during the conversation.

Even when the phone is not being used for a call, there is a continuous link between the mobile phone and the base stations through which it travels. An incoming call can thus be routed via the most appropriate base station to the mobile phone.

Cells with identical numbers use the same broadcasting frequencies. The diagram simplifies the situation as most cells are circular and there are many regions where they overlap.

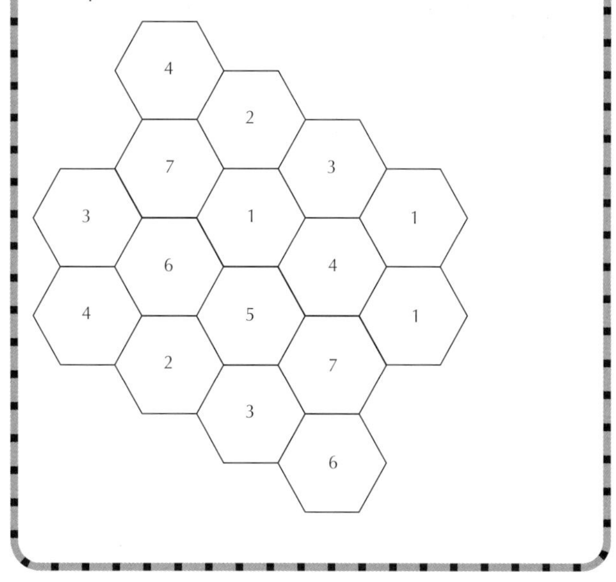

IB QUESTIONS – COMMUNICATIONS

1 **(a)** Explain the difference between amplitude modulation and frequency modulation. [4]

(b) Describe the relative advantages and disadvantages of AM and FM for radio broadcast and reception. [4]

(c) The block diagram below represents a simple amplitude modulated radio receiver. Three component blocks are labeled A, B and C.

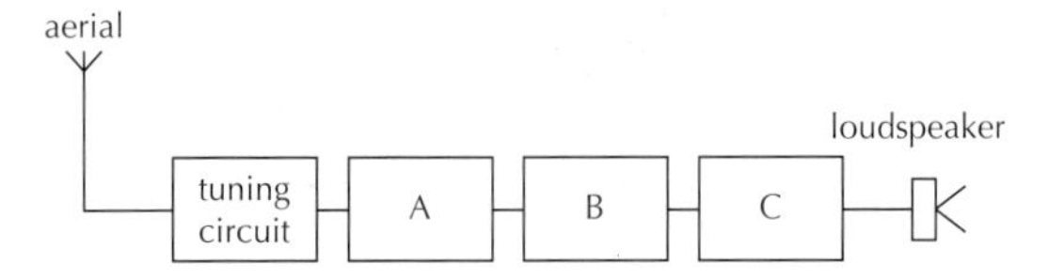

(i) Identify the component blocks represented by A, B and C. [3]

(ii) Draw labelled diagrams to show the shape of the waveform at the output of each of A, B and C. [3]

2 An optical fibre telephone system limits the range of frequencies of the audio signal that it carries to between 300 Hz and 3.4 kHz.

(a) Sketch a typical power spectrum diagram for these audio frequencies using a carrier wave of frequency 100 kHz for

(i) AM modulated signals
(ii) FM modulated signals. [4]

(b) Use your answer to **(c)** to explain the terms

(i) sideband frequencies
(ii) bandwidth. [4]

(c) Estimate the bandwidth for

(i) AM modulated signals
(ii) FM modulated signals. [4]

(d) The maximum frequency signal that can be sent down the optical fibre is 500 kHz. Estimate the number of channels available for

(i) AM modulated signals
(ii) FM modulated signals. [4]

3 Explain the difference between an analogue-to-digital converter and a serial-to-parallel converter. [4]

4 An audio analogue signal is transmitted along a wire between a reporter and a broadcasting unit. The signal transmitted down the wire is a serial digital signal but the technicians in the unit are able to listen to the audio signal on headphones.

(a) Describe, using block diagrams, the principles of the conversion, transmission and reception for this process. [4]

(b) Explain how time division multiplexing could be used to allow other audio signals to be broadcast down the same wire during the broadcast. [4]

5 Explain the difference between the effects of material and modal dispersion of optical fibre signals. [4]

6 A 15 km length of optical fibre has an attenuation of 4 dB km^{-1}. A 5 mW signal is sent along the wire using two amplifiers as represented by the diagram below.

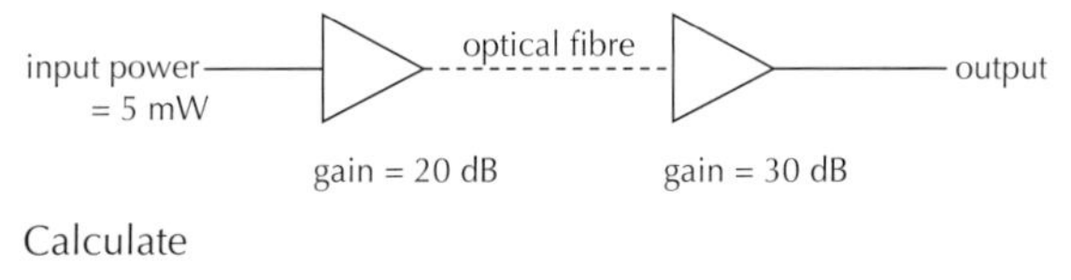

Calculate

(a) the overall gain of the system

(b) the output power. [2]

7 An ideal operational amplifier (op-amp) is powered by a ±15 V d.c. supply and connected in the circuit shown below.

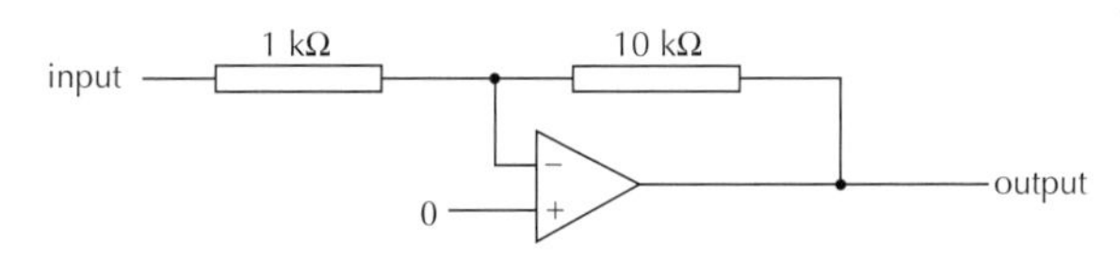

(a) State the properties of an ideal op-amp. [2]

(b) Calculate the output voltage when the input voltage is

(i) +0.7 V
(ii) −1.2 V
(iii) +5 V. [3]

(c) Sketch the output signal when a sinusoidal signal of amplitude 2 V is connected to the input. [3]

8 Describe the role of the cellular exchange and the public switched telephone network in communications using mobile phones. [3]

Dispersion

DISPERSION

White light is, in fact, a combination of all the visible frequencies. These different frequencies of light are perceived as different colours of the rainbow. The splitting up of white light into its component colours is called **dispersion**. A prism causes the dispersion of light because the refractive indexes are slightly different for each of the different colours.

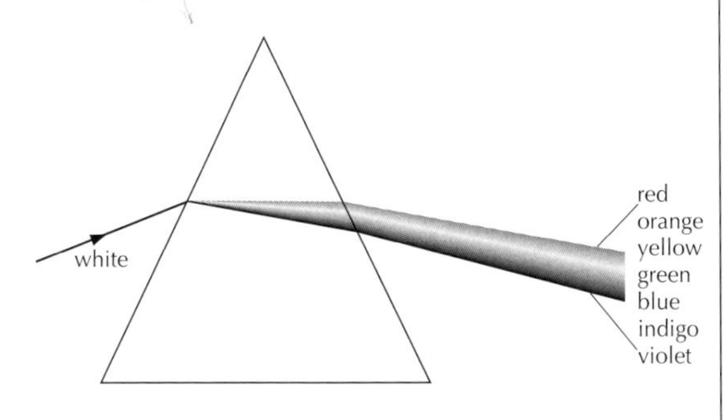

Note that

- the refraction takes place at both surfaces.
- red light is bent the least and blue light is bent the most. In other words the refractive index for red light is smaller than for blue.
- Dispersion can occur throughout the EM spectrum.

TRANSMISSION, ABSORPTION AND SCATTERING

When any electromagnetic wave passes through a medium (i.e. anything apart from a vacuum), it will be **transmitted**, **absorbed** or **scattered** to different extents depending on the wavelength and the medium involved.

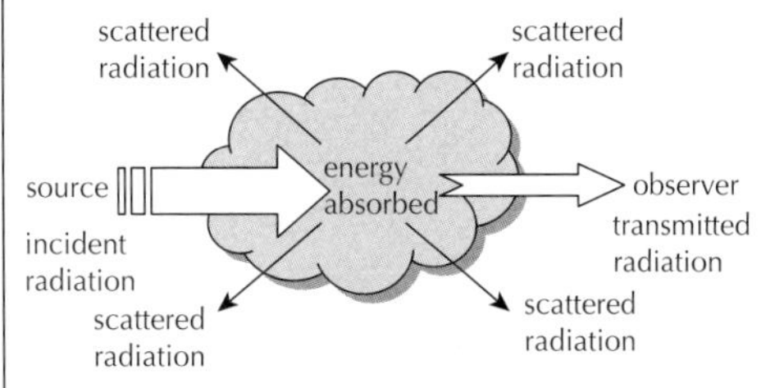

The energy that is received directly by an observer after the wave has travelled through the medium is the transmitted energy. The direction of energy transfer may be affected by refraction taking place when the energy entered the medium, but energy will be transmitted in straight lines.

The medium may absorb some of the energy causing its temperature to increase and decreasing the energy transmitted. The medium may re-emit some of this absorbed energy. Radiation can also be scattered by the medium.

Electromagnetic radiation from the Sun is incident on the Earth's atmosphere and provides an example of each of these processes:

- Blue light is scattered in all directions as a result of interaction with small dust particles in the atmosphere. This is the reason that the sky appears blue.
- The transmitted light will not contain the same amount of blue. The grazing incidence to the atmosphere at sunset and sunrise means that light from the sun travels a greater length through the atmosphere and sunsets or sunrises appear to be red.
- Harmful UV radiation is absorbed by the ozone layer in the atmosphere which would otherwise be harmful to creatures living on the surface of the Earth including humans.
- The atmosphere absorbs strongly in the infra-red radiation region. This process is involved in the greenhouse effect (see page 76). Increasing the carbon dioxide content of the atmosphere will increase the absorption and result in global warming.

NATURE OF ELECTROMAGNETIC WAVES

Most people know that light is an electromagnetic wave, but it is quite hard to understand what this actually means. A physical wave involves the oscillation of matter, whereas an electromagnetic wave involves the oscillation of electric and magnetic fields. The diagram below attempts to show this.

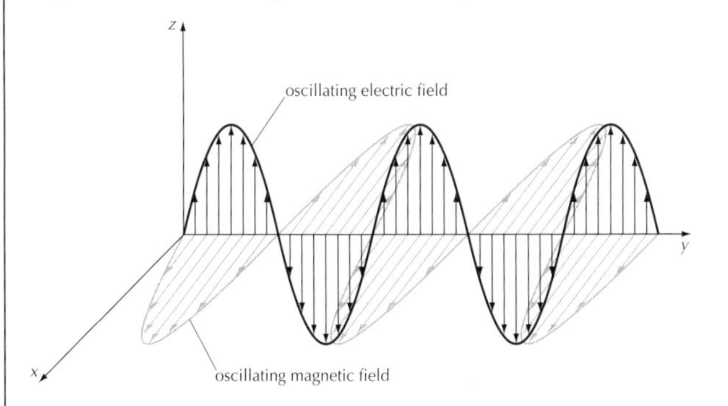

The changing electric and magnetic fields move through space – the technical way of saying this is that the fields **propagate** through space. The physics of how these fields propagate was worked out in the nineteenth century. The "rules" are summarised in four equations known as **Maxwell's equations**. The application of these equations allows the speed of all electromagnetic waves (including light) to be predicted. It turns out that this can be done in terms of the electric and magnetic constants of the medium through which they travel.

$$c = \sqrt{\frac{1}{\varepsilon_0 \mu_0}}$$

This equation does not need to be understood in detail. The only important idea is that the speed of light is **independent** of the velocity of the source of the light. In other words, a prediction from Maxwell's equations is that the speed of light in a vacuum has the same value for all observers. This is experimentaly verified.

ELECTROMAGNETIC SPECTRUM

The electromagnetic spectrum is shown on page 40. You should learn the regions and be able to identify a source and possible risks associated with each region. Health hazards associated with transmission lines are shown on page 102.

Lasers

MONOCHROMATIC

Laser light is **monochromatic**. It contains only a very very narrow band of frequencies. This is very different to the light given out by a light bulb, which contains a mixture of many different frequencies.

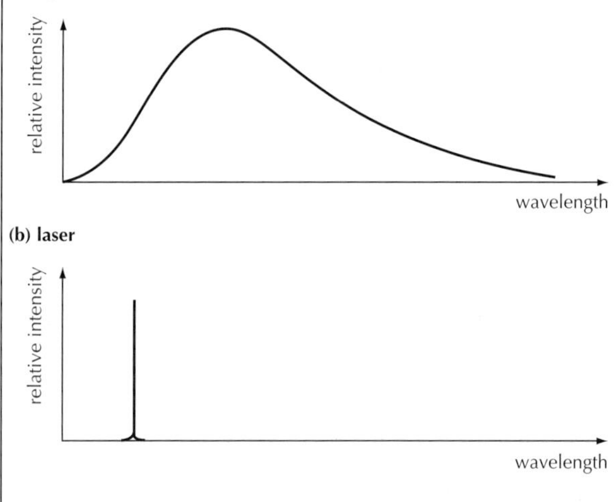

COHERENT

Laser light is **coherent**. The oscillations that make up the waves are linked together. Light is always emitted in 'packets' of energy called photons. Each photon in laser light is in phase with all the other photons that are emitted. This is very different to the light given out by a light bulb. Each atom that emits light acts independently from the other atoms. This means that every photon has an independent phase.

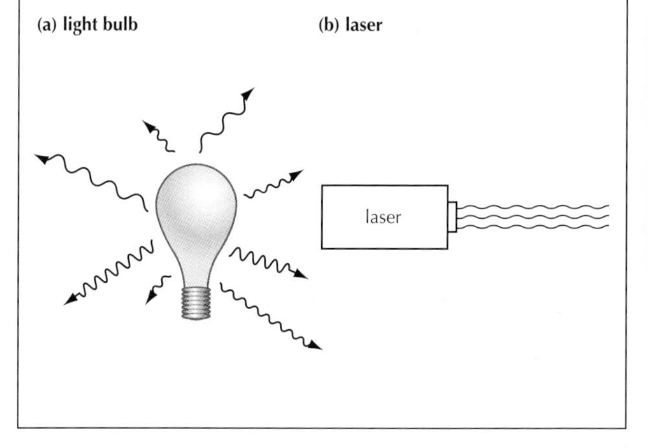

APPLICATIONS

The are many uses of laser in technology, industry and medicine. Some of these rely on the fact that laser light does not tend to diverge. This means that lasers can provide energy that is concentrated in a small region. When used medically, the energy supplied by the laser can be enough to heat and destroy a small area of tissue while leaving neighbouring areas virtually unaffected. The use of a laser in this way to cut through tissue is called laser surgery. It has the additional advantage that the blood vessels that are cut by the process also tend to be sealed at the same time. This means that there is less bleeding than when using a knife.

Other possible applications include:

- technology (bar-code scanners, laser discs)
- industry (surveying, welding and machining metals, drilling tiny holes in metals)
- medicine (destroying tissue in small areas, attaching the retina, corneal correction)
- communication
- production of CDs
- Reading and writing of CDs, DVDs etc.

PRODUCTION OF LASER LIGHT

Laser stands for "**L**ight **A**mplification by **S**timulated **E**mission of **R**adiation". Light photons are produced when an atomic electron falls from a higher energy level down to a lower energy level – see page 59. Normally electrons will always occupy the lowest available energy levels in an atom. The production of laser light involves a process that promotes (or **pumps**) a large number of electrons to a higher energy level – this is known as **population inversion**. These electrons are stimulated to fall down and emit light of a particular frequency.

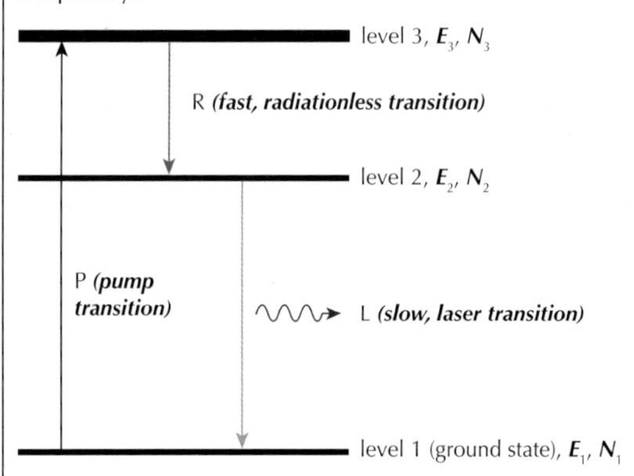

In the example above the pump transition takes the electrons to level 3, they quickly lose energy via collisions and fall in to level 2. They can be stimulated to fall into level 1.

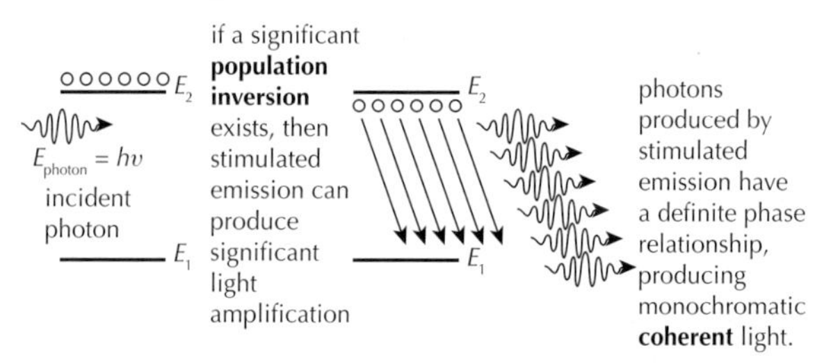

Light Amplification by Stimulated Emission of Radiation.

The production of laser light involves repeated reflections between carefully aligned mirrors.

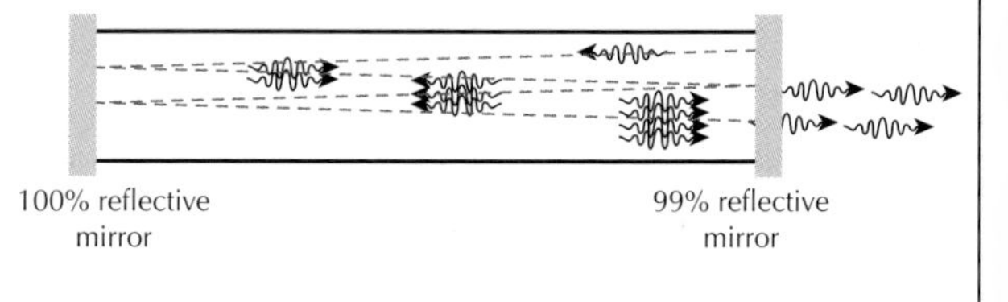

Image formation

RAY DIAGRAMS

If an object is placed in front of a plane mirror, an image will be formed.

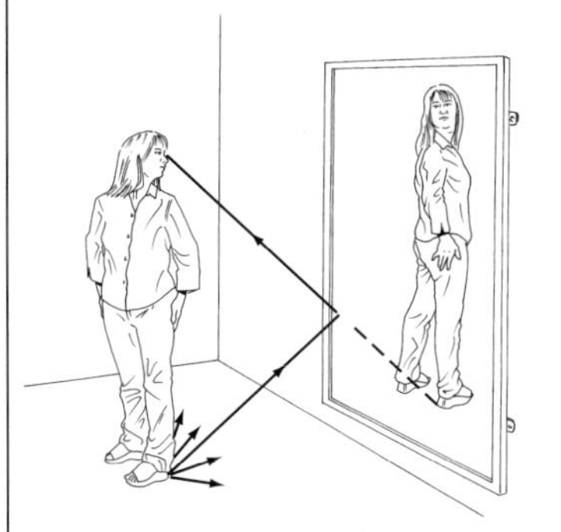

In order to find the location and nature of this image a ray diagram is needed.

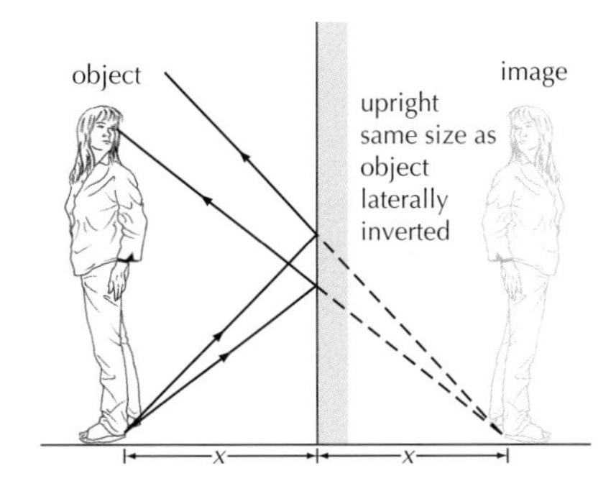

The process is as follows:

- Light sets off in all directions from every part of the object. (This is a result of diffuse reflections from a source of light.)
- Each ray of light that arrives at the mirror is reflected according to the law of reflection.
- These rays can be received by an observer.
- The location of the image seen by the observer arises because the rays are assumed to have travelled in straight lines.

The image formed by reflection in a plane mirror is always

- **The same distance behind the mirror as the object is in front.**
- **Upright** (as opposed to being inverted).
- **The same size as the object** (as opposed to being magnified or diminished).
- **Laterally inverted** (i.e. left and right are interchanged).
- **Virtual** (see below).

REAL AND VIRTUAL IMAGES

The image formed by reflection in a plane mirror is described as a **virtual image**. This term is used to describe images created when rays of light **seem** to come from a single point but in fact they do not pass through that point. In the example above, the rays of light seem to be coming from behind the mirror. They do not, of course, actually pass behind the mirror at all.

The opposite of a virtual image is a **real image**. In this case, the rays of light do actually pass through a single point. Real images cannot be formed by plane mirrors, but they can be formed by concave mirrors or by lenses. For example, if you look into the concave surface of a spoon, you will see an image of yourself. This particular image is

- **Upside down**
- **Diminished**
- **Real**

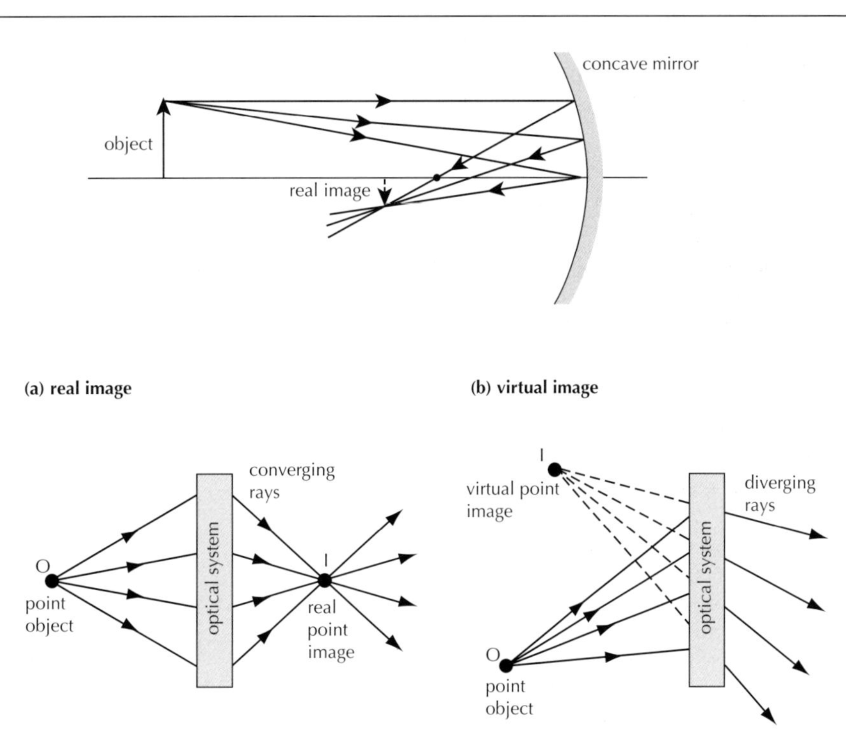

STICK IN WATER

The image formed as a result of the refraction of light leaving water is so commonly seen that most people forget that the objects are made to seem strange. A straight stick will appear bent if it is place in water. The brain assumes that the rays that arrive at one's eyes must have been travelling in a straight line.

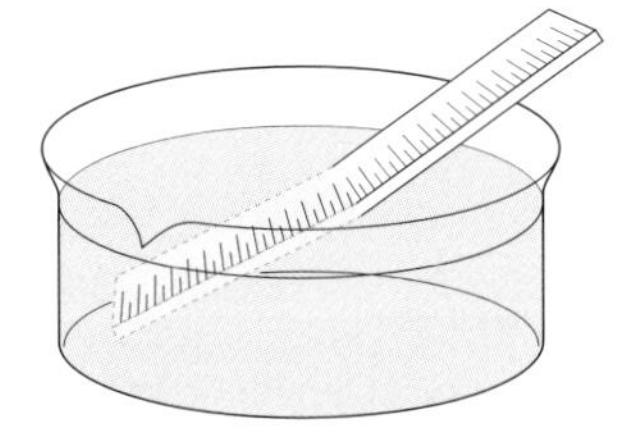

A straight stick appears bent when placed in water

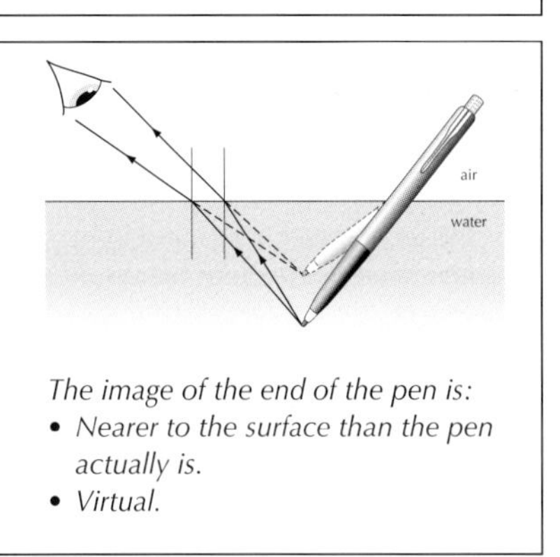

The image of the end of the pen is:

- *Nearer to the surface than the pen actually is.*
- *Virtual.*

Converging lenses

CONVERGING LENSES

A converging lens brings parallel rays into one focus point.

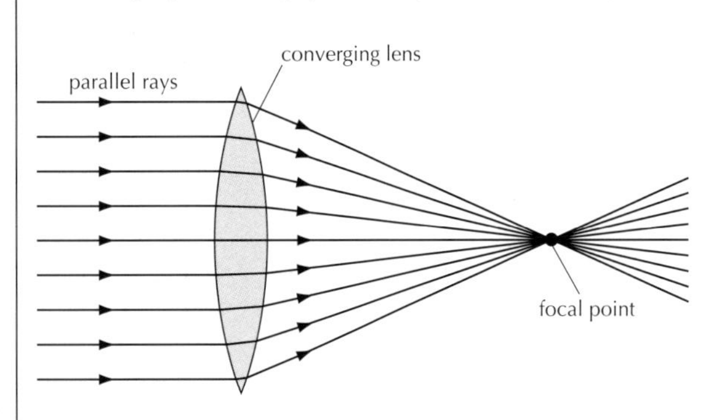

The reason that this happens is the refraction that takes place at both surfaces of the lens.

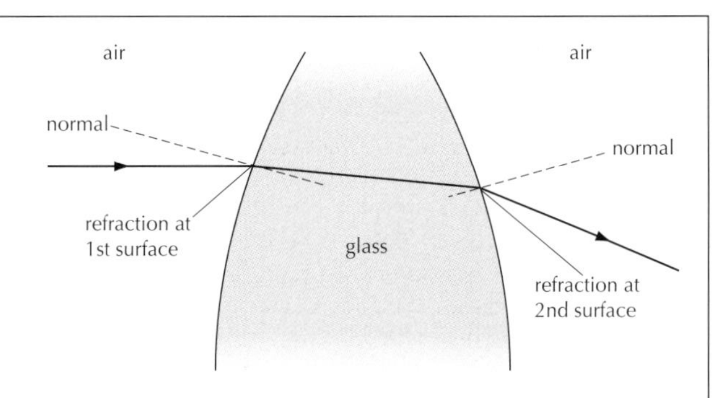

The rays of light are all brought together in one point because of the particular shape of the lens. Any one lens can be thought of as a collection of different-shaped glass blocks. It can be shown that any thin lens that has surfaces formed from sections of spheres will converge light into one focus point.

A converging lens will always be thicker at the centre when compared with the edges.

DEFINITIONS

When analysing lenses and the images that they form, some technical terms need to be defined.

- The curvature of each surface of a lens makes it part of a sphere. The **centre of curvature** for the lens surface is the centre of this sphere.
- The **principal axis** is the line going directly through the middle of the lens. Technically it joins the centres of curvature of the two surfaces
- The **focal point** (principal focus) of a lens is the point on the principal axis to which rays that were parallel to the principal axis are brought to focus after passing through the lens. A lens will thus have a focal point on each side.
- The **focal length** is the distance between the centre of the lens and the focal point.
- The **linear magnification** is the ratio between the size (height) of the image and the size (height) of the object. It has no units.

$$\text{linear magnification} = \frac{\text{image size}}{\text{object size}}$$

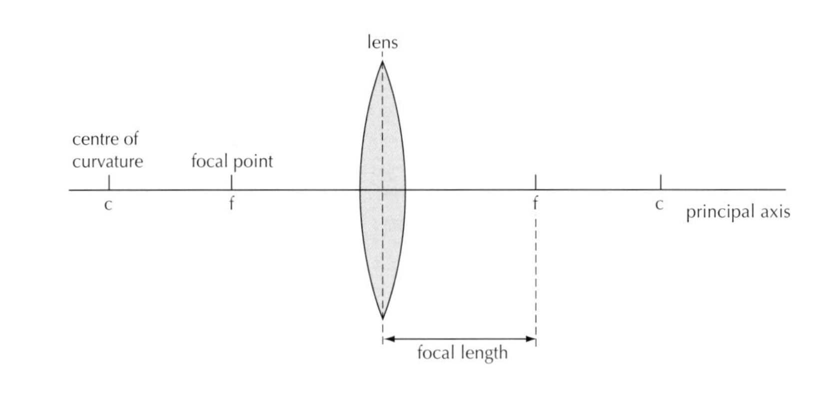

POWER OF A LENS

The power of a lens measures the extent to which light is bent by the lens. A higher power lens bends the light more and thus has a smaller focal length. The definition of the power of a lens, P, is the reciprocal of the focal length, f:

$$P = \frac{1}{f}$$

f is the focal length measured in m
P is the power of the lens measured in m^{-1} or **dioptres** (dpt)

A lens of power = +5 dioptre is converging and has a focal length of 20 cm. When two thin lenses are placed close together their powers approximately add.

Image formation in convex lenses

IMPORTANT RAYS

In order to determine the nature and position of the image created of a given object, we need to construct **ray diagrams** of the set-up. In order to do this, we concentrate on the paths taken by three particular rays. As soon as the paths taken by two of these rays have been constructed, the paths of all the other rays can be inferred. These important rays are described below.

Converging lens

1. Any ray that was travelling parallel to the principal axis will be refracted towards the focal point on the other side of the lens.
2. Any ray that travelled through the focal point will be refracted parallel to the principal axis.
3. Any ray that goes through the centre of the lens will be undeviated.

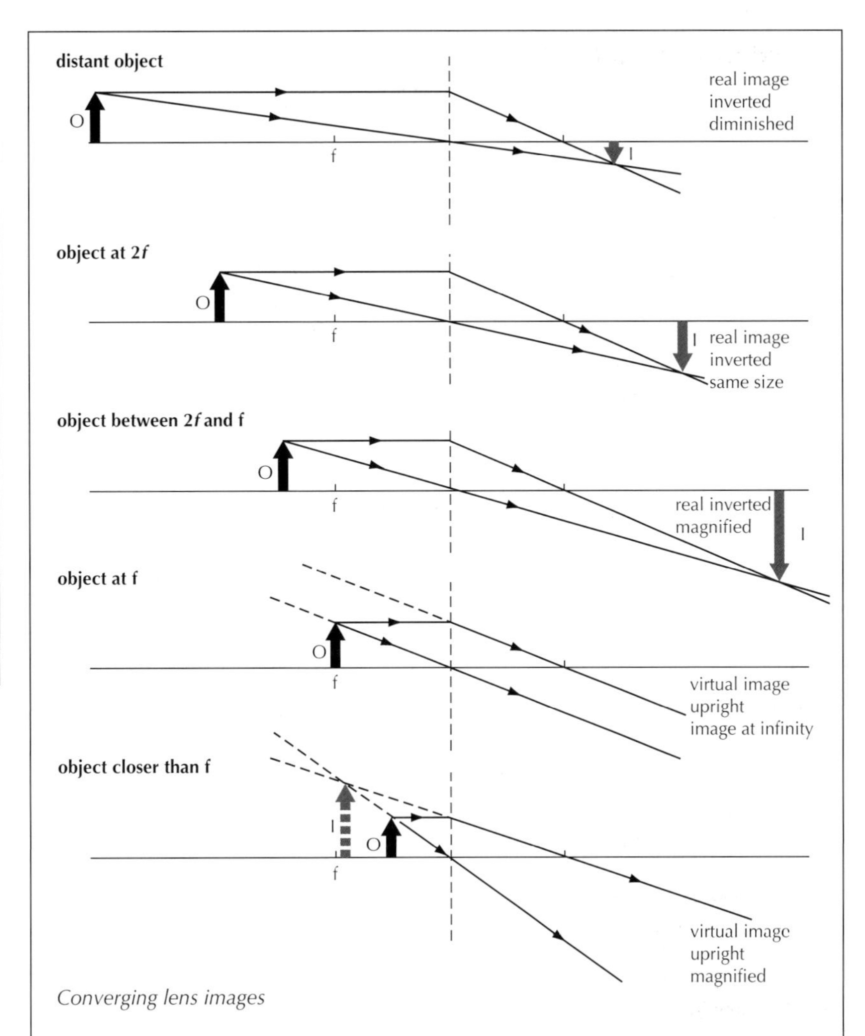

Converging lens images

POSSIBLE SITUATIONS

A ray diagram can be constructed as follows

- an upright arrow on the principal axis represents the object.
- the paths of two important rays from the top of the object are constructed.
- this locates the position of the top of the image.
- the bottom of the image must be on the principal axis directly above (or below) the top of the image.

A full description of the image created would include the following information

- if it is real or virtual.
- if it is upright or inverted.
- if it is magnified or diminished.
- its exact position.

It should be noted that the important rays are just used to locate the image. The real image also consists of all the other rays from the object. In particular, the image will still be formed even if some of the rays are blocked off.

Thin lens equation

LENS EQUATION

There is a mathematical method of locating the image formed by a lens. An analysis of the angles involved shows that the following equation can be applied to thin spherical lenses:

$$\frac{1}{u} + \frac{1}{v} = \frac{1}{f}$$

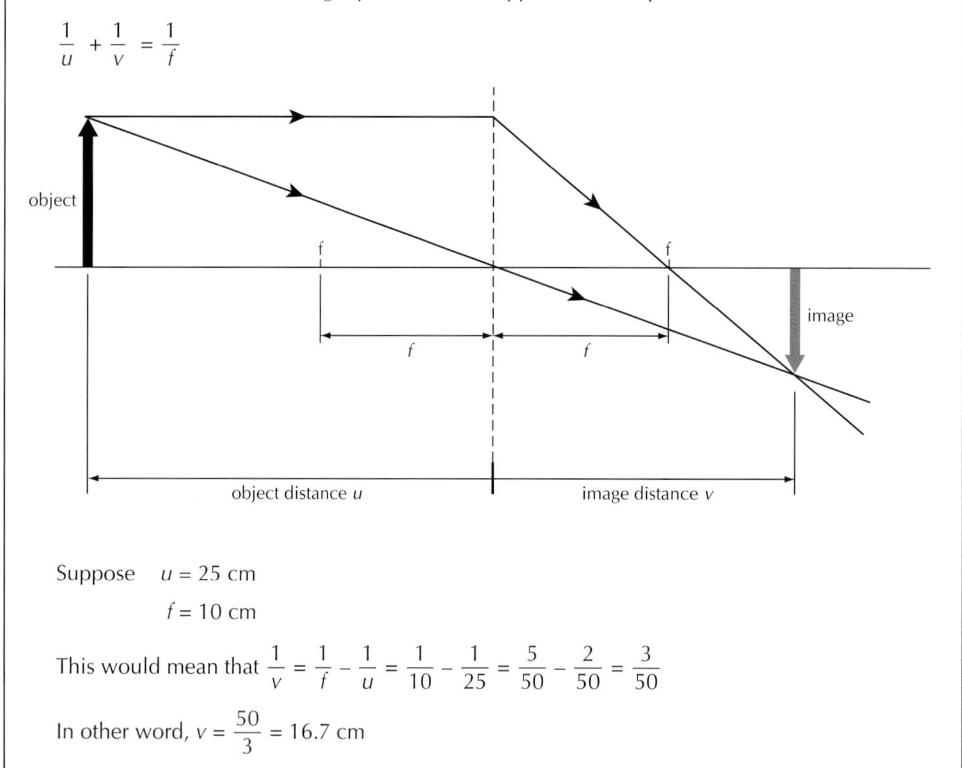

Suppose $u = 25$ cm

$f = 10$ cm

This would mean that $\frac{1}{v} = \frac{1}{f} - \frac{1}{u} = \frac{1}{10} - \frac{1}{25} = \frac{5}{50} - \frac{2}{50} = \frac{3}{50}$

In other word, $v = \frac{50}{3} = 16.7$ cm

REAL IS POSITIVE

Care needs to be taken with virtual images. The equation does work but for this to be the case, the following convention has to be followed:

- a virtual image is represented by a negative value for v – in other words, it will be on the same side of the lens as the object.

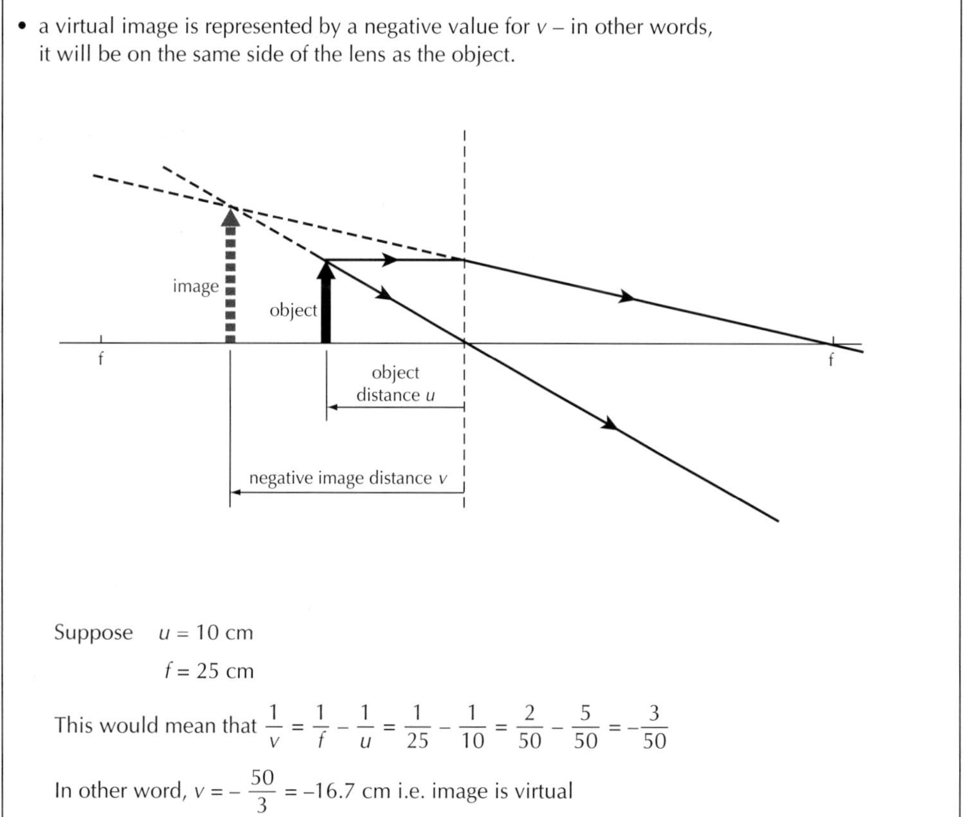

Suppose $u = 10$ cm

$f = 25$ cm

This would mean that $\frac{1}{v} = \frac{1}{f} - \frac{1}{u} = \frac{1}{25} - \frac{1}{10} = \frac{2}{50} - \frac{5}{50} = -\frac{3}{50}$

In other word, $v = -\frac{50}{3} = -16.7$ cm i.e. image is virtual

The simple magnifying glass

NEAR AND FAR POINT

The human eye can focus objects at different distances from the eye. Two terms are useful to describe the possible range of distances – the **near point** and the **far point distance**.

- The distance to the **near point** is the distance between the eye and the nearest object that can be brought into clear focus (without strain or help from, for example, lenses). It is also known as the "least distance of distinct vision". By convention it is taken to be 25 cm for normal vision.
- The distance to the **far point** is the distance between the eye and the furthest object that can be brought into focus. This is taken to be infinity for normal vision.

ANGULAR SIZE

If we bring an object closer to us (and our eyes are still able to focus on it) then we see it in more detail. This is because, as the object approaches, it occupies a bigger visual angle. The technical term for this is that the object **subtends** a larger angle.

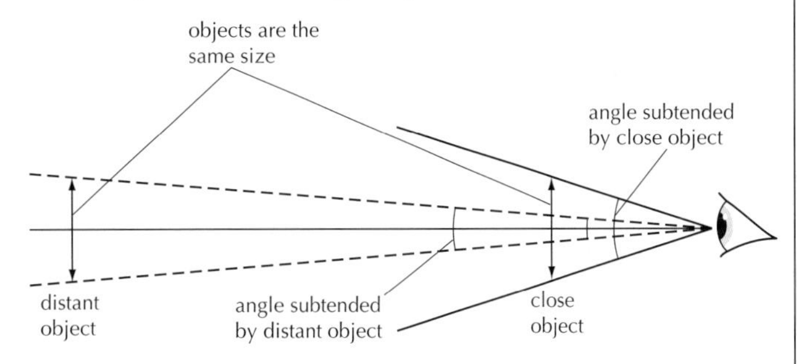

ANGULAR MAGNIFICATION

The angular magnification M of an optical instrument is defined as the ratio between the angle that an object subtends normally and the angle that its image subtends as a result of the optical instrument. The 'normal' situation depends on the context. It should be noted that the angular magnification is not the same as the linear magnification.

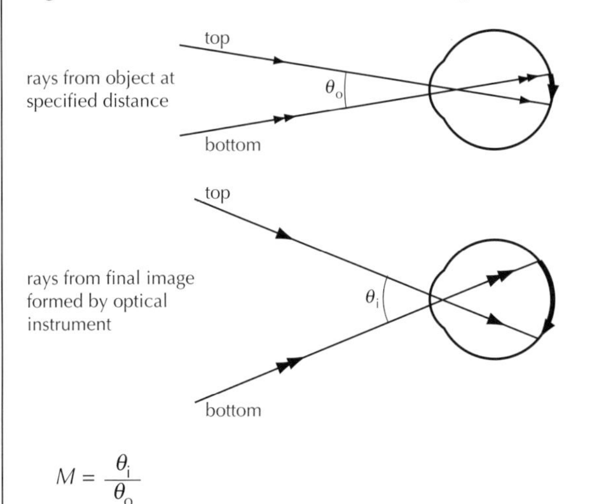

$$M = \frac{\theta_i}{\theta_o}$$

The largest visual angle that an object can occupy is when it is placed at the near point. This is often taken as the 'normal' situation.

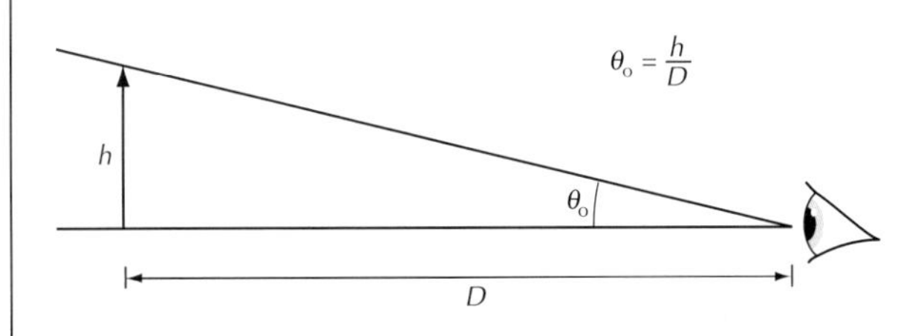

A simple lens can increase the angle subtended. It is usual to consider two possible situations.

1. Image formed at infinity

In this arrangement, the object is placed at the focal point. The resulting image will be formed at infinity and can be seen by the relaxed eye.

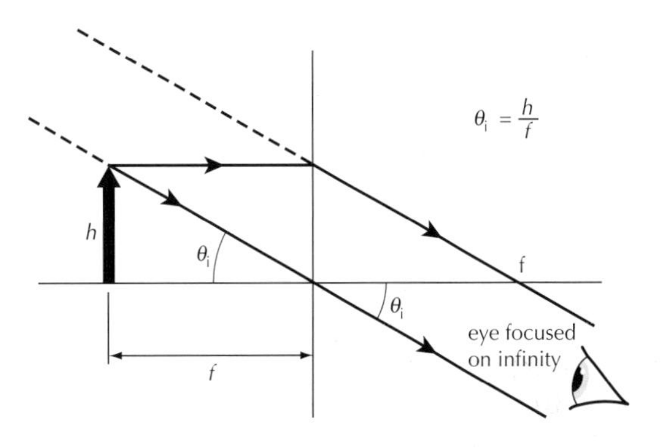

In this case the angular magnification would be

$$M = \frac{\theta_i}{\theta_o} = \frac{h/f}{h/D} = \frac{D}{f}$$

This is the smallest value that the angular magnification can be.

2. Image formed at near point

In this arrangement, the object is placed nearer to the lens. The resulting virtual image is located at the near point. This arrangement has the largest possible angular magnification.

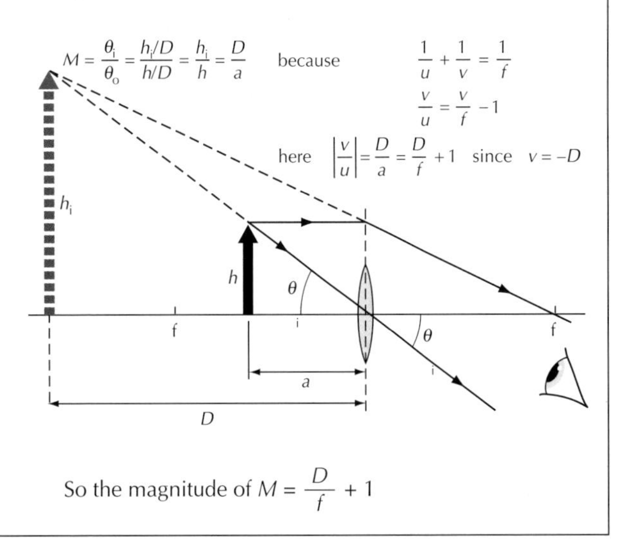

$M = \frac{\theta_i}{\theta_o} = \frac{h_i/D}{h/D} = \frac{h_i}{h} = \frac{D}{a}$
because
$\frac{1}{u} + \frac{1}{v} = \frac{1}{f}$
$\frac{v}{u} = \frac{v}{f} - 1$
here $\left|\frac{v}{u}\right| = \frac{D}{a} = \frac{D}{f} + 1$ since $v = -D$
h_i
h
θ_i
f
θ_i
f
a
D

So the magnitude of $M = \frac{D}{f} + 1$

The compound microscope and astronomical telescope

COMPOUND MICROSCOPE

A compound microscope consists of two lenses – the **objective lens** and the **eyepiece lens**. The first lens (the objective lens) forms a **real magnified** image of the object being viewed. This real image can then be considered as the object for the second lens (the eyepiece lens) which acts as a magnifying lens. The rays from this real image travel into the eyepiece lens and they form a **virtual magnified** image. In normal adjustment, this virtual image is arranged to be located at the near point so that maximum angular magnification is obtained.

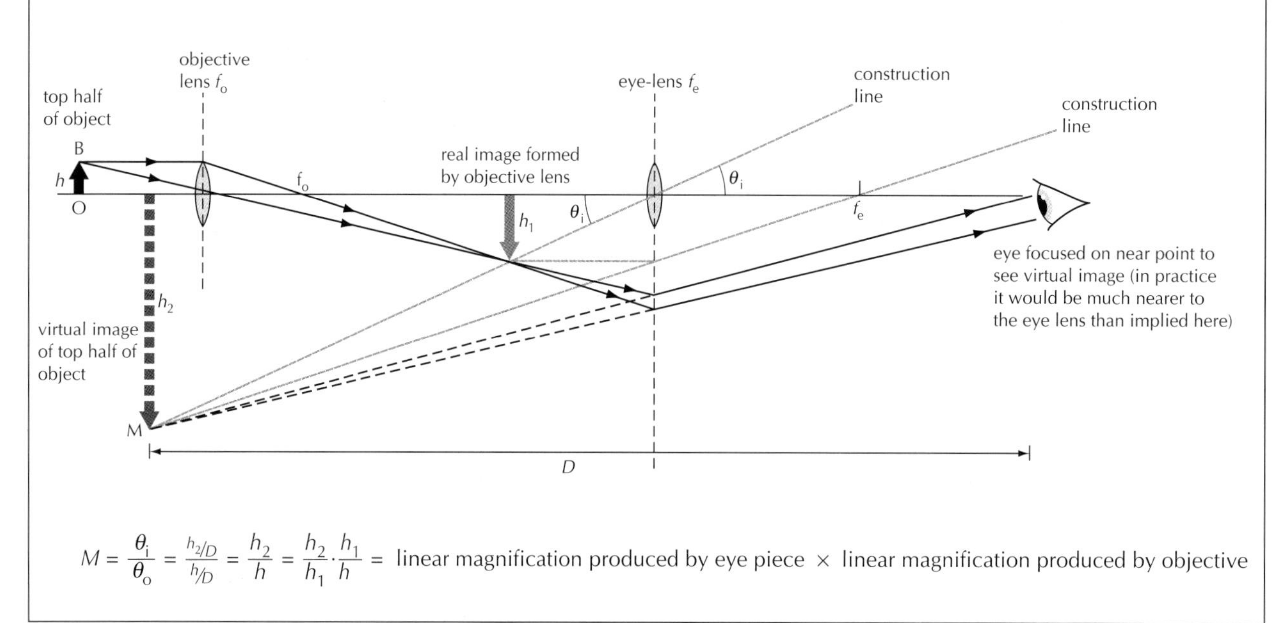

$$M = \frac{\theta_i}{\theta_o} = \frac{h_2/D}{h/D} = \frac{h_2}{h} = \frac{h_2}{h_1} \cdot \frac{h_1}{h} = \text{linear magnification produced by eye piece} \times \text{linear magnification produced by objective}$$

ASTRONOMICAL TELESCOPE

An astronomical telescope also consists of two lenses. In this case, the objective lens forms a **real** but **diminished** image of the distant object being viewed. Once again, this real image can then be considered as the object for the eyepiece lens acting as a magnifying lens. The rays from this real image travel into the eyepiece lens and they form a **virtual magnified** image. In normal adjustment, this virtual image is arranged to be located at infinity.

$$M = \frac{\theta_i}{\theta_o} = \frac{h_1/f_e}{h_1/f_o} = \frac{f_o}{f_e}$$

The length of the telescope $\approx f_o + f_e$.

Aberrations

SPHERICAL

A lens is said to have an **aberration** if, for some reason, a point object does not produce a perfect point image.
In reality, lenses that are spherical do not produce perfect images. **Spherical aberration** is the term used to describe the fact that rays striking the outer regions of a spherical lens will be brought to a slightly different focus point from those striking the inner regions of the same lens.

In general, a point object will focus into a small circle of light, rather than a perfect point. There are several possible ways of reducing this effect:

- the shape of the lens could be altered in such a way as to correct for the effect. The lens would, of course, no longer be spherical. A particular shape only works for objects at a particular distance away.
- the effect can be reduced for a given lens by decreasing the aperture. The technical term for this is **stopping down** the aperture. The disadvantage is that the total amount of light is reduced and the effects of diffraction (see page 95) would be made worse.

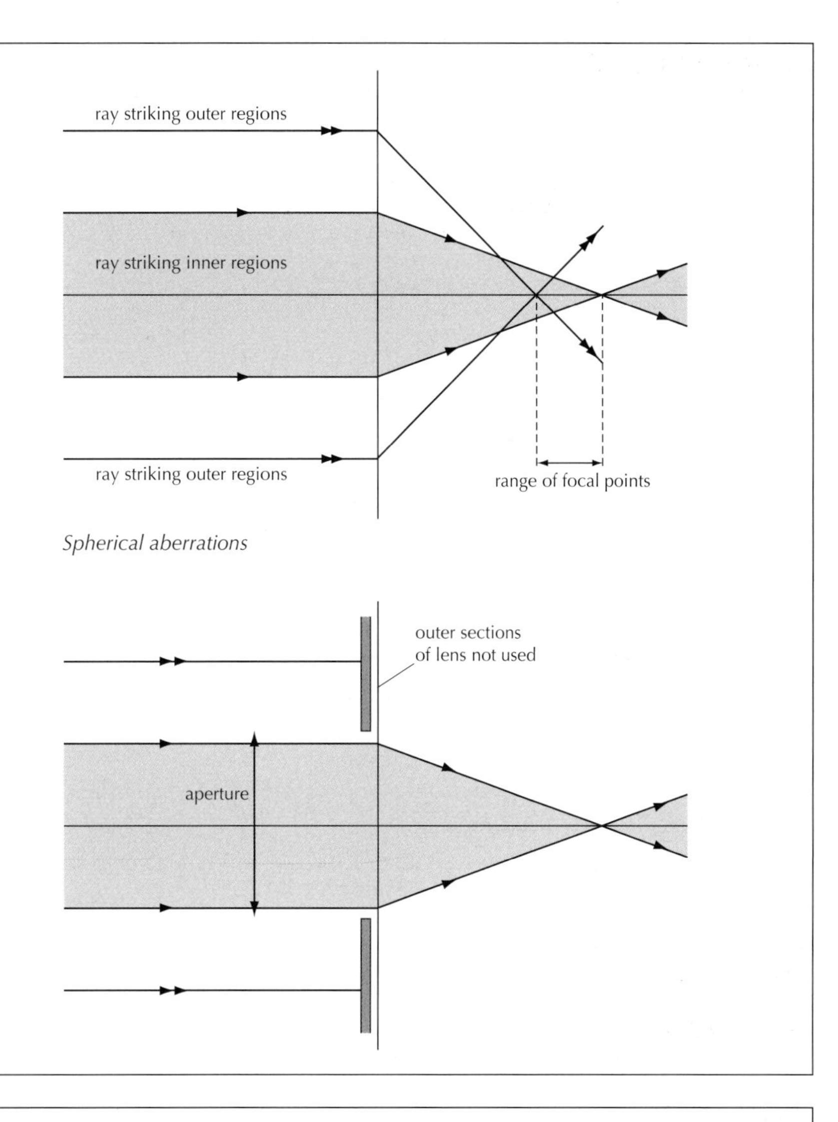

Spherical aberrations

CHROMATIC

Chromatic aberration is the term used to describe the fact that rays of different colours will be brought to a slightly different focus point by the same lens. The refractive index of the material used to make the lens is different for different frequencies of light.

A point object will produce a blurred image of different colors.

The effect can be eliminated for two given colours (and reduced for all) by using two different materials to make up a compound lens. This compound lens is called an **achromatic doublet**.
The two types of glass produce equal but opposite dispersion.

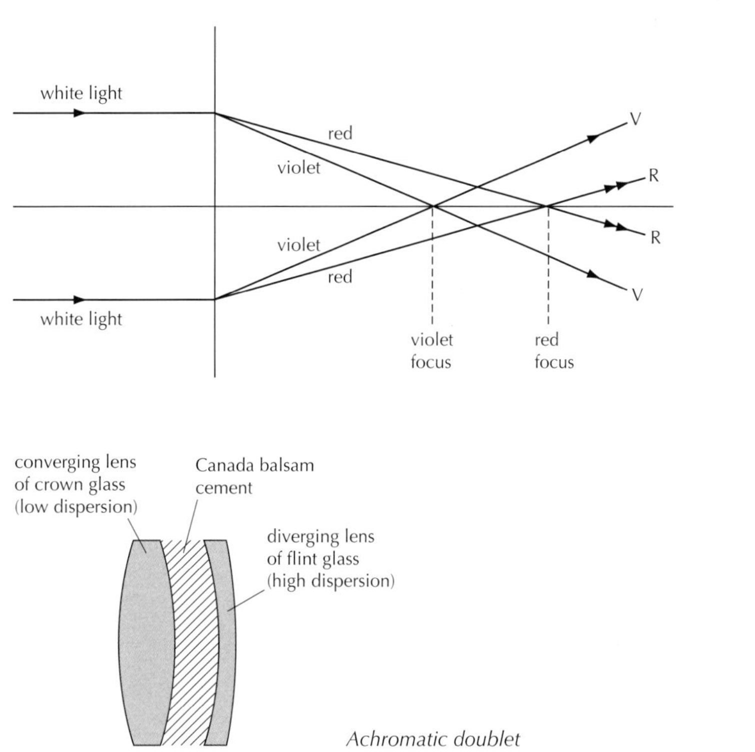

Achromatic doublet

Two-source interference of waves

PRINCIPLES OF THE TWO-SOURCE INTERFERENCE PATTERN

Two-source interference is simply another application of the principle of superposition, for two coherent sources having roughly the same amplitude.

Two sources are coherent if

- they have the same frequency.
- there is a constant phase relationship between the two sources.

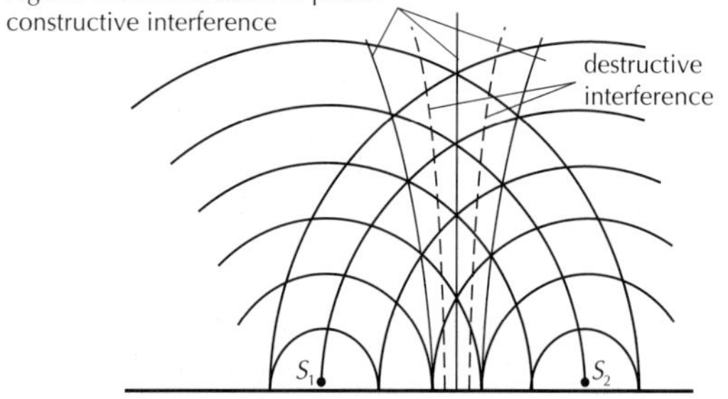

Two dippers in water moving together are coherent sources. This forms regions, of water ripples and other regions with no waves.

Two loudspeakers both connected to the same signal generator are coherent sources. This forms regions of loud and soft sound.

A set-up for viewing two-source interference with light is shown below. It is known as **Young's double slit** experiment. A **monochromatic** source of light is one that gives out only one frequency. Light from the twin slits (the sources) interferes and patterns of light and dark regions, called **fringes**, can be seen on the screen.

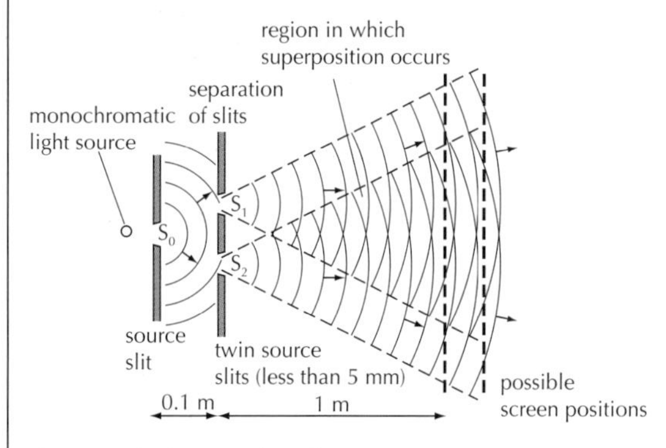

The use of a laser makes the set-up easier.

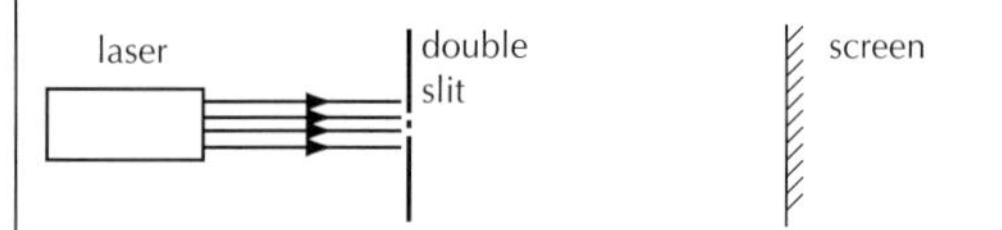

The experiment results in a regular pattern of light and dark strips across the screen as represented below.

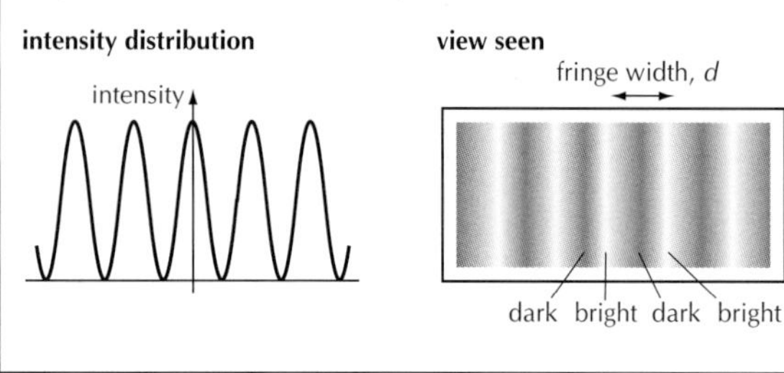

MATHEMATICS

The location of the light and dark fringes can be mathematically derived in one of two ways.

Method 1

The simplest way is to consider two parallel rays setting off from the slits as shown below.

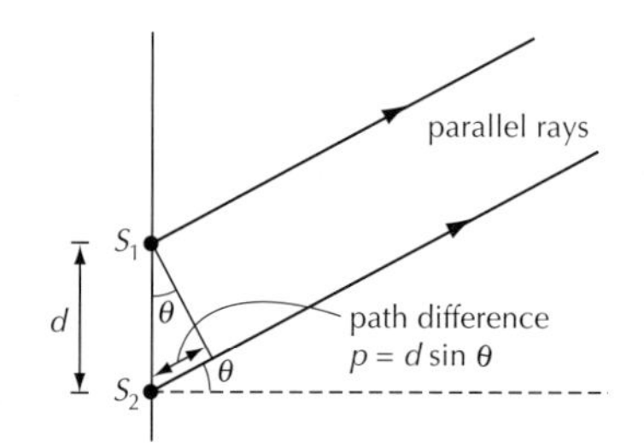

If these two rays result in a bright patch, then the two rays must arrive in phase. The two rays of light started out in phase but the light from source 2 travels an extra distance. This extra distance is called the **path difference**.

Constructive interference can only happen if the path difference is a whole number of wavelengths. Mathematically,

Path difference = $n\lambda$

[where n is an integer – e.g. 1, 2, 3 etc.]

From the geometry of the situation

Path difference = $d \sin\theta$

In other words $n\lambda = d \sin\theta$

Method 2

If a screen is used to make the fringes visible, then the rays from the two slits cannot be absolutely parallel, but the physical set-up means that this is effectively true.

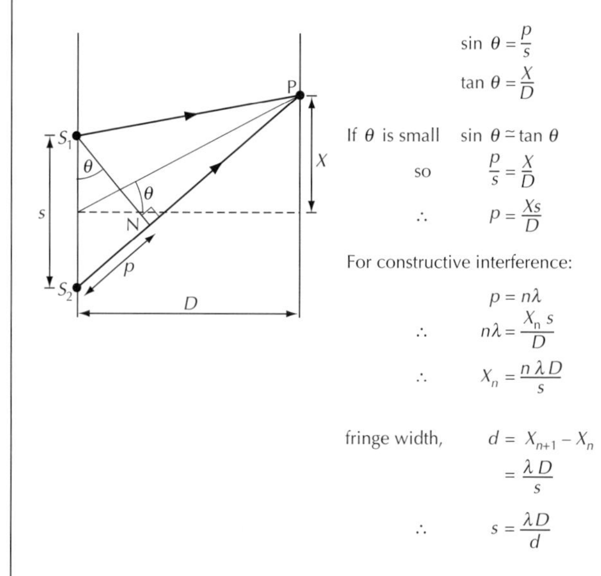

$$\sin\theta = \frac{p}{s}$$

$$\tan\theta = \frac{X}{D}$$

If θ is small $\quad \sin\theta \simeq \tan\theta$

so $\quad \frac{p}{s} = \frac{X}{D}$

$\therefore \quad p = \frac{Xs}{D}$

For constructive interference:

$$p = n\lambda$$

$\therefore \quad n\lambda = \frac{X_n s}{D}$

$\therefore \quad X_n = \frac{n\lambda D}{s}$

fringe width, $\quad d = X_{n+1} - X_n$

$$= \frac{\lambda D}{s}$$

$\therefore \quad s = \frac{\lambda D}{d}$

This equation only applies when the angle is small.

Example

Laser light of wavelength 450 nm is shone on two slits that are 0.1 mm apart. How far apart are the fringes on a screen placed 5.0 m away?

$$d = \frac{\lambda D}{s} = \frac{4.5 \times 10^{-7} \times 5}{1.0 \times 10^{-4}} = 0.0225 \text{ m} = 2.25 \text{ cm}$$

Multiple slit diffraction

THE DIFFRACTION GRATING

The diffraction that takes place at an individual slit affects the overall appearance of the fringes in Young's double slit experiment (see page 164 for more details). This section considers the effect on the final interference pattern of adding further slits. A series of parallel slits (at a regular separation) is called a **diffraction grating**.

Additional slits at the same separation will not effect the condition for constructive interference. In other words, the angle at which the light from slits adds constructively will be unaffected by the number of slits. The situation is shown below.

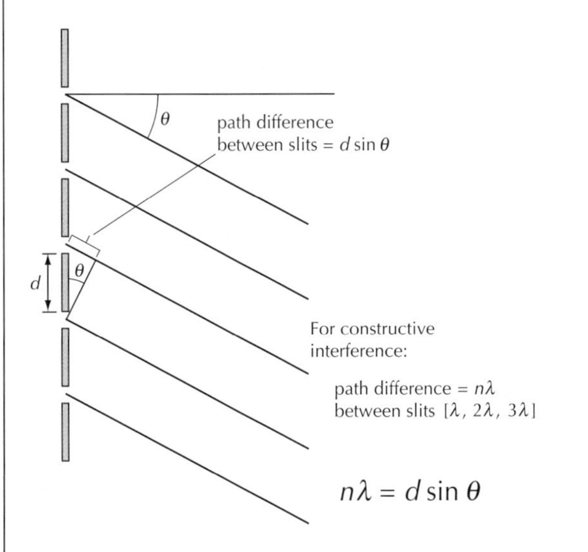

$$n\lambda = d \sin \theta$$

This formula also applies to the Young's double slit arrangement. The difference between the patterns is most noticeable at the angles where perfect constructive interference does not take place. If there are only two slits, the maxima will have a significant angular width. Two sources that are just out of phase interfere to give a resultant that is nearly the same amplitude as two sources that are exactly in phase.

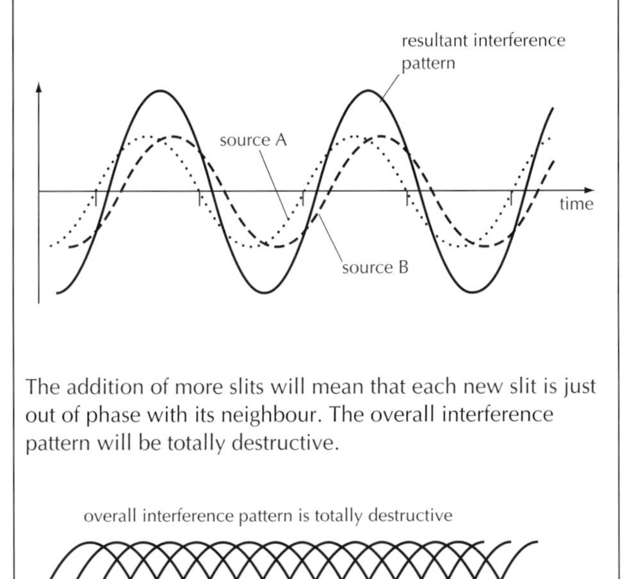

The addition of more slits will mean that each new slit is just out of phase with its neighbour. The overall interference pattern will be totally destructive.

overall interference pattern is totally destructive

time

The addition of further slits at the same slit separation has the following effects:

- the principal maxima maintain the same separation.
- the principal maxima become much sharper.
- the overall amount of light being let through is increased, so the pattern increases in intensity.

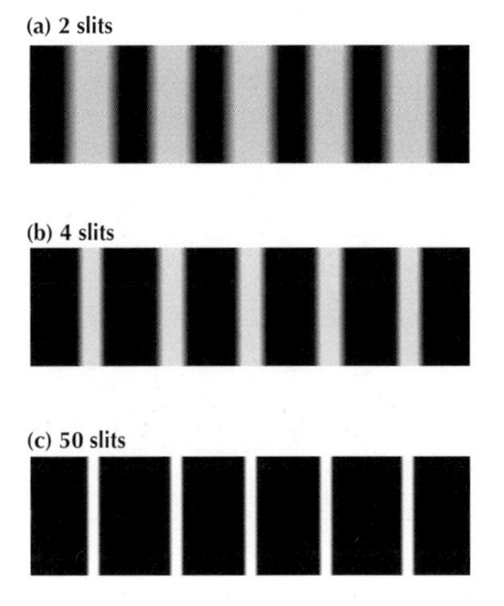

Grating patterns

USES

One of the main uses of a diffraction grating is the accurate experimental measurement of the different wavelengths of light contained in a given spectrum. If white light is incident on a diffraction grating, the angle at which constructive interference takes place depends on wavelength. Different wavelengths can thus be observed at different angles. The accurate measurement of the angle provides the experimenter with an accurate measurement of the exact wavelength (and thus frequency) of the colour of light that is being considered. The apparatus that is used to achieve this accurate measurement is called a **spectrometer**.

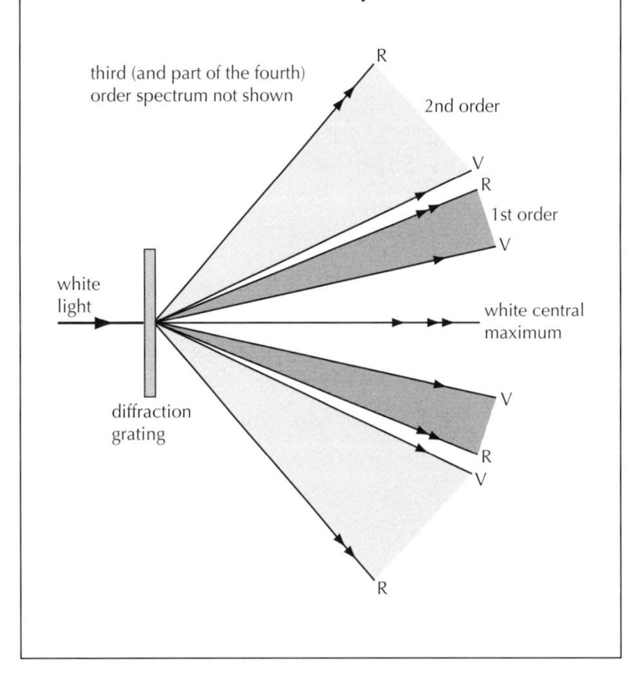

X-rays

PRODUCTION OF X-RAYS

At one time or another in their lives, most people have an X-ray photograph taken of some part of their body. The experimental procedure used for the production of X-rays relies on electrons.

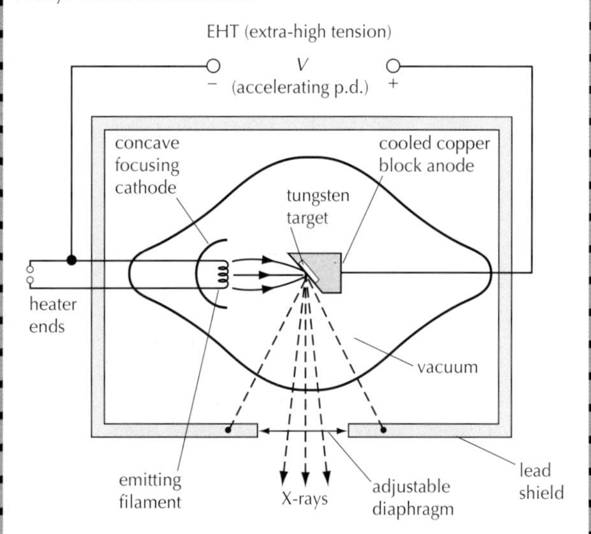

A Coolidge X-ray tube

- Electrons are accelerated by travelling through a potential difference, V. The kinetic energy gained is thus Ve.
- The fast moving electrons collide with the metal target and the collision results in X-rays. There is also a great deal of heat generated in the target. This needs to be kept cool by being kept rotating in oil.
- The adjustable diaphragm is used to define the beam of X-rays.
- Increasing the heater current to the cathode increases its temperature thus the number of electrons emitted and thus the intensity of the X-rays.
- The **hardness** of an X-ray beam measures its penetration power. Higher frequency (lower wavelength) X-rays are harder.
- A filter is often used to absorb the soft X-rays that are emitted by the target

EXAMPLE

Calculate the minimum wavelength of X-rays produced by an electron beam accelerated through a potential difference of 30 keV.

$$\text{Energy} = Ve$$
$$= 30\,000 \times 1.6 \times 10^{-19}$$
$$= 4.8 \times 10^{-15}\text{ J}$$
$$\therefore\ hf = 4.8 \times 10^{-15}$$
$$f = \frac{4.8 \times 10^{-15}}{6.63 \times 10^{-34}}\text{ Hz}$$
$$= 7.24 \times 10^{18}\text{ Hz}$$
$$\lambda_{min} = \frac{c}{f}$$
$$= \frac{3 \times 10^{8}}{7.24 \times 10^{18}}$$
$$= 4.1 \times 10^{-11}\text{ m}$$

General equation: $\lambda_{min} = \dfrac{hc}{eV}$

TYPICAL X-RAY SPECTRUM

The X-rays emitted from the target contain a range of wavelengths – all with different relative amplitudes. A graph of a typical X-ray spectrum is

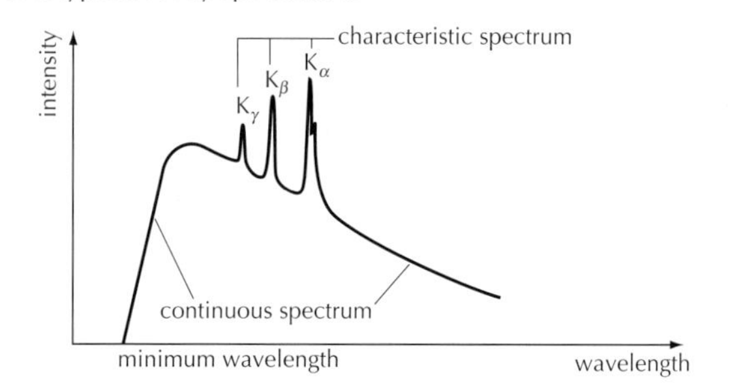

There are two different process taking place in order for this spectrum to be produced

- **Continuous features**
  As the incoming electrons collide with the target atoms, they are decelerated. This deceleration means that X-rays are emitted. The energy of the X-ray photon depends on the energy lost in the collisions. The maximum amount of energy that can be lost is all the initial kinetic energy of the electrons. The maximum energy available means that there is a maximum frequency of X-rays produced. This corresponds to a minimum wavelength limit shown on the graph.
- **Characteristic features**
  In some circumstances, the collisions between the incoming electrons and the target atoms can cause electrons from the inner orbital of the target atom to be promoted up to higher energy levels. When these electrons fall back down they emit X-rays of a particular frequency which is fixed by the energy levels available.

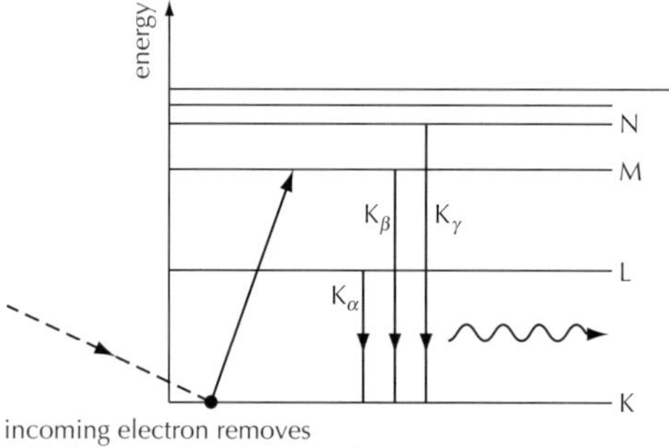

These two processes combine to give the overall spectrum. Changes to this spectrum can thus be predicted.

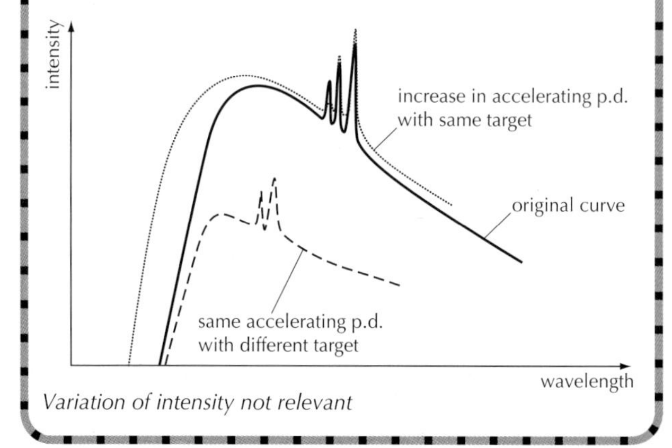

Variation of intensity not relevant

X-ray diffraction

PRINCIPLES OF X-RAY DIFFRACTION

When X-rays are incident on a regular structure (e.g. a crystal) the majority of X-rays will pass through the material. At particular angles from the 'straight through' direction, high intensity X-ray signals are recorded. These angles correspond to points of constructive interference of the X-rays scattered from different planes of atoms in the crystal.

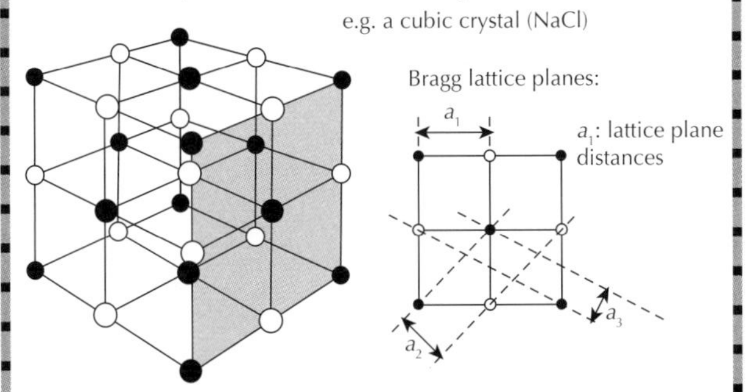

A regular crystal structure contains many different lattice planes that can cause the interference so typically a crystal structure will give rise to many different constructive interference positions. A powdered sample of crystal will contain every orientation of these planes so the resulting X-ray diffraction picture will contain circles rather than points:

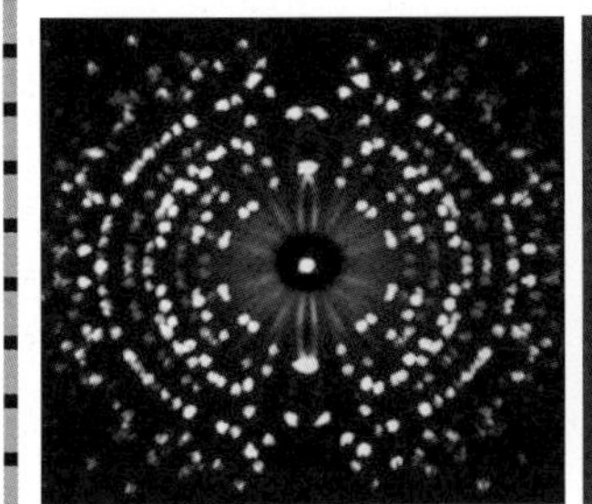

X-ray diffraction with single individual crystal

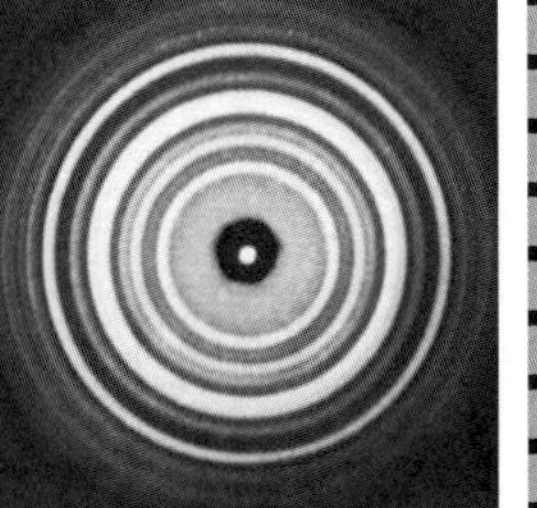

X-ray diffraction with powdered crystal

BRAGG SCATTERING EQUATION

The Bragg scattering equation relates the path length difference between two lattice planes to the wavelength of the X-rays:

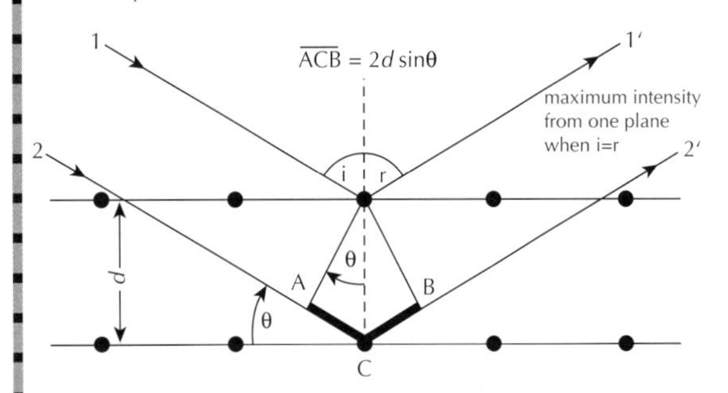

Constructive interference will take place at the angle θ if the path difference between the upper and lower plane ($\overline{ACB}$) is equal to a whole number of wavelengths. The distance between the atomic planes is d.

Path difference $\overline{ACB} = 2\,d \sin\theta$

For constructive interference, path difference $= n\,\lambda$

n is an integer 1, 2, 3 etc

λ is the wavelength of the X-rays used.

$\therefore\ 2\,d \sin\theta = n\,\lambda$ (beam deviated by 2θ)

USE OF X-RAY DIFFRACTION TO DETERMINE STRUCTURE

Using the Bragg equation it is possible to

- measure lattice plane distances by using X-rays of known wavelength
- use cubic crystals of known lattice plane distances to measure wavelength of X-rays

Many complex structures (including DNA) have been determined by detailed analysis of X-ray diffraction patterns.

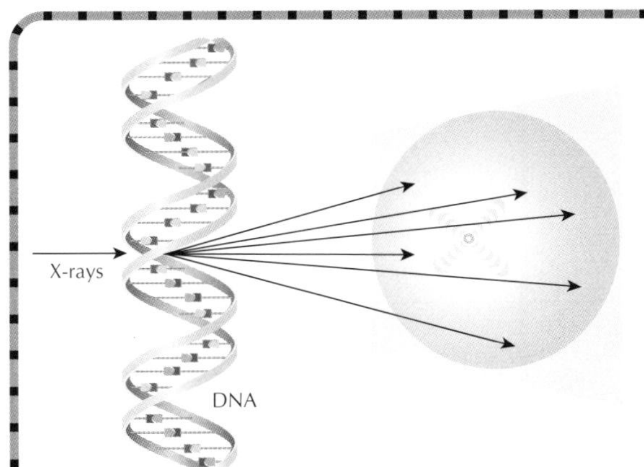

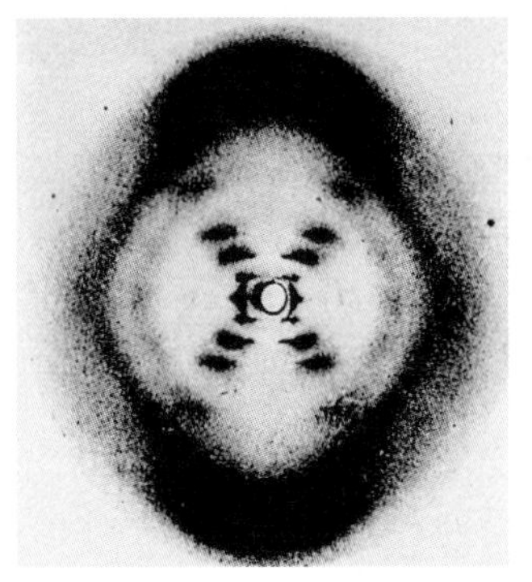

X-rays passing through a helix diffract at angles perpendicular to helix making an "X" pattern

EXAMPLE

A powder X-ray diffraction photograph of copper(I) oxide (Cu_2O) using X-rays of wavelength 1.54×10^{-10} m has the first diffraction maximum at an angle of deviation of 29°. Calculate the associated lattice plane distance.

$$d = \frac{n\lambda}{2\sin\theta} = \frac{1.54 \times 10^{-10}}{2\sin(14.5^\circ)} = 3.08 \times 10^{-10}\ \text{m}$$

The above calculation has assumed that $n = 1$. Higher values of n will correspond to larger angles and the question says this angle was the first diffraction maximum.

Wedge films

THEORY

If two glass plates are at a small angle to one another, the gap between the two is called an air wedge. An example of this set-up would be two microscope slides with a piece of paper between them at one end.

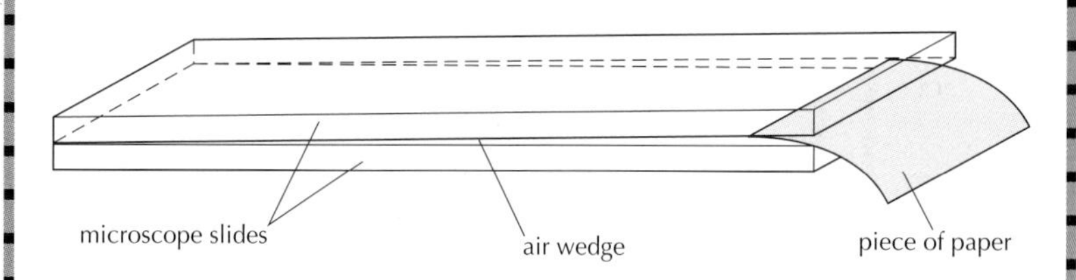

There is a path difference between the rays of light reflecting from the top and from the bottom surfaces of the air wedge. This results in parallel lines of equally spaced constructive and destructive interference fringes.

If the arrangement involves parallel rays they are called **equal inclination** fringes

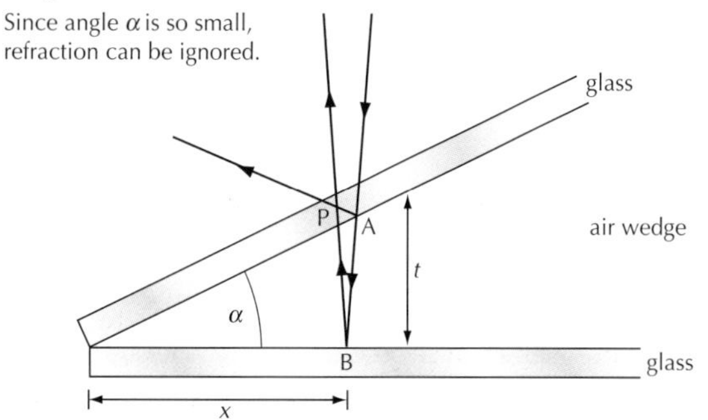

Exaggerated diagram of air wedge

$\therefore$ path difference due to air wedge $\approx 2t$

A phase change happens at B, but not A (see following page).

$$\therefore \text{path difference} = 2t + \frac{\lambda}{2} = m\lambda, \text{ where } m = 0,1,2,\ldots$$

$$\text{or } 2t = (m + \tfrac{1}{2})\lambda$$

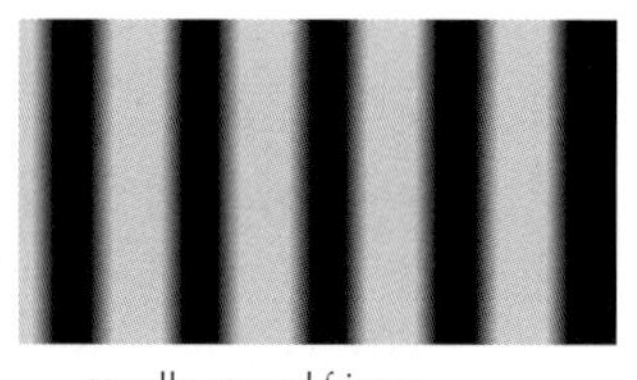

equally spaced fringes

If the wedge has refractive index n

constructive interference occurs when: $2nt = (m + \frac{1}{2})\lambda$

destructive interference occurs when: $2nt = m\lambda$

USE

If the separation of the fringes is measured and the wavelength of light used is known, the angle and thus the thickness of the wedge at any point can be calculated.

This method can be used for measuring very small distances. A possible application includes measuring the thickness of the film of moisture (tear film) on the surface of the eye. Any small change in distance will result in a movement of the fringe pattern. Using this technique, a distance change of the order of about 10^{-7} m can be recorded.

A region that is designed to be completely smooth is sometimes known as an **optical flat**. Any deviation from complete smoothness will affect the fringes formed as a result of reflection from the surface.

Parallel films

PHASE CHANGES

There are many situations when interference can take place that also involve the reflection of light. When analysing in detail the conditions for constructive or destructive interference, one needs to take any **phase changes** into consideration. A phase change is the inversion of the wave that can take place at a reflection interface, but it does not always happen. It depends on the two media involved.

The technical term for the inversion of a wave is that it has 'undergone a phase change of π'.

- When light is reflected back from an optically denser medium there is a phase change of π.
- When light is reflected back from an optically less dense medium there is no phase change.

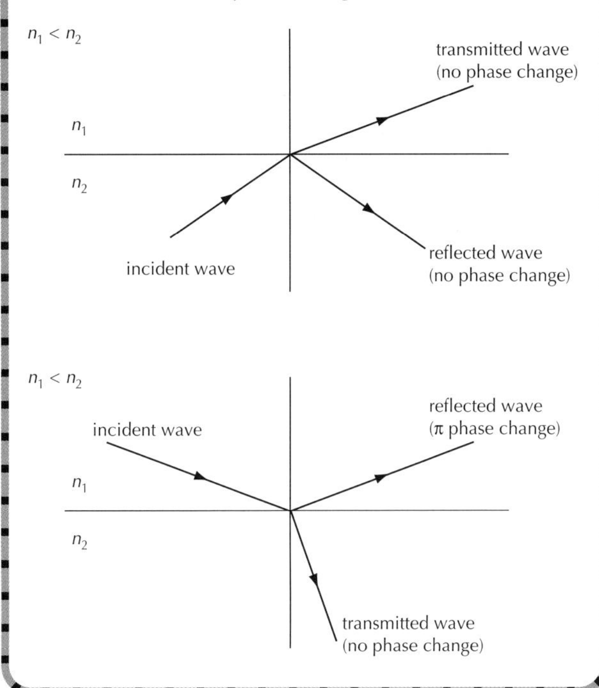

EXAMPLE

The above equations work out the angles for which constructive and destructive interference take place for a given wavelength. If the source of light is an extended source, the eye receives rays leaving the film over a range of values for θ.

If white light is used then the situation becomes more complex. Provided the thickness of the film is small, then one or two colours may reinforce along a direction in which others cancel. The appearance of the film will be bright colours, such as can be seen when looking at

- an oil film on the surface of water or
- soap bubbles.

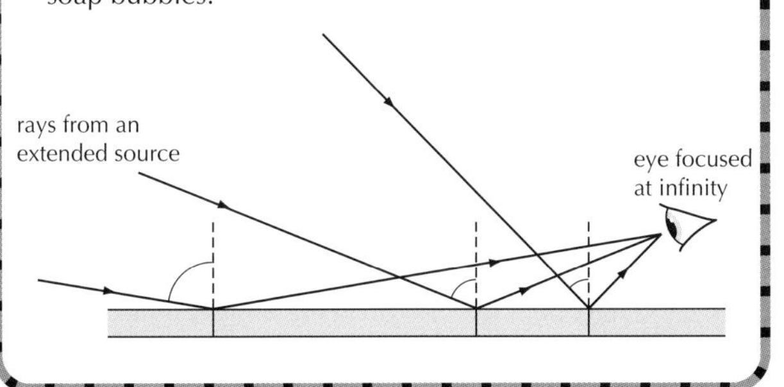

CONDITIONS FOR INTERFERENCE PATTERNS

A parallel-sided film can produce interference as a result of the reflections that are taking place at both surfaces of the film.

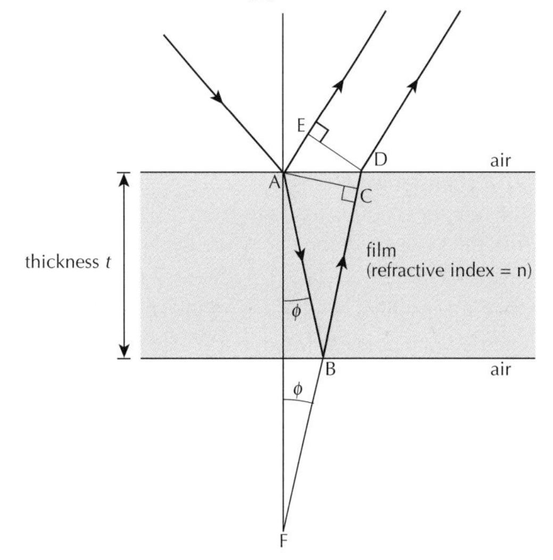

From point A, there are two possible paths:

1. along path AE in air
2. along ABCD in the film

These rays then interfere and we need to calculate the optical path difference.

The path AE in air is equivalent to CD in the the film
So path difference = (AB + BC) in the film.

In addition, the phase change at A is equivalent to $\frac{\lambda}{2}$ path difference.

So total path difference = (AB + BC) in film + $\frac{\lambda}{2}$

$= n(AB + BC) + \frac{\lambda}{2}$

By geometry:

$(AB + BC) = FC$

$= 2t\cos\phi$

$\therefore$ path difference $= 2nt\cos\phi + \frac{\lambda}{2}$

if $2nt\cos\phi = m\lambda$: destructive

if $2nt\cos\phi = \left(m + \frac{1}{2}\right)\lambda$: constructive

$m = 0,1,2,3,4$

APPLICATIONS

Applications of parallel thin films include:

- The design of non-reflecting radar coatings for military aircraft. If the thickness of the extra coating is designed so that radar signals destructively interfere when they reflect from both surfaces, then no signal will be reflected and an aircraft could go undetected.
- Measurements of thickness of oil slicks caused by spillage. Measurements of the wavelengths of electromagnetic signals that give constructive and destructive interference (at known angles) allow the thickness of the oil to be calculated.
- Design of non-reflecting surfaces for lenses (blooming), solar panels and solar cells. A strong reflection at any of these surfaces would reduce the amount of energy being usefully transmitted. A thin surface film can be added so that destructive interference takes place for a typical wavelength and thus maximum transmittance takes place at this wavelength.

IB QUESTIONS – OPTION G – ELECTROMAGNETIC WAVES

1 A student is given two converging lenses, A and B, and a tube in order to make a telescope.

(a) Describe a simple method by which she can determine the focal length of each lens. [2]

(b) She finds the focal lengths to be as follows:

Focal length of lens A 10 cm
Focal length of lens B 50 cm

Draw a diagram to show how the lenses should be arranged in the tube in order to make a telescope. Your diagram should include:

(i) labels for each lens;
(ii) the focal points for each lens;
(iii) the position of the eye when the telescope is in use. [4]

(c) On your diagram, mark the location of the intermediate image formed in the tube. [1]

(d) Is the image seen through the telescope upright or upside-down? [1]

(e) Approximately how long must the telescope tube be? [1]

2 An elderly lady buys a 'magnifying glass' to read small print in the telephone directory. To her surprise she finds that if she holds the convex lens fairly close to the page she gets one kind of image, while if she holds it fairly far from the page she gets quite another kind of image.

(a) *Lens close to the page*. For the lens quite close to the page she draws the ray diagram below.

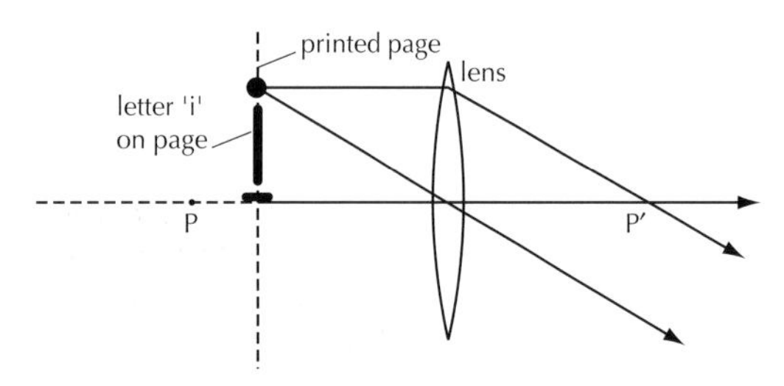

(i) Where should her eye be located in order to see the image of the letter 'i'? Tick the correct answer below. [1]

To the left of the lens ☐
Anywhere to the right of the lens ☐
To the right of the focal point P' ☐

(ii) If she looks at letters on a page in this way, how will they appear to her?

Right way up or upside down?
Enlarged or diminished?
Behind the lens or in front of the lens?

Should she be able to read the telephone directory using the lens this way? [2]

(b) *Lens further from the page.* The lady now moves the page further from the lens. The diagram below represents the situation where the page is more than twice the focal distance from the lens.

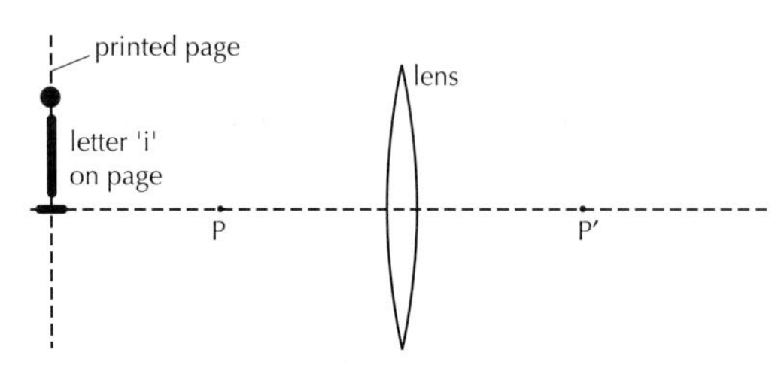

(i) Locate the image of the 'i' by tracing suitable rays on the diagram. [4]

(ii) Where should the lady's eye be located in order to see the image? Tick the correct answer below. [1]

To the left of the lens ☐
Anywhere to the right of the lens ☐
Between the lens and P' ☐
To the right of the image ☐

(iii) If she looks at letters on a page in this way, how will they appear to her?

Right way up or upside-down?
Enlarged or diminished?
Behind the lens or in front of the lens?
Nearer or further away than the page?

Will she be able to read the telephone directory using the lens this way? [3]

HL

3 This question is about the formation of coloured fringes when white light is reflected from thin films.

(a) Name the wave phenomenon that is responsible for the formation of regions of different colour when white light is reflected from a thin film of oil floating on water. [1]

(b) A film of oil of refractive index 1.45 floats on a layer of water of refractive index 1.33 and is illuminated by white light at normal incidence.

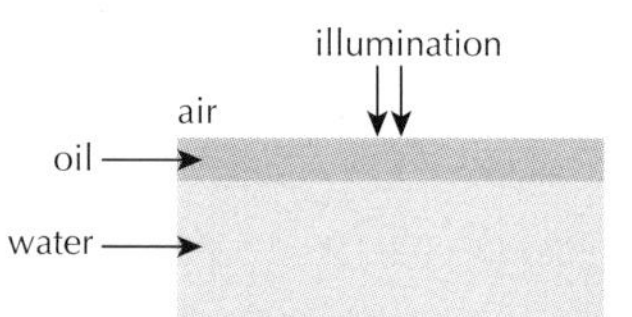

When viewed at near normal incidence a particular region of the film looks red, with an average wavelength of about 650 nm. An equation relating this dominant average wavelength λ, to the minimum film thickness of the region t, is $\lambda = 4nt$.

(i) State what property n measures and explain why it enters into the equation. [2]
(ii) Calculate the minimum film thickness. [1]
(iii) Describe the change to the conditions for reflection that would result if the oil film was spread over a flat sheet of glass of refractive index 1.76, rather than floating on water. [2]

4 The graph below shows a typical X-ray spectrum produced when electrons are accelerated through a potential difference and are then stopped in a metal target.

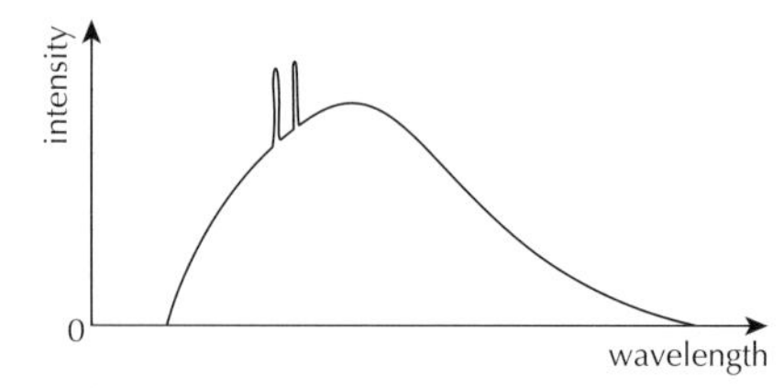

Which feature of the graph enables this potential difference to be determined?

A The maximum wavelength of the X-rays produced
B The minimum wavelength of the X-rays produced
C The wavelength of the peaks on the graph
D The maximum intensity of the X-rays produced

Frames of reference

OBSERVERS AND FRAMES OF REFERENCE

The proper treatment of large velocities involves an understanding of Einstein's theory of relativity and this means thinking about space and time in a completely different way. The reasons for this change are developed in the following pages, but they are surprisingly simple. They logically follow from two straightforward assumptions. In order to see why this is the case we need to consider what we mean by an object in motion in the first place.

A person sitting in a chair will probably think that they are at rest. Indeed from their point of view this must be true, but this is not the only way of viewing the situation. The Earth is in orbit around the Sun, so from the Sun's point of view the person sitting in the chair must be in motion. This example shows that an object's motion (or lack of it) depends on the observer.

The calculation of relative velocity was considered on page 13. This treatment, like all the mechanics in this book so far, assumes that the velocities are small enough to be able to apply Newton's laws to different frames of reference.

Is this person at rest...

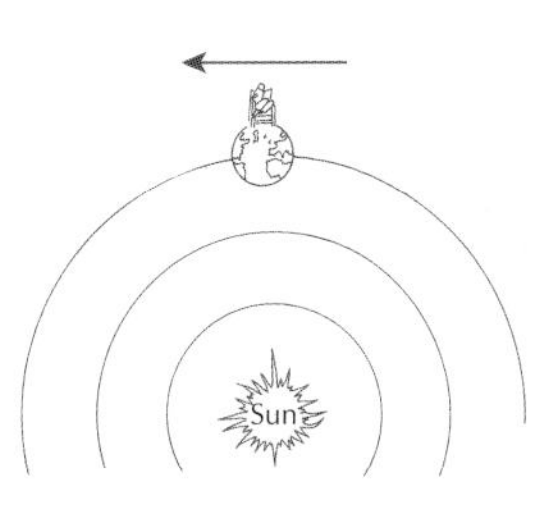

...or moving at great velocity?

GALILEAN TRANSFORMATIONS

It is possible to formalise the relationship between two different frames of reference. The idea is to use the measurement in one frame of reference to work out the measurements that would be recorded in another frame of reference. The equations that do this without taking the theory of relativity into consideration are called **Galilean transformations**.

The simplest situation to consider is two frames of reference (S and S′) with one frame (S′) moving past the other one (S) as shown below.

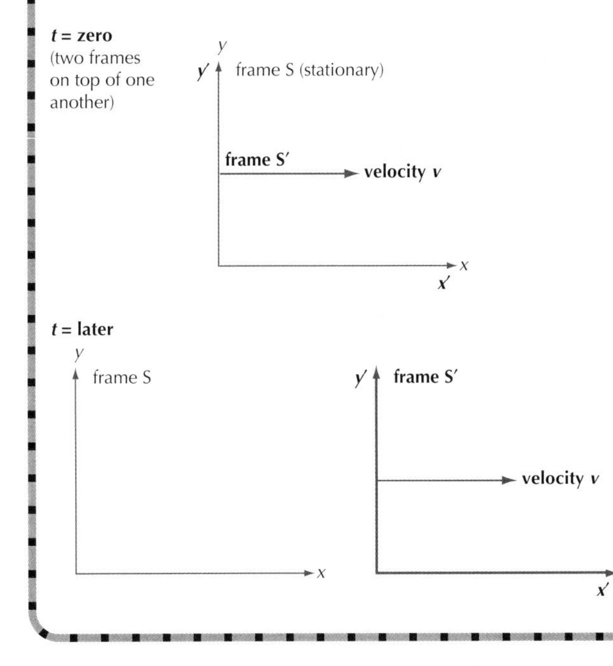

Each frame of reference can record the position and time of an event. Since the relative motion is along the x-axis, most measurements will be the same:

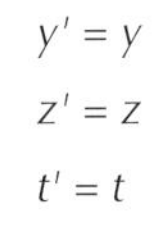

$$y' = y$$

$$z' = z$$

$$t' = t$$

If an event is stationary according to one frame, it will be moving according to the other frame – the frames will record different values for the x measurement. The transformation between the two is given by

$$x' = x - v\,t$$

We can use these equations to formalise the calculation of velocities. The frames will agree on any velocity measured in the y or z direction, but they will disagree on a velocity in the x-direction. Mathematically,

$$u' = u - v$$

For example, if the moving frame is going at 4 m s^{-1}, then an object moving in the same direction at a velocity of 15 m s^{-1} as recorded in the stationary frame will be measured as travelling at 11 m s^{-1} in the moving frame.

Constancy of the speed of light

FAILURE OF GALILEAN TRANSFORMATION EQUATIONS

If the speed of light has the same value for all observers then the Galilean transformation equations cannot work for light.

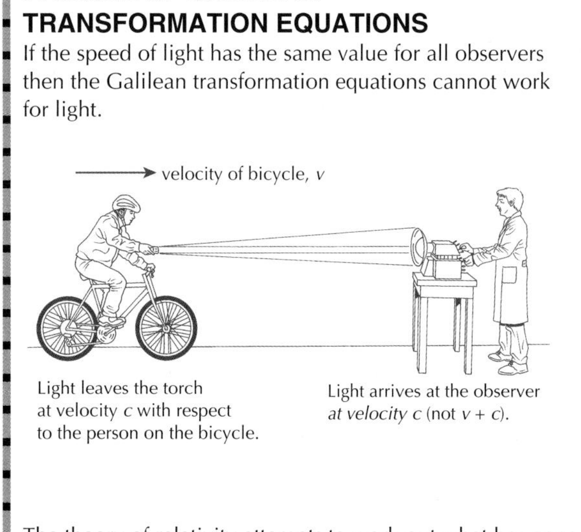

Light leaves the torch at velocity c with respect to the person on the bicycle.

Light arrives at the observer at velocity c (not $v + c$).

The theory of relativity attempts to work out what has gone wrong.

PION DECAY EXPERIMENTS

In 1964 an experiment at the European Centre for Nuclear Reseach (CERN) measured the speed of gamma-ray photons that had been produced by particles moving close to the speed of light and found these photons also to be moving at the speed of light. This is consistent with the speed of light being independent of the speed of its source, to a high degree of accuracy.

The experiment analysed the decay of a particle called the neutral pion into two gamma-ray photons. Energy considerations meant that the pions were known to be moving faster than 99.9% of the speed of light and the speed of the photons was measured to be $2.9977 \pm 0.004 \times 10^8$ m s^{-1}.

Gravitational and inertial mass

MASS

General relativity attempts to include gravitational effects in the ideas developed by the special theory of relativity.

The mass of an object is a single quantity but it can be thought of as being two very different properties. In order to distinguish between the two they are sometimes given different names – **gravitational mass** and **inertial mass**.

1. Inertial mass $m_{(i)}$

Inertial mass is the property of an object that determines how it responds to a given force – whatever the nature of the force.

A push of 700 N on a car of mass 1400 kg gives it an acceleration of 0.5 m s^{-2}.

The same push on a toy car would give it a much greater acceleration!

Inertial mass – different masses have different accelerations when a force acts on them

Newton's second law has already given us a rule for this ($F = ma$). This can be used to describe inertial mass. 'Inertial mass is the ratio of resultant force to acceleration'.

$$\text{Inertial mass } m_{(i)} = \frac{F}{a}$$

The units of inertial mass would be N m^{-1} s^{2}.

2. Gravitational mass $m_{(g)}$

Gravitational mass is the property of an object that determines how much gravitational force it feels when near another object.

$$\text{Gravitational force} \propto m_{(g)}$$

It also determines the gravitational force that the other object feels.

The pull of gravity (from the Earth) on an elephant is large.

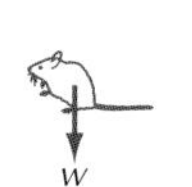

The pull of gravity (from the Earth) on a mouse is much less.

Gravitational mass – different masses have different gravitational forces acting between them

As well as developing his laws of motion, Newton also developed a law of gravitational attraction. See page 51 for details.

These two concepts (inertial mass and gravitational mass) are very different. The surprise is that they turn out to be the equivalent. In other words, an object's gravitational mass is equal to its inertial mass. The fact that different objects have the same value for free-fall acceleration shows this.

A mouse and an elephant would fall together (if air friction were negligible)

Concepts and postulates of special relativity

POSTULATES OF THE SPECIAL THEORY OF RELATIVITY

The special theory of relativity is based on two fundamental assumptions or **postulates**. If either of these postulates could be shown to be wrong, then the theory of relativity would be wrong. When discussing relativity we need to be even more than usually precise with our use of technical terms.

One important technical phrase is an **inertial frame of reference**. This means a frame of reference in which the laws of inertia (Newton's laws) apply. Newton's laws do not apply in accelerating frames of reference so an inertial frame is a frame that is either stationary or moving with constant velocity.

An important idea to grasp is that there is no fundamental difference between being stationary and moving at constant velocity. Newton's laws link forces and accelerations. If there is no resultant force on an object then its acceleration will be zero. This could mean that the object is at **rest** or it could mean that the object is **moving at constant velocity**.

The two postulates of special relativity are:

- the speed of light in a vacuum is the same constant for all inertial observers.
- the laws of physics are the same for all inertial observers.

The first postulate leads on from Maxwell's equations and can be experimentally verified. The second postulate seems completely reasonable – particularly since Newton's laws do not differentiate between being at rest and moving at constant velocity. If both are accepted as being true then we need to start thinking about space and time in a completely different way. If in doubt, we need to return to these two postulates.

SIMULTANEITY

One example of how the postulates of relativity disrupt our everyday understanding of the world around us is the concept of simultaneity. If two events happen together we say that they are **simultaneous**. We would normally expect that if two events are simultaneous to one observer, they should be simultaneous to all observers – but this is not the case! A simple way to demonstrate this is to consider an experimenter in a train.

The experimenter is positioned **exactly** in the middle of a carriage that is moving at constant velocity. She sends out two pulses of light towards the ends of the train. Mounted at the ends are mirrors that reflect the pulses back towards the observer. As far as the experimenter is concerned, the whole carriage is at rest. Since she is in the middle, the experimenter will know that:

- the pulses were sent out simultaneously.
- the pulses hit the mirrors simultaneously.
- the pulses returned simultaneously

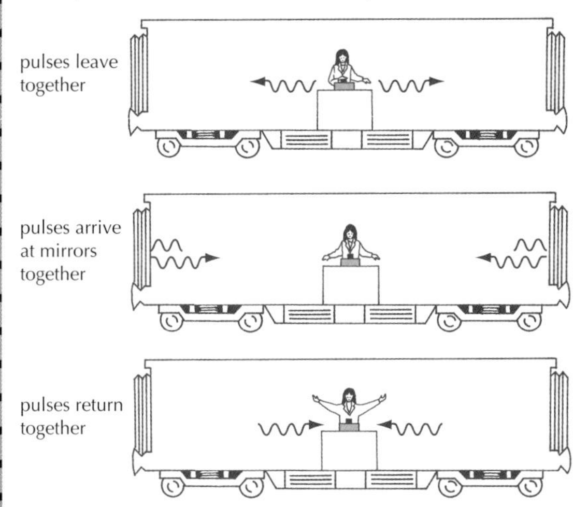
pulses leave together
pulses arrive at mirrors together
pulses return together

The situation will seem very different if watched by a stationary observer (on the platform). This observer knows that light must travel at constant speed – both beams are travelling at the same speed as far as he is concerned, so they must hit the mirrors at different times. The left-hand end of the carriage is moving towards the beam and the right hand end is moving away. This means that the reflection will happen on the left-hand end first.

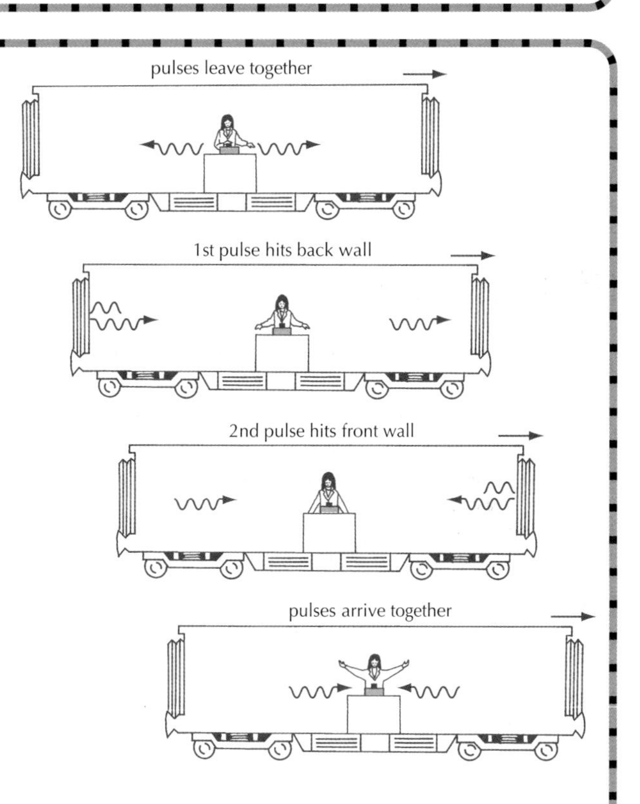
pulses leave together
1st pulse hits back wall
2nd pulse hits front wall
pulses arrive together

Interestingly, the observer on the platform does see the beams arriving back at the same time. The observer on the platform will know that:

- the pulses were sent out simultaneously.
- the left-hand pulse hit the mirror before the right hand pulse.
- the pulses returned simultaneously.

In general, simultaneous events that take place at the same point in space will be simultaneous to all observers whereas events that take place at different points in space can be simultaneous to one observer but not simultaneous to another!

Do not dismiss these ideas because the experiment seems too fanciful to be tried out. The use of a pulse of light allowed us to rely on the first postulate. This conclusion is valid whatever event is considered.

Time dilation

LIGHT CLOCK

A **light clock** is an imaginary device. A beam of light bounces between two mirrors – the time taken by the light between bounces is one 'tick' of the light clock.

As shown in the derivation the path taken by light in a light clock that is moving at constant velocity is longer. We know that the speed of light is fixed so the time between the 'ticks' on a moving clock must also be longer. This effect – that moving clocks run slow – is called **time dilation**.

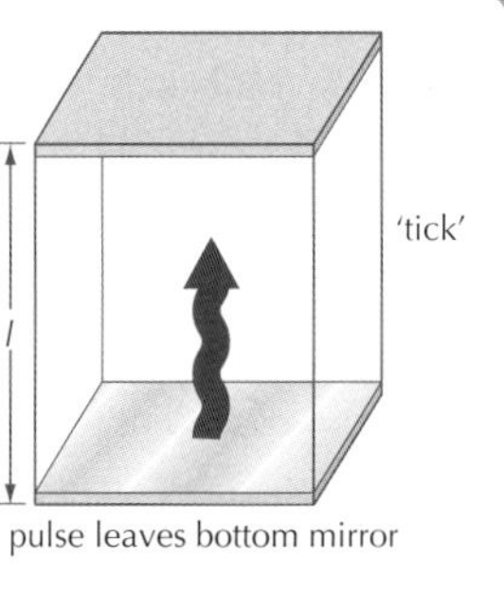

pulse leaves bottom mirror

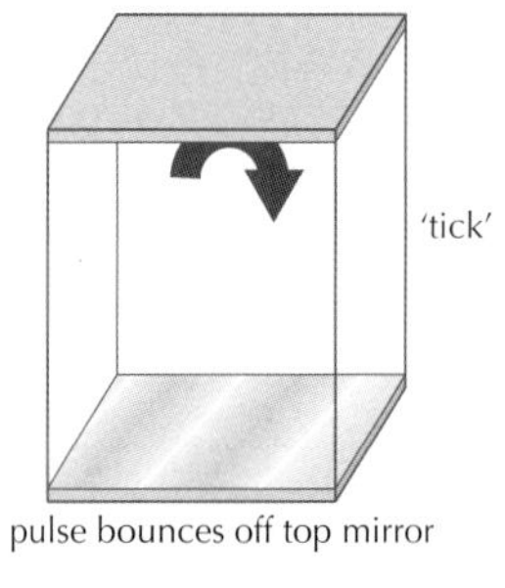

pulse bounces off top mirror

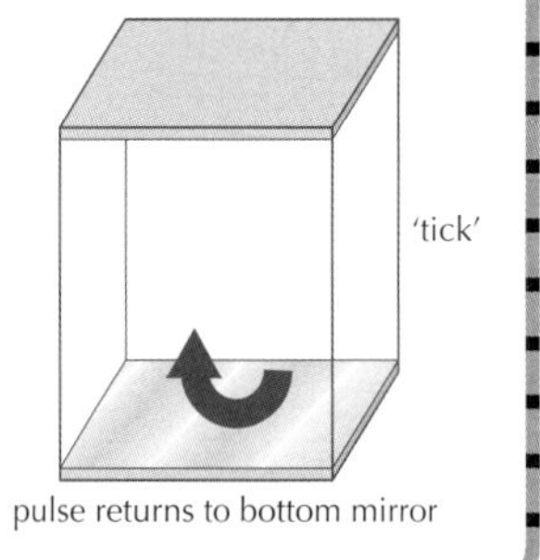

pulse returns to bottom mirror

PROPER TIME

When expressing the time taken between events (for example the length of time that a firework is giving out light), the **proper time** is the time as measured in a frame where the events take place at the same point in space. It turns out to be the shortest possible time that any observer could correctly record for the event.

measuring how long a firework lasts

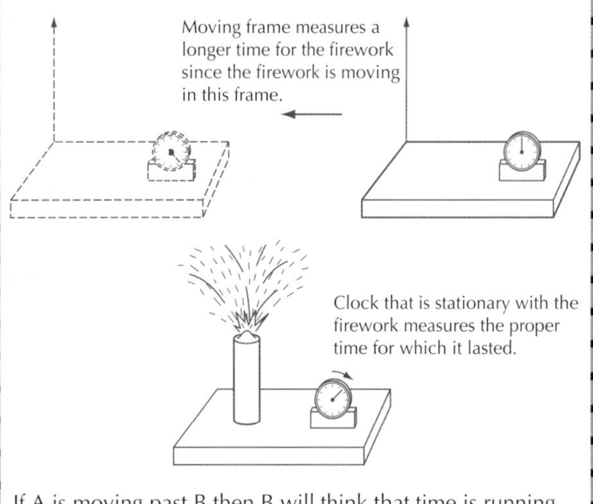

Moving frame measures a longer time for the firework since the firework is moving in this frame.
Clock that is stationary with the firework measures the proper time for which it lasted.

If A is moving past B then B will think that time is running slowly for A. From A's point of view, B is moving past A. This means that A will think that time is running slowly for B. Both views are correct!

DERIVATION OF THE EFFECT

If we imagine a stationary observer with one light clock then t is the time between 'ticks' on their stationary clock. In **this stationary frame**, a moving clock runs **slowly** and t' is the time between 'ticks' on the moving clock: t' is greater than t.

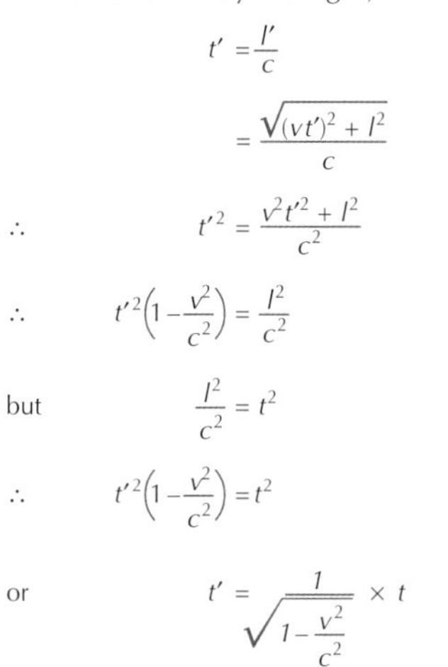

In the time t',

the clock has moved on a distance $= v\,t'$

Distance travelled by the light, $l' = \sqrt{((vt')^2 + l^2)}$

$$t' = \frac{l'}{c}$$

$$= \frac{\sqrt{(vt')^2 + l^2}}{c}$$

$$\therefore \quad t'^2 = \frac{v^2t'^2 + l^2}{c^2}$$

$$\therefore \quad t'^2\left(1 - \frac{v^2}{c^2}\right) = \frac{l^2}{c^2}$$

$$\text{but} \quad \frac{l^2}{c^2} = t^2$$

$$\therefore \quad t'^2\left(1 - \frac{v^2}{c^2}\right) = t^2$$

$$\text{or} \quad t' = \frac{1}{\sqrt{1 - \frac{v^2}{c^2}}} \times t$$

This equation is true for all measurements of time, whether they have been made using a light clock or not.

LORENTZ FACTOR

The time dilation formula expresses the time according to a stationary observer, Δt, in terms of the time measured on a moving clock, Δt_0

$$\Delta t = \frac{1}{\sqrt{1 - \frac{v^2}{c^2}}} \times \Delta t_0$$

We call the **Lorentz** factor, $\gamma = \frac{1}{\sqrt{1 - \frac{v^2}{c^2}}}$

so that $\Delta t = \gamma \Delta t_0$

At low velocities, the Lorentz factor is approximately equal to one – relativistic effects are negligible. It approaches infinity near the speed of light.

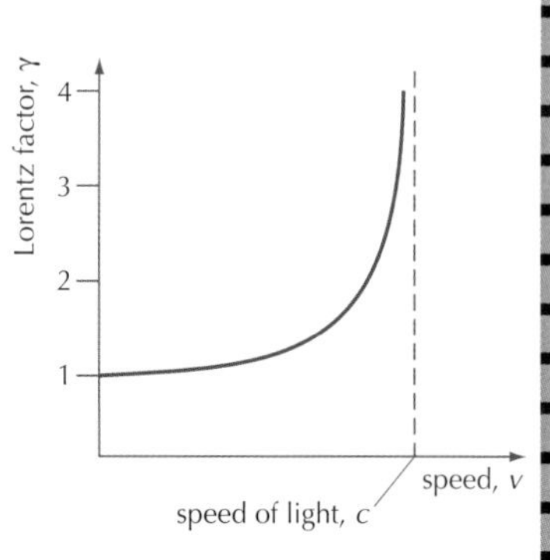

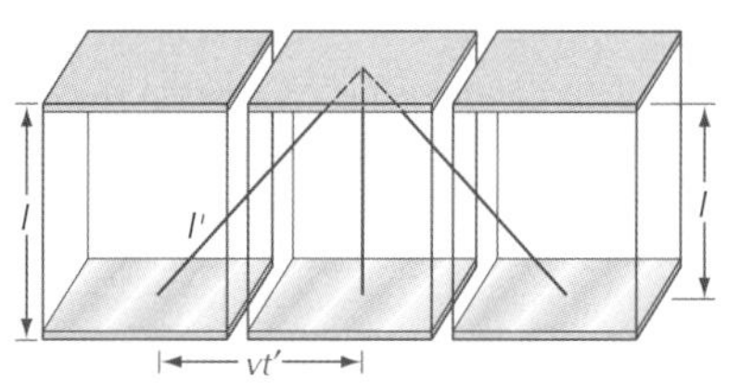

Length contraction

EFFECT

Time is not the only measurement that is affected by relative motion. There is another relativistic effect called **length contraction**. According to a (stationary) observer, the separation between two points in space contracts if there is relative motion in that direction. The contraction is in the same direction as the relative motion.

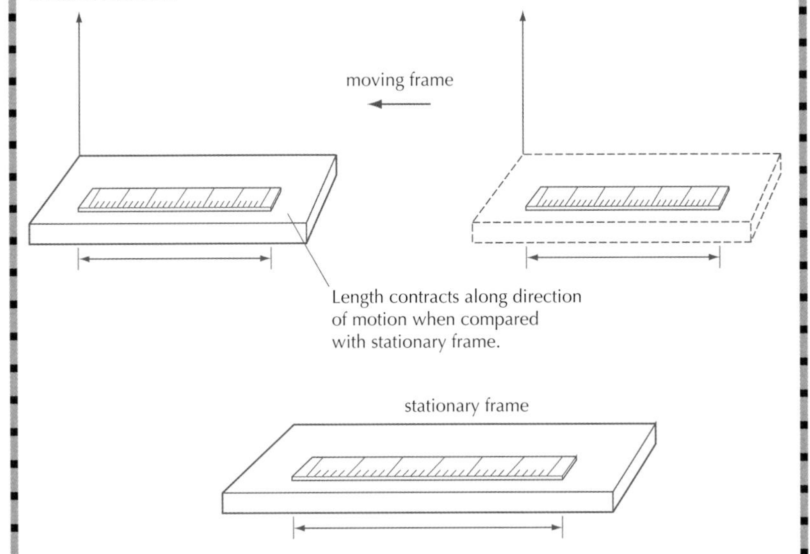
moving frame
Length contracts along direction of motion when compared with stationary frame.
stationary frame

Length contracts by the same proportion as time dilates – the Lorentz factor is once again used in the equation, but this time there is a division rather than a multiplication.

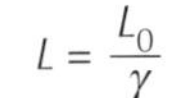

$$L = \frac{L_0}{\gamma}$$

PROPER LENGTH

As before different observers will come up with different measurements for the length of the same object depending on their relative motions. The **proper length** of an object is the length recorded in a frame where the object is at rest.

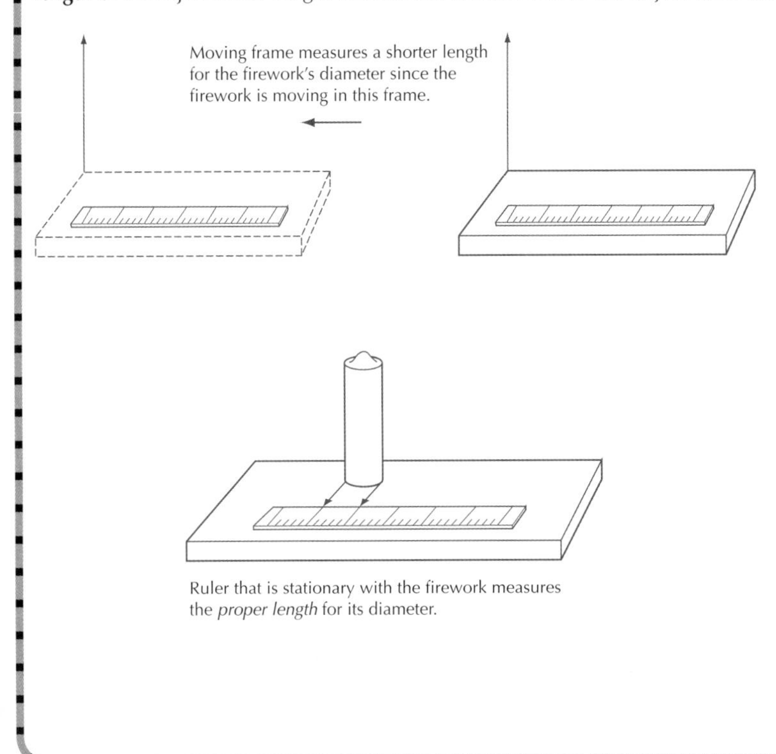
Moving frame measures a shorter length for the firework's diameter since the firework is moving in this frame.

Ruler that is stationary with the firework measures the *proper length* for its diameter.

EXAMPLE

An unstable particle has a life time of 4.0×10^{-8} s in its own rest frame. If it is moving at 98% of the speed of light calculate:

a) Its life time in the laboratory frame.

b) The length travelled in both frames.

a) $\gamma = \sqrt{\frac{1}{1 - (0.98)^2}}$

$= 5.025$

$\Delta t = \gamma \Delta t_0$

$= 5.025 \times 4.0 \times 10^{-18}$

$= 2.01 \times 10^{-7}$s

b) In the laboratory frame, the particle moves

Length = speed × time

$= 0.98 \times 3 \times 10^8 \times 2.01 \times 10^{-7}$

$= 59.1$ m

In the particle's frame, the laboratory moves

$\Delta l = \frac{59.1}{\gamma}$

$= 11.8$ m

(alternatively: length = speed × time

$= 0.98 \times 3 \times 10^8 \times 4.0 \times 10^{-8}$

$= 11.8$ m)

The twin paradox

As mentioned on page 174, the theory of relativity gives no preference to different inertial observers – the time dilation effect (moving clocks run slowly) is always the same. This leads to the '**twin paradox**'. In this imaginary situation, two identical twins compare their views of time. One twin remains on Earth while the other twin undergoes a very fast trip out to a distant star and back again.

As far as the twin on the Earth is concerned the other twin is a moving observer. This means that the twin that remains on the Earth will think that time has been running slowly for the other twin. When they meet up again, the returning twin should have aged less.

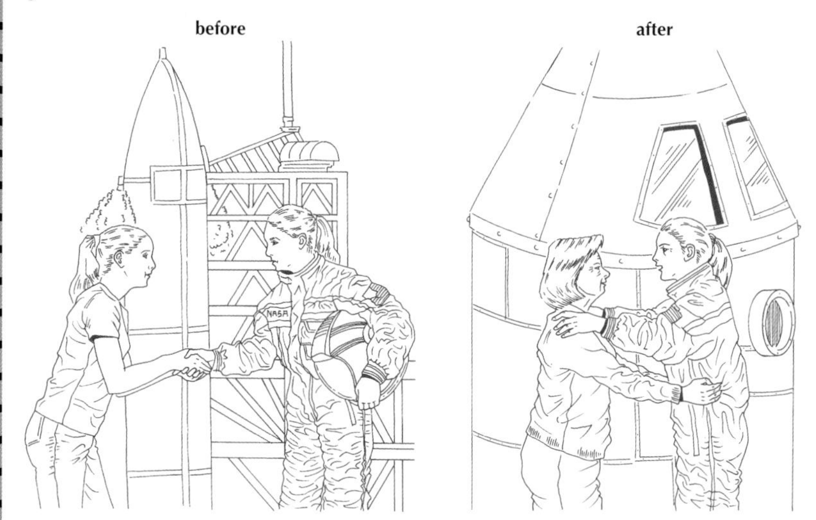

This seems a very strange prediction, but it is correct according to the time dilation formula. Remember that:

- this is a relativistic effect – time is running at different rates because of the **relative velocity** between the two twins and **not** because of the **distance** between them.
- the difference in ageing is relative. Neither twin is getting younger; as far as both of them are concerned, time has been passing at the normal rate. It's just that the moving twin thinks that she has been away for a shorter time than the time as recorded by the twin on the Earth.

The paradox is that, according to the twin who made the journey, the twin on the Earth was moving all the time and so the twin left on the Earth should have aged less. Who's version of time is correct?

The solution to the paradox comes from the realisation that the equations of special relativity are only symmetrical when the two observers are in constant relative motion. For the twins to meet back up again, one of them would have to turn around. This would involve external forces and acceleration. If this is the case then the situation is no longer symmetrical for the twins. The twin on the Earth has not accelerated so her view of the situation must be correct.

HAFELE-KEATING EXPERIMENT

In 1971 Hafele and Keating tested the predictions of time dilation. Atomic clocks were put into aircraft and flown, both eastwards and westwards, around the world. Before and after the flights, the times on the clocks were compared with clocks that remained fixed in the same location on the surface of the Earth.

An observer in the centre of the Earth would describe the clock flying eastwards as moving the fastest, the clock that is on the same location on the Earth's surface as also moving eastwards (due to the rotation of the Earth) but not as fast as the clock in the airplane, and the clock flying westwards as moving the slowest.

The results of their experiment agreed with the predictions of special relativity within the uncertainties of the experimental procedure. It should be noted that general relativity also predicts a mechanism that will additionally cause the clocks to register different times. This is a result of the clocks being at different heights in a gravitational field, see page 184. The full analysis of the experimental data also took this effect into account.

Velocity addition

MATHEMATICAL RELATIONSHIP

When two observers measure each other's velocity, they will always agree on the value. The calculation of relative velocity is not, however, normally straightforward. For example, an observer might see two objects approaching one another, as shown below.

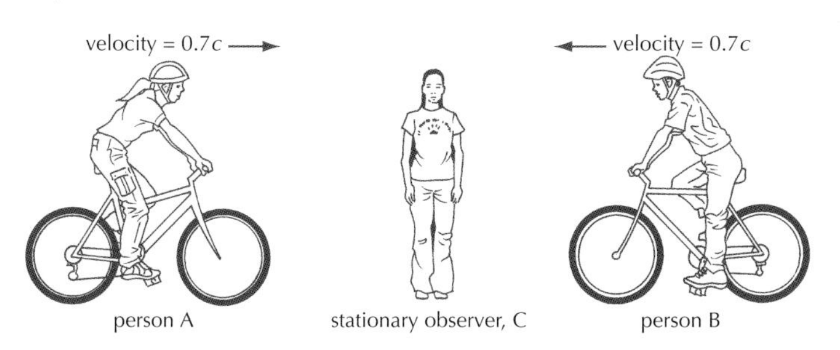

If each object has a relative velocity of 0.7 c, the Galilean transformations would predict that the relative velocity between the two objects would be 1.4 c. This cannot be the case as the Lorentz factor can only be worked out for objects travelling at less than the speed of light.

The situation considered is one frame moving relative to another frame at velocity v.

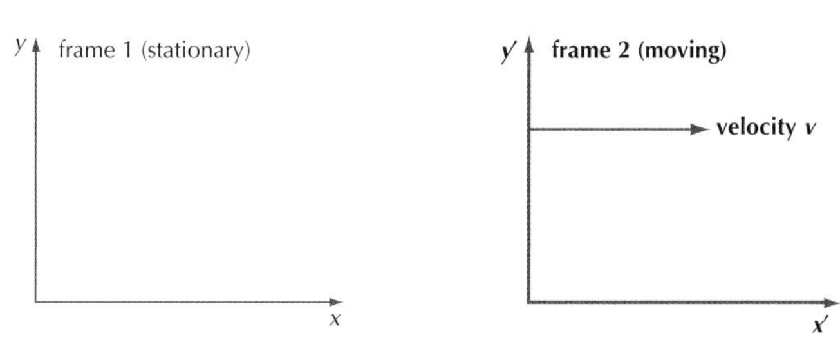

The equation used to move between frames is given below

$$u_x' = \frac{u_x - v}{1 - \frac{u_x v}{c^2}}$$

u_x' – the velocity in the x-direction as measured in the second frame

u_x – the velocity in the x-direction as measured in the first frame

v – the velocity of the second frame as measured in the first frame

In each of these cases, a positive velocity means motion along the positive x-direction. If something is moving in the negative x-direction then a negative velocity should be substituted into the equation.

Example

In the example above, two objects approached each other with 70% of the speed of light. So

u_x' = relative velocity of approach – to be calculated

$u_x = 0.7\ c$

$v = -0.7\ c$

$$u_x' = \frac{1.4\ c}{(1 + 0.49)}$$ *note the sign in the brackets*

$$= \frac{1.4\ c}{1.49}$$

$$= 0.94\ c$$

COMPARISON WITH GALILEAN EQUATION

The top line of the relativistic addition of velocities equation can be compared with the Galilean equation for the calculation of relative velocities.

$$u_x' = u_x - v$$

At low values of v these two equations give the same value. The Galilean equation only starts to fail at high velocities.

At high velocities, the Galilean equation can give answers of greater than c, while the relativistic one always gives a relative velocity that is less than the speed of light.

Mass and energy

$E = mc^2$

The most famous equation in all of physics is surely Einstein's mass-energy relationship $E = mc^2$, but where does it come from? By now it should not be a surprise that if time and length need to be viewed in a different way, then so does mass. To be absolutely precise there are two different aspects to the quantity that we call mass. In this section we shall consider what is called the **inertial mass** – (see page 172 for more details).

According to Newton's laws, a constant force produces a constant acceleration. If this was always true then any velocity at all should be achievable – even faster than light. All we have to do is apply a constant force and wait.

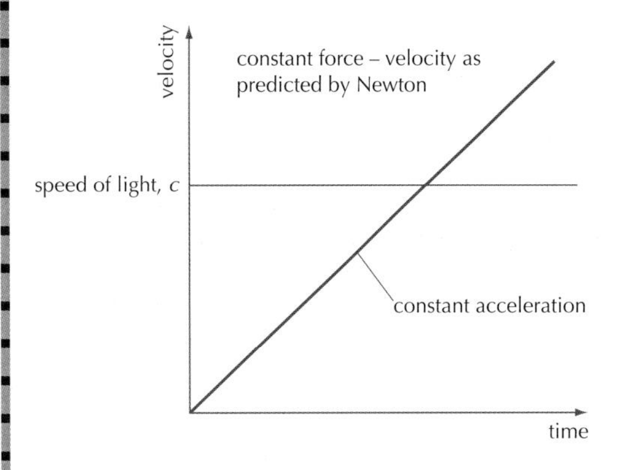

In practice, this does not happen. As soon as the speed of an object starts to approach the speed of light, the acceleration gets less and less even if the force is constant.

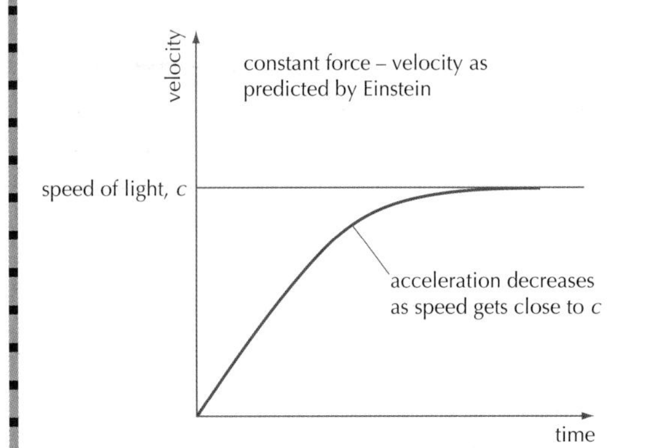

Since the force is constant and the acceleration is decreasing, the mass must be increasing. The Lorentz factor can be used to calculate the new mass. The ratio between the mass when moving with a constant velocity and the mass when the object appears at rest is equal to the Lorentz factor, γ.

$$m = \gamma m_0$$

Where has this extra mass come from? The force is still doing work (= force×distance). Therefore the object must be gaining in energy. The form of energy that it gains is mass-energy. Mass is just another form of energy. The equation that works out how energy and mass are related is

$$E = mc^2$$

energy in joules | mass in kilograms | speed of light in m s^{-1}

See pages 63 and 179 for more details on how to apply this equation in given situations.

REST MASS

The measurement of mass depends on relative velocity. Once again it is important to distinguish the measurement taken in the frame of the object from all other possible frames. The **rest mass** of an object is its mass as measured in a frame where the object is at rest. A frame that is moving with respect to the object would record a higher mass.

WHY NOTHING CAN GO FASTER THAN THE SPEED OF LIGHT

The relativistic mass increase formula shows how the Lorentz factor can be used to predict the increase in mass of an object. The greater the speed of the object, the greater its mass. A greater mass means that a fixed force would produce less acceleration. If you want the acceleration to remain constant, you would need to increase the force.

As the object approaches the speed of light, its mass approaches infinite values. This means that an infinite force would be needed to accelerate it even more. Nothing can get round this absolute speed limit. The speed of light is the maximum speed that anything can ever have. Objects with mass can never actually attain this speed.

MASS-ENERGY

Mass and energy are equivalent. This means that energy can be converted into mass and vice versa. Einstein's mass-energy equation can always be used, but one needs to be careful about how the numbers are substituted. Newtonian equations (such as $KE = \frac{1}{2}mv^2$ or momentum = mv) will take different forms when relativity theory is applied.

The energy needed to create a particle at rest is called the rest energy E_0 and can be calculated from the rest mass:

$$E_0 = m_0c^2$$

If this particle is given a velocity, it will have a greater total energy. This means that its mass will have increased. Einstein's mass-energy equation still applies.

$$E = mc^2$$

but in this case the mass m is greater than the rest mass m_0 ($m = \gamma m_0$).

EXAMPLE

An electron is accelerated through a p.d. of 1.0×10^6 V. Calculate its velocity.

$$\text{Energy gained} = 1.0 \times 10^6 \times 1.6 \times 10^{-19}\text{ J}$$
$$= 1.6 \times 10^{-13}\text{ J}$$
$$E_0 = m_0c^2 = 9.11 \times 10^{-31} \times (3 \times 10^8)^2$$
$$= 8.2 \times 10^{-14}\text{ J}$$
$$\therefore \text{Total energy} = 2.42 \times 10^{-13}\text{ J}$$
$$\therefore \gamma = \frac{2.42 \times 10^{-13}}{8.2 \times 10^{-14}} = 2.95$$
$$\text{velocity} = \sqrt{1 - \frac{1}{\gamma^2}}\, c$$
$$= 0.94\, c$$

Relativistic momentum and energy

EQUATIONS

The connection between mass and energy was introduced on page 178:

$$E_0 = m_0 c^2$$

$$E = mc^2$$

So kinetic energy E_k is

$$E_k = (\gamma - 1)m_0 c^2$$

Other equations can be derived for relativistic situations. Physical laws that apply in Newtonian situations also apply in relativistic situations. However the concepts often have to be refined to take into account the new ways of viewing space and time.

For example, in Newtonian mechanics, momentum p is defined as the product of mass and velocity.

$$p = mv$$

In relativity it has a similar form, but the Lorentz factor needs to be taken into consideration.

$$p = \gamma m_0 v$$

The momentum of an object is related to its total energy. In relativistic mechanics, the relationship can be stated as

$$E^2 = p^2c^2 + m_0^2 c^4$$

In Newtonian mechanics, the relationship between energy and momentum is

$$E = \frac{p^2}{2m}$$

Do not be tempted to use the standard Newtonian equations – if the situation is relativistic, then you need to use the relativistic equations.

UNITS

SI units can be applied in these equations. Sometimes, however, it is useful to use other units instead.

At the atomic scale, the joule is a huge unit. Often the electronvolt (eV) is used. One electronvolt is the energy gained by one electron if it moves through a potential difference of 1 volt. Since

$$\text{Potential difference} = \frac{\text{energy difference}}{\text{change}}$$

$$1 \text{ eV} = 1 \text{ V} \times 1.6 \times 10^{-19} \text{ C}$$

$$= 1.6 \times 10^{-19} \text{ J}$$

In fact the electronvolt is too small a unit, so the standard SI multiples are used

1 keV = 1000 eV

1 MeV = 10^6 eV   etc.

Since mass and energy are equivalent, it makes sense to have comparable units for mass. The equation that links the two ($E = mc^2$) defines a new unit for mass – the MeV c^{-2}. The speed of light is included in the unit so that no change of number is needed when switching between mass and energy – If a particle of mass of 5 MeV c^{-2} is converted completely into energy, the energy released would be 5 MeV. It would also be possible to used keV c^{-2} or GeV c^{-2} as a unit for mass.

In a similar way, the easiest unit for momentum is the MeV c^{-1}. This is the best unit to use if using the equation which links relativistic energy and momentum.

EXAMPLE

The Large Electron / Positron (LEP) collider at the European Centre for Nuclear Research (CERN) accelerates electrons to total energies of about 90 GeV. These electrons then collide with *positrons* moving in the opposite direction as shown below. Positrons are identical in rest mass to electrons but carry a positive charge. The positrons have the same energy as the electrons.

Electron →	← Positron
Total energy = 90 GeV	Total energy = 90 GeV

(a) Use the equations of special relativity to calculate,

(i) the velocity of an electron (with respect to the laboratory);

Total energy = 90 GeV = 90 000 MeV

Rest mass = 0.5 MeVc^{-2} $\therefore \gamma$ = 18 000 (huge)

$\therefore v \simeq c$

(ii) the momentum of an electron (with respect to the laboratory).

$p^2 c^2 = E^2 - m_0^2 c^4$

$\simeq E^2$

$p \simeq 90$ GeVc^{-1}

(b) For these two particles, estimate their relative velocity of approach.

since γ so large

relative velocity $\simeq c$

(c) What is the total momentum of the system (the two particles) before the collision?

zero

(d) The collision causes new particles to be created.

(i) Estimate the maximum total rest mass possible for the new particles.

Total energy available = 180 GeV

$\therefore$ max total rest mass possible = 180 GeVc^{-2}

(ii) Give one reason why your answer is a *maximum*.

Above assumes that particles were created at rest

Evidence to support special relativity

THE MUON EXPERIMENT

Muons are leptons (see page 206) – they can be thought of as a more massive version of an electron. They can be created in the laboratory but they quickly decay. Their average lifetime is 2.2×10^{-6} s as measured in the frame in which the muons are at rest.

Muons are also created high up (10 km above the surface) in the atmosphere. Cosmic rays from the Sun can cause them to be created with huge velocities – perhaps 0.99 c. As they travel towards the Earth some of them decay but there is still a detectable number of muons arriving at the surface of the Earth.

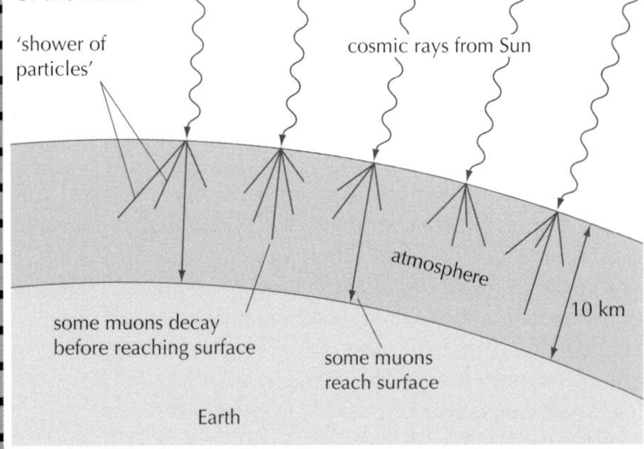

Without relativity, no muons would be expected to reach the surface at all. A particle with a lifetime of 2.2×10^{-6} s which is travelling near the speed of light (3×10^{8} m s^{-1}) would be expected to travel less than a kilometre before decaying ($2.2 \times 10^{-6} \times 3 \times 10^{8} = 660$ m).

The moving muons are effectively moving 'clocks'. Their high speed means that the Lorentz factor is high.

$$\gamma = \frac{1}{\sqrt{1 - 0.99^2}} = 7.1$$

Therefore an average lifetime of 2.2×10^{-6} s in the muons' frame of reference will be time dilated to a longer time as far as a stationary observer on the Earth is concerned. From this frame of reference they will last, on average, 7.1 times longer. Many muons will still decay but some will make it through to the surface – this is exactly what is observed.

In the muons' frame they exist for 2.2×10^{-6} s on average. They make it down to the surface because the atmosphere (and the Earth) is moving with respect to the muons. This means that the atmosphere will be length-contracted. The 10 km distance as measured by an observer on the Earth will only be $\frac{10}{7.1} =$ 1.4 km. A significant number of muons will exist long enough for the Earth to travel this distance.

THE MICHELSON–MORLEY EXPERIMENT

With hindsight, the **Michelson–Morley experiment** provides crucial support for the first postulate of special relativity – the constancy of the speed of light. The aim of the experiment was to measure the speed of the Earth through space. Using the technical language of the time, they were trying to measure the speed of the Earth through the Aether.

The experiment involved two beams of light travelling down two paths at right angles to one another. Having travelled different paths, the light was brought together where it interfered and produced fringes of constructive and destructive interference.

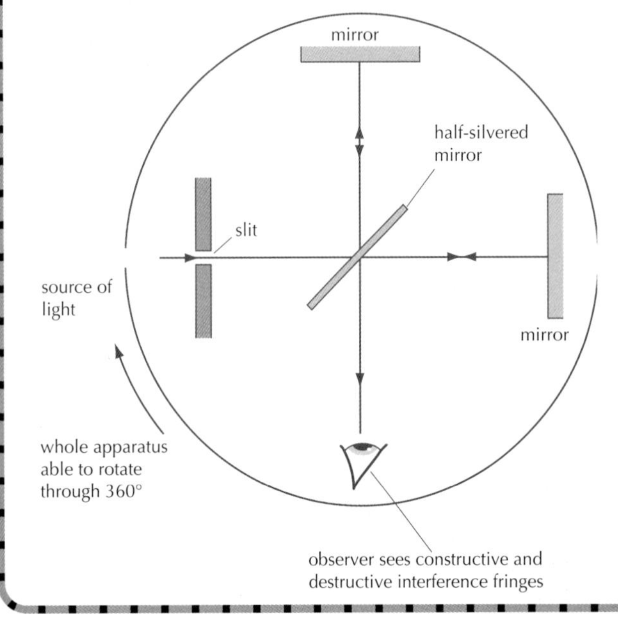

From the Galilean point of view, the motion of the Earth through space would affect the time of travel for the two beams by different amounts. Suppose the Earth was moving through the Aether in a direction that was parallel to one arm; the speed of light down that beam would be $(c + v)$ and on the way back the speed would be $(c - v)$. This is different to the speed of light in the arm that is at right angles to the motion. Here the light would have a speed of $\sqrt{(c^2 + v^2)}$.

If the whole apparatus were rotated around, the speed down the paths would change. This would move the interference pattern. The idea was to measure the change and thus work out the speed of the Earth through the Aether.

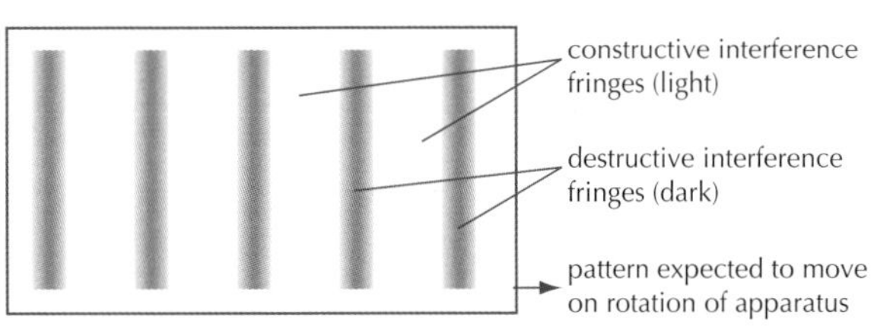

The experiment was tried but the rotation of the apparatus did not produce any observable change in the interference pattern. This null result can be easily understood from the first postulate of relativity – the constancy of the speed of light. The interference pattern does not change because the speed of light along the paths is always the same. It is unaffected by the motion of the Earth.

General relativity – the equivalence principle

PRINCIPLE OF EQUIVALENCE

One of Einstein's 'thought experiments' considers how an observer's view of the world would change if they were accelerating. The example below considers an observer inside a closed spaceship.

There are two possible situations to compare.

- the rocket could be far away from any planet but accelerating forwards.
- the rocket at rest on the surface of a planet.

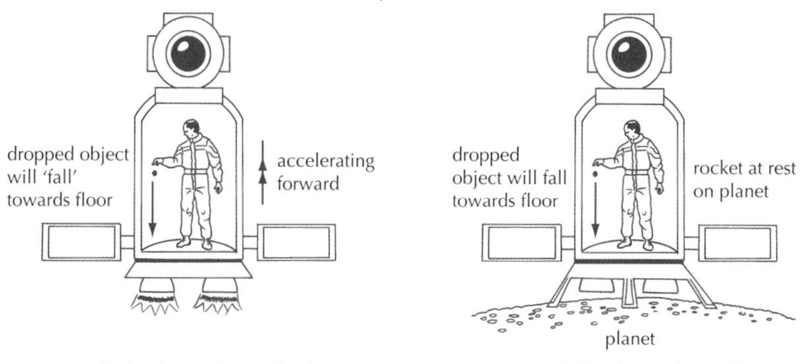

dropped object will 'fall' towards floor
accelerating forward
dropped object will fall towards floor
rocket at rest on planet
planet

astronaut feels a force when rocket is accelerating forward

astronaut feels a force when rocket is at rest on the surface of a planet

Although these situations seem completely different, the observer **inside** the rocket would interpret these situations as being identical.

This is Einstein's 'principle of equivalence' – a postulate that states that there is no difference between an accelerating frame of reference and a gravitational field.

From the principle of equivalence, it can be deduced that light rays are bent in a gravitational field (see below) and that time slows down near a massive body (see page 182).

BENDING OF LIGHT

Einstein's principle of equivalence suggests that a gravitational field should bend light rays! There is a small window high up in the rocket that allows a beam of light to enter.

In both of the cases to the right, diagrams 1 and 2, the observer is an **inertial** observer and would see the light shining on the wall at the point that is exactly opposite the small window. If, however, the rocket was accelerating upwards (see below, diagram 3) then the beam of light would hit a point on the wall **below** the point that is opposite the small window.

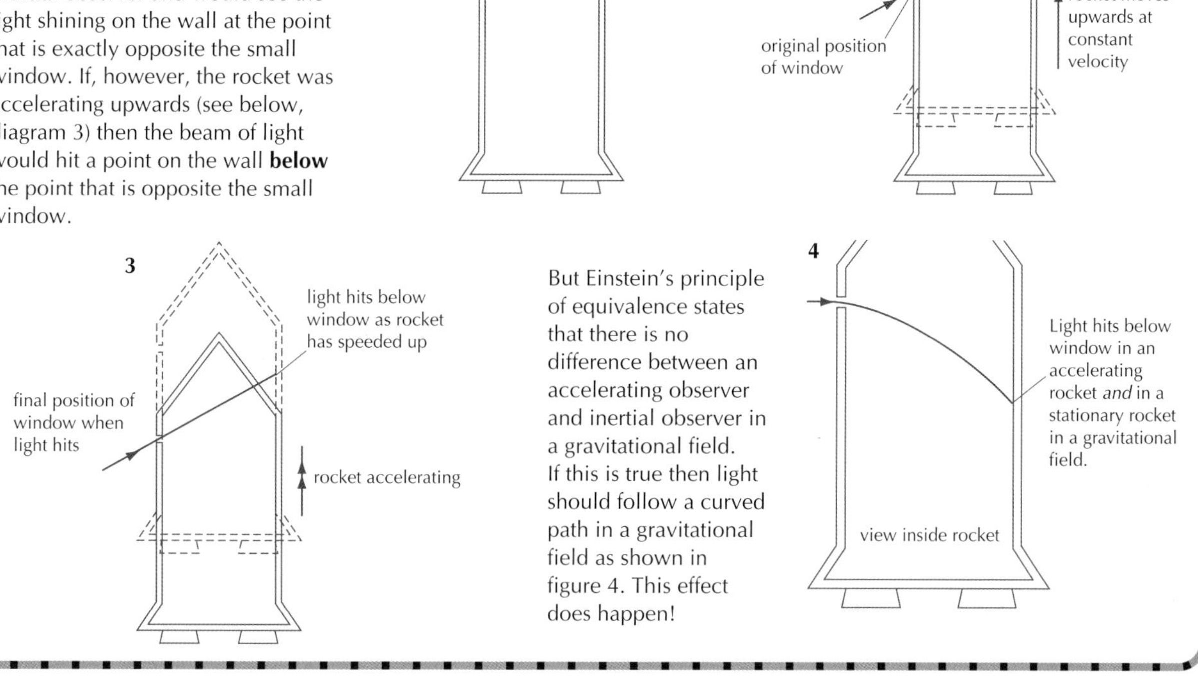

1 rocket at rest in space
window
light hits wall opposite window
2 rocket moving with constant velocity
final position of window when light hits
light hits wall opposite window
original position of window
rocket moves upwards at constant velocity
3
light hits below window as rocket has speeded up
final position of window when light hits
rocket accelerating
4
Light hits below window in an accelerating rocket and in a stationary rocket in a gravitational field.
view inside rocket

But Einstein's principle of equivalence states that there is no difference between an accelerating observer and inertial observer in a gravitational field. If this is true then light should follow a curved path in a gravitational field as shown in figure 4. This effect does happen!

Space–time

SPACE-TIME

Relativity has shown that our Newtonian ideas of space and time are incorrect. Two inertial observers will generally disagree on their measurements of space and time but they will agree on a measurement of the speed of light. Is there anything else upon which they will agree?

In relativity, a good way of imagining what is going on is to consider everything as different 'events' in something called space-time. From one observer's point of view, three co-ordinates (x, y and z) can define a position in space. One further 'coordinate' is required to define its position in time (t). An event is a given point specified by these four co-ordinates (x, y, z, t).

As a result of length contraction and time dilation, another observer would be expected to come up with totally different numbers for all of these four measurements – (x', y', z', t'). The amazing thing is that these two observers will agree on something. This is best stated mathematically:

$$x^2 + y^2 + z^2 - c^2t^2 = (x')^2 + (y')^2 + (z')^2 - c^2(t')^2$$

On normal axes, Pythagoras's theorem shows us that the quantity $\sqrt{(x^2 + y^2 + z^2)}$ is equal to the length of the line from the origin, so $(x^2 + y^2 + z^2)$ is equal to (the length of the line)^2. In other words, it is the separation in space.

$$(\text{Separation in space})^2 = (x^2 + y^2 + z^2)$$

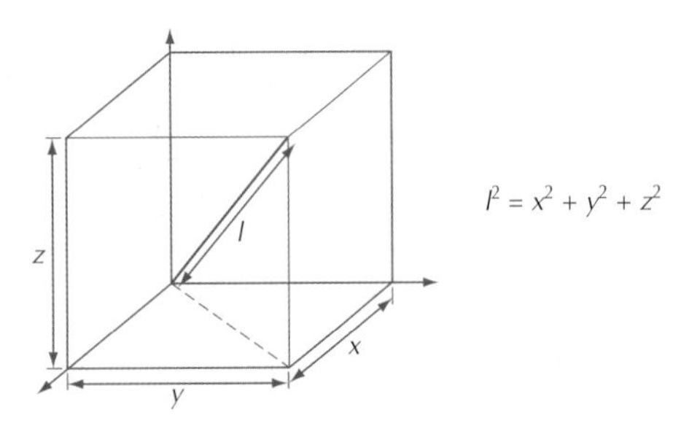

The two observers agree about something very similar to this, but it includes a co-ordinate of time. This can be thought of as the separation in imaginary four- dimensional space-time.

$$(\text{Separation in space-time})^2 = (x^2 + y^2 + z^2 - c^2t^2)$$

We cannot represent all four dimensions on the one diagram, so we usually limit the number of dimensions of space that we represent. The simplest representation has only one dimension of space and one of time as shown below.

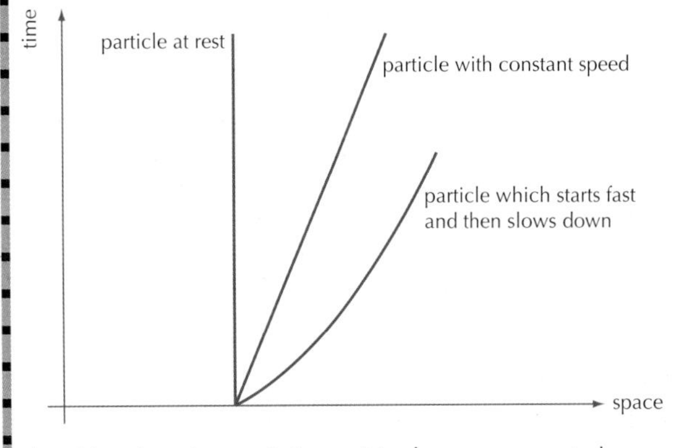

An object (moving or stationary) is always represented as a line in space-time.

EFFECT OF GRAVITY ON SPACE-TIME

The Newtonian way of describing gravity is in terms of the forces between two masses. In general relativity the way of thinking about gravity is not to think of it as a force, but as changes in the shape (warping) of space-time. The warping of space-time is caused by mass. Think about two travellers who both set off from different points on the Earth's equator and travel north.

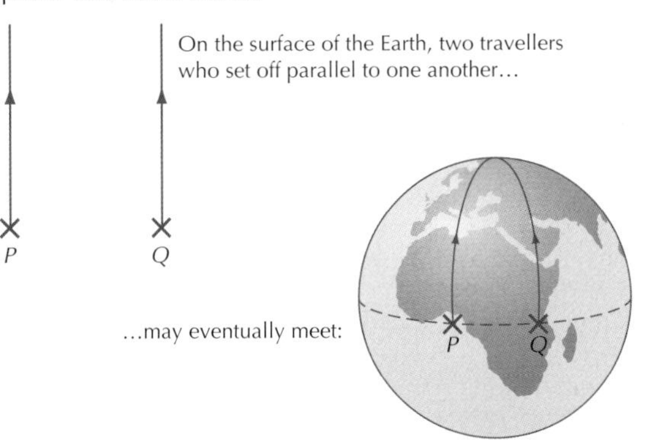

As they travel north they will get closer and closer together. They could explain this coming together in terms of a force of attraction between them or they could explain it as a consequence of the surface of the Earth being curved. The travellers have to move in straight lines across the surface of the Earth so their paths come together.

Einstein showed how space-time could be thought of as being curved by mass. The more matter you have, the more curved space-time becomes. Moving objects follow the curvature of space-time or in other words, they take the shortest path in space-time. As has been explained, it is very hard to imagine the four dimensions of space-time. It is easier to picture what is going on by representing space-time as a flat two-dimensional sheet.

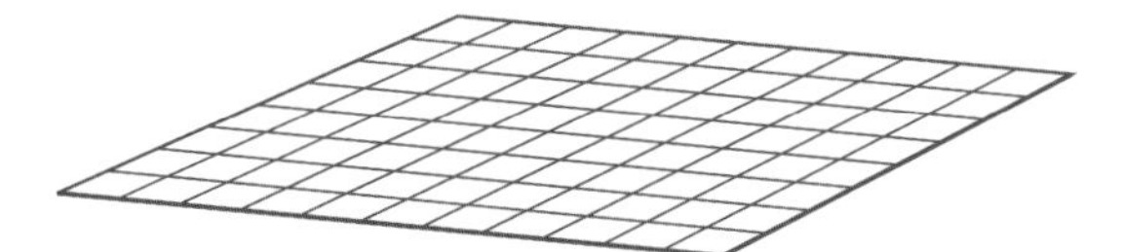

space–time represented by flat sheet

Any mass present warps (or bends) space-time. The more mass you have the greater the warping that takes place. This warping of space can be used to describe the the orbit of the Earth around the Sun. The diagram below represents how Einstein would explain the situation. The Sun warps space-time around itself. The Earth orbits the Sun because it is travelling along the shortest possible path in space-time. This turns out to be a curved path.

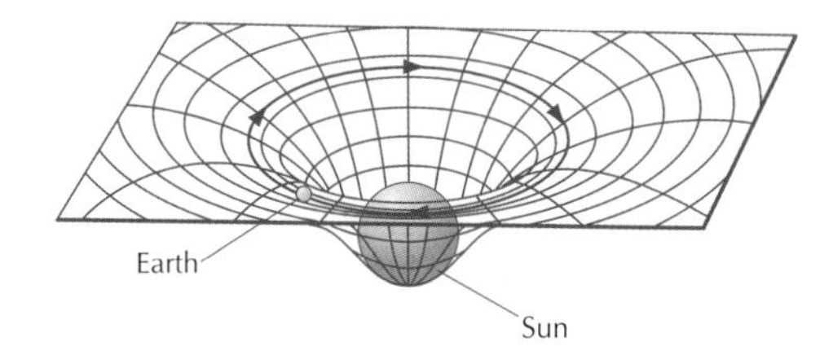

- mass 'tells' space how to curve.
- space 'tells' matter how to move.

Black holes

DESCRIPTION

When a star has used up all of its nuclear fuel, the force of gravity makes it collapse down on itself (see the astrophysics section for more details). The more it contracts the greater the density of matter and thus the greater the gravitational field near the collapsing star. In terms of general relativity, this would be described in terms of the space near a collapsing star becoming more and more curved. The curvature of space becomes more and more severe depending on the mass of the collapsing star.

If the collapsing star is less than about 1.4 times the mass of the Sun, then the electrons play an important part in eventually stopping this contraction. The star that is left is called a white dwarf. If the collapsing star is greater than this, the electrons cannot halt the contraction. A contracting mass of up to three times the mass of the Sun can also be stopped – this time the neutrons play an important role and the star that is left is called a neutron star. The curvature of space-time near a neutron star is more extreme than the curvature near a white dwarf.

At masses greater than this we do not know of any process that can stop the contraction. Space-time around the mass becomes more and more warped until eventually it becomes so great that it folds in over itself. What is left is called a black hole. All the mass is concentrated into a point – the **singularity**.

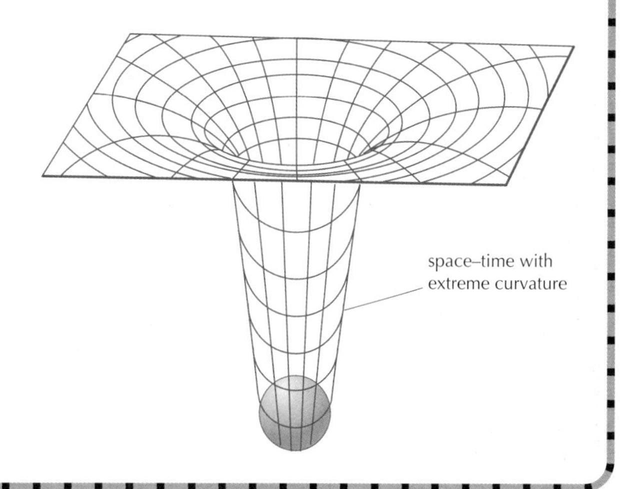

SCHWARZCHILD RADIUS

The curvature of space-time near a black hole is so extreme that nothing, not even light, can escape. Matter can be attracted into the hole, but nothing can get out since nothing can travel faster than light. The gravitational forces are so extreme that light would be severely deflected near a black hole.

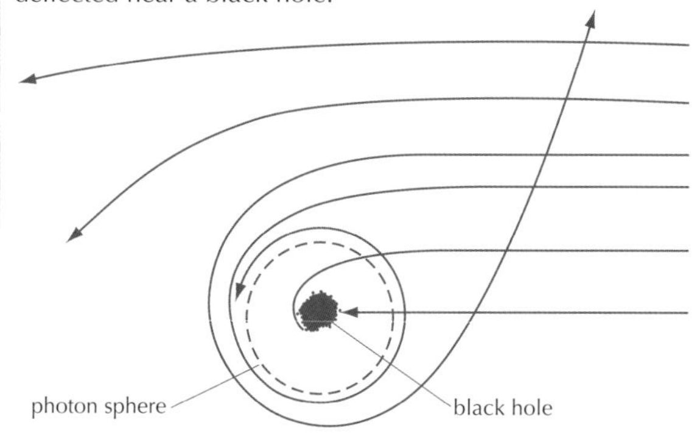

If you were to approach a black hole, the gravitational forces on you would increase. The first thing of interest would be the **photon sphere**. This consists of a very thin shell of light photons captured in orbit around the black hole. As we fall further in, the gravitational forces increase and so the escape velocity at that distance also increases.

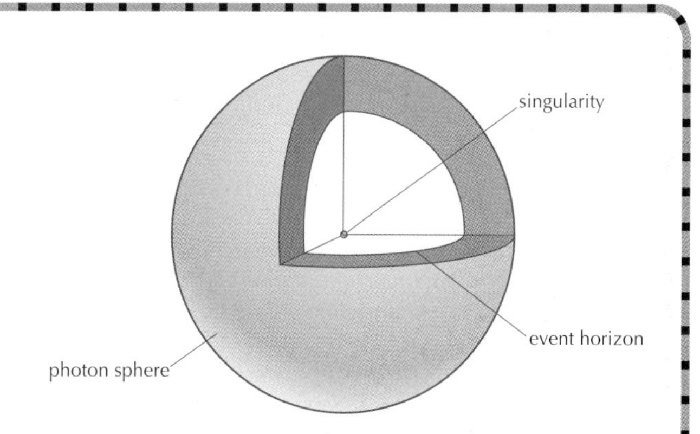

At a particular distance from the centre, called the **Schwarzchild radius**, we get to a point where the escape velocity is equal to the speed of light. Newtonian mechanics predicts that the escape velocity v from a mass M of radius r is given by the formula

$$v = \sqrt{\frac{2GM}{r}}$$

If the escape velocity is the speed of light, c, then the Schwarzchild radius would be given by

$$R_{\text{Sch}} = \frac{2GM}{c^2}$$

It turns out that this equation is also correct if we use the proper equations of general relativity. If we cross the Schwarzchild radius and get closer to the singularity, we would no longer be able to communicate with the Universe outside. For this reason crossing the Schwarzchild radius is sometimes called crossing the **event horizon**. An observer watching an object approaching a black hole would see time slowing down for the object.

The time dilation is worked out from $\Delta t = \dfrac{\Delta t_0}{\sqrt{1-\dfrac{R_{\text{Sch}}}{r}}}$

where r is the distance from the black hole.

EXAMPLE

Calculate the size of a black hole that has the same mass as our sun (1.99×10^{30} kg).

$$R_{\text{Sch}} = \frac{2 \times 6.67 \times 10^{-11} \times 1.99 \times 10^{30}}{(3 \times 10^{8})^2}$$

$$= 2949.6 \text{ m}$$

$$= 2.9 \text{ km}$$

Gravitational red shift

CONCEPT

The general theory of relativity makes other predictions that can be experimentally tested. One such effect is **gravitational redshift** – clocks slow down in a gravitational field. In other words a clock on the ground floor of a building will run slowly when compared with a clock in the attic – the attic is further away from the centre of the Earth.

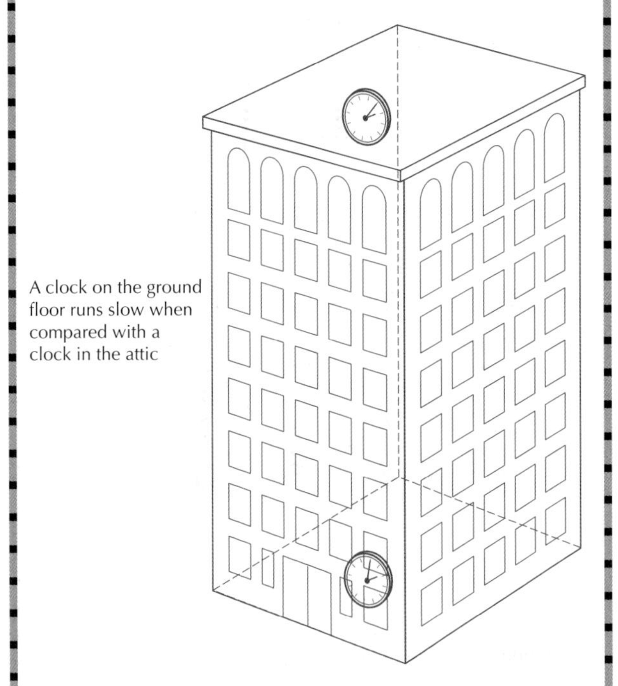

A clock on the ground floor runs slow when compared with a clock in the attic

The same effect can be imagined in a different way. We have seen that a gravitational field affects light. If light is shone away from a mass (for example the Sun), the photons of light must be increasing their gravitational potential energy as they move away. This means that they must be decreasing their total energy. Since frequency is a measure of the energy of a photon, the observed frequency away from the source must be less than the emitted frequency.

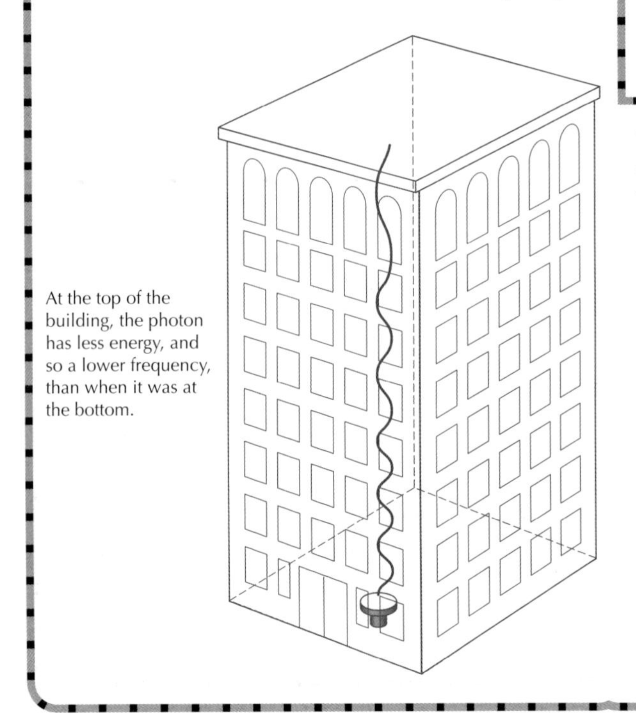

At the top of the building, the photon has less energy, and so a lower frequency, than when it was at the bottom.

The oscillations of the light can be imagined as the pulses of a clock. An observer at the top of the building would perceive the clock on the ground floor to be running slowly.

MATHEMATICS

This gravitational time dilation effect can be mathematically worked out for a uniform gravitational field g. The change in frequency Δf is given by

$$\frac{\Delta f}{f} = \frac{g\Delta h}{c^2}$$

where

- f – is the frequency emitted at the source
- g – is the gravitational field strength (assumed to be constant)
- Δh – is the height difference and
- c – is the speed of light.

EXAMPLE

A UFO travels at such a speed to remain above one point on the Earth at a height of 200 km above the Earth's surface. A radio signal of frequency of 110 MHz is sent to the UFO.

(i) What is the frequency received by the UFO?

(ii) If the signal was reflected back to Earth, what would be the observer frequency of the return signal? Explain your answer.

(i) $f = 1.1 \times 10^8$ Hz

$g = 10 \text{ m s}^{-2}$

$\Delta h = 2.0 \times 10^5$m

$$\therefore \; \Delta f = \frac{10 \times 2.0 \times 10^5}{(3 \times 10^8)^2} \times 1.1 \times 10^8 \text{ Hz}$$

$= 2.4 \times 10^{-3}$ Hz

$\therefore \; f$ received $= 1.1 \times 10^8 - 2.4 \times 10^{-3}$

$= 109999999.998$ Hz

$\approx 1.1 \times 10^8$ Hz

(ii) The return signal will be gravitationally blueshifted. Therefore it will arrive back at **exactly** the same frequency as emitted.

Supporting evidence

EVIDENCE TO SUPPORT GENERAL RELATIVITY

Bending of star light

The predictions of general relativity, just like those of special relativity, seem so strange that we need strong experimental evidence. One main prediction was the bending of light by a gravitational field. One of the first experiments to check this effect was done by a physicist called Arthur Eddington in 1919.

The idea behind the experiment was to measure the deflection of light (from a star) as a result of the Sun's mass. During the day, the stars are not visible because the Sun is so bright. During a solar eclipse, however, stars are visible during the few minutes when the Moon blocks all of the light from the Sun. If the positions of the stars during the total eclipse were compared with the positions of the same stars recorded at different time, the stars that appeared near the edge of the Sun would appear to have moved.

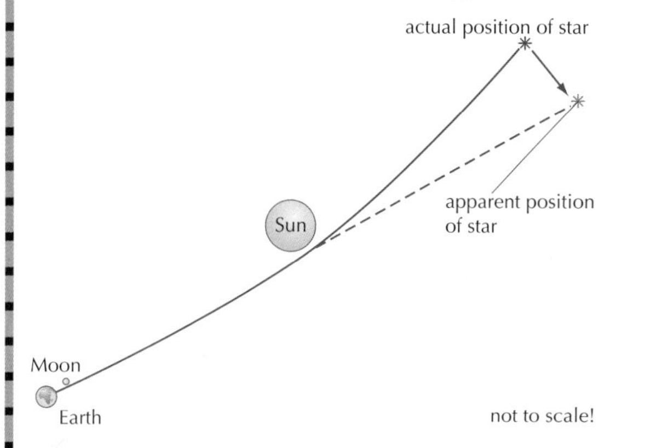

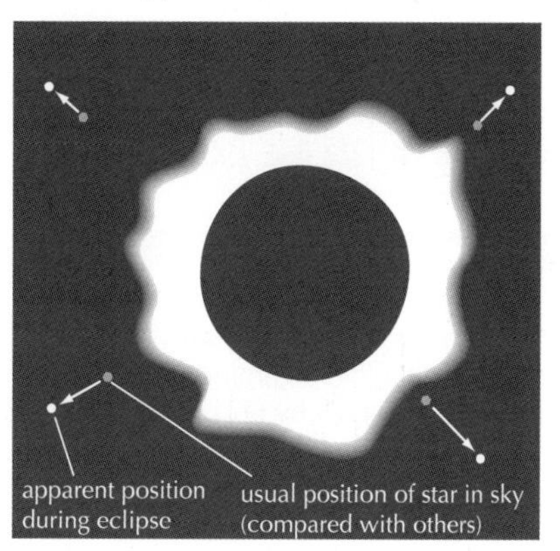

The angle of the shift of these stars turned out to be exactly the angle as predicted by Einstein's general theory of relativity.

Gravitational lensing

The bending of the path of light or the warping of space-time (depending on which description you prefer) can also produce some very extreme effects. Massive galaxies can deflect the light from quasars (or other very distance sources of light) so that the rays bend around the galaxy as shown below.

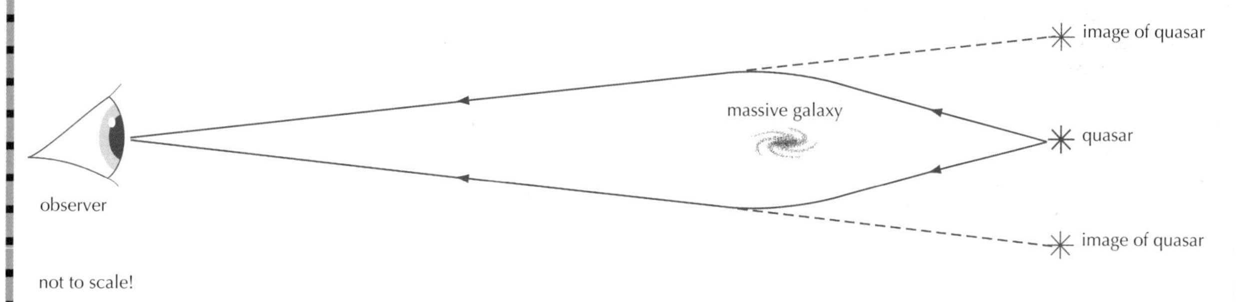

In this strange situation, the galaxy is acting like a lens and we can observe multiple images of the distant quasar.

EVIDENCE TO SUPPORT GRAVITATIONAL REDSHIFT

Pound-Rebka

The decrease in the frequency of a photon as it climbs out of a gravitational field can be measured in the laboratory. The measurements need to be very sensitive, but they have been successfully achieved on many occasions. One of the experiments to do this was done in 1960 and is called the **Pound-Rebka** experiment. The frequencies of gamma-ray photons were measured after they ascended or descended Jefferson Physical Laboratory Tower at Harvard University.

Atomic clock frequency shift

Because they are so sensitive, comparing the difference in time recorded by two identical atomic clocks can provide a direct measurement of gravitational redshift. One of the clocks is taken to high altitude by a rocket, whereas a second one remains on the ground. The clock that is at the higher altitude will run faster.

Shapiro time delay experiments

The time taken for a radar pluse to travel to another nearby planet (e.g. Venus or Mercury) and back can be accurately recorded. The gravitational field of the Sun can affect the time taken. The extent of the effect depends on the orientation of the planets and the Sun. The experiment was first performed in the 1960s and the results confirmed the predictions of general relativity.

IB QUESTIONS – OPTION H – RELATIVITY

1 *Relativity and simultaneity*

(a) State two postulates of the special theory of relativity. [2]

Einstein proposed a 'thought experiment' along the following lines. Imagine a train of proper length 100 m passing through a station at half the speed of light. There are two lightning strikes, one at the front and one at the rear of the train, leaving scorch marks on both the train and the station platform. Observer S is standing on the station platform midway between the two strikes, while observer T is sitting in the middle of the train. Light from each strike travels to both observers.

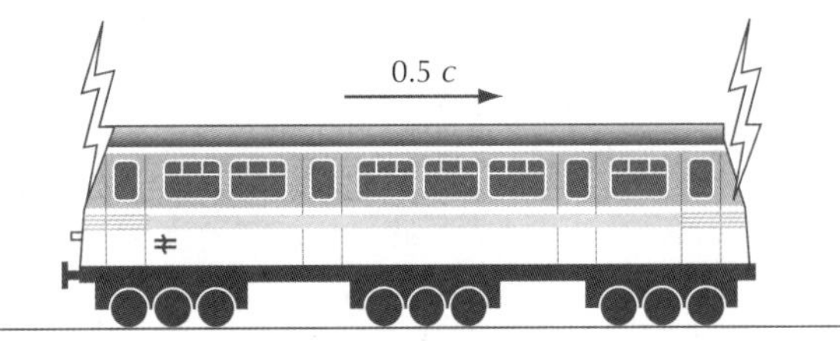

(b) If observer S on the station concludes from his observations that the two lightning strikes occurred simultaneously, explain why observer T on the train will conclude that they did **not** occur simultaneously. [4]

(c) Which strike will T conclude occurred first? [1]

(d) What will be the distance between the scorch marks on the *train*, according to T and according to S? [3]

(e) What will be the distance between the scorch marks on the *platform*, according to T and according to S? [2]

2 An electron is travelling at a constant speed in a vacuum. A laboratory observer measures its speed as 95% of the speed of light and the length of its journey to be 100 m.

(a) Show that for these electrons, $\gamma = 3.2$.

(b) What is the length of the journey in the **electron's** frame of reference? [1]

(c) What is the time taken for this journey in the **electron's** frame of reference? [2]

(d) What is the mass of the electron according to the **laboratory** observer? [2]

(e) Use the axes below to show how the observed mass of the electron will change with velocity as measured by the laboratory observer. There is no need to do any further calculations. [3]

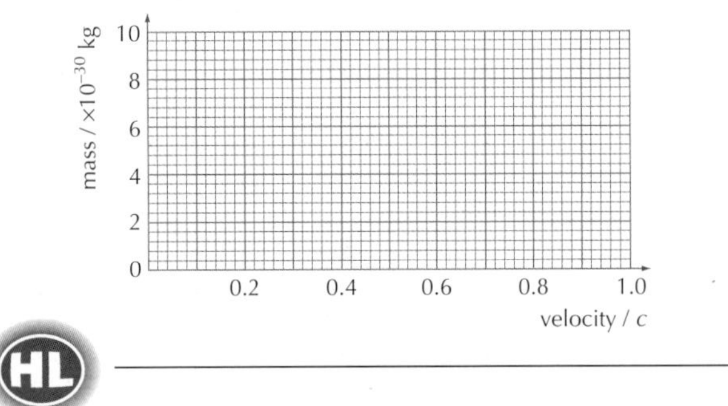

3 In a laboratory experiment two identical particles (P and Q), each of rest mass m_0, collide. In the **laboratory frame of reference**, they are both moving at a velocity of 2/3 *c*. The situation before the collision is shown in the diagram on the right.

Before:

(a) In the laboratory frame of reference,

(i) what is the **total momentum** of P and Q? [1]

(ii) what is the **total energy** of P and Q? [3]

The same collision can be viewed according to **P's frame of reference** as shown in the diagrams below.

velocity = *v*

P (rest) Q

(b) In P's frame of reference,

(i) what is Q's velocity, *v*? [3]

(ii) what is the **total momentum** of P and Q? [3]

(iii) what is the **total energy** of P and Q? [3]

(c) As a result of the collision, many particles and photons are formed, but the total energy of the particles depends on the frame of reference. Do the observers in each frame of reference agree or disagree on the number of particles and photons formed in the collision? Explain your answer. [2]

4 Two space travellers Lee and Anna are put into a state of hibernation in a ventilated capsule in a spaceship, for a long trip to find another habitable planet. They eventually awake, but do not know whether the ship is still travelling or whether they have landed. They feel attracted toward the floor of the capsule, an experience rather like weak gravity. Lee says the spaceship must have landed on a planet and they are experiencing its gravitational attraction. Anna says the spaceship must be accelerating and the capsule floor is pushing on them.

(a) Hoping to decide which of them is right, they try an experiment. They release a hammer in mid air, and it accelerates straight to the floor. Does this observation help them decide? How would *each* of them explain the motion of the hammer? [4]

(b) They propose another experiment, namely to shine monochromatic light from the floor to the ceiling of the capsule and use sensitive apparatus to detect any change in frequency.

(i) How would Lee explain how a redshift arises, viewing the radiation as photons moving in a gravitational field? [2]

(ii) How would Anna explain how a redshift arises, viewing the radiation in terms of wavefronts arriving at a detector whose speed is increasing? [2]

(c) Can Lee and Anna perform *any* experiment in the capsule which could distinguish whether they are on the surface of a planet or accelerating in space? State why. [1]

(d) Later they notice that the gravitational-like sensation starts diminishing gradually, until they eventually 'float weightless' in the capsule. Lee suggests that they must have taken off from the planet, and as they got further away its gravitational attraction diminished until it was negligible. Anna suggests that the spaceship must have gradually reduced its thrust and acceleration to zero. Which explanation is feasible, or is there no way to tell who is right? Explain. [3]

HL The ear and the mechanism of hearing

THE EAR

The ear converts sounds into electrical signals in the brain. There are three main sections – the **outer**, the **middle** and the **inner** ear.

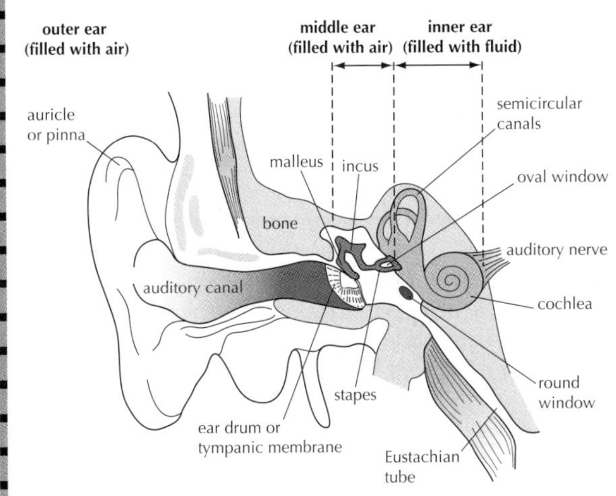

- The shape of the outer ear allows air vibrations to arrive at the ear drum (**tympanic membrane**).
- The middle ear converts the oscillations of air (and so oscillations of the ear drum) into oscillations in the fluid in the inner ear at the **oval window**. Three small bones that are called the **malleus**, **incus** and **stapes** (or hammer, anvil and stirrup) achieve this. Collectively they are known as the **ossicles**.
- The inner ear (and in particular the **cochlea**) converts the oscillations in the fluid into electrical signals that are sent along the auditory nerve to the brain.

Part of the reason for the complexity is the need to transmit (rather than reflect) as much as possible of the sound from the air into the cochlear fluid. Technically achieving this transmission of energy is known as **impedance matching**.

- The arrangement of the ossicles is such that it achieves a **mechanical advantage** of about 1.5. This means that the bones act as levers and pistons and the forces on the eardrum are increased by 50% by the time they are transmitted to the oval window of the inner ear.
- The area of the oval window is about 15 times smaller than the area of the eardrum. This means that the pressure on the oval window will be greater than the pressure on the eardrum.

These two processes result in larger pressure variations in the cochlear fluid as compared to the pressure variations on the eardrum.

Other parts of the ear's structure include the **semicircular canals** and the **Eustachian tube**.

- The semicircular canals in the inner ear are not used for hearing sounds. They are involved in detecting movement and keeping the body balanced.
- The Eustachian tubes connect the middle ear to the mouth and allow the pressures on either side of the eardrum to be equalised. Although the tube is normally closed, it opens during swallowing, yawning or chewing.

PITCH

The pitch of a sound corresponds to the frequency of the wave – the higher the frequency of the sound, the higher the pitch.

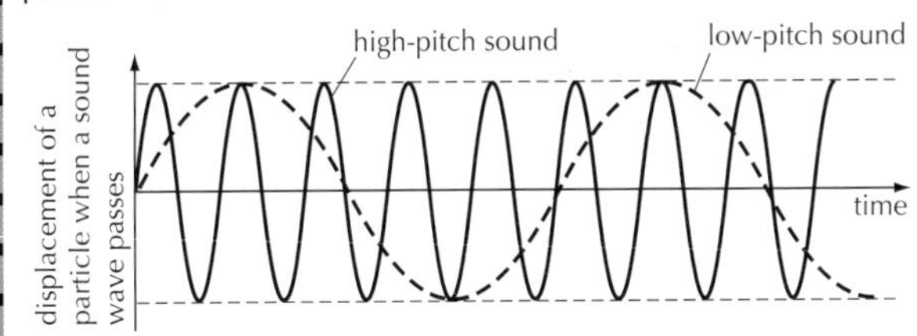

A normal human ear can hear sounds in the range 20 Hz up to 20 000 Hz (20 kHz).

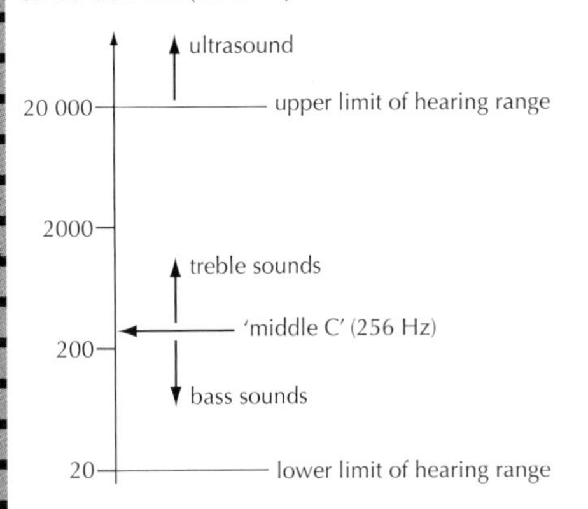

Most sounds are not just one pure frequency – they are a mixture of different frequencies. The particular mixture used makes different sounds unique. The different frequencies present in the sound wave are distinguished in the cochlea.

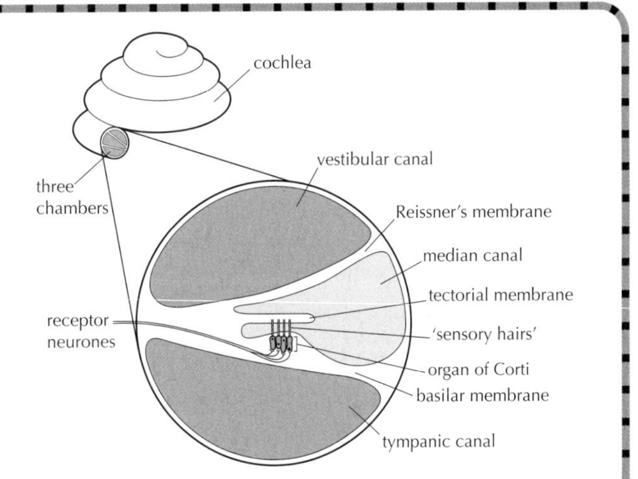

The cochlea is a spiral that contains three chambers. The pressure wave starts from the oval window and passes to the top of the spiral along the **vestibular chamber**. It then returns via the **tympanic chamber** and ends up being absorbed at the round window. As the pressure wave travels along these two chambers, 'sensors' in the middle chamber (**cochlear duct**) convert the variations in pressure into electrical signals in the auditory nerve. The operation of these 'sensors' is not fully understood, but they involve small hair-like structures in something called the **organ of Corti**. As these hair-like structures are moved back and forth, electrical impulses are sent to the brain. The hair-cells are set in motion by movements in the **basilar membrane**. This consists of fibres that change in length and tension as one travels along the chambers. Different frequencies are detected by the different sized structures along the cochlea.

HL Sound intensity and the dB scale

SOUND INTENSITY

The **sound intensity** is the amount of energy that a sound wave brings to a unit area every second. The units of sound intensity are W m^{-2}.

It depends on the amplitude of the sound. A more intense sound (one that is louder) must have a larger amplitude.

Intensity $\propto$ (amplitude)^2

This relationship between intensity and amplitude is true for all waves.

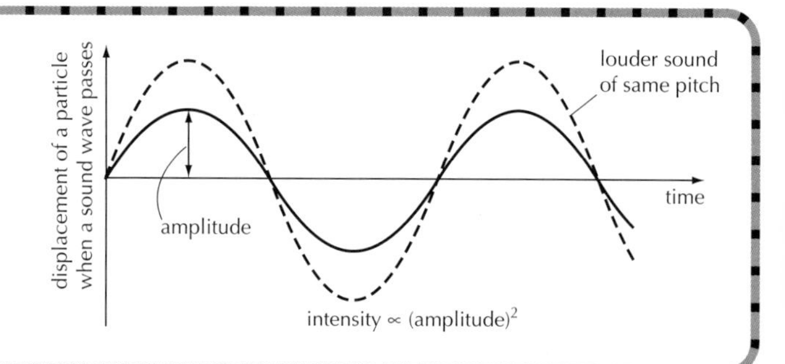

LOUDNESS

Intensity is a measurable quantity whereas loudness is subjective and depends on the listener.

Different people can describe different intensity sounds as appearing to have the same loudness – the frequency of the sound is an important factor. The following diagram shows the intensity – frequency diagram for a person with normal hearing.

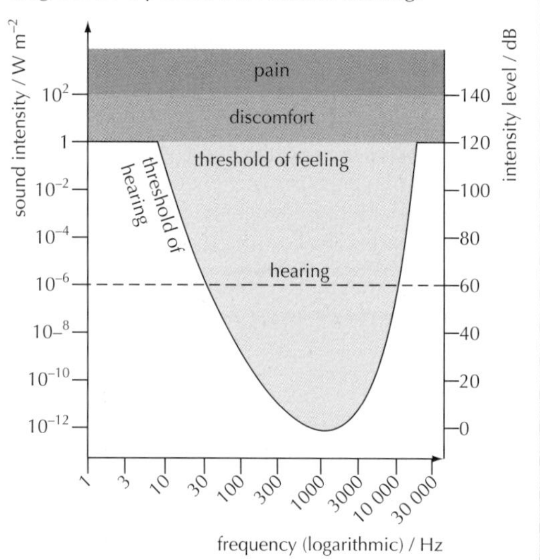

Diagram of intensity versus frequency for the average ear

The following points should be noted:

- the line represents sounds that are perceived to be the same loudness.
- below the threshold of hearing line, nothing is audible.
- above the threshold of feeling line, the sound is becoming uncomfortable.
- the ear is most sensitive to sounds at a frequency of about 2 kHz.
- a standard frequency for measurement is 1kHz. At this frequency, the threshold of hearing corresponds to an intensity of 10^{-12} W m^{-2}.
- this diagram represents the response of a normal ear. Medical problems or simple ageing would alter this diagram.

DECIBEL SCALE

Sound intensity levels are measured on the decibel scale (dB). As its name suggests, the decibel unit is simply one tenth of a base unit that is called the bel (B). Human hearing can respond to a huge range of different sound intensities. The decibel scale is logarithmic. The scale compares any given sound intensity with intensity at the threshold of hearing (the weakest sound that a person can normally hear). This threshold value is taken to be exactly 1 picowatt per square metre or 10^{-12} W m^{-2}.

Mathematically

$$\text{Intensity level in bels} = \log_{10} \frac{\text{intensity of sound}}{\text{intensity at the threshold of hearing}}$$

$$\text{or Intensity level in bels} = \log_{10} \frac{I}{I_0} \quad (\text{where } I_0 = 10^{-12} \text{ W m}^{-2})$$

Since 1 bel = 10 dB

$$\text{Intensity level in dB} = 10 \times \log_{10} \frac{I}{I_0}$$

sound intensity level (dB)	
150	permanent ear damage ↑
140	
130	THRESHOLD OF PAIN
120	near aeroplane engine
110	near pneumatic drill
100	near loud car horn
90	heavy traffic
80	door slamming
70	loud conversation
60	quiet motor at 1 m
50	quiet conversation
40	quiet street
30	watch ticking at 1 m
20	quiet room
10	virtual silence
0	THRESHOLD OF HEARING

EXAMPLE

The intensity next to an aircraft that is taking off is about 1 W m^{-2}. This means that the sound intensity level is given by

$$I = 10 \log \left(\frac{1}{10^{-12}}\right) \text{ dB}$$

$$= 10 \log (10^{12}) \text{ dB}$$

$$= 10 \times 12 \text{ dB}$$

$$= 120 \text{ dB}$$

Hearing tests

EXPERIMENTAL PROCEDURES

The aim of a basic hearing test is to see how hearing ability varies with frequency. To gain a rough idea of how a person's hearing responds to frequency it is possible to use tuning forks as the source of known frequencies. If the hearing is poor in a particular frequency range, then a tuning fork would need to be brought closer to the ear before it was heard.

For proper diagnosis, a complete record of the variation of hearing with frequency is required. This is called an **audiogram**. A complete audiogram tests for both **air conduction** and **bone conduction** as a comparison between the two can help to find out what part of the hearing mechanism has gone wrong. In each case, the patient is placed in a soundproofed room and the quietest sound that can be heard is recorded at a given frequency. The test is then repeated at different frequencies and then for the other ear.

- Air conduction is usually tested using headphones. At any one given frequency, sounds of varying intensity are fed into the ear and the threshold of hearing is established. This process is repeated at different frequencies.
- Bone conduction is tested by placing a vibrator on the bone behind the ear. It is held in place by a small band stretching over the top of the head. The vibrations are carried through the bones and tissues of the skull directly to the cochlea. This process allows the tester to bypass the outer ear and the middle ear. It tests the sensitivity of the inner ear directly.

In both tests, the amount of hearing loss is measured at different frequencies. A typical set of data would include the following frequencies:

125, 250, 500, 1000, 2000, 4000 and 8000 Hz

Before the start of the test, 'normal' levels of hearing have been established for the different frequencies used. The difference between this normal level and the recorded level is what is recorded on the audiogram using the dB scale. Different symbols are used to represent air conduction and bone conduction for the left and for the right ear. Air conduction thresholds are often represented by circles or crosses and bone conduction by triangles

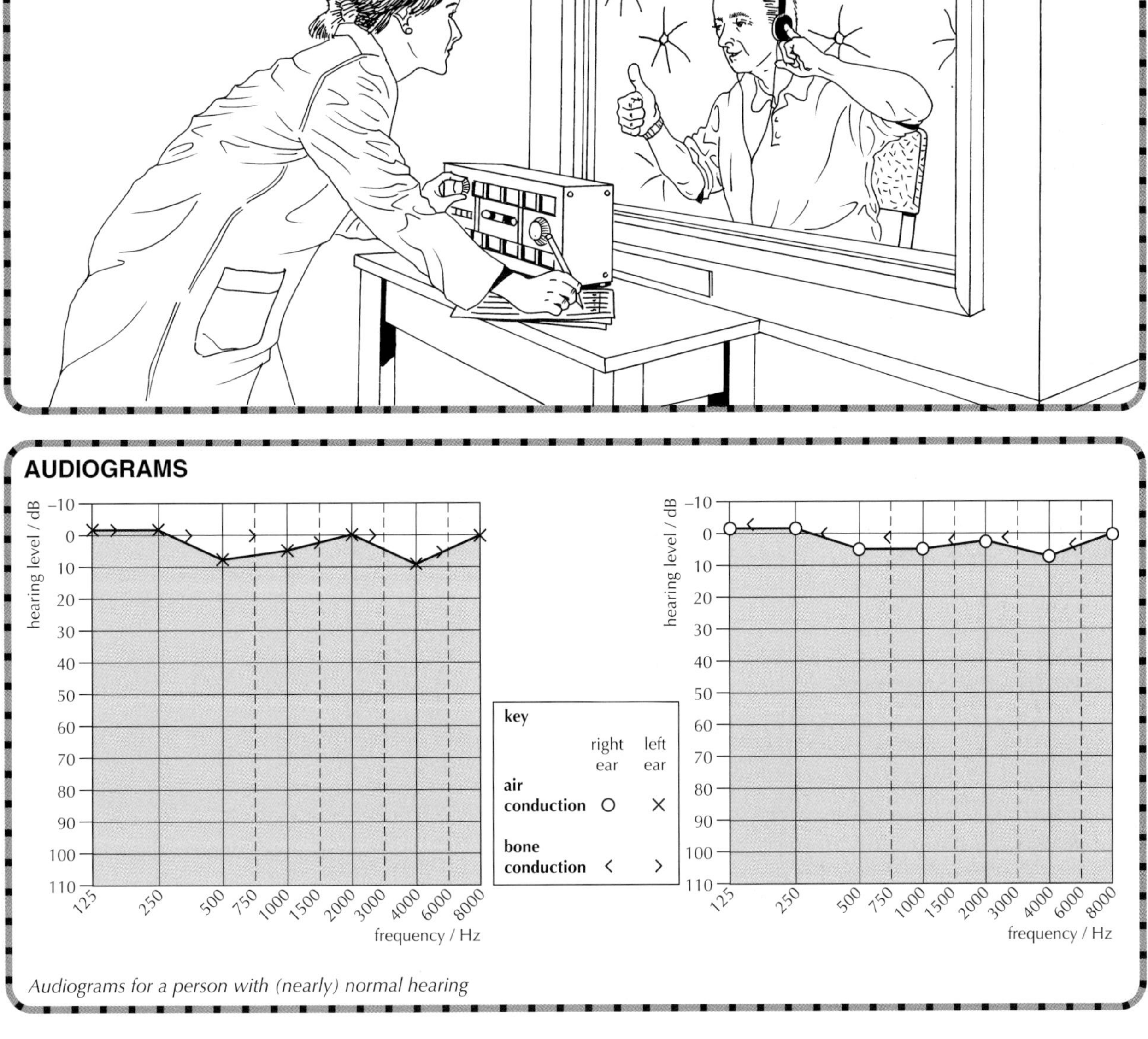

Audiograms for a person with (nearly) normal hearing

Hearing defects and correction

CONDUCTIVE LOSS

If the air conduction thresholds show a hearing loss but the bone conduction thresholds are normal, this is called a **conductive loss** of hearing. The sounds are being processed correctly in the inner ear, but the vibrations are not reaching it. This can sometimes be corrected by surgery. The main causes for conductive losses are:

- blockages – build up of wax or fluid.
- accidents – the eardrum can be ruptured or the middle ear could be damaged.
- diseases – the bones in the middle ear (and the oval window) can be prevented from moving.
- age – with increasing age, the bones in the middle ear (and the oval window) tend to become solidified.

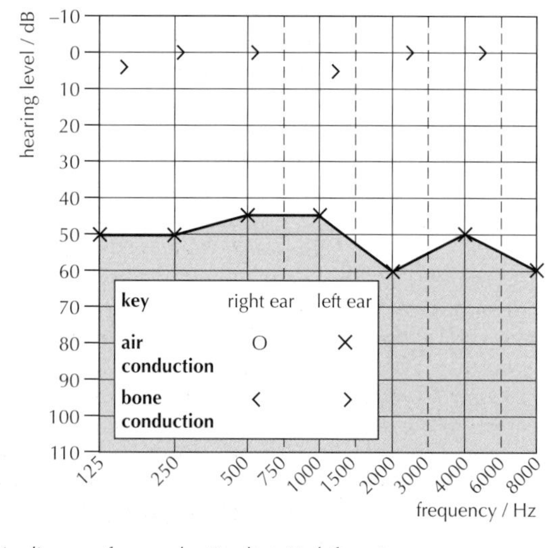

Audiogram for conductive loss (in left ear)

SENSORY LOSS

If both air conduction and bone conduction thresholds show the same amount of hearing loss, we call it **sensory** (**sensorineural**) hearing loss. It is also possible to have a greater hearing loss in air conduction as compared to bone conduction. This would be called a **mixed** hearing loss.

Sensory loss can be caused by ageing or the exposure to excessive noise over periods of time.

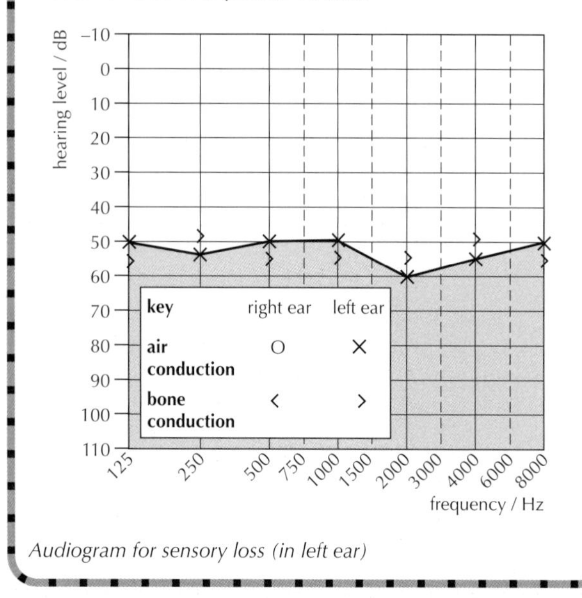

Audiogram for sensory loss (in left ear)

SELECTIVE FREQUENCY LOSS

All of the previous audiograms show a similar hearing loss at all frequencies. The audiogram below shows conductive loss particularly in the low and mid frequency range. This could lead to loss in speech discrimination.

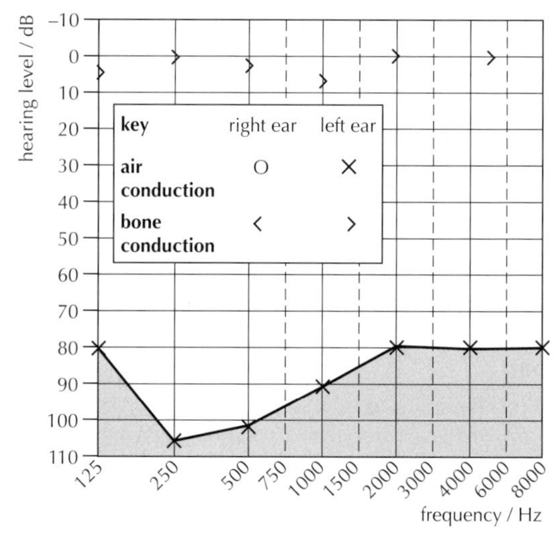

Audiogram for selective frequency loss (in left ear)

EFFECTS OF EXPOSURE TO NOISE

There are short-term and long-term effects on human hearing that result from exposure to noise and/or loud sounds. Common causes include people's working environment (e.g. a noisy factory or proximity to aircraft/heavy machinery) and their music listening habits (e.g. regular loud concerts or listening to music on headphones with the volume left too high).

In the short-term, exposure to loud sounds (e.g. the music from a loud concert) can cause temporary deafness across the frequency range and/or the sensation of "ringing" in the ears or **tinnitus**. This can often disappear with time, but a significant proportion of the population suffer from permanent tinnitus, the onset of which can be associated with ageing. Tinnitus can be untreatable.

Long-term and permanent damage can also occur resulting in permanent deafness. Some hearing aids provide very effective support for those with hearing loss, but the type of hearing aid used needs to be matched to the cause of the problem.

X-ray imaging

INTENSITY, QUALITY AND ATTENUATION

The details of X-ray production are shown on page 166. The effects of X-rays on matter depend on two things, the **intensity** and the **quality** of the X-rays.

- The intensity, I, is the amount of energy per unit area that is carried by the X-rays.
- The quality of the X-ray beam is the name given to the spread of wavelengths that are present in the beam. The incoming energy of the X-rays arrives in 'packets' of energy called photons (see page 104 for more details). A typical X-ray spectrum is shown on page 166.

If the energy of the beam is absorbed, then it is said to be **attenuated**. If there is nothing in the way of an X-ray beam, it will still be attenuated as the beam spreads out. Two processes of attenuation by matter, **simple scattering** and the **photoelectric effect** are the dominant ones for low-energy X-rays.

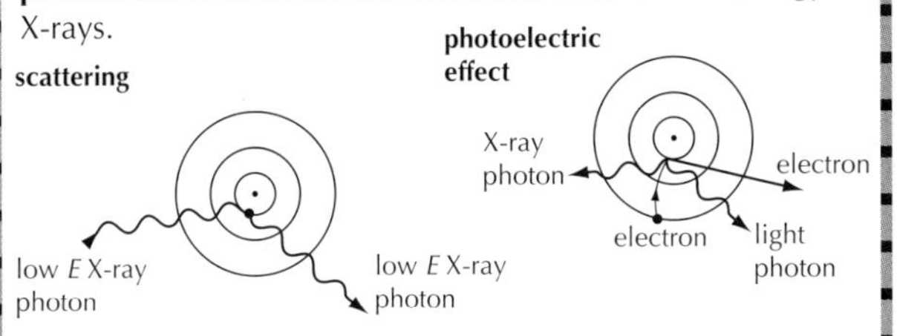

- Simple scattering affects X-ray photons that have energies between zero and 30 keV.
- In the photoelectric effect, the incoming X-ray has enough energy to cause one of the inner electrons to be ejected from the atom. It will result in one of the outer electrons 'falling down' into this energy level. As it does so, it releases some light energy. This process affects X-ray photons that have energies between zero and 100 keV.

Both attenuation processes result in a near exponential transmission of radiation as shown in the diagram below. For a given energy of X-rays and given material there will be a certain thickness that reduces the intensity of the X-ray by 50%. This is known as the **half-value thickness**.

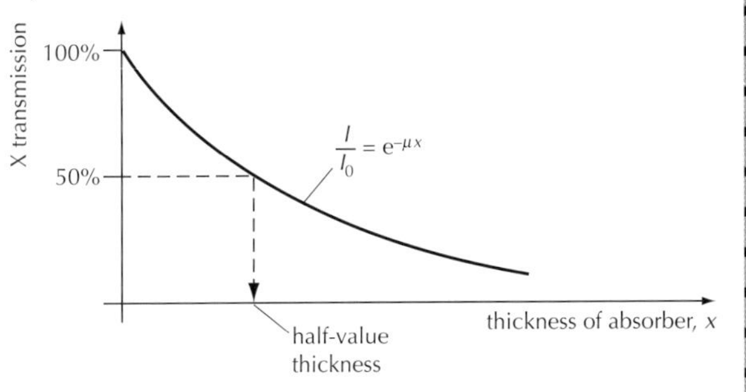

The **attenuation coefficient** μ is a constant that mathematically allows us to calculate the intensity of the X-rays given any thickness of material. The equation is as follows:

$$I = I_0 e^{-\mu x}$$

The relationship between the attenuation coefficient and the half-value thickness is

$$x_{\frac{1}{2}} = \frac{\ln 2}{\mu}$$

$x_{\frac{1}{2}}$ The half-value thickness of the material (in m)

ln 2 The natural log of 2. This is the number 0.6931

μ The attenuation coefficient (in m^{-1})

BASIC X-RAY DISPLAY TECHNIQUES

The basic principle of X-ray imaging is that some body parts (for example bones) will attenuate the X-ray beam much more than other body parts (for example skin and tissue). Photographic film darkens when a beam of X-rays is shone on them so bones show up as white areas on an X-ray picture.

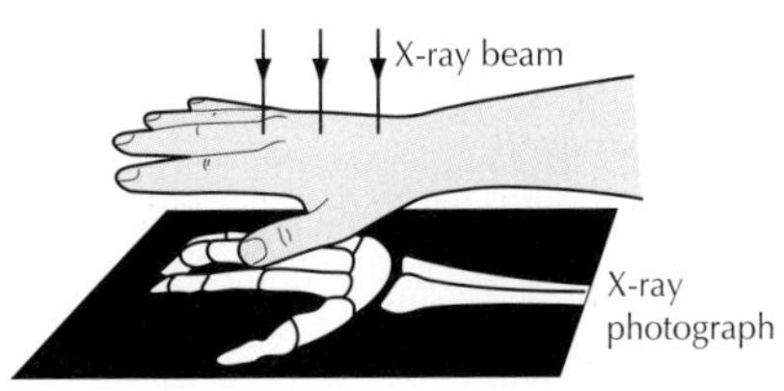

Since X-rays cause ionisations, they are dangerous. This means that the intensity used needs to be kept to an absolute minimum. This can be done by introducing something to **intensify** (to enhance) the image. There are two simple techniques of **enhancement**:

- When X-rays strike an intensifying screen the energy is re-radiated as visible light. The photographic film can absorb this extra light. The overall effect is to darken the image in the areas where X-rays are still present.
- In an image-intensifier tube, the X-rays strike a fluorescent screen and produce light. This light causes electrons to be emitted from a photocathode. These electrons are then accelerated towards an anode where they strike another fluorescent screen and give off light to produce an image.

X-RAY IMAGING TECHNIQUES INCLUDING COMPUTER TOMOGRAPHY

With a simple X-ray photograph it is hard to identify problems within soft tissue, for example in the gut. There are two general techniques aimed at improving this situation.

- In a **barium meal**, a dense substance is swallowed and its progress along the gut can be monitored.
- **Tomography** is a technique that makes the X-ray photograph focus on a certain region or 'slice' through the patient. All other regions are blurred out of focus. This is achieved by moving the source of X-rays and the film together.

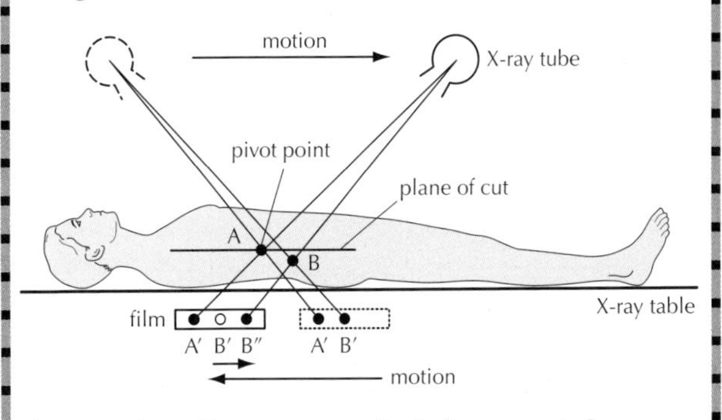

An extension of basic tomography is the **computed tomography scan** or **CT scan**. In this set-up a tube sends out a pulse of X-rays and a set of sensitive detectors collects information about the levels of X-radiation reaching each detector. The X-ray source and the detectors are then rotated around a patient and the process is repeated. A computer can analyse the information recorded and is able to reconstruct a 3-dimensional 'map' of the inside of the body in terms of X-ray attenuation.

Ultrasonic imaging

ULTRASOUND

The limit of human hearing is about 20 kHz. Any sound that is of higher frequency than this is known as **ultrasound**. Typically ultrasound used in medical imaging is just within the MHz range. The velocity of sound through soft tissue is approximately 1500 m s^{-1} meaning that typical wavelengths used are of the order of a few millimetres.

Unlike X-rays, ultrasound is not ionising so it can be used very safely for imaging inside the body – with pregnant women for example. The basic principle is to use a probe that is capable of emitting and receiving pulses of ultrasound. The ultrasound is reflected at any boundary between different types of tissue. The time taken for these reflections allows us to work out where the boundaries must be located.

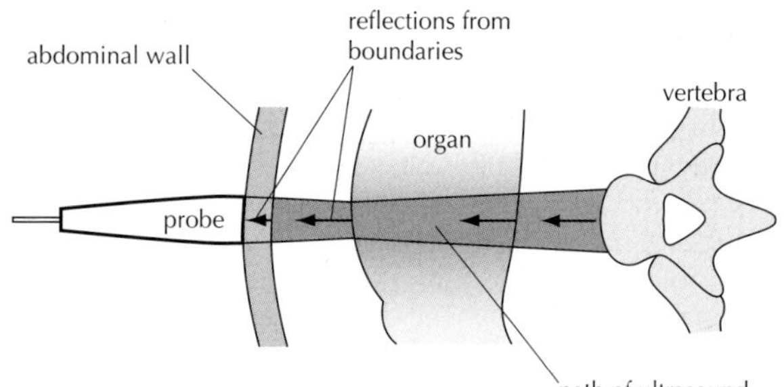

Acoustic impedance

The acoustic impedance of a substance is the product of the density, ρ, and the speed of sound, c.

$$Z = \rho c$$

unit of Z = kg m^{-2} s^{-1}

Very strong reflections take place when the boundary is between two substances that have very different acoustic impedances. This can cause some difficulties.

- In order for the ultrasound to enter the body in the first place, there needs to be no air gap between the probe and the patient's skin. An air gap would cause almost all of the ultrasound to be reflected straight back. The transmission of ultrasound is achieved by putting a gel or oil (of similar density to the density of tissue) between the probe and the skin.
- Very dense objects (such as bones) can cause a strong reflection and multiple images can be created. These need to be recognised and eliminated.

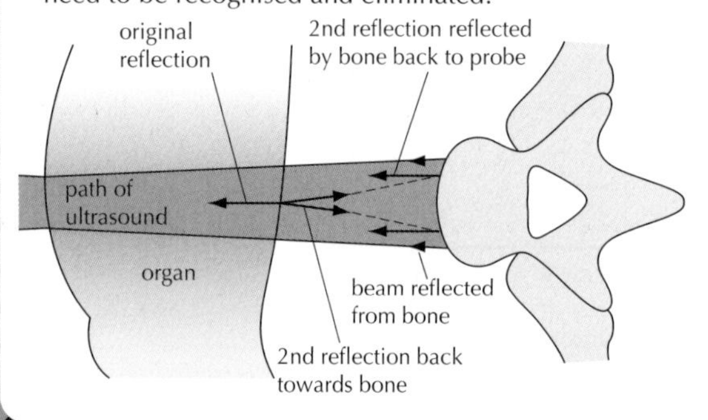

PIEZOELECTRIC CRYSTALS

These quartz crystals change shape when an electric current flows and can be used with an alternating p.d. to generate ultrasound. They also generate p.d.s when receiving sound pressure waves so one crystal is used for generation and detection.

A- AND B-SCANS

There are two ways of presenting the information gathered from an ultrasound probe, the **A-scan** or the **B-scan**. The A-scan (amplitude-modulated scan) presents the information as a graph of signal strength versus time. The B-scan (brightness-modulated scan) uses the signal strength to affect the brightness of a dot of light on a screen.

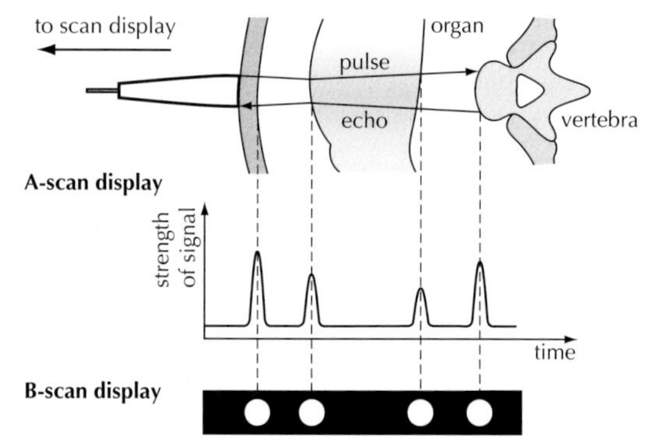

Pulse display by amplitude (A-scan) and brightness (B-scan)

A-scans are useful where the arrangement of the internal organs is well known and a precise measurement of distance is required. If several B-scans are taken of the same section of the body at one time, all the lines can be assembled into an image which represent a section through the body. This process can be achieved using a large number of transducers.

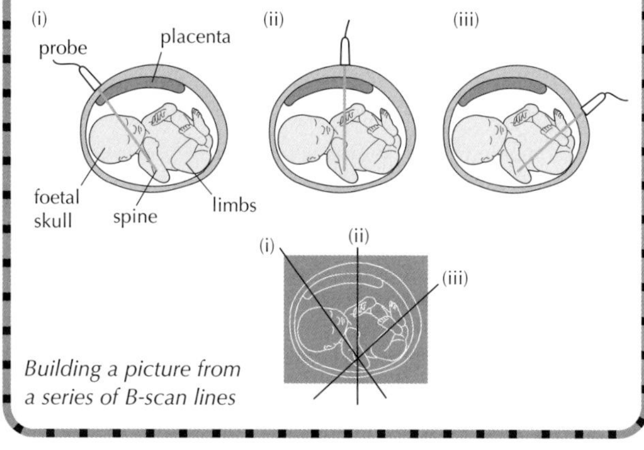

Building a picture from a series of B-scan lines

CHOICE OF FREQUENCY

The choice of frequency of ultrasound to use can be seen as the choice between resolution and attenuation.

- Here, the resolution means the size of the smallest object that can be imaged. Since ultrasound is a wave motion, diffraction effects will be taking place. In order to image a small object, we must use a small wavelength. If this was the only factor to be considered, the frequency chosen would be as large as possible.
- Unfortunately attenuation increases as the frequency of ultrasound increases. If very high frequency ultrasound is used, it will all be absorbed and none will be reflected back. If this was the only factor to be considered, the frequency chosen would be as small as possible.

On balance the frequency chosen has to be somewhere between the two extremes. It turns out that the best choice of frequency is often such that the part of the body being imaged is about 200 wavelengths of ultrasound away from the probe.

Other imaging techniques

NMR

Nuclear Magnetic Resonance (NMR) is a very complicated process but one that is extremely useful. It can provide detailed images of sections through the body without any invasive or dangerous techniques. It is of particular use in detecting tumours in the brain. It involves the use of a non-uniform magnetic field in conjuction with a large uniform field.

In outline, the process is as follows:

- the nuclei of atoms have a property called spin.
- the spin of these nuclei means that they can act like tiny magnets.
- these nuclei will tend to line up in a strong magnetic field.

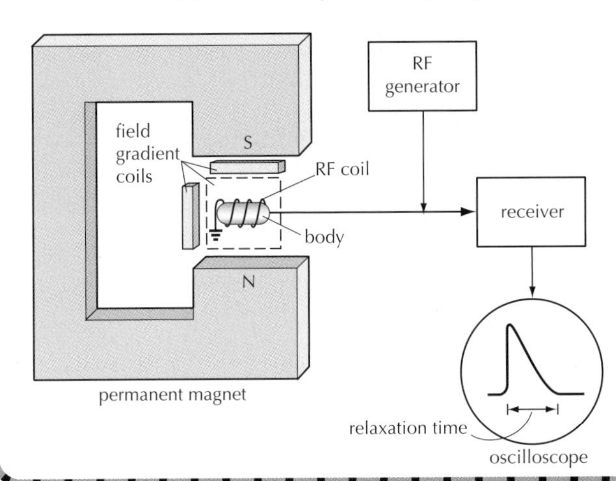

- they do not, however, perfectly line up – they oscillate in a particular way that is called **precession**. This happens at a very high frequency – the same as the frequency of radio waves.
- the particular frequency of precession depends on the magnetic field and the particular nucleus involved. It is called the **Larmor frequency**.
- if a pulse of radio waves is applied at the Larmor frequency, the nuclei can absorb this energy in a process called **resonance**.
- after the pulse, the nuclei return to their lower energy state by emitting radio waves.
- the time over which radio waves are emitted is called the **relaxation time**.
- the radio waves emitted and their relaxation times can be processed to produce the NMR scan image.

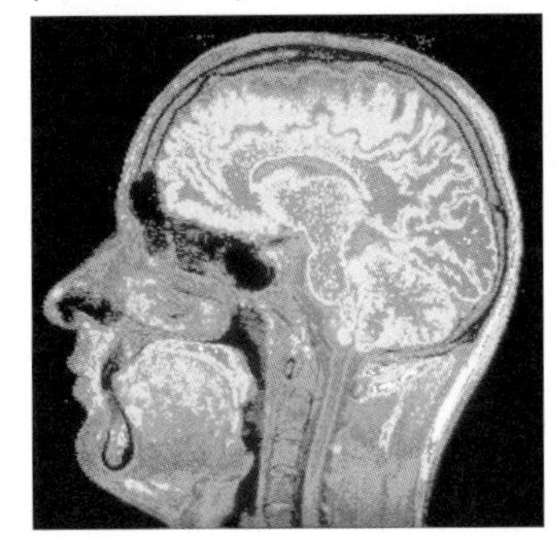

NMR scan image of the head

LASERS

Laser have a wide range of applications in both clinical diagnosis and therapy. Some examples include:

Pulse oximetry

Red and infrared laser light is shone through a thin part of a patient's anatomy, for example an earlobe or fingertip. The amount of absorbance of each wavelength will depend on many factors but a changing relative absorbance between the two wavelengths can be used to determine the ratio of blood cells with oxygen (which are bright red) and those without oxygen (which are dark red). This allows the overall oxygen content of the blood to be determined without having to take a sample of blood.

Endoscopes

An endoscope provides the ability to look inside the human body without invasive surgery. A tube is inserted into the body. Within the tube there is a collection of optical fibers that allows illumination from a outside source to reach the end of the tube and illuminate the region under investigation (e.g. a section of a patient's bowels). Reflected light is collected using a lens system and further optical fibres are used to allow an image to be viewed. Many endoscopes included additional instruments including a laser.

Laser as a scalpel and as a coagulator

Lasers can be used to accurately scan objects but are also commonly used to provide therapy both as a scalpel and a coagulator.

A laser focused on a small region can increase its temperature so high that it cuts through tissue like a scalpel. The laser heating ensures the site is kept free of germs and thus the risk of infection is reduced. In addition, blood vessels and nerves are automatically sealed off. A defocused laser beam can stop bleeding by stimulating the blood to form a clot.

Biological effects of radiation, dosimetry and radiation safety

DOSIMETRY

Three quantities are of particular note – **exposure, absorbed dose** and **dose equivalent**.

- **Exposure** is a measure of the total amount of ionisation produced. It is defined as

$$X = \frac{Q}{m}$$

exposure (X); total charge of one sign produced (Q); mass of air (m)

The units of exposure X are coulombs per kilogram (C kg^{-1})

- The **absorbed dose** is the energy absorbed per unit mass of tissue.

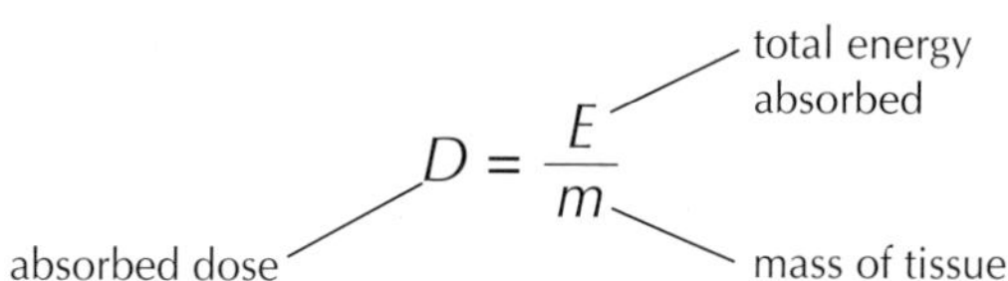

The units of absorbed dose D are joules per kilogram (J kg^{-1}). This unit is called the gray, Gy.

- The **dose equivalent** is an attempt to measure the radiation damage that actually occurs in tissues.

  The dose equivalent is defined as

$$H = QD$$

dose equivalent (H); quality factor (Q); absorbed dose (D)

The units of dose equivalent H are once again joules per kilogram (J kg^{-1}).
In order to differentiate between this and the absorbed dose, the unit is called the sievert (Sv).

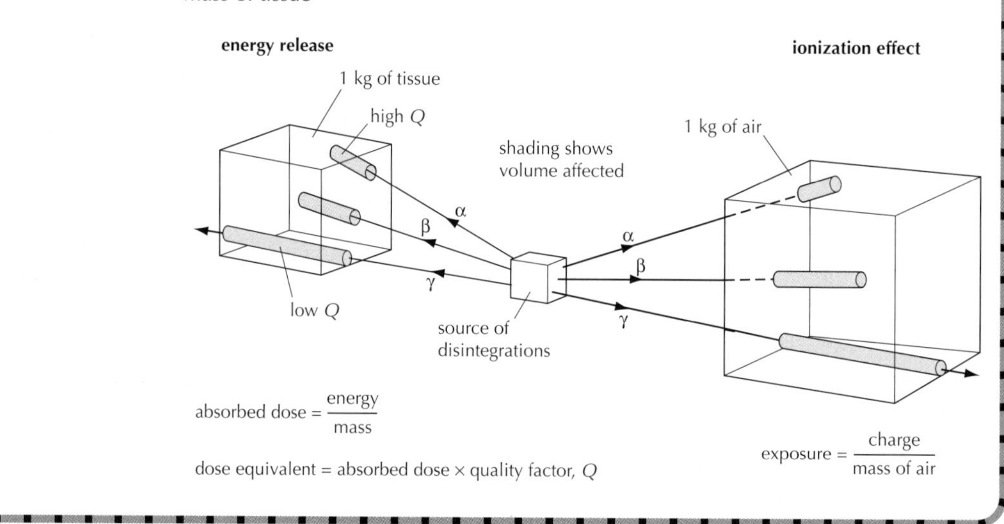

PRECAUTIONS TAKEN

The three fundamental factors used to reduce individual's exposure to ionising radiation are time of exposure, shielding distance from the source and shielding. These were introduced on page 61. An important procedure used to monitor the exposure received by those who work with radiation is the wearing and regular checking of a **film badge**.

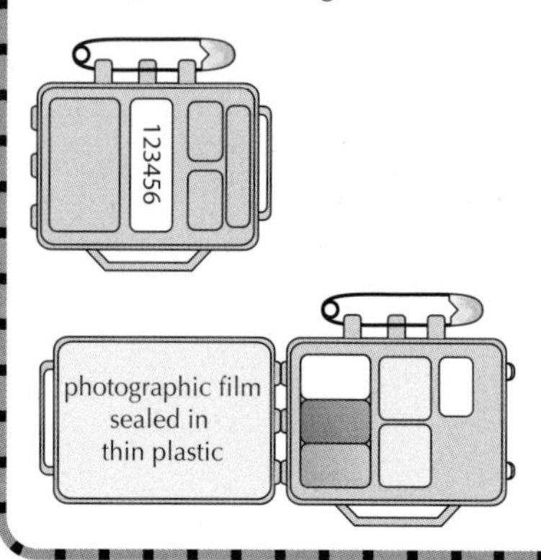

The photographic film is in a plastic container and not exposed to any light. Any ionising radiation that makes it way to the film will cause the film to darken in the areas that have been exposed. Different filters are placed either side of the film so as to be able to determine the nature of the ionizing radiation that has affected the film (and its wearer).

Typically the filters may include:

- An open window to allows all incident radiation that can penetrate the film wrapping to interact with the film.
- A thin plastic filter that attenuates beta radiation but passes all other radiations.
- A thick plastic filter that passes all but the lowest energy photon radiation and absorbs all but the highest energy beta radiation.
- A range of filters of varying thickness to allow typical photon energies to be determined.

BALANCED RISK

There is no such thing as a safe dose of radiation. Radioisotopes have many medical applications both in terms of diagnosis and in terms of treatment, and the risk of not treating a particular condition needs to be weighed against any extra risk that involves using the ionising radiation. In general all additional exposure needs to be **as** **l**ow **as** can be **r**easonably **a**chieved (ALARA) and needs to show a positive overall benefit to the patient.

Radiation sources in diagnosis and therapy (1)

RADIOACTIVE TRACERS

Radioactive tracers can be used as 'tags'. If they are introduced into the body, their progress around the body can be monitored from outside (so long as they are gamma emitters). This can give information about how a specific organ is (or is not) functioning as well as being used to analyse a whole body system (for example the circulation of blood around the body).

There are many factors that affect the choice of a particular radioisotope for a particular situation. Some of these are listed below:

- the radioisotope should be able to be 'taken up' by the organ in question in its usual way. In other words it needs to have specific chemical properties.
- the quantity of radioisotope needs to be as small as possible so as to minimise the harmful ionising radiation received by the body.
- the lifetime of the tracer needs to be matched to the time scale of the process being studied.

Organ/tissue	Tracers	Uses
general body composition	^{3}H, ^{24}Na, ^{42}K, ^{82}Br	Used to measure volumes of body fluids and estimate quantities of salts (e.g. of sodium, potassium, chlorine).
blood	^{32}P, ^{51}Cr, ^{125}I, ^{131}I, ^{132}I	Used to measure volumes of blood and the different components of blood (plasma, red blood cells) and the volumes of blood in different organs. Also used to locate internal bleeding sites.
bone	^{45}Ca, ^{47}Ca, ^{85}Sr, ^{99m}Tc	Used to investigate absorption of calcium, location of bone disease and how bone metabolises minerals.
cancerous tumours	^{32}P, ^{60}Co, ^{99m}Tc, ^{131}I	Used to detect, locate, and diagnose tumours. ^{60}Co is used to treat tumours.
heart and lungs	^{99m}Tc, ^{131}I, ^{133}Xe	Used to measure cardiac action: blood flow, volume, and circulation. Labelled gases used in investigations of respiratory activity.
liver	^{32}P, ^{99m}Tc, ^{131}I, ^{198}Au	Used in diagnosing liver disease and disorders in hepatic circulation.
muscle	^{201}Tl	Diagnosis in organs; in particular, heart muscle.

PET SCANS

Positron Emission Tomography (PET) is a technique that uses an isotope of carbon, carbon-11, as the radioactive tracer to measure the changes in blood flow within the brain. This particular isotope can be introduced into the body by breathing in a small sample of carbon monoxide. The radioactive carbon atoms will be easily taken up by the red blood cells and flow around the body.

When the carbon-11 decays, it does so by positive beta decay. A positron is emitted. When the positron meets an electron, they will destroy each other and two gamma rays will be emitted. These two gamma rays will essentially be travelling in opposite directions. Detectors around the patient can pick up these two gamma rays and calculate with great accuracy the position of the source of the gamma rays.

Current researchers can use this technique to image the functioning of the brain. A patient is given a specific task (such as reading or drawing) and the PET scan can identify the areas of the brain that receive more blood as a result of doing the task.

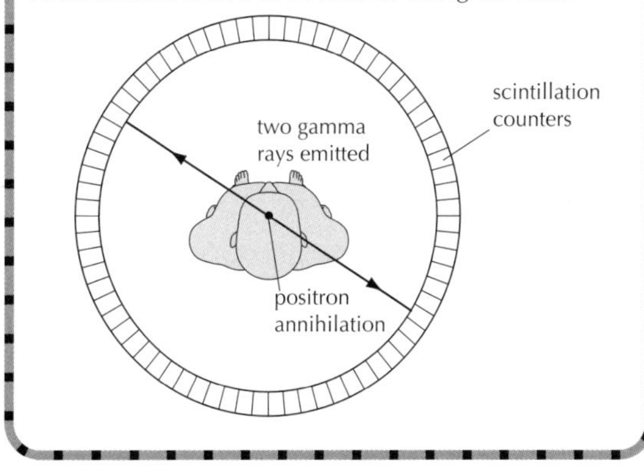

RADIATION THERAPY

So far, we have considered only the applications of ionising radiation involved with diagnostic medicine (identifying diseases), but they can also be used for therapy (treatment). The aim of radiotherapy is to target malignant cells in preference to normal healthy cells.

The dose that is used is critical. Too high a dose and too many healthy cells are killed. Too low a dose and the cancer is not destroyed. The dose needs to be as high as possible in the region of the cancer and as low as possible elsewhere. This can be achieved by one of two techniques:

- A radioactive source can be placed in the tumour itself. This can be done chemically or physically.
- Overlapping beams of radiation can be used. Where they overlap the dose will be high, elsewhere the dose will be lower.

The source of the ionising radiation can be a radioactive element, but these days it is also common to use high-energy X-rays or gamma rays from particle accelerators. High-energy protons can also be used directly. Malignant cells, are slightly more susceptible to damage from radiation compared with normal healthy cells.

CHOICE OF RADIOISOTOPE

Important factors include:

- the nature of the radiation – alpha, beta and gamma have different penetrating powers.
- the energy of the radiation.
- the physical half-life of the radioisotope. A short half-life means that the radioisotope will decay quickly, but sometimes a constant source is required.
- the biological half-life.
- the chemical properties – some chemicals are 'taken up' more readily than others.
- whether the radioisotope is solid, liquid or gas.
- the availability of the radioisotope.

Some common radioisotopes are listed below:

Radioisotope	Comment
Cobalt-60	Source of high-energy gamma rays used for radiotherapy. Long half-life so does not need to be replaced often.
Technetium-99m	This is a gamma emitter with a short half-life that can be produced relatively easily. It can be used in different forms to study blood flow, liver function, and bone growth.
Iodine-123	A gamma emitter that is readily taken up by the thyroid. It can also be used to study the functioning of the liver.
Iodine-131	A gamma emitter than is sometimes used to treat cancer of the thyroid.
Xenon-133	A gamma emitter in gaseous form which can be inhaled and used to study lung function.

Radiation sources in diagnosis and therapy (2)

DIFFERENT TYPES OF HALF-LIFE

A radioactive source will decay in the laboratory. If the source is introduced into a living body, there is the extra complication of the biological processes that are taking place. These will tend to remove the source from the body and this means that the **effective half-life** is reduced.

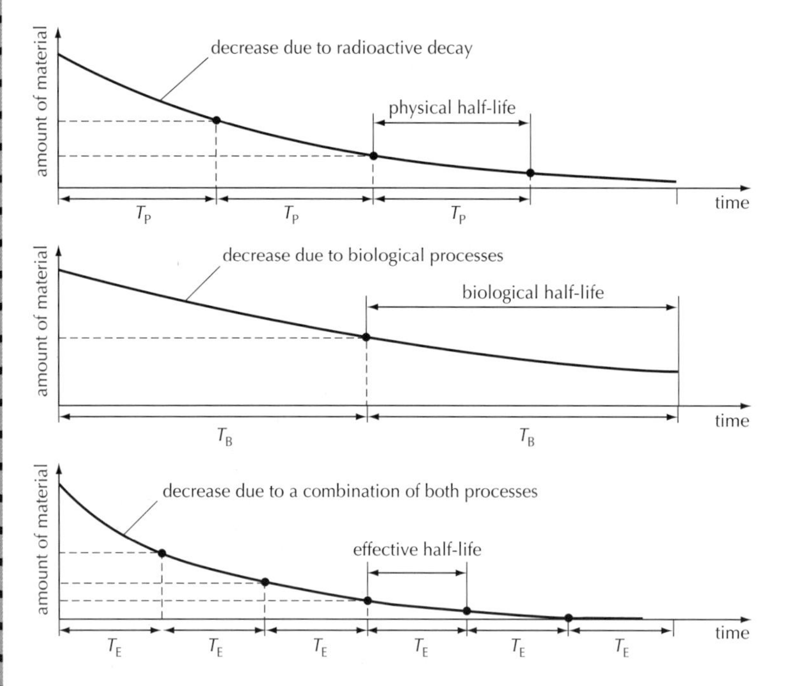

The effective half-life T_E of an isotope is the time taken for the number of radioactive nuclei of that isotope **in the body** to halve. This can be seen as the result of two processes. One process is radioactive decay – this will have a certain **physical half-life** T_P. The other process is the chemical removal of the isotope from the body and this will have a certain **biological half-life** T_B. The relationship between these values is

$$\frac{1}{T_E} = \frac{1}{T_P} + \frac{1}{T_B}$$

EXAMPLE

If the physical half-life of a radioisotope is 10 days and the biological half-life is 15 days, what fraction will remain after 30 days?

$$\frac{1}{T_E} = \frac{1}{10} + \frac{1}{15} = \frac{1}{6}$$

$\therefore\ T_E = 6$

30 days = $5 \times T_E$

$\therefore$ fraction remaining = $\left(\frac{1}{2}\right)^5 = 3.1\%$

EXAMPLE OF RADIOACTIVE PROCEDURE

A possible procedure to determine the total red blood cell volume would be as follows:

- Sample of patient's blood taken
- The red blood cells in the sample are radioactively labelled with chromium-51.
- This sample is re-injected into the patient.
- Radioactively labelled blood is allowed to circulate and mix with the rest of the patient's blood.
- After a suitable time interval, further blood samples are taken.
- The radioactive levels of chromium-51 in the new samples are measured using a gamma counter.
- A comparison between the original sample and the new samples allows the dilution level to be calculated and hence the total red blood cell volume can be calculated.

IB QUESTIONS – OPTION I – MEDICAL PHYSICS

1 The diagram below shows how the typical threshold of hearing varies with frequency for a normal young person.

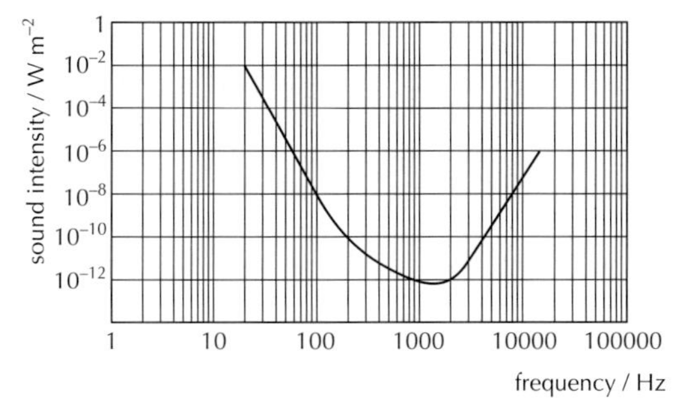

(a) Outline how the data for this graph could be obtained. [3]

(b) To what approximate frequency of sound is the ear most sensitive? [1]

(c) Over what range of frequencies is a sound of intensity 10^{-10} W m^{-2} audible? [1]

2 This question is about human hearing.

(a) Explain what is meant by *conductive* hearing loss. [1]

(b) A person with conductive hearing loss who uses an effective hearing aid will suffer little loss of hearing when taking part in a conversation. However, explain why such a person should be advised **not** to spend a lot of money on a 'Hi-Fi' music system. [3]

(c) A person has hearing loss of 40 dB at a frequency of 1000 Hz. Estimate the least intensity of sound that the person can detect at this frequency. [2]

3 Radioisotopes can be introduced into the body for **imaging** or for **therapy**. One common radioisotope is iodine-131.

(a) Explain the difference between **biological** half-life and **physical** half-life of a radioisotope. [2]

(b) A sample of a compound of iodine-131 is administered to a patient. The physical half-life of iodine-131 is 8 days whereas the biological half-life of this compound is about 20 days. What percentage activity will remain after 40 days? [3]

(c) If a patient receives the same **absorbed dose** from two different sources they would not necessarily receive the same **dose equivalent**. Explain what is meant by the terms **absorbed dose** and **dose equivalent**. Explain how they are related. [3]

(d) Outline **two** precautions necessary when introducing radioisotopes into the body. [2]

4 This question is about ultrasound scanning.

(a) State a typical value for the frequency of ultrasound used in medical scanning. [1]

The diagram below shows an ultrasound transmitter and receiver placed in contact with the skin.

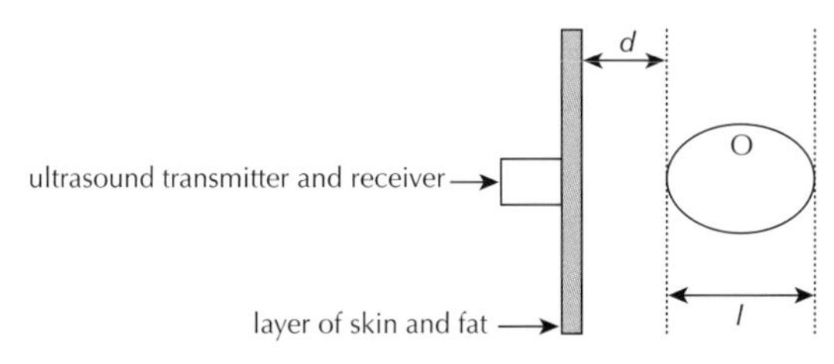

The purpose of this particular scan is to find the depth d of the organ labelled O below the skin and also to find its length, l.

(b) **(i)** Suggest why a layer of gel is applied between the ultrasound transmitter/receiver and the skin. [2]

On the graph below the pulse strength of the reflected pulses is plotted against the time lapsed between the pulse being transmitted and the time that the pulse is received, t.

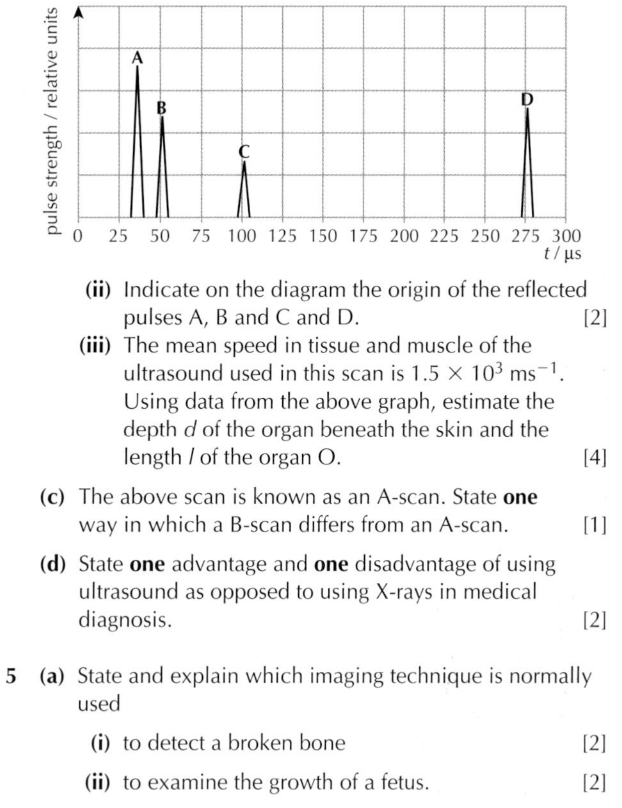

(ii) Indicate on the diagram the origin of the reflected pulses A, B and C and D. [2]

(iii) The mean speed in tissue and muscle of the ultrasound used in this scan is 1.5×10^3 ms^{-1}. Using data from the above graph, estimate the depth d of the organ beneath the skin and the length l of the organ O. [4]

(c) The above scan is known as an A-scan. State **one** way in which a B-scan differs from an A-scan. [1]

(d) State **one** advantage and **one** disadvantage of using ultrasound as opposed to using X-rays in medical diagnosis. [2]

5 **(a)** State and explain which imaging technique is normally used

(i) to detect a broken bone [2]

(ii) to examine the growth of a fetus. [2]

The graph below shows the variation of the intensity I of a parallel beam of X-rays after it has been transmitted through a thickness x of lead.

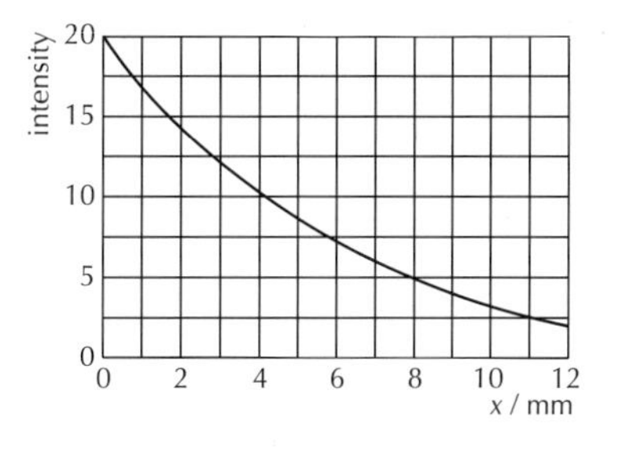

(b) **(i)** Define *half-value thickness*, $x_{1/2}$. [2]

(ii) Use the graph to estimate $x_{1/2}$ for this beam in lead. [2]

(iii) Determine the thickness of lead required to reduce the intensity transmitted to 20% of its initial value. [2]

(iv) A second metal has a half-value thickness $x_{1/2}$ for this radiation of 8 mm. Calculate what thickness of this metal is required to reduce the intensity of the transmitted beam by 80%. [3]

Description and classification of particles

CLASSIFICATION OF PARTICLES

Particle accelerator experiments identify many, many "new" particles. Two original classes of particles were identified – the **leptons** (= "light") and the **hadrons** (= "heavy"). The hadrons were subdivided into **mesons** and **baryons**. Another class of particles is involved in the mediation of the interactions between the particles. These were called **gauge bosons** or 'exchange bosons'.

Particles are called **elementary** if they have no internal structure, that is, they are not made out of smaller constituents. The classes of elementary particles are quarks, leptons and the exchange particles. Another theoretical particle, the Higgs particle, could be elementary (see page 206). Combinations of elementary particles are called **composite** particles. An important method of classification of all particles is a concept called **spin**.

CONSERVATION LAWS

Not all reactions between particles are possible. The study of the reactions that did take place gave rise to some experimental conservation laws that applied to particle physics. Some of these laws were simply confirmation of conservation laws that were already known to physicists – charge, momentum (linear and angular) and mass-energy. On top of these fundamental laws there appeared to be other rules that were never broken e.g. the law of conservation of baryon number. If all baryons were assigned a "baryon number" of 1 (and all antibaryons were assigned a baryon number of –1) then the total number of baryons before and after a collision was always the same. A similar law of conservation of lepton number applies.

Other reactions suggested new and different particle properties that were often, but not always, conserved in reactions. "Strangeness" and "charm" are examples of two such properties. Strangeness is conserved in all strong interactions.

All particles, whether they are elementary or composite, can be specified in terms of their mass and the various quantum numbers that are related to the conservation laws that have been discovered. The quantum numbers that are used to identify particles include:

- electric charge
- spin (see box below)
- strangeness
- charm
- color (this property is not the same as an object's actual colour – see page 205)
- lepton number
- baryon number.

Every particle has its own **antiparticle**. An antiparticle has the same mass as its particle but all its quantum numbers (including charge, spin etc) are opposite. There are some particles (e.g. the photon) that are their own antiparticle.

SPIN

A very important particle property is the quantum-mechanical phenomenon of spin angular momentum or **spin**. In any reaction the total angular momentum of the particles must be conserved and particles' individual spins contribute to the overall total. Particle spin is quantized, which means that only discrete values are possible.

A quantum of spin is $\frac{h}{2\pi}$ where h is Planck's constant (6.63×10^{-34} J s). This is often written as $\hbar$ Using this notation, an electron's spin will either be $+\frac{\hbar}{2}$ or $-\frac{\hbar}{2}$. The $\hbar$ is often neglected when talking about spin, so these two states are known as spin $+\frac{1}{2}$ (up) or spin $-\frac{1}{2}$ (down).

A classical way of interpreting of an electron's spin would be to imagine the electron as a massive charged sphere spinning on its own axis. This spin will generate a magnetic field and in turn the magnetic field of an atom would be the addition of the fields generated by the electrons orbiting around the nucleus and their own magnetic field. This model, however, does not agree with experimental observations.

The spin of an electron can be measured by, for example, sending a beam of electrons into a region of weak magnetic field. Some of the spins arrange themselves in the direction of the magnetic field and the others will be in exactly the opposite direction. The result is that the electron spins will be in one of two possible states however we perform the experiment. Whatever direction of magnetic field we apply, we will find that the electrons' spin is either parallel or antiparallel to the direction chosen.

In fact all spin $\frac{1}{2}$ particles always have only two possible orientations of spin – parallel or antiparallel to the chosen axis. A spin 1 particle will have three allowed orientations and a spin $\frac{3}{2}$ would have four.

Particles with non-integer spin ($\pm\frac{\hbar}{2}$, $\pm\frac{3\hbar}{2}$ etc.) are called **fermions** and when they interact in large numbers they obey a set of statistical laws called Fermi–Dirac statistics. A very important property of all fermions is that they obey the Pauli exclusion principle (see below). All 'everyday' particles of matter (proton, neutron, electrons, etc) are fermions. Specifically, leptons and baryons are fermions

Particles with zero or integer spin (0, $\pm\hbar$, etc) are called **bosons** and they obey a set of statistical laws called Bose–Einstein statistics. They do not obey the Pauli exclusion principle. Mesons are all bosons as are photons and other exchange bosons (e.g. gluons).

Fundamental interactions

CLASSIFICATION OF INTERACTIONS-EXCHANGE PARTICLES

There are only four fundamental interactions that exist: Gravity, Electromagnetic, Strong and Weak.

- At the end of the nineteenth century, Maxwell showed that the electrostatic force and the magnetic force were just two different aspects of the more fundamental electromagnetic force.
- Friction is simply a result of the forces between atoms and this must be governed by electromagnetic interactions.
- The strong and the weak interaction only exist over nuclear ranges.
- The strong force binds the nucleus together.
- The weak force explains radioactive β decay.
- The electromagnetic force and the weak nuclear force are now considered to be aspects of a single electroweak force.
- All four interactions can be thought of as being mediated by an exchange of particles. Each interaction has its own exchange particle or particles. The bigger the mass of the exchange boson, the smaller the range of the force concerned.

Interaction	Relative strength	Range (m)	Exchange particle
strong	1	$\sim 10^{-15}$	8 different gluons
electromagnetic	10^{-2}	infinite	photon
weak	10^{-13}	$\sim 10^{-18}$	W^+, W^-, Z^0
gravity	10^{-39}	infinite	graviton

The electromagnetic interaction involves charged matter.

The gravitational and the weak interactions apply to all matter.

The strong interaction only applies to hadrons (baryons and mesons). Leptons and bosons are unaffected by the strong force.

THE PAULI EXCLUSION PRINCIPLE

No two fermions can occupy the same quantum state. In other words, no two fermions in the same quantum system can have the same set of quantum numbers as each other.

Electrons are fermions so this must apply to all electrons inside any given atom. There are four quantum numbers that apply in an atom. A difference in quantum number identifies a different possible orbital state. The Pauli exclusion principle means that no two electrons can be in the same orbital state.

The four quantum numbers are:

- The principle quantum number n, which gives the main energy level (1, 2, 3 etc.).
- The orbital quantum number l which is an integer from 0, 1, 2 etc up to $(n - 1)$. (Also often denoted by the letters s, p, d etc.).
- The magnetic quantum number m which is an integer from $-l$ to $+l$ (e.g. −1, 0, +1 for the p orbital, also sometimes called p_x, p_y, p_z).
- The spin quantum number m_s which is always $\pm\frac{1}{2}$ for electrons.

Each quantum number designates an available orbital. It is possible for two electrons to have the same quantum numbers for n, l and m only if they have opposite spins. A third electron must occupy a different orbital.

EXCHANGE PARTICLES AND THE UNCERTAINTY PRINCIPLE

Any force can be imagined as being mediated by the exchange of virtual particles.

The exchange results in repulsion between the two particles

From the point of view of quantum mechanics, the energy needed to create these virtual particles, ΔE is available so long as the energy of the particle does not exist for a longer time Δt than is proscribed by the uncertainty principle (see page 108).

$$\Delta E\,\Delta t \geq \frac{h}{4\pi}$$

h is Planck's constant = 6.63×10^{-34} J s

The greater the mass of the exchange particle, the smaller the time for which it can exist. The range of the weak interaction is small as the masses of its exchange particles (W^+, W^- and Z^0) are large.

In particle physics, all real particles can be thought of as being surrounded by a cloud of virtual particles that appear and disappear out of the surrounding vacuum. The lifetime of these particles is inversely proportional to their mass. The interaction between two particles takes place when one or more of the virtual particles in one cloud is absorbed by the other particle.

Feynman diagrams (1)

RULES FOR DRAWING FEYNMAN DIAGRAMS

Feynman diagrams can be used to represent possible particle interactions. The diagrams are used to calculate the overall probability of an interaction taking place. In quantum mechanics, in order to find out the overall probability of an interaction, it is necessary to add together all the possible ways in which an interaction can take place. Used properly they are a mathematical tool for calculations but, at this level, they can be seen as a simple pictorial model of possible interactions.

Feynman diagrams are space-time diagrams. Typically the *x*-axis represents time going from left to right and the *y*-axis represents space (some books reverse these two axes). Some simple rules help in the construction of correct diagrams:

- Each junction in the diagram (vertex) has an arrow going in and one going out. These will represent a lepton-lepton transition or a quark–quark transition.
- Quarks or leptons are solid straight lines.
- Exchange particles are either wavy (photons, $W^{\pm}$ or Z) or curly (gluons).
- Time flows from left to right. Arrows from left to right represent particles travelling forward in time. Arrows from right to left represent antiparticles travelling forward in time.
- The labels for the different particles are shown at the end of the line.
- The junctions will be linked by a line representing the exchange particle involved.

EXAMPLES

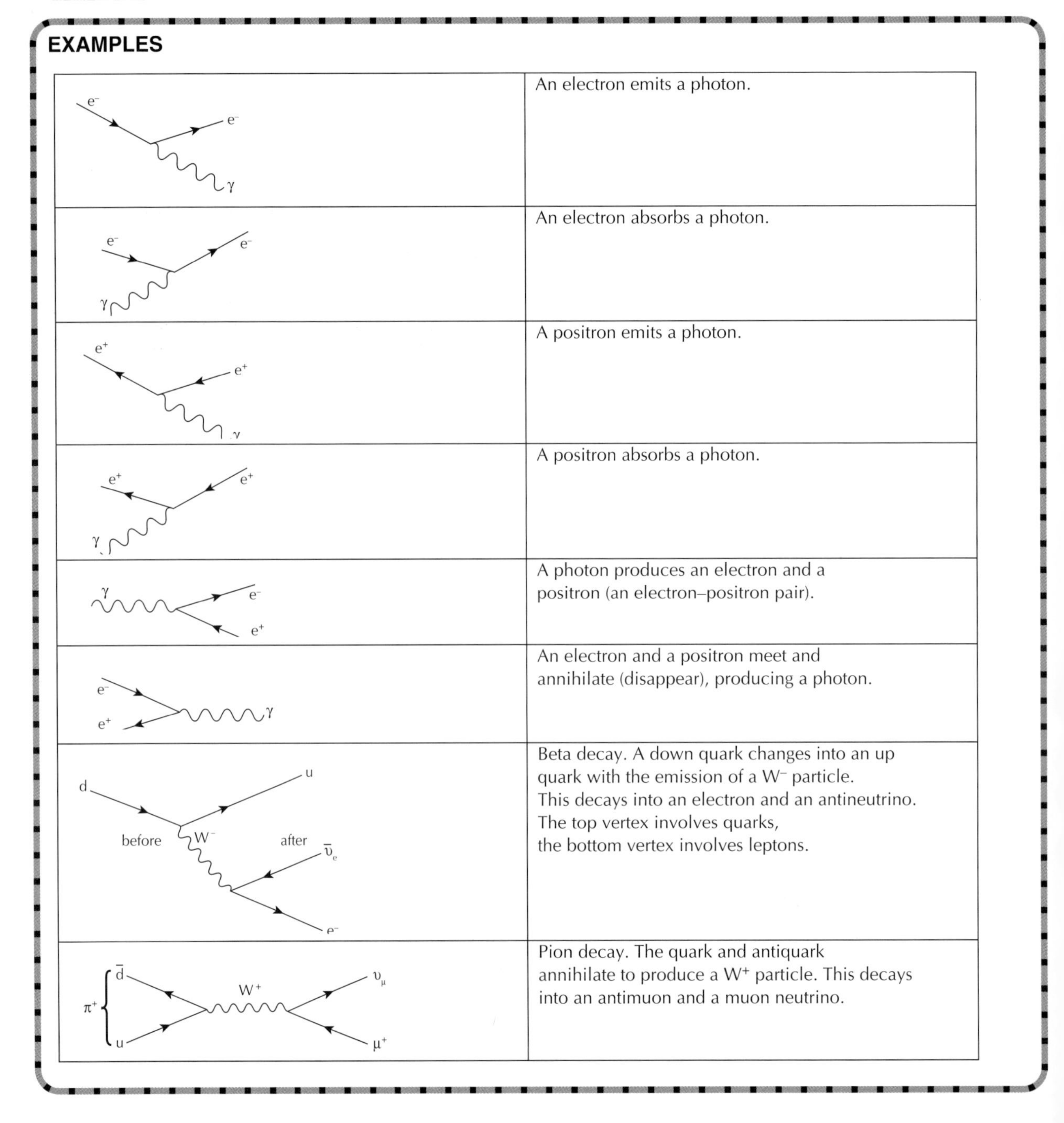

e⁻
e⁻
γ
An electron emits a photon.
An electron absorbs a photon.
A positron emits a photon.
A positron absorbs a photon.
A photon produces an electron and a positron (an electron–positron pair).
An electron and a positron meet and annihilate (disappear), producing a photon.
d
u
before
W⁻
after
ῡe
e⁻
Beta decay. A down quark changes into an up quark with the emission of a W⁻ particle.
This decays into an electron and an antineutrino.
The top vertex involves quarks,
the bottom vertex involves leptons.
π⁺
d̄
u
W⁺
υμ
μ⁺
Pion decay. The quark and antiquark annihilate to produce a W⁺ particle. This decays into an antimuon and a muon neutrino.

Feynman diagrams (2)

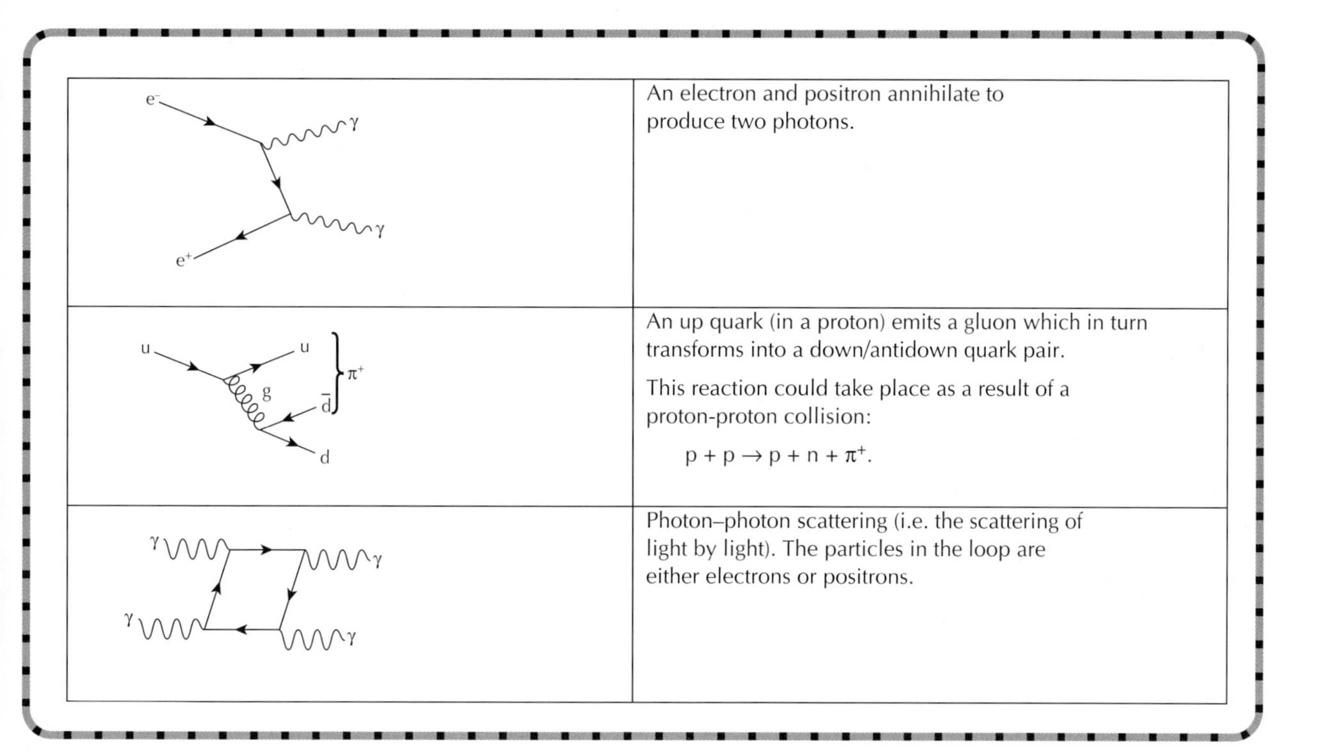

	An electron and positron annihilate to produce two photons.
	An up quark (in a proton) emits a gluon which in turn transforms into a down/antidown quark pair. This reaction could take place as a result of a proton-proton collision: $p + p \rightarrow p + n + \pi^+$.
	Photon–photon scattering (i.e. the scattering of light by light). The particles in the loop are either electrons or positrons.

USES OF FEYNMAN DIAGRAMS

Once a possible interaction has been identified with a Feynman diagram, it is possible to use it to calculate the probabilities for certain fundamental processes to take place. Each line and vertex corresponds to a mathematical term. By adding together all the terms, the probability of the interaction can be calculated using the diagram.

More complicated diagrams with the same overall outcome need to be considered in order to calculate the overall probability of a chosen outcome. The more diagrams that are included in the calculation, the more accurate the answer.

In a Feynman diagram, lines entering or leaving the diagram represent real particles and must obey relativistic mass, energy and momentum relationships. Lines in intermediate stages in the diagram represent virtual particles and do not have to obey energy conservation providing they exist for a short enough time for the uncertainty relationship to apply. Such virtual particles cannot be detected.

RANGE OF INTERACTIONS

A formula exists to calculate the range of an interaction from the mass of the exchange particle:

$$R \approx \frac{h}{4\pi mc}$$

R is the approximate range of the interaction in m
h is Planck's constant = 6.63×10^{-34} J s
m is the rest mass of the exchange particle in kg
c is the speed of light in m s^{-1}

The range of the weak interaction is approximately 10^{-18} m. This means the mass of the exchange particles involved is given by:

$$m \approx \frac{h}{4\pi cR} = \frac{6.63 \times 10^{-34}}{4\pi \times 3 \times 10^{8} \times 10^{-18}} = 1.8 \times 10^{-25}\text{kg} \approx 10^{2}\ \text{GeV c}^{-2}$$

The accepted mass of the $W^{\pm}$ is 80 GeV c^{-2} and the mass of the Z is 91 GeV c^{-2}.

Particle accelerators (1)

NEED FOR HIGH ENERGIES

In a particle accelerator particles are accelerated up to very high energies and then collided. New particles can be created in accordance with Einstein's mass-energy relation. The tracks of the particles are recorded and the curvature of the path (as a result of a magnetic field) is used to calculate the mass and the velocity and the charge of the particle.

The high energies involved are needed for two reasons:

- The energy for the creation of new particles comes from the kinetic energy of the collisions. In order to produce particles of large mass, large energies of collisions are required.
- In order to resolve particles of a small size, a large energy is required. Wave-particle duality (see page 105) means that every moving particle will have a wavelength associated with it given by de Broglie equation: $\lambda = \frac{h}{p}$.

  If the particle's wavelength is large compared with the size of the object that is being imaged, then diffraction effects will mean that the object cannot be resolved. In order for a object of size d to be resolved, λ needs to be the same magnitude as d. Large energy means large momentum and a small wavelength.

LINEAR ACCELERATOR

In a linear accelerator, charged particles are accelerated along a horizontal evacuated tube. A radio frequency alternating potential difference is connected along a series of hollow electrodes. As it moves between two electrodes, the charged particle is accelerated by the electric field between the electrodes.

Once inside the electrode, the charged particle does not experience any electric field and the particle drifts along at constant velocity. During the time in the electrode, one half period of the alternating potential difference has taken place. As the particle emerges from the electrode, it is once again accelerated to the next electrode. Each time the charged particle moves between electrodes, the field is arranged to give it another burst of acceleration. The length of the electrodes gets longer down the tube as the particles accelerates, so that the frequency of the alternating p.d. can remain constant.

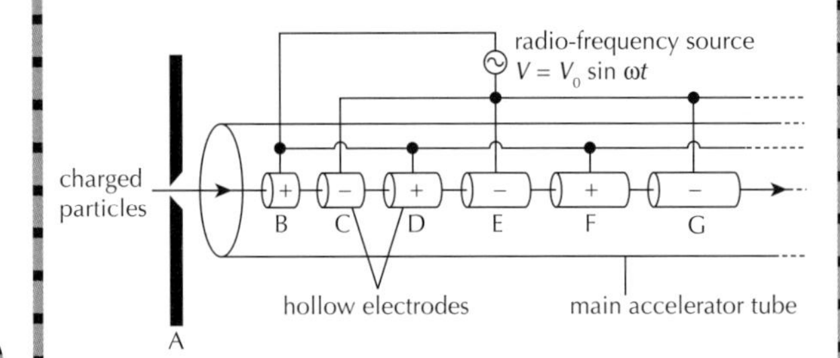

A negative particle will accelerate from A to B. During its time in B the potentials are reversed (B becomes negative and C becomes positive) and it will again be accelerated from B to C. The potential is reversed again while the particle is inside C and so on.

The final particle energy depends on the length of the tube and very long accelerators would be difficult to build. Linear accelerators can accelerate electrons to energies of around 30 GeV.

When a collision takes place, some of the energy will be lost as radiation as a result of the charged particle decelerating. This is known as **bremsstrahlung** (**braking radiation**). This process accounts for the continuous features of a typical X-ray spectrum (see page 166).

CYCLOTRON

The charged particles being accelerated in a cyclotron travel in an outward spiral in a vacuum. The curved motion of the particles is achieved by a fixed magnetic field. The particles move inside two hollow electrodes called **dees** that have a gap between them. The particles are accelerated as they move between the dees. An alternating potential difference between the dees is arranged so that each time the particles arrive at the gap, the potential accelerates the charged particles across the gap.

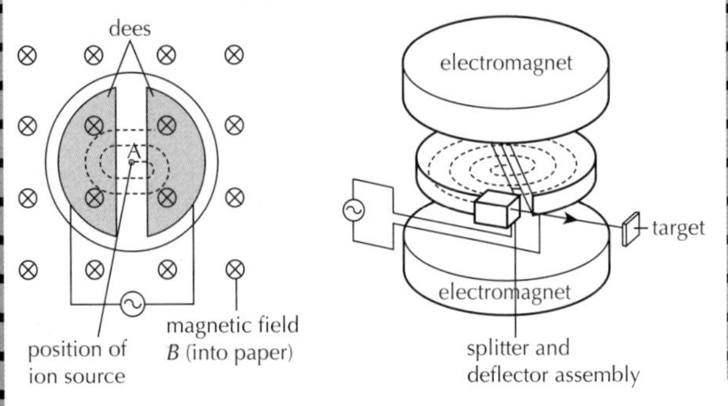

The magnetic force provides the centripetal force required:

$$Bqv = \frac{mv^2}{r}$$

The time taken for one revolution

$$T = \frac{2\pi r}{v} = \frac{2\pi m}{Bq}$$

So long as the velocities remain non-relativistic, this time is independent of the speed of the charged particle. As the particle accelerates, it moves in a circle of larger radius but arrives at the dees in the same time. This means that the frequency of the alternating potential difference remains constant. As the particles approach the speed of light either the alternating frequency or the magnetic field needs to be adjusted.

Frequency of alternating potential difference

$$f = \frac{1}{T} = \frac{Bq}{2\pi m}$$

Particle energies in the order of 100 MeV are possible. Higher energies would require too large a cyclotron.

Particle accelerators (2)

SYNCHROTRON

Synchrotrons are able to accelerate particles to extremely high energies of approximately 1000 GeV (= 1TeV). The cost of construction and maintenance is extremely high and collaborations of institutions in many different countries often jointly fund the few examples that exist worldwide (CERN, DESY, SLAC, Fermilab and Brookhaven).

The accelerated particles move around a roughly circular path of large radius enclosed in an evacuated pipe. Typically particles travel around the ring in bunches. It is possible to arrange for particles and their antiparticles (of the same energy) to both orbit the ring at the same time but in opposite directions. As particles move around, they encounter four different types of components:

- **Radio-frequency cavities** – where the particles are accelerated as they travel around the ring.
- **Bending magnets** – the strength of the field is modified so as to keep the particles travelling in the ring as their velocities are increased.
- **Focusing magnets** – the particles in the beam are kept close together in a bunch.
- **Experimental regions** – the particles are arranged to collide with other particles at these points.

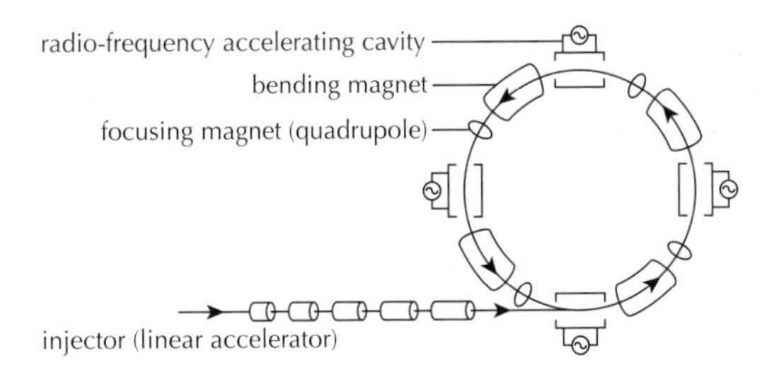

Whenever a charged particle changes its direction it must have accelerated. All accelerated charges radiate energy. Thus a charged particle going around a circular track will emit synchrotron radiation. Synchrotrons have large radius rings to minimize the energy lost in synchrotron radiation as this radiated energy must be continually replaced.

PARTICLE PRODUCTION

The energy, E, needed to create a particle (of rest mass m) at rest is given by Einstein's mass-energy relation, $E = mc^2$. A moving particle will have an additional kinetic energy, E_K, so in general the total energy of a particle is:

$$E = mc^2 + E_K$$

When a moving particle collides with a stationary target, not all of the input energy is available for particle production. As the incoming particle carried momentum with it, the law of conservation of momentum requires that some of the created particles must also be moving and thus have kinetic energy.

m v M

accelerated particle with energy E — target particle (at rest)

There is a linear momentum in the system before the collision, so the particles must be moving after the collision – momentum must be conserved.

The calculation of the energy available in any given situation is not trivial because the high velocities of the particles mean that the equations of special relativity must be used to move between the laboratory frame (in which the target is stationary) and the centre of mass frame. This gives the following relationship:

$$E_a^2 = 2Mc^2E + (Mc^2)^2 + (mc^2)^2$$

E_a is the energy available for the formation of particles as a result of the collision
M is the rest mass of the target particle
m is the rest mass of the incoming particle
c is the speed of light
E is the total energy of the incoming particle (i.e. rest energy + kinetic energy)

The left hand side of the equation is the total energy of the system in the centre of mass frame and the right hand side of the equation is in the laboratory frame.

EXAMPLE

A π^- particle is fired towards a proton. At high enough energies, the following reaction can take place:

$$\pi^- + p \rightarrow K^0 + \Lambda^0$$

In order to calculate the minimum kinetic energy of the π^- particle, we need to know the rest masses of the particles:

π^- 140 MeV c^{-2}
p 938 MeV c^{-2}
K^0 498 MeV c^{-2}
Λ^0 1116 MeV c^{-2}

For the reaction to happen,
total rest mass of products = 498 + 1116 MeV c^{-2} = 1614 MeV c^{-2}

$\therefore\ E_a = 1614$ MeV

substituting, $M = 938$ MeV and $m = 140$ MeV gives:

$(1614)^2 = 2 \times 938 \times E + 938^2 + 140^2$

$\therefore\ 1876\,E = 170\,552$

$\therefore\ E = 909.1$ MeV

This is the minimum total energy of the π^- particle, so

minimum KE of π^- = total energy – rest energy = 909.1 – 140 MeV = 769.1 MeV

The difference between the rest-mass energies before and after the collision is 536 MeV but we need to provide at least 769.1 MeV for the reaction to take place.

In order to ensure the maximum energy is available for particle production, it is better to arrange for a collision between two particles moving in opposite directions than for a moving particle to strike a stationary target:

M v — v M

accelerated particle — accelerated particle

If the masses and the velocities of the particles are identical, the total linear momentum before the collision is zero. Momentum must be conserved and thus the particles after the collision do not have to have significant KE. A higher proportion of the initial energy can go into particle production.

Particle detectors

BUBBLE CHAMBER

The bubble chamber is a tank of liquid hydrogen whose temperature is kept just below its boiling point. A sudden expansion of the chamber causes the pressure to be reduced. The liquid hydrogen is said to be **superheated** and will begin to boil. Any high-energy particle that enters the tank when the hydrogen is superheated will cause ionisations along its path. These ionisations will preferentially cause the formation of the initial bubbles along the particle's path. Cameras record the bubbles that have been formed and the pressure is returned to normal so that the process can be repeated.

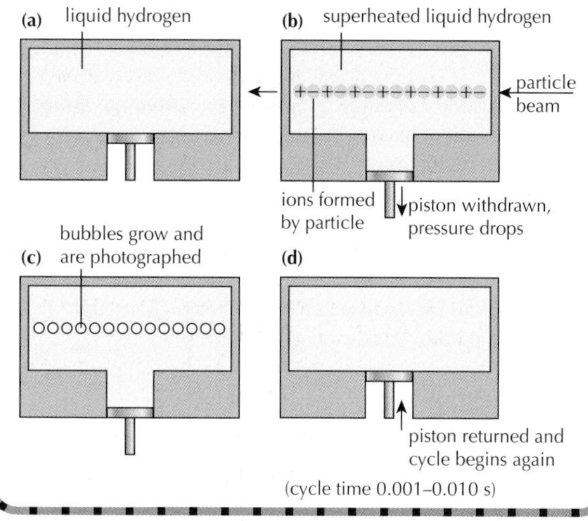

PHOTOMULTIPLIER

Individual particles can cause a photon to be emitted from certain materials, such as crystals of sodium iodide. A photomultiplier allows the single photon that is emitted to be readily detected. The photon strikes a photocathode and causes an electron to be emitted as a result of the photoelectric effect. This photoelectron is accelerated towards a curved electrode called a **dynode**. When it arrives it causes several electrons to be emitted. A chain of dynodes is arranged so that a large number of electrons (an avalanche) arrives at the anode as is recorded as a pulse of current.

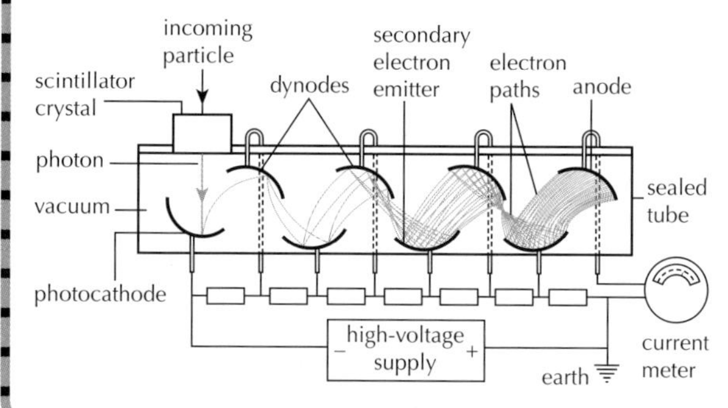

ISSUES

There are very few large-scale particle research facilities worldwide because the costs of building and maintaining such facilities are exceptionally high. There are arguments both for and against allocating funding to fundamental particle research.

Arguments for:

- Understanding the nature of the universe is one of the most fundamental, interesting and important areas for mankind as a whole and it therefore deserves to be properly resourced.
- All fundamental research has the potential to give rise to new technology that may eventually improve the quality of life for many.
- International particle research provides a forum for citizens of different countries to collaborate and work together. For example even at the height of the cold war, Western and Soviet scientists collaborated in the field of particle physics.

Arguments against:

- The enormous amounts of money required to do particle physics research could be more usefully spent providing food, shelter and medical care to the many millions of people who are suffering from hunger, homelessness and disease around the world.
- If money is to be allocated to research, it is much more worthwhile to invest limited resources into medical research. This offers the immediate possibility of saving lives and improving the quality of life for some sufferers.
- It might be more beneficial to fund a great deal of small diverse pieces of research rather than concentrating all funding into one expensive area.
- Is the information gained really worth the cost?

WIRE CHAMBER

The wire chamber consists of a 'sandwich' of many very fine wires held a few millimetres apart from one another with a potential difference of several thousand volts between the wires. When a charged particle passes into the chamber it will cause an ionisation. As a result, a number of electrons (some as a result of collisions) will be produced near one of the wires. The position of the wire that senses this pulse of current identifies one point along the particle's path. Many wire chambers can be used together to record a particle's path.

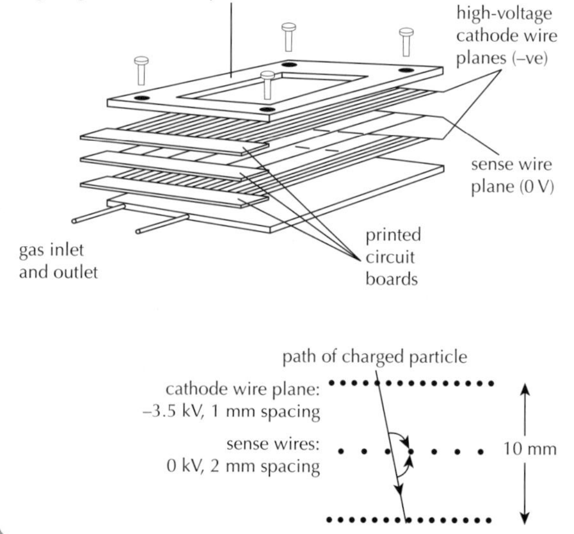

Quarks (1)

QUARK MODEL

All hadrons are made up from different combinations of fundamental particles called **quarks**. There are six different types of quark and six types of antiquark. This very neatly matches the six leptons that are also known to exist. Quarks are affected by the strong force (see below), whereas leptons are not. The weak interaction can change one type of quark into another.

	Electric charge	'Generation' 1	2	3
Quarks	$+\frac{2}{3}$	u (up) $M = 5$ MeV c^{-2}	c (charm) $M = 1500$ MeV c^{-2}	t (top) $M = 174\,000$ MeV c^{-2}
	$-\frac{1}{3}$	d (down) $M = 10$ MeV c^{-2}	s (strange) $M = 200$ MeV c^{-2}	b (bottom) $M = 4700$ MeV c^{-2}
		All quarks and antiquarks have spin $\frac{1}{2}$ All quarks have a baryon number of $+\frac{1}{3}$, All antiquarks have a baryon number of $-\frac{1}{3}$		

Isolated quarks cannot exist. They can exist only in twos or threes. Mesons are made from two quarks (a quark and an antiquark) whereas baryons are made up of a combination of three quarks (either all quarks or all antiquarks).

	Name of particle	Quark structure
Baryons	proton (p)	u u d
	neutron (n)	u d d
Mesons	π^-	d $\bar{u}$
	π^+	u $\bar{d}$

The force between quarks is still the strong interaction but the full description of this interaction is termed QCD theory – quantum chromodynamics. A quark can be one of three different 'colors' – red, green or blue. Only the 'white' combinations are possible. The force between quarks is sometimes called the color force. Eight different types of gluon mediate it.

CONSERVATION LAWS

All particle reactions involving hadrons must conserve electric charge, total energy and momentum. Particle reactions can result in the creation of different hadrons (baryons or mesons) but 'free' quarks do not exist on their own. Whenever a quark is created, an antiquark is also created.

In all reactions (electromagnetic, strong and weak) the **conservation of baryon number** applies. All baryons contain three quarks so their baryon number is +1. All antibaryons contain three antiquarks, to their baryon number is –1. Mesons contain a quark and an antiquark so their baryon number is 0.

Strange quarks have a strangeness of –1. Antistrange quarks have a strangeness of +1. All other quarks have a strangeness of 0. The **conservation of strangeness** will thus apply in electromagnetic and strong interactions. Since weak interactions change the nature of quarks, strangeness will not necessarily be conserved in weak interactions.

See page 199 for additional rules involving leptons.

SPIN STRUCTURE OF HADRONS

Each baryon is made up of three quarks. For example the proton is the bound state of two up quarks and one down quark (uud). Each quark has a spin $\frac{1}{2}$ so we could have two of the quarks with spins parallel and one antiparallel. Overall the spin of the proton is $\frac{1}{2}$. It is possible for baryons to have spin $\frac{3}{2}$ if their spins are all aligned. The Δ^+ particle quark content is also (uud) but it always has spin $\frac{3}{2}$. This arrangement is more energetic and so less stable.

Each meson is made up of a quark–antiquark pair. The quarks and antiquarks have spins in opposite senses (one parallel and one antiparallel). Overall most mesons have spin 0. Is also possible to form mesons with spin 1 if the quark–antiquark pair has parallel spin.

COLOR AND THE PAULI EXCLUSION PRINCIPLE

The Pauli exclusion principle states that no two fermions can occupy the same quantum state. The quarks are fermions (they have spin $\frac{1}{2}$) so they cannot be in the same quantum state when bound together inside hadrons. The quantum difference between the quarks is a property called color. All quarks can be red (r), yellow (y) or blue (b). Antiquarks can be antired ($\bar{r}$), antiyellow ($\bar{y}$) or antiblue ($\bar{b}$). The two up quarks in a proton are not identical because they have different colors. This means their spins can be parallel to one another.

Only white (**color neutral**) combinations are possible. Baryons must contain r, y & b quarks (or $\bar{r}$, $\bar{y}$, $\bar{b}$) where as mesons contain a color and the anticolor (e.g. r and $\bar{r}$ or b and $\bar{b}$ etc.)

Quarks (2)

QUANTUM CHROMODYNAMICS

The interaction between objects with color is called the color interaction and is explained by a theory called quantum chromodynamics. The force-carrying particle is called the gluon. There are eight different types of gluon each with zero mass and spin 1. Each gluon carries a combination of color and anticolor and their emission and absorption by different quarks causes the color force.

As the gluons themselves are colored, there will be a color interaction between gluons themselves as well as between quarks. The overall effect is that they bind quarks together. The force between quarks increases as the separation between quarks increases. Isolated quarks and gluons cannot be observed. If sufficient energy is supplied to a hadron in order to attempt to isolate a quark, then more hadrons will be produced rather than isolated quarks. This is known as **quark confinement**.

The six color-changing gluons are: $G_{r\bar{b}}$, $G_{r\bar{y}}$, $G_{b\bar{y}}$, $G_{b\bar{r}}$, $G_{y\bar{b}}$, $G_{y\bar{r}}$.

For example when a blue up quark emits the gluon $G_{b\bar{r}}$, it loses its blue color and becomes a red up quark (the gluon contains antired, so red color must be left behind). A red down quark absorbing this gluon will become a blue down quark.

There are two additional color-neutral quarks: G_0 and $G_0{}'$, making a total of eight gluons.

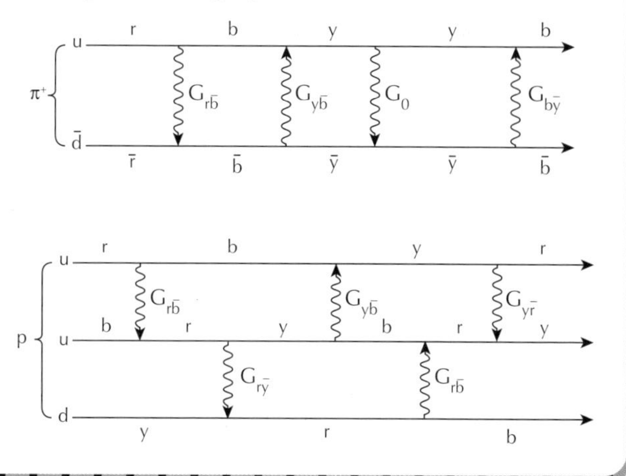

STRONG INTERACTION

The color interaction and the strong interaction are essentially the same thing. Properly, the color interaction is the fundamental force that binds quarks together into baryons and mesons. It is mediated by gluons. The **residual strong interaction** is the force that binds color-neutral particles (such as the proton and neutron) together in a nucleus. The overall effect of the interactions between all the quarks in the nucleons is a short-range interaction between color-neutral nucleons.

The particles mediating the strong interaction can be considered to involve the exchange composite particles (π mesons) whereas the fundamental color interaction is always seen as the exchange of gluons.

Leptons and the standard model

THE STANDARD MODEL

There are six different leptons and six different antileptons. The six leptons are considered to be in three different generations or families in exactly the same way that there are considered to be three different generations of quarks.

	Electric charge	'Generation' 1	2	3
Lepton	0	ν_e (electron-neutrino) M = 0 or almost 0	ν_μ (muon-neutrino) M = 0 or almost 0	ν_τ (tau-neutrino) M = 0 or almost 0
	–1	e (electron) $M = 0.511$ MeV c^{-2}	μ (muon) $M = 105$	τ (tau) $M = 1784$ MeV

All leptons and antileptons have spin $\frac{1}{2}$

The electron and the electron neutrino have a lepton (electron family) number of +1. The antielectron and the antielectron neutrino have a lepton (electron family) number of –1.

Similar principles are used to assign lepton numbers to the muon and the tau family members.

The **standard model** of particle physics is the theory that says that all matter is considered to be composed of combinations of six types of quark and six types of lepton. This is the currently accepted theory. Each of these particles is considered to be fundamental, which means they do not have any deeper structure. Gravity is not explained by the standard model.

CONSERVATION LAWS

In addition to the rules introduced on page 199, **lepton family number** is also conserved in all reactions. For example, whenever a muon is created, an antimuon or an antimuon neutrino must also be created so that the total number of leptons in the muon-family is always conserved.

HIGGS BOSON

The Higgs boson is an additional theoretical boson. It has been proposed in order to explain the process by which particles can acquire mass. If it exists, a particle's mass may be the result of interactions involving the Higgs boson. Experiments to look for its existence are already being constructed.

Experimental evidence for the quark and standard models

DEEP INELASTIC SCATTERING

Deep inelastic scattering experiments involve scattering leptons off hadrons when large amounts of energy (compared to their rest masses) and momentum are transferred to the hadrons. Charged constituents of the hadron can be identified.

For example, if a beam of electrons has sufficient energy, the de Bröglie wavelength associated with the electrons can become small enough to probe inside neutrons and protons.

$$\lambda = \frac{h}{p}$$

λ is the de Bröglie wavelength in m
h is Planck's constant (= 6.63×10^{-34} J s)
p is the momentum in kg m s^{-1}

As a result, some electrons are seen to deflect through large angles, suggesting an interaction with a tiny, dense, charged particle within the proton. This interaction cannot be via the strong interaction as leptons do not feel the strong force – it must be electromagnetic hence this observation provides evidence for the existence of centres of charge within the proton.

The result of electron scattering can be compared with similar experiments involving the scattering of neutrinos (by the weak interaction). At large energies, a large number of particles can be produced, including hadrons emitted in a narrow beam, or jet. The jet of particles is produced at a large angle to the direction of the incident neutrinos. The results are interpreted to show that the neutrino collides with a quark and as it is expelled, new quarks and antiquarks are created, forming a jet of hadrons all moving in approximately the same direction.

The difference between the two experiments allows the charge of the scattering objects to be calculated as $+\frac{2}{3}e$ and $-\frac{1}{3}e$ which agrees with the predictions of the quark model.

These experiments provide direct evidence not only for quarks in hadrons but also in mesons and for the existence of gluons and color.

It is possible to analyse the scattering results to measure the amount of momentum carried by each of the charged constituent parts of the hadron if it is moving. This turns out to be about one half of the total momentum in the case of a meson and about a third in the case of a baryon. These numbers agree well with the theoretical predictions of the quark compositions of these particles. In addition, the total momentum carried by the charged constituents does not add up to the momentum of the hadron, implying the existence of other neutral constituents (thought to be the gluons) inside the hadron.

It should be noted that the individual quarks and gluons behave as though they were 'free' particles while confined within the hadron. When quarks come very close together, the color force is very weak. This is called **asymptotic freedom** – the quarks inside a hadron are essentially free within the bounds of their confinement.

In deep inelastic scattering experiments, the energy transfer is very large. The strength of the strong interaction between the quarks decreases as the energy available for the interaction increases and so the quarks behave almost as free particles.

NEUTRAL CURRENT

Beta decay occurs by weak interaction. The overall process involves a down quark changing into an up quark and thus a neutron changes into a proton. The force-carrying particle is a W^- particle which decays into an electron and an electron antineutrino.

The reverse process is also possible: when an electron neutrino is fired at a neutron it can cause it to convert into a proton with the emission of an electron. In this situation the force-carrying particle is the W^+ particle.

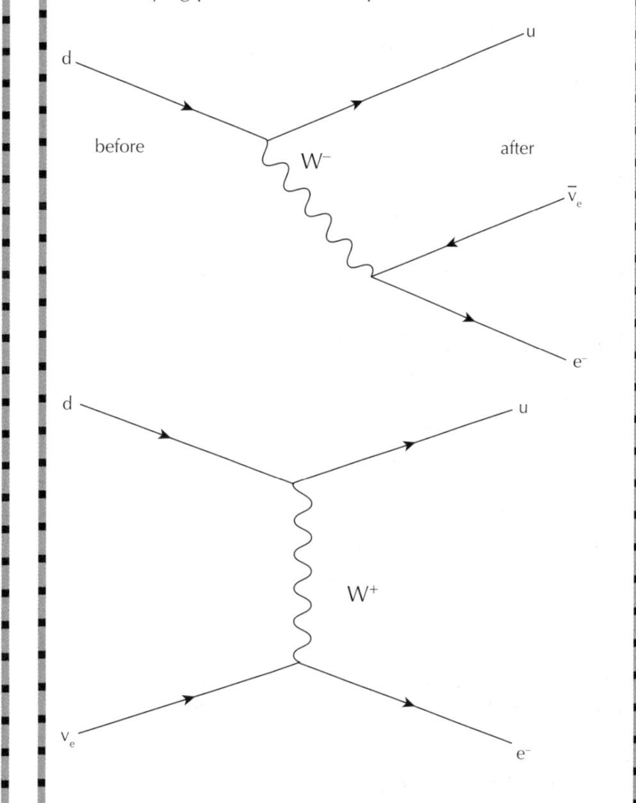

In both these cases the weak interaction involves a transfer of charge. One of the predictions of standard model was the unification of the weak and electromagnetic forces. This suggested that neutral weak interactions could take place involving the neutral Z^0 boson. If these reactions, called **neutral currents**, could be detected then this would provide evidence for the standard model.

Typically a neutral Z^0 boson may decay into an electron–antielectron or a muon–antimuon pair. For example the collision of a high energy neutrino with a proton may result in the scattering of the neutrino and the disintegration of the proton into several hadrons. The total charge of the hadrons must be +1 indicating that the boson involved was neutral. The Z^0 boson has been observed and its mass verified by experiments.

Cosmology and strings

TEMPERATURE OF THE UNIVERSE

The current model for the creation of the universe is called the **Big Bang** model. The universe is thought to have started life at extremely high temperature and density. Following the Big Bang, the universe expanded and cooled down. Important points in the development of the early universe are shown on page 135. Calculations working backwards from the current average temperature of universe of 2.7 K estimate that the temperature the universe was 10^{32} K very soon (10^{-43} s) after the Big Bang.

MATTER AND ANTIMATTER IN THE UNIVERSE

The Big Bang model predicts that matter and antimatter would be created from photon–photon collisions in equal amounts and the early universe contained almost equal numbers of particles and antiparticles. The current universe, however, appears to be predominately matter, with very limited amounts of antimatter existing. This means that the early universe must have contained slightly more matter than antimatter particles.

Theorists are researching possible reasons for this so-called **asymmetry** in the reactions that took place in the early universe that meant that slightly more matter than antimatter was created. Whatever the reason, once the temperature fell below about 10^{10} K (see above) the photons would not have had enough energy to create particle–antiparticle pairs and the predominance of matter over antimatter would be fixed for ever.

STRING THEORY

The standard model of particle physics does not include an explanation of the gravitational force and so far all attempts to reconcile gravitation with quantum theories have not been successful. Current theories of the universe have introduced the idea that all elementary particles might be regarded not as points but as strings. This idea has been developed into a set of theories called string theories.

The lengths of these strings are extremely small but all the quantum properties correspond to different modes of vibration. This is similar to the harmonics of an ordinary vibrating string in, for example, a violin where different frequencies correspond to different musical notes.

The standard model is formulated in four dimensions (three space and one time) whereas string theories typically require many extra dimensions.

PARTICLE INTERACTIONS IN THE EARLY UNIVERSE

The average kinetic energy per particle E_K of a system of interacting particles is related to its absolute temperature T by the following relationship:

$$E_K = \frac{3}{2}kT$$

k is Boltzmann's constant = 1.38×10^{-23} J K^{-1}

The early universe is believed to have started as a hot, high-energy photon gas. Photons interacted to created particle–antiparticle pairs and the reverse process also took place.

The equation can be used to work out the minimum temperature below which the production of proton–antiproton becomes impossible:

Rest mass of a proton = 938 MeV c^{-2}

∴ energy needed for proton–antiproton pair

$= 1876 \text{ MeV} = 1876 \times 1.6 \times 10^{-13} = 3.0 \times 10^{-10}$ J

$$T = \frac{2E}{3k} = \frac{2 \times 3.0 \times 10^{-10}}{3 \times 1.38 \times 10^{-23}} \approx 10^{13} \text{ K}$$

The minimum temperature for the production of electron–positron pairs will be lower as the rest mass of the electron is lower.

Rest mass of an electron = 0.511 MeV c^{-2}

∴ energy needed for electron–positron pair

$= 1.022 \text{ MeV} = 1.022 \times 1.6 \times 10^{-13} = 1.6 \times 10^{-13}$ J

$$T = \frac{2E}{3k} = \frac{2 \times 1.6 \times 10^{-13}}{3 \times 1.38 \times 10^{-23}} \approx 10^{10} \text{ K}$$

Nuclei can be formed if protons and neutrons can combine together. If the temperature is too high, nuclei will not be stable and will break apart. If the temperature is too low, then protons will not be able to overcome their mutual electrostatic repulsion. The energy for two protons to collide together can be calculated from the electrostatic potential energy:

$$\text{PE required} = \frac{q^2}{4\pi\varepsilon_0 r} \text{ where } q \text{ is the charge on the proton}$$

and r is the approximate diameter of a proton. Using 10^{-15} m for the required distances gives energies of the order of 1 MeV, so the temperature is approximately 10^{10} K. Approximately 100 s after the Big Bang, nucleosynthesis stopped with only the nuclei of the lightest elements having been formed.

Matter and photons continued to interact with one another until the temperature became low enough for the nuclei to capture electrons and for atoms of hydrogen and helium to be formed. The ionisation energy of atoms is of the order of 10 eV, which corresponds to a temperature of approximately 10 000K for the beginning of the process. After the atoms had been formed, the universe is said to have become transparent to radiation.

IB QUESTIONS – OPTION J – PARTICLE PHYSICS

1 Two protons within a nucleus can feel an electrostatic repulsion while at the same time feeling an attraction due to the strong force. If their separation is increased to 10^{-14} m, the magnitudes of both forces change.

(a) Explain how each force (strong and electrostatic) can be viewed in terms of the exchange of virtual particles. [3]

(b) Explain how each force varies as the separation is increased. [2]

2 This question is about radioactive decay and fundamental forces.

(a) The nucleus of manganese-54 ($^{54}_{25}$Mn) undergoes **positive** beta decay to form a nucleus of chromium (Cr). Complete the following equation for this decay process.

$$^{54}_{25}\text{Mn} \rightarrow \text{Cr} + \beta^+ + \qquad [3]$$

(b) Positive beta decay of a nucleus involves the weak nuclear interaction (force). State the name of the **exchange** particle involved in the weak nuclear interaction. [1]

(c) State the name of

(i) the **interaction** involved when a nucleus undergoes **alpha** particle decay [1]

(ii) an **exchange** particle involved with **alpha** particle decay. [1]

3 The reaction

$$n \rightarrow p + e^-$$

never occurs because it violates the law of conservation of

A baryon number

B lepton number

C electric charge

D baryon number and electric charge.

4 Which **one** of the following lists the three classes of fundamental particle?

A	leptons	quarks	exchange bosons
B	hadrons	quarks	exchange bosons
C	leptons	mesons	baryons
D	hadrons	baryons	mesons.

5 Possible particle reactions are given below. They **cannot** take place because they violate one or more conservation laws. For each reaction identify **one** conservation law that is violated.

(a) $\mu^- \rightarrow e^- + \gamma$ [1]

(b) $p + n \rightarrow p + \pi^0$ [1]

(c) $p \rightarrow \pi^+ + \pi^-$ [1]

6 This question is about deducing the quark structure of a nuclear particle.

When a K^- meson collides with a proton, the following reaction can take place.

$$K^- + p \rightarrow K^0 + K^+ + X$$

X is a particle whose quark structure is to be determined.

The quark structure of mesons is given below.

particle	**quark structure**
K^-	$s\bar{u}$
K^+	$u\bar{s}$
K^0	$d\bar{s}$

(a) State and explain whether the original K^- particle is a hadron, a lepton **or** an exchange particle. [2]

(b) State the quark structure of the proton. [2]

(c) The quark structure of particle X is sss. Show that the reaction is consistent with the theory that hadrons are composed of quarks. [2]

7 Which particles among the following are leptons?

A Protons and neutrons.

B Electrons and photons.

C Electrons and neutrinos.

D Quarks and bosons.

8 Which of the following is **not** conserved when and electron and its antiparticle undergo mutual annihilation?

A Lepton number.

B Electric charge.

C Linear momentum.

D Kinetic energy.

9 For each of the situations listed below, state which fundamental interaction is involved

(a) a person's weight

(b) three quarks are held together to make up a proton

(c) the push up from a chair on the person sitting on it

(d) a compass needle pointing toward geographic north. [4]

10 A collision between a proton and an electron produces a neutron and a neutrino. Draw a Feynman diagram to illustrate this process. [3]

11 A moving proton collides with a stationary proton and produces a neutral meson π^0 according to the reaction

$$p + p \rightarrow p + p + \pi^0$$

The rest mass of the proton is 938 MeV c^{-2} and rest mass of the π^0 is 135 MeV c^{-2}.

(a) Calculate the **minimum** kinetic energy, E_k, of the moving proton, in order to produce the π^0.

(b) π^0 particles can also be produced when a moving proton collides with another proton with the same kinetic energy moving in the opposite direction. State and explain whether the total energy required to produce a π^0 particle in this way will be greater, smaller or the same as your answer in **(a)**. [3]

12 (a) Explain what is meant by the standard model of particle physics. [3]

(b) Outline one piece of experimental evidence that supports the existence of quarks. [3]

13 (a) State the order of magnitude for the ionization energy of atoms. [1]

(b) Use your answer to **(a)** to estimate the temperature when the universe became transparent to radiation. [1]

Answers to questions

PHYSICS AND PHYSICAL MEASUREMENT *(page 11)*
1. B **2.** B **3.** B **4.** C **5.** A **6.** B **7.** B **8.** C
9. (a) (i) 0.5 × acceleration down slope **(iv)** 0.36 ms^{-1} **10.** C
11. D **12.** D **13.** A **14. (b) (i)** –3 **(ii)** 2.6×10^{-4} Nm^{-3}
15. (b) 2.4 ± 0.1 s **(c)** 2.6 ± 0.2 ms^{-2}

MECHANICS *(page 26)*
1. C **2.** D **3.** B **4.** B **5.** C **6. (a)** equal **(b)** left **(c)** 20 km h^{-1}
(e) car driver **(f)** no **7. (a)** yes, in towards the centre of the circle
(c) 9.6 ms^{-1}

THERMAL PHYSICS *(page 32)*
1. B **2.** D **3.** D **4.** D **5. (a) (i)** Length = 20m; depth = 2m; width = 5m; temp = 25°C **(ii)** $464 **(b) (i)** 84 days

OSCILLATIONS AND WAVES *(page 45)*
1. C **2.** C **3.** C **4. (b)** 1200 ms^{-1} **(d)** 10 kHz **(e)** 0.12 m
(f) none **5.** B **6. (a)** 4.55 rad s^{-1} **(b)** 4.15 N m^{-1} **(c)** 14.6 cm s^{-1}
(d) 66.3 cm s^{-2}

ELECTRIC CURRENTS *(page 50)*
1. D **2.** D **3.** A **4.** A **5.** B **6. (b)** equal **(c)** A & C brighter; B dimmer **(d)** 1.08 W

FIELDS AND FORCES *(page 57)*
1. C **2.** A **3. (c) (ii)** 7.2×10^{15} ms^{-2}
(iv) 100 V **4.** B **5. (a) (i)** →← **(ii)**

(b) velocity increases; acceleration increases; the wires get closer together **6.** C **7.** C

ATOMIC AND NUCLEAR PHYSICS *(page 65)*
1. B **2.** D **3.** A **4.** D **5.** B **6.** C **7.** D
8. (b) (i) ${}^{2}_{1}H + {}^{26}_{12}Mg \rightarrow {}^{24}_{11}Na + {}^{4}_{2}He$ **9. (a)** ${}^{12}_{6}C \rightarrow {}^{12}_{7}N + {}^{0}_{-1}\beta$
(b) (ii) 11 600 years **10. (a) (i)** 3 **(b) (i)** 1.72×10^{19}

ENERGY, POWER AND CLIMATE CHANGE *(page 79)*
1. (c) 15 MW **(d) (i)** 20% **4. (a)** 1000 MW **(b)** 1200 MW
(c) 17% **(d)** 43 kg s^{-1} **5. (c)** 1.8 MW

MOTION IN FIELDS *(page 85)*
1. A **2.** C **3.** D **4.** C **5. (a) (i)** 1.9×10^{11} J **(ii)** 7.7 km s^{-1}
(iii) 2.2×10^{12} J **(c)** 2.6 h **7. (b) (i)** 3.9 ms^{-1}

THERMAL PHYSICS *(page 91)*
1. D **2.** A **3.** C **4.** A **5. (a)** no **(b)** equal **(c)** 300 J
(d) – 500 J **(e)** 500 J **(f)** 150 J **(g)** 16% **6. (b)** 990 K
(c) (i) 1 **(ii)** 2 and 3 **(iii)** 3 **7. (b) (i)** 170 (166) **(ii)** 1.0×10^{26}
(c) (i) 2.0×10^{-28} m^3 **(ii)** 5.8×10^{-10} m

WAVE PHENOMENA *(page 99)*
1. D **2.** B **3.** C **4. (b)** 27.5 ms^{-1} **5.** 45° **6.** 56.3°
7. (c) 330 ms^{-1} **8. (c)** 2.0×10^{11} m **9. (b)** 3 ± 2 mm

ELECTROMAGNETIC INDUCTION *(page 103)*
1. D **3.** B **4.** D **5.** D **6. (b)** 0.7 V

QUANTUM AND NUCLEAR PHYSICS *(page 111)*
1. D **2.** C **3.** B **4. (b)** ln *R* and *t* **(c)** yes **(e)** 0.375 h^{-1}
(h) 1.85 h **5. (c)** 2×10^{-10} m **6. (b) (i)** $6.9 \pm 0.3 \times 10^{-34}$ Js
(ii) $3.3 \pm 0.5 \times 10^{-19}$ J **7.** 4.5×10^4 Bq

DIGITAL TECHNOLOGY *(page 117)*
1. (i) 6 bits (ASCII uses 8) **(ii)** 154 × 6 = 924 **2.** 5.76×10^7
4. 125 nm **6. (a)** 7.2×10^{-4} C **(b)** 2.9×10^{-3} s **(c) (ii)** no
7. (a) 5×10^{-6} **(b) (i)** 5×10^{-5} m **(ii)** 1×10^6

ASTROPHYSICS *(page 137)*
1. (a) Aldebaran **(c)** Aldebaran **(e)** smaller **(f)** – 2.9 **(g)** 200 pc
2. (a) 5800 K **3. (e)** 499.83 nm

COMMUNICATIONS *(page 154)*
2. (c) (i) 6.8 kHz **(ii)** 34 kHz **(d) (i)** 73 **(ii)** 14 **6. (a)** –10 dB
(b) 0.5 mW **7. (b) (i)** 7 V **(ii)** –12 V **(iii)** +13 V (positive saturation)

ELECTROMAGNETIC WAVES *(page 170)*
1. (d) upside-down **(e)** 60 cm **2. (a) (i)** anywhere to the right of the lens **(ii)** right way up; enlarged; behind the lens; yes **(b) (ii)** to the right of the image **(iii)** upside-down; diminished; in front; nearer; no
3. (b) (ii) 110 nm **4.** B

RELATIVITY *(page 186)*
1. (c) front **(d)** T: 100 m, S: 87 m **(e)** T: 75 m, S: 87 m **2. (b)** 31 m
(c) 1.1×10^{-7} s **(d)** 2.9×10^{-30} kg **3. (a) (i)** zero **(ii)** 2.7 m_0c^2
(b) (i) 0.923 c **(ii)** 2.4 m_0c **(iii)** 3.6 m_0c^2 **(c)** agree **4. (c)** no
(d) Anna

MEDICAL PHYSICS *(page 197)*
1. (b) 1500 Hz **(c)** 200 → 4300 Hz **2. (c)** 10^{-8} W m^{-2} **3. (b)** 0.8%
4. (a) 1 → 20 MHz **(b) (i)** 130 mm **2. (b) (i)** 50% **(ii)** 4 mm

PARTICLE PHYSICS *(page 209)*
3. B **4.** A **6. (b)** uud **11. (a)** 280 MeV **13. (a)** 10 eV **(b)** 10^4 K

Origin of individual questions

The questions detailed below are all taken from past IB examination papers and are all © IB. The questions are from the May (M) or November (N), 1998 (98), 1999 (99), 2000 (00), 2001 (01), Specimen (Sp), Standard level (S1/2/3) or Higher level (H1/2/3) papers with question numbers in brackets.

PHYSICS AND PHYSICAL MEASUREMENT
1. M99SpS1(1) **2.** M99S1(1) **3.** N99S1(1) **4.** M00S1(1)
5. M00S1(2) **6.** N00S1(1) **7.** N01S1(1) **8.** N01S1(2) **9.** N99S2(S2)
10. M98H1(5) **11.** N98H1(5) **12.** M99H1(3) **13.** M99H1(4)
14. N98H2(A1) **15.** M98SpH2(A2)

MECHANICS
1. M98S1(2) **2.** M98S1(4) **3.** M98S1(8) **4.** M98S1(9) **5.** N98S1(4)
6. N00H2(B2) **7.** N00H2(B1)

THERMAL PHYSICS
1. N99H1(15) **2.** N99H1(16) **3.** N99H1(17) **5.** M99SpS2(B3)
6. M99S2(B3)

OSCILLATIONS AND WAVES
1. M01H1(14) **2.** M00H1(19) **3.** M98H1(24) **4.** N99H2(B2)

ELECTRIC CURRENTS
1. N99H1(29) **2.** M99H1(30) **3.** M98H1(29) **4.** M98H1(30)
5. N99SpH1(23) **6.** M99H2(A3)

ATOMIC AND NUCLEAR PHYSICS
1. N98S1(29) **2.** M99S1(29) **3.** M99S1(30) **4.** M98SpS1(29)
5. M98SpS1(30) **6.** M98S1(29) **7.** M98S1(30) **8.** M98S2(A3)
9. M99S2(A3) **10.** M99H2(B4)

ENERGY, POWER AND CLIMATE CHANGE
1. N01S3(C1) **2.** M99S3(C1) **3.** M98SpS3(C3) **4.** M98SpS3(C2)
5. M98S3(C2) **6.** N98S3(C2)

MOTION IN FIELDS
3. N00H1(7) **4.** M01H1(1) **5.** N98H2(B4) **6.** N01H2(A3)

THERMAL PHYSICS
1. M99H1(18) **2.** M00H1(16) **3.** M00H1(18) **4.** M01H1(25)
5. N01H2(B1) **6.** N98H2(A2)

WAVE PHENOMENA
1. M99H1(23) **2.** N01H1(24) **3.** N00H1(24) **4.** N98H2(A5)

ELECTROMAGNETIC INDUCTION
1. N00H1(31) **3.** M98H1(33) **4.** M99H1(27) **5.** N99H1(34)
6. N98H2(A4)

QUANTUM AND NUCLEAR PHYSICS
1. N98H1(32) **2.** M98H1(39) **3.** M01H1(35) **4.** N00H2(A1)

ASTROPHYSICS
1. N01H3(F1) **2.** N01H3(F2) **3.** N98H3(F2) **4.** N00H3(F2)

ELECTROMAGNETIC WAVES
1. N00H3(H1) **2.** M00H3(H1)

RELATIVITY
1. M00H3(G1) **2.** N00H3(G2) **3.** N01H3(G2) **4.** M00H3(G2)

MEDICAL PHYSICS
1. N01S3(D1) **2.** M99H3(D3) **3.** N01H3(D3)

PARTICLE PHYSICS
1. M98H3(E4)

Index

D

E